AF893667

Ecological Studies, Vol. 185

Analysis and Synthesis

Edited by

M.M. Caldwell, Logan, USA
G. Heldmaier, Marburg, Germany
R.B. Jackson, Durham, USA
O.L. Lange, Würzburg, Germany
H.A. Mooney, Stanford, USA
E.-D. Schulze, Jena, Germany
U. Sommer, Kiel, Germany

Ecological Studies

Volumes published since 2001 are listed at the end of this book.

M. Kappelle (Ed.)

Ecology and Conservation of Neotropical Montane Oak Forests

With 62 Figures and 64 Tables

Springer

Dr. Maarten Kappelle
The Nature Conservancy (TNC)
Apartado 230-1225
San José
Costa Rica

Cover illustration: Landscape mosaic of the oak forest zone along the Savegre River at about 2,300 m elevation near San Gerardo de Dota, Costa Rica. This landscape is made up of old-growth montane oak forest along the mountain crests, recovering forests at the lower forest edges, pastures with isolated oak and Buddleja trees, living fences of cypress trees, and orchards with young apple trees. The photo was taken by Maarten Kappelle in 1992.

ISSN 0070-8356
ISBN-10 3-540-28908-9 Springer Berlin Heidelberg New York
ISBN-13 978-3-540-28908-1 Springer Berlin Heidelberg New York

This work is subject to copyright. All rights are reserved, whether the whole or part of the material is concerned, specifically the rights of translation, reprinting, reuse of illustrations, recitation, broadcasting, reproduction on microfilm or in any other way, and storage in data banks. Duplication of this publication or parts thereof is permitted only under the provisions of the German Copyright Law of September 9, 1965, in its current version, and permissions for use must always be obtained from Springer-Verlag. Violations are liable for prosecution under the German Copyright Law.

Springer is a part of Springer Science+Business Media
springeronline.com

© Springer-Verlag Berlin Heidelberg 2006
Printed in Germany

The use of general descriptive names, registered names, trademarks, etc. in this publication does not imply, even in the absence of a specific statement, that such names are exempt from the relevant protective laws and regulations and therefore free for general use.

Editor: Dr. Dieter Czeschlik, Heidelberg, Germany
Desk editor: Dr. Andrea Schlitzberger, Heidelberg, Germany
Cover design: *design & production* GmbH, Heidelberg, Germany
Typesetting and production: Friedmut Kröner, Heidelberg, Germany

31/3152 YK – 5 4 3 2 1 0 – Printed on acid free paper

The editor dedicates this book to his sons Derk Frederik and Bernard Floris, and to all other children living in and near the highland oak forests of the American Tropics. Today, these magnificent forests suffer severely from climate change, land use change, and ultimately, biodiversity loss. If we want our children – and their children and grandchildren – to enjoy the numerous, economically valuable environmental goods and services that these forests provide us, we need to pay for their conservation and sustainable use. Only then will we be able to ensure that human society continues to obtain the benefits of Earth's natural capital as expressed in unique ecosystems such as the Neotropical montane oak forests. Only then will we assure the conditions for a decent, healthy, and secure life for our children and those to come.

Preface

Today, mid- and high-elevation belts in the American Tropics still support montane evergreen broad-leaved oak (*Quercus*) forests. They range from relatively dry woodlands to extremely wet cloud forests, and may occur either as pure monotypic stands – sometimes with giant oaks up to 60 m tall – or as mixed-species systems in which oak co-occurs with other predominant genera such as pine (*Pinus*) and sweetgum (*Liquidambar*). They are found throughout southern Mexico, Central America and the Colombian Andes, and form a major component of the American Tropics ecoregions, biodiversity hotspots, and centers of plant diversity.

Their biological richness, expressed in the large variety of trees, shrubs, epiphytic orchids and bromeliads, ferns, bryophytes, lichens and fungi, is indeed striking. Even animal life is astonishing: the avifauna is among the greatest worldwide, with the mythical Resplendent Quetzal as its most beautiful representative. Large mammals such as jaguar, puma, tapir, peccary and deer still roam around in considerable quantities. In terms of biogeochemical cycling, most of these forests, and especially the oak cloud forests filter large air masses. They capture and incorporate water and nutrients from mist and fog into their cycles, providing nascent rivers with clear fresh water.

Originally, these montane oak forests were widely distributed. However, since the early 1800s, large oak forest areas in the highland Neotropics have made way for coffee plantations and pastures. Today, only few intact blocks remain while most forests are fragmented, suffering from severe disturbance. Remnant patches of forest and woodland are under increasing threat as they are cut for timber, charcoal and fuelwood, or converted into grasslands for cattle.

The importance of this kind of forest for humanity has recently been recognized by various scholars. Experts have noted their key role in providing society with drinking and irrigation water, supplying large urban and rural populations in and near mayor cities in Mesoamerica and the Colombian Andes (e.g., Guatemala City, San José and Bogotá). However, the destructive anthropogenic forces that cause oak forest fragmentation and degradation ultimately lead to species extinction, and loss of environmental goods and services on which human society so strongly depends.

Over the last 20 years, neotropical montane oak forests have been studied intensively by numerous scientists. In recent years, a considerable amount of scientific knowledge on this forest system has become available. To date, however, this knowledge has mainly appeared in a scattered fashion, often only in gray literature. So far, no publication has addressed this ecosystem in a coherent and integrated manner, oriented to a wider audience. Certainly, such a comprehensive volume, providing a thorough understanding of forest patterns and processes in a synthetic and holistic manner, is particularly important for sustainable forest management and lasting biodiversity conservation.

In view of this growing demand, the editor has assembled, in close cooperation with 67 authors from ten countries, the existing body of knowledge on these magnificent oak forests into one comprehensive scientific volume. It is the first state-of-the-art regional account that treats such diverse aspects as the paleo-ecology, biogeography, structure, composition, biodiversity, population dynamics, ecosystem dynamics, fragmentation and recovery, and conservation and sustainable use of natural and managed oak-dominated forests in the highlands of the American Tropics.

It is expected that this volume will be useful to tropical and temperate biologists alike, to biogeographers, plant ecologists, conservation biologists, foresters, policy makers, site practitioners, researchers, lecturers, tutors, and all others with an interest in tropical oak forest ecology and conservation. The editor is confident that this work will help advance scientific knowledge, vitally needed for conserving, restoring and sustainably using the rich oak forests still present in the highland tropics of the New World.

At Springer Verlag in Heidelberg, I would like to gratefully acknowledge Andrea Schlitzberger for initial encouragement to prepare the book and for guiding it to completion. Dieter Czeschlik supported the project throughout its development. Monique Delafontaine and Friedmut Kröner did an excellent job eopy-editing and production-editing the chapters, respectively. Ernst-Detlef Schulze, Series Editor in Jena, suggested many improvements to the original manuscript. Finally, I can never thank enough my beloved wife – and co-author of one of the chapters – Marta E. Juárez, for her moral support and encouragement during the gestation of this book.

Maarten Kappelle

San José, Costa Rica
October 2005

Contents

Contributors

Alcántara-Ayala, Othón

Departamento de Biología Evolutiva, Universidad Nacional Autónoma de México (UNAM), Apartado Postal 70-399, Ciudad Universitaria, México 04510 DF, Mexico

Asbjornsen, Heidi

Department of Natural Resource Ecology and Management, Iowa State University, 234 Science II, Ames, IA 50011, USA, e-mail: hasbjorn@iastate.edu

Ashton, P. Mark S.

School of Forestry and Environmental Studies, Yale University, 360 Prospect Street, New Haven, CT 06511, USA

aus der Beek, Robin

Regional Community Forestry Training Center, Kasetsart University, P.O. Box 1111, Bangkok 10903, Thailand. Current address: c/o SNV Bhutan, P.O. Box 825, Langjophakha, Timphu, Bhutan, e-mail: robinadb@druknet.bt

Barrantes Moreno, Gerardo

Fundación Instituto de Políticas para la Sostenibilidad (IPS), Apartado Postal 900-3000, Heredia, Costa Rica, e-mail: gerardo@ips.or.cr

Bonfil, Consuelo

Departamento de Ecología y Recursos Naturales, Facultad de Ciencias, Universidad Nacional Autónoma de México (UNAM), Circuito Exterior, Ciudad Universitaria, México DF 04510, Mexico, e-mail: cbs@fciencias.unam.mx

Brown, Nick

Department of Plant Sciences, Oxford Forestry Institute, University of Oxford, South Parks Road, Oxford OX1 3RB, UK

Camacho-Cruz, Angélica

Departamento Interuniversitario de Ecología, Facultad de Biología, Universidad Complutense, 28040 Madrid, Spain

Carranza, Julieta

School of Biology, University of Costa Rica (UCR), San Pedro de Montes de Oca, Costa Rica

Cavelier, Jaime

Departamento de Ciencias Biológicas, Universidad de los Andes, Bogotá, Colombia, and The Gordon and Betty Moore Foundation, 1747 Connecticut Avenue NW, Washington, DC 20009, USA

Chaverri, Adelaida †

School of Environmental Sciences, Universidad Nacional (UNA), Heredia, Costa Rica

Cleef, Antoine M.

Institute for Biodiversity and Ecosystem Dynamics (IBED), University of Amsterdam, P.O. Box 94062, 1090 GB Amsterdam, The Netherlands

Contreras-Medina, Raúl

Departamento de Biología Evolutiva, Universidad Nacional Autónoma de México (UNAM), Apartado Postal 70-399, Ciudad Universitaria, México 04510 DF, Mexico

Cortés-S, Sandra P.

Instituto de Ciencias Naturales, Universidad Nacional de Colombia, Bogotá, Colombia

Cuevas-Reyes, Pablo

Facultad de Biología, Universidad Michoacana de San Nicolás de Hidalgo, Morelia, Michoacán, Mexico

Figueroa-Rangel, Blanca L.

Departamento de Ecología y Recursos Naturales, IMECBIO, Centro Universitario de la Costa Sur, Universidad de Guadalajara, Apartado Postal # 108, Autlán de Navarro, CP 48900 Jalisco, Mexico, and School of Geography and the Environment, University of Oxford, Mansfield Road, Oxford OX1 3TB, UK

Flamenco-S., Alejandro

El Colegio de la Frontera Sur (ECOSUR), Apartado Postal 63, San Cristóbal de Las Casas, Chiapas C.P. 29200, Mexico

Galindo-Jaimes, Luis

Departamento Interuniversitario de Ecología, Facultad de Biología, Universidad Complutense, 28040 Madrid, Spain

García-Rojas, Michael

Programa Regional de Manejo de Vida Silvestre (PRMVS), Universidad Nacional Costa Rica, P.O. Box 1350-3000 Heredia, Costa Rica. Current address: Instituto Monteverde, P.O. Box 69-5655, Monteverde, Puntarenas, Costa Rica, e-mail: mgarcia@mvinstitute.org

González-Espinosa, Mario

Departamento de Ecología y Sistemática Terrestres, El Colegio de la Frontera Sur (ECOSUR), Carretera Panamericana y Periférico Sur s/n, 29290 San Cristóbal de Las Casas, Chiapas, Mexico, e-mail: mgonzale@sclc.ecosur.mx

Groot, Thomas V.M.

Institute for Biodiversity and Ecosystem Dynamics (IBED), University of Amsterdam, P.O. Box 94062, 1090 GB Amsterdam, The Netherlands

Guariguata, Manuel R.

Centro Agronómico Tropical de Investigación y Enseñanza (CATIE), 7170 Turrialba, Costa Rica. Current address: United Nations Environment Program (UNEP), Secretariat of the Convention on Biological Diversity (SCBD), 413 Rue St. Jacques, Suite 800, Montréal H2Y 1N9, Canada, e-mail: manuel.guariguata@biodiv.org

Halling, Roy E.

Institute of Systematic Botany, The New York Botanical Garden, Bronx, 10458-5126 NY, USA

Herrera, Bernal

Tropical Agricultural Research and Education Center (CATIE), Turrialba 7170, Costa Rica, and University of Costa Rica (UCR), San Pedro de Montes de Oca, Costa Rica. Current address: The Nature Conservancy (TNC), Apartado 230-1225, San José, Costa Rica, e-mail: bherrera@tnc.org

Hertel, Dietrich

Department of Plant Ecology, Albrecht-von-Haller-Institute of Plant Sciences, University of Göttingen, Untere Karspüle 2, 37073 Göttingen, Germany, e-mail: dhertel@gwdg.de

Hölscher, Dirk

Department of Tropical Silviculture, Institute of Silviculture, University of Göttingen, Büsgenweg 1, 37077 Göttingen, Germany

Holz, Ingo

Universität Greifswald, Botanisches Institut und Botanischer Garten, Grimmer Str. 88, 17487 Greifswald, Germany,
e-mail: ingo.holz@uni-greifswald.de

Hooghiemstra, Henry

Institute for Biodiversity and Ecosystem Dynamics (IBED), University of Amsterdam, P.O. Box 94062, 1090 GB Amsterdam, The Netherlands,
e-mail: hooghiemstra@science.uva.nl

Islebe, Gerald A.

El Colegio de la Frontera Sur (ECOSUR), Unidad Chetumal, Herbarium, AP 424, CP 77000 Chetumal, Quintana Roo, Mexico, e-mail: gerald@ecosur-qroo.mx

Juárez, Marta E.

Apartado 549-1260, Plaza Colonial, Escazú, Costa Rica

Kappelle, Maarten

The Nature Conservancy (TNC), Apartado 230-1225, San José, Costa Rica,
e-mail: mkappelle@tnc.org

Köhler, Lars

Department of Plant Ecology, Albrecht-von-Haller-Institute of Plant Sciences, University of Göttingen, Untere Karspüle 2, 37073 Göttingen, Germany, e-mail: koehlerlars@hotmail.com

Leuschner, Christoph

Department of Plant Ecology, Albrecht-von-Haller-Institute of Plant Sciences, University of Göttingen, Untere Karspüle 2, 37073 Göttingen, Germany

Lira-Torres, Iván

Universidad del Mar, Puerto Escondido, Oaxaca 71980, Mexico

López-Barrera, Fabiola

Departamento de Ecología Funcional, Instituto de Ecología, A.C., km 2.5 Carretera Antigua a Coatepec No. 351, Congregación el Haya Xalapa, Veracruz 91070, Mexico, e-mail: fabiola@ecologia.edu.mx

Luna-Vega, Isolda

Departamento de Biología Evolutiva, Universidad Nacional Autónoma de México (UNAM), Apartado Postal 70-399, Ciudad Universitaria, México 04510 DF, Mexico, e-mail: ilv@hp.fciencias.unam.mx

Manson, Robert H.

Departamento de Ecología Funcional, Instituto de Ecología, A.C., km 2.5 Carretera Antigua a Coatepec No. 351, Congregación el Haya Xalapa, Veracruz 91070, Mexico, e-mail: manson@ecologia.edu.mx

Mata, Milagro

Instituto Nacional de Biodiversidad (INBio), Apartado 22-3100, Santo Domingo de Heredia, Costa Rica

Meave, Jorge A.

Departamento de Ecología y Recursos Naturales, Facultad de Ciencias, Universidad Nacional Autónoma de México (UNAM), Ciudad Universitaria, México 04510 DF, Mexico, e-mail: jamdc@fciencias.unam.mx

Mondragón-Sánchez, Maria L.

Instituto Tecnológico de Morelia, Morelia, Michoacán, Mexico

Mueller, Gregory M.

Department of Botany, Field Museum of Natural History, 1400 S. Lake Shore Drive, Chicago, 60605 IL, USA, e-mail: gmueller@fieldmuseum.org

Naranjo, Eduardo J.

Departamento de Ecología y Sistemática Terrestres, El Colegio de la Frontera Sur (ECOSUR), Carretera Panamericana y Periférico Sur s/n, 29290 San Cristóbal de Las Casas, Chiapas, Mexico

Nixon, Kevin C.

L.H. Bailey Hortorium, Department of Plant Biology, Cornell University, Ithaca, NY 14853, USA, e-mail: kcn2@cornell.edu

Olvera-Vargas, Miguel

Departamento de Ecología y Recursos Naturales, IMECBIO, Centro Universitario de la Costa Sur, Universidad de Guadalajara, Apartado Postal # 108, Autlán de Navarro, CP 48900 Jalisco, Mexico, and Department of Plant Sciences, Oxford Forestry Institute, University of Oxford, South Parks Road, Oxford OX1 3RB, UK, e-mail: molvera@cucsur.udg.mx

Oostermeijer, J. Gerard B.

Institute for Biodiversity and Ecosystem Dynamics (IBED), University of Amsterdam, P.O. Box 94062, 1090 GB Amsterdam, The Netherlands

Oyama, Ken

Centro de Investigaciones en Ecosistemas, Universidad Nacional Autónoma de México (UNAM), Antigua Carretera a Pátzcuaro No. 8701, Col. Ex-Hacienda de San José de la Huerta, Morelia, 58190 Michoacán, Mexico, e-mail: akoyama@oikos.unam.mx

Pedroni, Lucio

Centro Agronómico Tropical de Investigación y Enseñanza (CATIE), 7170 Turrialba, Costa Rica

Pulido, María T.

Departamento de Ciencias Biológicas, Universidad de los Andes, Bogotá, Colombia, and Jardín Botánico, Universidad Nacional Autónoma de México (UNAM), México DF 04510, Mexico, e-mail: mtpulido@yahoo.com

Ramírez-Marcial, Neptalí

Departamento de Ecología y Sistemática Terrestres, El Colegio de la Frontera Sur (ECOSUR), Carretera Panamericana y Periférico Sur s/n, 29290 San Cristóbal de Las Casas, Chiapas, Mexico, e-mail: nramirezm@sclc.ecosur.mx

Rincón, Armando

Departamento de Ecología y Recursos Naturales, Facutad de Ciencias, Universidad Nacional Autónoma de México (UNAM), Ciudad Universitaria, México 04510 DF, Mexico

Romero-Romero, Marco A.

Departamento de Ecología y Recursos Naturales, Facultad de Ciencias, Universidad Nacional Autónoma de México (UNAM), Ciudad Universitaria, México 04510 DF, Mexico

Ruiz-Jiménez, Carlos A.

Departamento de Biología Evolutiva, Universidad Nacional Autónoma de México (UNAM), Apartado Postal 70-399, Ciudad Universitaria, México 04510 DF, Mexico

Sáenz, Grace P.

Centro Agronómico Tropical de Investigación y Enseñanza (CATIE), 7170 Turrialba, Costa Rica

Scareli-Santos, Claudia

Centro de Investigaciones en Ecosistemas, Universidad Nacional Autónoma de México (UNAM), Antigua Carretera a Pátzcuaro No. 8701, Col. Ex-Hacienda de San José de la Huerta, Morelia, 58190 Michoacán, Mexico

Schmit, John P.

Center for Urban Ecology, 4598 Macarthur Blvd. Nw, Washington, DC 20007, USA

Sipman, Harrie J.M.

Botanic Garden & Botanical Museum, Koenigin-Luise-Str. 6-8, 14191 Berlin, Germany, e-mail: h.sipman@bgbm.org

Stift, Marc

Institute for Biodiversity and Ecosystem Dynamics (IBED), University of Amsterdam, P.O. Box 94062, 1090 GB Amsterdam, The Netherlands

ten Hoopen, Martijn

Centro Agronómica de Investigación y Enseñanza (CATIE), 7170 Turrialba, Costa Rica

Tobler, Mathias W.

Botanical Research Institute of Texas, 509 Pecan Street, Fort Worth, TX 76102, USA, e-mail: matobler@gmx.net

Tovar-Sánchez, Efrain

Centro de Investigaciones en Ecosistemas, Universidad Nacional Autónoma de México (UNAM), Antigua Carretera a Pátzcuaro No. 8701, Col. Ex-Hacienda de San José de la Huerta, Morelia, 58190 Michoacán, Mexico

van den Bergh, Maurits B.

Institute for Biodiversity and Ecosystem Dynamics (IBED), University of Amsterdam (UvA), P.O. Box 94062, 1090 GB Amsterdam, The Netherlands

van Uffelen, Jan-Gerrit

Hessenweg 59, 7771 RD Hardenberg, The Netherlands

Vázquez-López, José M.

Departamento de Ecología y Recursos Naturales, IMECBIO, Centro Universitario de la Costa Sur, Universidad de Guadalajara, Apartado Postal # 108, Autlán de Navarro, CP 48900 Jalisco, Mexico

Venegas, Geoffrey

Centro Agronómico Tropical de Investigación y Enseñanza (CATIE), 7170 Turrialba, Costa Rica

Vogt, Kristina A.

College of Forest Resources, University of Washington, Seattle, WA 98195, USA

WILMS, JOOST J.A.M.

Jaboncillos, San Gerardo de Dota, Costa Rica

WOLF, JAN H.D.

Institute for Biodiversity and Ecosystem Dynamics (IBED), Universiteit van Amsterdam, P.O. Box 94062, 1090 GB Amsterdam, The Netherlands, e-mail: wolf@science.uva.nl

Part I
Introduction to Neotropical Montane Oak Forests

1 Global and Neotropical Distribution and Diversity of Oak (genus *Quercus*) and Oak Forests

K.C. Nixon

1.1 Introduction

The genus *Quercus* is one of the most important clades of woody angiosperms in the northern hemisphere in terms of species diversity, ecological dominance, and economic value. Oaks are dominant members of a wide variety of habitats, including temperate deciduous forest, temperate and subtropical evergreen forest, subtropical and tropical savannah, subtropical woodland, oak-pine forest, oak-'piñon'-juniper woodlands, various kinds of 'cloud forest', tropical premontane forest, tropical montane forest, matorral (summer rain chaparral), and a variety of Mediterranean climate vegetations, including chaparral (French: maqui), oak woodland, and evergreen oak forest (Nixon 1993a, b, 1997b, 2002; Kappelle et al. 1995). Oaks also enter, and are important, along the margins of various other vegetation types, such as coniferous forests, prairies, tropical grasslands, desert and semi-desert scrublands, dry (deciduous) tropical forest, and in some evergreen tropical forests (Barbour and Billings 1999).

Although many species of *Quercus* are exceptionally large, dominant overstory trees (Kappelle et al. 1995, Chaps. 8–11 and 14–17), perhaps an almost equal number of species are shrubs or small trees, particularly in drier habitats such as chaparral, in edaphically challenging environments, and in some higher elevation forests. Oaks also occur as 'specialists' in a diversity of edaphically distinct habitats, such as serpentine, sandy barrens, and swamps. However, in wetter forests oaks are often among the largest trees of the region, particularly when compared to other angiosperms. In the Americas, this is true both in the temperate deciduous forests of the eastern USA and in the evergreen oak forests of Mexico and Central America. Oaks also occur in the Himalayas and Southeast Asia (Indonesia).

The economic importance of *Quercus* in the northern hemisphere is widely known. Various species are sources of high-quality lumber, and it is the

Ecological Studies, Vol. 185
M. Kappelle (Ed.) Ecology and Conservation of Neotropical Montane Oak Forests
© Springer-Verlag Berlin Heidelberg 2006

preferred firewood in many areas, particularly as a cooking/heating fuel throughout the highlands of Mexico and Central America. Because of the dominance of oak in many forests, it is the subject of a vast number of ecological studies that focus on interactions between oaks and fungi (Chap. 5), plants (Chaps. 18, 19 and 23) and animals (Chaps. 24–27).

1.2 Higher-Level Taxonomy

The genus *Quercus* in the broad sense is a member of the family Fagaceae (excluding *Nothofagus*), which also includes *Fagus* (beeches), *Castanea* (the true chestnuts), other 'castaneoid' genera (*Chysolepis*, *Castanopsis*, and *Lithocarpus*), and three monotypic tropical genera (*Trigonobalanus*, *Formanodendron*, and *Colombobalanus*). In the New World, in addition to *Quercus* we have *Fagus* (two spp.), *Chrysolepis* (one spp.), *Lithocarpus* (one sp.). *Castanea* (two spp.), and *Colombobalanus* (one spp.; Nixon and Crepet 1989; Nixon 1997a, 2003). The family Fagaceae sensu stricto (excluding *Nothofagus*) is monophyletic, based on both morphological and molecular analyses (Nixon 1989; Manos et al. 1999). In the recent literature, oaks are treated either as a single genus with two subgenera (*Quercus* and *Cyclobalanopsis*; Nixon 1993b), or as two distinct genera (*Quercus* and *Cyclobalanopsis*). The evidence at this point, based on molecular data, is equivocal as to whether *Quercus* and *Cyclobalanopsis* form a monophyletic group (P.S. Manos, personal communication). In the Flora of China, the two lineages were separated as distinct genera, with 35 species recognized for *Quercus*, and 69 species of *Cyclobalanopsis* within China (Chengyiu et al. 1999). Within the New World, only *Quercus* sensu stricto occurs (Nixon 1997b; Nixon and Muller 1997), so the issue of whether to recognize one or two genera (thankfully) does not affect the nomenclature in this region.

Within New World *Quercus*, there have been traditionally recognized three distinct groups – the white oaks (section *Quercus*, sometimes referred to as subgenus or section *Leucobalanus* or *Lepidobalanus*), the red or black oaks (section *Lobatae*; also sometimes referred to as subgenus or section *Erythrobalanus*), and the intermediate or golden oaks (section *Protobalanus*; Nixon 1993a, b, 1997b; Manos 1997). A fourth group, section *Cerris*, is restricted to Eurasia and North Africa. Sections *Quercus* and *Lobatae* are widespread in the Americas and relatively diverse, whereas section *Protobalanus* is a small clade of ca. six species restricted to the southwestern USA and northern Mexico, including some islands near the west coasts of both countries (the Channel Islands, Guadalupe Island, and Cedros Island; see Manos 1997). Section *Quercus* is widespread in the northern hemisphere of the Old World in addition to the Americas, whereas section *Lobatae* and *Protobalanus* are both endemic to the New World.

The genus *Quercus* first appears in the fossil record in the Early Tertiary of North America about 50–55 million years ago (Crepet and Nixon 1989a, b), although the oldest evidence for the family Fagaceae is in the Late Cretaceous, about 90 million years ago (Crepet et al. 2004). Although in both cases the earliest records are North American, this is likely due to sampling error, and the biogeographic origins of both *Quercus* and Fagaceae remain equivocal at this point (Nixon 1989). By the mid to late Tertiary, *Quercus* fossils are among the most common found at numerous localities in western North America, suggesting that widespread (evergreen) oak forests occurred over wide areas in the northern hemisphere, particularly in the Miocene. In Chapter 2, Hooghiemstra provides information (from fossil pollen records) on the immigration of *Quercus* in the Colombian Andes.

From the perspective of oak taxonomy and systematics, several aspects of the genus are important. For one, the oaks are considered exceptional for their apparent ability to hybridize within species groups. This is based mostly on observations that species are highly variable, often with isolated individuals and occasionally with significant populations showing morphological variability that encompasses characteristics of more than one recognized species. Numerous studies have attempted to document and characterize hybridization among oak species, and in the last part of the 20th Century, several studies employed genetic and/or molecular markers to address questions of hybridization. In addition to documenting obvious cases of morphological introgression, some studies also found that cryptic hybridization could be present, as evidenced by the distribution of plastid types that seemed to be independent of species boundaries, but correlated instead with geographic proximity of populations. For example, within European white oaks, it was found that *Quercus robur* and *Q. petraea* populations in close proximity shared the same chloroplast genome, whereas they differed from conspecific populations from more distant localities. This was also found in at least one US study (Whittemore and Schaal 1991).

More recently, based on similar kinds of observations, it has been suggested that *Quercus* species may accomplish at least some dispersal solely through pollen transport by wind; pollen reaching relative populations of a related species might produce hybrids, and eventually through repeated backcrossing and selection, the 'invading' species emerges and produces its own morphologically distinct populations, similar to those that produced the pollen. Such scenarios might explain the pattern of morphological variation that is seen throughout the range of *Quercus*, and in some specific cases, putative hybrids (morphological intermediates) are well-known outside the geographic or ecologic range of one of the parents (Nixon 1993b). Whether these cases are due to past contact of populations followed by swamping of one of the parents, or rather to pollen dispersal over long distances remains to be seen. It is important to note that natural hybrids have been documented only between species within the same section. Although there have been scattered reports of artificial hybridization between species from different sections, these have not been verified with genetic or molecular data.

When considering the nature of *Quercus* forests in the Neotropics, and the possible history of the group, all of the above life history factors must be considered. Although less well-known taxonomically, the oaks of the Mexican and Central American forests appear to exhibit similar patterns of morphological variation and life history as do oaks in the forests of the USA and Europe, which have been more intensively studied. That said, there also clearly are differences in the Neotropical species of *Quercus* in terms of life history factors. The most obvious of these is a less precise seasonality in Neotropical oaks.

1.3 Distribution and Species Diversity

On the American continent, species of the genus *Quercus* (oak) occur in Canada, the USA, Mexico, Belize, Guatemala, El Salvador, Honduras, Nicaragua, Costa Rica, Panama and Colombia. Figure 1.1 shows the distribution of *Quercus* in the Americas.

Estimates of species diversity within new World *Quercus* have changed over the years, but we now have a fairly accurate estimate based on recent floras and broad treatments (Nixon 1993a, 1997b, 2003), and work in progress (Flora Mesoamericana, Nixon, unpublished data). The overall number of species in the New World, including Latin America, the United States and Canada, is probably around 220 species. Estimates of the total number of oak species that occur, along with endemics, in American countries in which *Quercus* is naturally found are as follows: four in Canada, 91 in the USA, one in Cuba, 160–165 in Mexico, nine in Belize, 25–26 in Guatemala, 8–10 in El Salvador, 14–15 in Honduras, 14 in Nicaragua, 14 in Costa Rica, 12 in Panama, and one (*Quercus humboldtii*) in Colombia.

The greatest species diversity for the genus *Quercus* in the New World occurs in the mountains of southern Mexico (Nixon 1993a). Another center of diversity occurs in the southeastern United States, but not particularly associated with the Appalachian Mountains. The Rocky Mountain region is depauperate in oak species, as is the Pacific Northwest.

Traveling southeast from Mexico into Central America, one notes a gradual reduction of oak species diversity. Eventually, when one reaches Colombia, there is a single species of oak (*Q. humboldtii*, subdivided into 2–3 species by some authors; see Chap. 11).

Seasonality in the temperate and subtropical oaks, including those of North America and Europe, results in relatively consistent patterns of flowering and fruit production. Most temperate species have a characteristic flowering time in the spring months (usually somewhere between February and

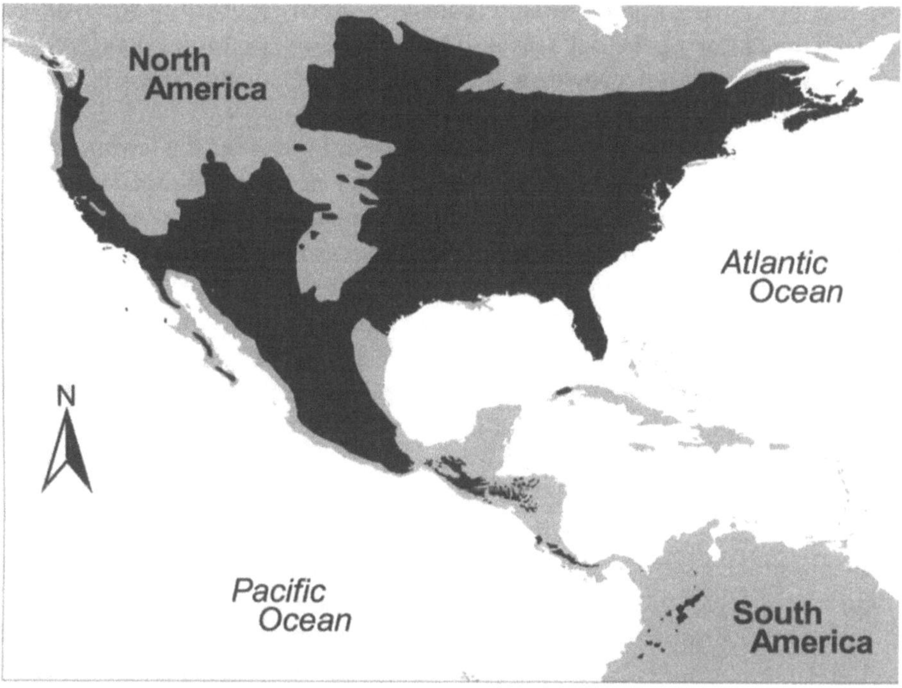

Fig. 1.1. Map showing the outer limits of the distribution of oak (genus *Quercus*) on the American continent. The genus ranges from southern Canada to southern Colombia, and is found in the northwest corner of Cuba. In Mesoamerica and Colombia, it is found mainly in mountainous areas above 1,000 m elevation. Map prepared by Marco V. Castro Campos at The Nature Conservancy

June, depending on the species and latitude), and a fairly fixed fall fruit production period – acorns fall mostly in the months of September–November, with the greatest production in October. This is true throughout much of montane Mexico as well, with most species flowering in March–April, and producing fruit in October–November.

The tropical and montane tropical oaks from southern Mexico to Colombia, however, present a different pattern of flowering and fruit production, which in some species is less predictable. The majority of oak species in the Mesoamerican region flower in the 'dry season', varying from October to February, with a peak fruiting time during the rainy season in June–July. To date, there have been very few studies of the exact phenology of these tropical species (Céspedes 1991, Chap. 19), and even less is known on the mechanisms by which flowering synchronization might occur. In Costa Rica, Céspedes (1991) observed during a year of observation a strong periodicity in leaf flushing, leaf fall, flowering and fruiting in *Quercus seemannii* at 1,700 m altitude. He noted that leaf fall occurred practically all year around, but was more pronounced during the dry season (February–April). Flowering and fruiting also occurred during the dry season (October–May), reaching a maximum in March, whereas shoot growth was more important during the wet season (May–October).

Such phenological patterns are also apparent in most of the lowland tropical species of the Pacific slope of Mexico as far north as Jalisco, and on the Atlantic slope during drier phases in the forests of Veracruz and Oaxaca (e.g., *Quercus sapotiifolia* Liebm.). However, some widespread lowland species, such as *Q. insignis*, produce mature fruits in June–July in the southern parts of their ranges in Central America, and produce mature fruits in October in the northernmost populations (e.g., near Jalapa in Veracruz). We do not yet have sufficient data to determine if this is a gradual, clinal shift in phenology from north to south, or if climatic factors are correlated with the differences.

Several ecological correlates of oak distributions in Central America are apparent, and are clearly seen in distributions within Costa Rica and Panama. Oaks are much rarer on the wet Atlantic slope of these countries, except at higher elevations, and reach their greatest abundance both in dominance and species on the drier Pacific slopes. The composition of the oak forests on the wetter Atlantic slopes also differs (Kappelle et al. 1992). This oak distribution may be more indicative of temporal distribution of rainfall and the occurrence of significant periods of drought on the Pacific slope, as opposed to exact amounts of precipitation. Thus, there is generally a more pronounced dry season, and overall seasonality, on the Pacific slope throughout Central America than on the wetter Atlantic slope. These differences in diversity seem to diminish as one travels northward, and particularly north of the Isthmus of Tehuantepec in Mexico, oaks seem to occur abundantly on both the Pacific and Atlantic slopes at lower elevations. Thus, some widespread lowland species that are restricted to the Pacific slope in Costa Rica occur on both coasts in Mexico (e.g., *Q. sapotiifolia* and *Q. elliptica*).

Overall, lowland oak distributions indicate increased diversity and dominance in the more seasonally dry forests of Mexico and on the Pacific slope of Central America. Of course, oaks are also dominant in the seasonally dry and cooler forests of montane Mexico, which might be classified as subtropical or mild temperate seasonal forests. Again, in these forests, the phenology of most oak species is suggestive of the phenology of their northern counterparts, with flowering generally occurring in the 'spring' (January–April), and mature fruits falling in the 'fall' (September–November, with a peak in October).

Thus, if we wish to contrast the oaks of Central America south of Mexico with those of Mexico, we can generally say that phenologically/ecologically, we have two groups – the montane Mexican group (generally above 1,500 m), and the tropical group (generally below 1,500 m). The latter group occurs at low elevations in the north, and at all elevations in the south (from near sea level to ca. 3,000 m). Life history correlates of these two groups include the almost complete lack of biennial fruit maturation in lowland tropical red oaks, and montane tropical red oaks from Central America (one notable exception being *Q. acatenangensis*/*Q. ocoteifolia*, which extends as far south as El Salvador).

By contrast, all red oaks from strongly temperate areas have biennial fruit maturation, and only a few subtropical species (e.g., *Q. pumila* in the SE USA, *Q. agrifolia* in California) are annual. The proportion of red oak species with annual maturation increases as one follows a gradient from north to south. Given that the condition of biennial maturation appears to be plesiomorphic (primitive) within the genus and the family, this suggests that the red oak group radiated from a relatively temperate or montane subtropical ancestor with biennial maturation into more tropical areas, where fruit maturation independently shifted to annual in more than one lineage. Such speculation, however, awaits confirmation from more precise phylogenetic analyses of species-level relationships within American *Quercus*.

In terms of growth form and habit, oaks have generally been divided into evergreen vs. deciduous species. However, such a classification is too simplistic. Most of the supposedly evergreen oaks of the southern and western USA and montane Mexico are actually sub-evergreen - holding their leaves approximately one full year, and losing their leaves either simultaneously with bud break in the 'spring' (or whenever they flush), or soon thereafter. By contrast, some species are truly evergreen, holding individual leaves for 18 months or longer, sometimes up to 3–4 years. One such species occurs in the western USA (*Q. sadleriana*), and a few in montane Mexico (e.g., the *Q. crassifolia* complex), but the majority of 'evergreen' species from the USA and montane Mexico are not truly evergreen. The most striking examples of these are *Q. agrifolia* (the California Coast Live Oak) and *Q. virginiana* (the Southern Live Oak). Both of these species drop all or most of their leaves in the spring; in some cases, they are bare for a few days, but in general the leaf drop occurs immediately after the new leaves emerge, so that the trees are never completely bare. In most of Central America, including Costa Rica and Panama, most oaks are truly evergreen.

1.4 Species Diversity in Central America

Overall, species diversity in *Quercus* in Central America diminishes as one heads to the southeast. For the Flora Mesoamericana project, which extends from Chiapas to Colombia, I estimate ca. 40 species of *Quercus* (Nixon, unpublished data). This is in contrast to southeastern Mexico (roughly including states from San Luis Potosi in the north to Chiapas in the south) where I have previously estimated the number of species to be as many as 75 (Nixon

1993b). In northwestern Mexico, in the states of Sonora and Sinaloa alone, the previous estimate is 41 species, and in southwestern Mexico, there are about 50 species (Nixon 1993b). Thus, Central America is not particularly diverse in terms of *Quercus*; more telling is the fact that there are relatively few species that are endemic to the region south of Mexico, and the majority of species that occur in the countries of Central America are found also in Mexico (probably only about five endemics of ca. 40 species). However, the lower species diversity of Central American oak forests does not diminish their ecological importance, and some of the forests, particularly in Costa Rica, are among the tallest for *Quercus* in the New World (Kappelle et al. 1992, 1995, Chaps. 4 and 10).

Within the oak forests of Central America, both red oaks (section *Lobatae*) and white oaks (section *Quercus*) commonly occur. The most commonly encountered associations of oak species in Costa Rica are *Quercus costaricensis–Q. bumelioides* (*Q. copeyensis* of authors in Chaps. 4, 7, 10, 15–18, 21–27, and 30–32) in upper montane forests, and *Q. seemannii–Q. bumelioides* in lower montane forests (Kappelle et al. 1995, Chaps. 4 and 10). The foliage of *Q. costaricensis* and *Q. bumelioides* is strikingly similar, typically being elliptic with impressed venation and usually, but not always, with a conspicuous tomentum on the lower surface. These two species belong to different sections (*Q. costaricensis* to *Lobatae*, and *Q. bumelioides* to *Quercus*) and are not closely related, and the similarities must be attributed to parallelism in response to similar environments. *Quercus costaricensis* seems to more predictably produce large crops of acorns, whereas acorn production seems to be more sporadic in *Q. bumelioides*. Both species tend to produce acorns from June to July, during the rainy season, and there appears to be no dormancy, the nuts germinating soon after falling. The seed dormancy that is seen in northern species of the red oak group (e.g., *Q. rubra*), and requires stratification (cold treatment) to break dormancy, is not known in the Neotropical species of either section of *Quercus*.

The *Q. seemannii* complex has been problematic particularly in Panama and Costa Rica, with various names applied by different authors. The problems with this complex are related to an apparent lack of easily tractable characters (the leaves are typically non-descript, glabrous at maturity, lanceolate and entire), compounded by a large degree of plasticity and a propensity to hybridize. Thus, in the Flora Costaricensis by Burger (1977), the complex was treated as including two species, *Q. seemannii* and *Q. rapurahuensis*, and he considered *Q. eugeniifolia* to be a synonym of *Q. seemannii*. Although never formally published, Breedlove later annotated numerous specimens formerly called *Q. seemannii* or *Q. eugeniifolia* in herbaria as *Q. salicifolia* Nee, a species that occurs on the west coast of Mexico in Jalisco and Oaxaca, in relatively dry, lowland forests. Based on my own examinations of type materials and of populations in the field in Mexico, Costa Rica, and Panama, I contend that the Mexican populations of *Q. salicifolia* should not be considered conspecific with what has been called *Q. seemannii* or *Q. eugeniifolia* in Costa Rica and Panama. I also recognize the distinction between *Q. eugeniifolia* and *Q. seemannii*, and prefer to treat those as separate species, too. Unfortunately, there is considerable morphological gradation between what would be called typical *Q. seemannii* and typical *Q. eugeniifolia* in both western Panama and Costa Rica. *Quercus eugeniifolia* typically occurs at lower elevations (upper premontane to lower montane), and is distinguished by extremely short petioles, a more acutely tapered leaf base, and secondary veins that are more numerous and leave the mid-vein at angles close to 90°. Of the two species, *Q. eugeniifolia* is perhaps the most similar to *Q. salicifolia* of Mexico, but differs in lacking the glandular trichomes of the latter, in venation features, and in fruit characters. In Mexico, *Q. eugeniifolia* is common on the wetter Atlantic slope (e.g., Veracruz and Oaxacan cloud forest), as one would expect, given its distribution in the relatively wet lower montane forests of Costa Rica.

The 'real' *Quercus seemannii* is common in Costa Rica and Panama, but is lacking from Mexico and Guatemala (its distribution in intervening countries is still under study). It typically has a short but distinct petiole, a broader, more rounded leaf base, and fewer veins that leave the mid-vein at a more acute angle. *Quercus rapurahuensis*, which occurs at lower elevations (lower montane and upper premontane), has similar but larger leaves, and much larger fruit. Although Breedlove has lumped *Q. rapurahuensis* with the more northern *Q. benthamii* in herbarium annotations, that species is strikingly different in having very large, tomentose buds, which are not seen in any of the Costa Rican or Panamanian material called *Q. rapurahuensis*. Thus, again following Burger (1975, 1977), it is best to recognize *Q. rapurahuensis* as distinct from *Q. benthamii*.

In northern Central America, from Oaxaca and Chiapas south to El Salvador, a common element of the montane oak forest is the red oak *Quercus acatenangensis*. Although superficially resembling some phases of *Q. seemannii*, with long entire glossy leaves, *Q. acatenangensis* belongs to a different complex of species that have biennial fruit maturation, unlike the annual fruit maturation found in all of the species from Costa Rica and Panama, including *Q. seemannii*. In turn, *Q. acatenangensis* appears to be the southern component of a complex that includes *Q. laurina* and *Q. affinis*, two of the most common high-elevation oaks of southern and eastern Mexico.

1.5 Conclusions

The Neotropics, particularly southern Mexico, harbors the greatest diversity of oak species in the New World. These oaks are often among the tallest trees in the forest areas in which they occur, ranging from low-elevation to high montane forests. The clearest relationships of Central American oaks are with lower-elevation Mexican oaks. Several species are widespread from Mexico to Costa Rica and Panama. In addition, at least one distinctive clade of red oaks with annual maturation is common in Central America and Colombia, and includes members of the *Q. seemannii* complex (including *Q. rapurahuensis* and *Q. gulielmi-treleasei*), *Q. costaricensis* and *Q. eugeniifolia*, and, in Colombia, *Q. humboldtii*.

References

Barbour MG, Billings WD (eds) (1999) North American terrestrial vegetation, 2nd edn. Cambridge Univ Press, New York, NY

Burger W (1975) The species concept in *Quercus*. Taxon 24(1):45–50

Burger W (1977) *Quercus*. Flora Costaricensis. Field Bot 40:59–82

Céspedes R (1991) Fenología de *Quercus seemannii* Lieb. (Fagaceae), en Cartago, Costa Rica. Rev Biol Trop 39(2):243–248

Chengjiu H, Yongtian Z, Bartholomew B (1999) Fagaceae. In: Zheng-Yi W, Raven P (eds) Flora of China, vol 4. Science Press, Beijing, pp 300–400

Crepet WL, Nixon KC (1989a) Earliest megafossil evidence of Fagaceae: phylogenetic and biogeographic implications. Am J Bot 76:842–855

Crepet WL, Nixon KC (1989b) Extinct transitional Fagaceae from the Oligocene and their phylogenetic implications. Am J Bot 76:1493–1505

Crepet WL, Nixon KC, Gandolfo MA (2004) Fossil evidence and phylogeny: the age of major angiosperm clades based on mesofossil and macrofossil evidence from Cretaceous deposits. Am J Bot 91:1666–1682

Kappelle M, Cleef AM, Chaverri A (1992) Phytogeography of Talamanca montane *Quercus* forests, Costa Rica. J Biogeogr 19(3):299–315

Kappelle M, Van Uffelen JG, Cleef AM (1995) Altitudinal zonation of montane *Quercus* forests along two transects in Chirripó National Park, Costa Rica. Vegetatio 119:119–153

Manos PS (1997) *Quercus* section *Protobalanus*. In: Flora of North America Editorial Committee (eds) Flora of North America, North of Mexico, vol 3. Oxford Univ Press, New York, pp 470–471

Manos PS, Doyle JJ, Nixon KC (1999) Phylogeny, biogeography, and processes of molecular differentiation in *Quercus* subg. *Quercus* (Fagaceae). Mol Phylogen Evol 12:333–349

Nixon KC (1989) Origins of Fagaceae. In: Crane PR, Blackmore S (eds) Evolution, systematics, and fossil history of the Hamamelidae. Syst Assoc Spec 40B(2):23–43

Nixon KC (1993a) The genus *Quercus* in Mexico. In: Ramamoorthy TP, Bye R, Lot A, Fa J (eds) Biological diversity of Mexico: origins and distribution. Oxford Univ Press, Oxford, UK, pp 447–458

Nixon KC (1993b) Infrageneric classification of *Quercus* (Fagaceae) and typification of sectional names. Ann Sci Forest 50 Suppl 1:25s–34s

Nixon KC (1997a) Fagaceae. In: Flora of North America Editorial Committee (eds) Flora of North America, North of Mexico, vol 3. Oxford Univ Press, New York, pp 436–437

Nixon KC (1997b) *Quercus*. In: Flora of North America Editorial Committee (eds) Flora of North America, North of Mexico, vol 3. Oxford Univ Press, New York, pp 445–447

Nixon KC (2002) The oak *(Quercus)* biodiversity of California and adjacent regions. USDA Forest Service, Gen Tech Rep PSW-GTR-184

Nixon KC (2003) Fagaceae. In: Smith N, Mori SA, Henderson A, Stevenson DW, Helad S (eds) Families of Neotropical flowering plants. Princeton Univ Press, Princeton, NJ, pp 156–158

Nixon KC, Crepet WL (1989) *Trigonobalanus* (Fagaceae): taxonomic status and phylogenetic relationships. Am J Bot 76:826–841

Nixon KC, Muller CH (1997) *Quercus* section *Quercus*. In: Flora of North America Editorial Committee (eds) Flora of North America, North of Mexico, vol 3. Oxford Univ Press, New York, pp 445–447

Whittemore AT, Schaal BA (1991) Interspecific gene flow in sympatric oaks. Proc Natl Acad Sci USA 88:2540–2544

Part II
Paleo-Ecology and Biogeography

2 Immigration of Oak into Northern South America: a Paleo-Ecological Document

H. Hooghiemstra

2.1 Introduction

In this chapter, a short overview is presented of the paleo-ecological aspects of Neotropical oak forests. On a long time scale, it is shown how oak forest migrated from the north into the Neotropics during Neogene and Pleistocene times. Evidence comes from long marine and terrestrial pollen records that show how the distribution area of *Quercus* extended southward through Central America, and eventually covered a small area in the northwestern part of South America. On a time scale of the last glacial-interglacial cycle, the dynamic history of oak forest is shown on the basis of a pollen record from Colombia. The aim of this chapter is to place the present-day distribution area, and the ecological requirements of oak forest, in a long-term perspective, in which evolution and speciation within the genus *Quercus*, the changing paleo-geographical setting of the Neotropics since the Neogene, and migration played an important role.

Understanding of the long-term paleo-ecological history can lead to a better understanding of the ecology of modern Neotropical oak forest, with positive feedback to conservation and restoration issues.

2.2 Miocene Central American Oak Forest and Oak Migration into South America During the Late Pleistocene

Oak forest is principally a northern hemisphere type of forest (Chap. 1). The distribution of *Quercus* ranges from Southeast Asia, west into the region of the Caspian and Black seas, via Europe, to North and Central America (Walter and Straka 1970). For a long time, Central America was a 'dead-ending' part of the distribution area of oak forest. The marine pollen record of core DSDP site

Ecological Studies, Vol. 185
M. Kappelle (Ed.) Ecology and Conservation of Neotropical Montane Oak Forests
© Springer-Verlag Berlin Heidelberg 2006

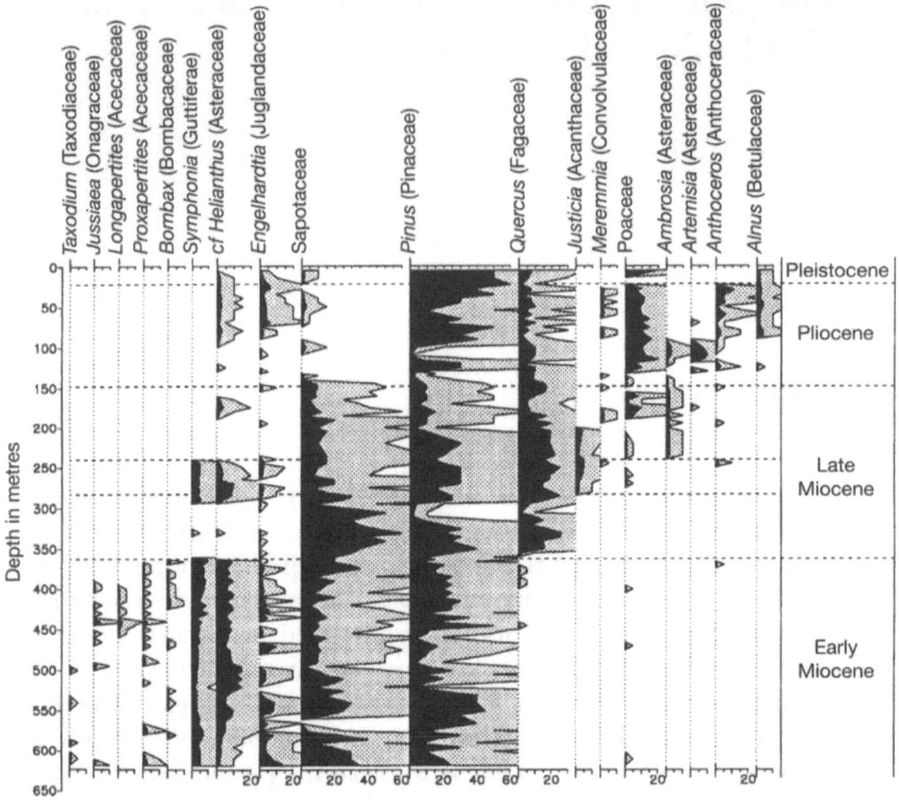

Fig. 2.1. Arrival of *Quercus* in late Miocene times in the region of Central Mexico, as documented by marine pollen record 'DSDP site 493' collected offshore the Mexican Pacific coast (modified after Fournier 1982, in Morley 2000). The curves covering *black areas* represent changing percentages per depth. Those covering *shaded areas* represent the same data but with scales exaggerated 5×, in order to emphasize low-value changes

493, located offshore the Mexican Pacific coast, shows clearly that *Quercus* arrived in the area of central west Mexico at the start of the late Miocene, about 10 million years ago (10 Ma BP; BP=before present), when there was a general transition from wet evergreen forest to drier semi-deciduous forest (Fournier 1982; Morley 2000; Fig. 2.1).For a long time, oak resided in Central America at the southern periphery of its distribution area.

During Pliocene times, between 5 and 3.5 Ma ago in particular, the Panamanian Isthmus gradually closed (Keigwin 1978; Webb and Rancy 1996), giving way to an enormous interchange of floral and faunal elements between the two continents (e.g., Stehli and Webb 1985). This is substantiated by the first appearance date (FAD) of two trees, *Alnus* (alder) and *Quercus* (oak), in long Colombian pollen records (Fig. 2.2). The Funza-1 pollen record from the basin of Bogotá shows that *Alnus* immigrated into this area around 1.1 Ma BP,

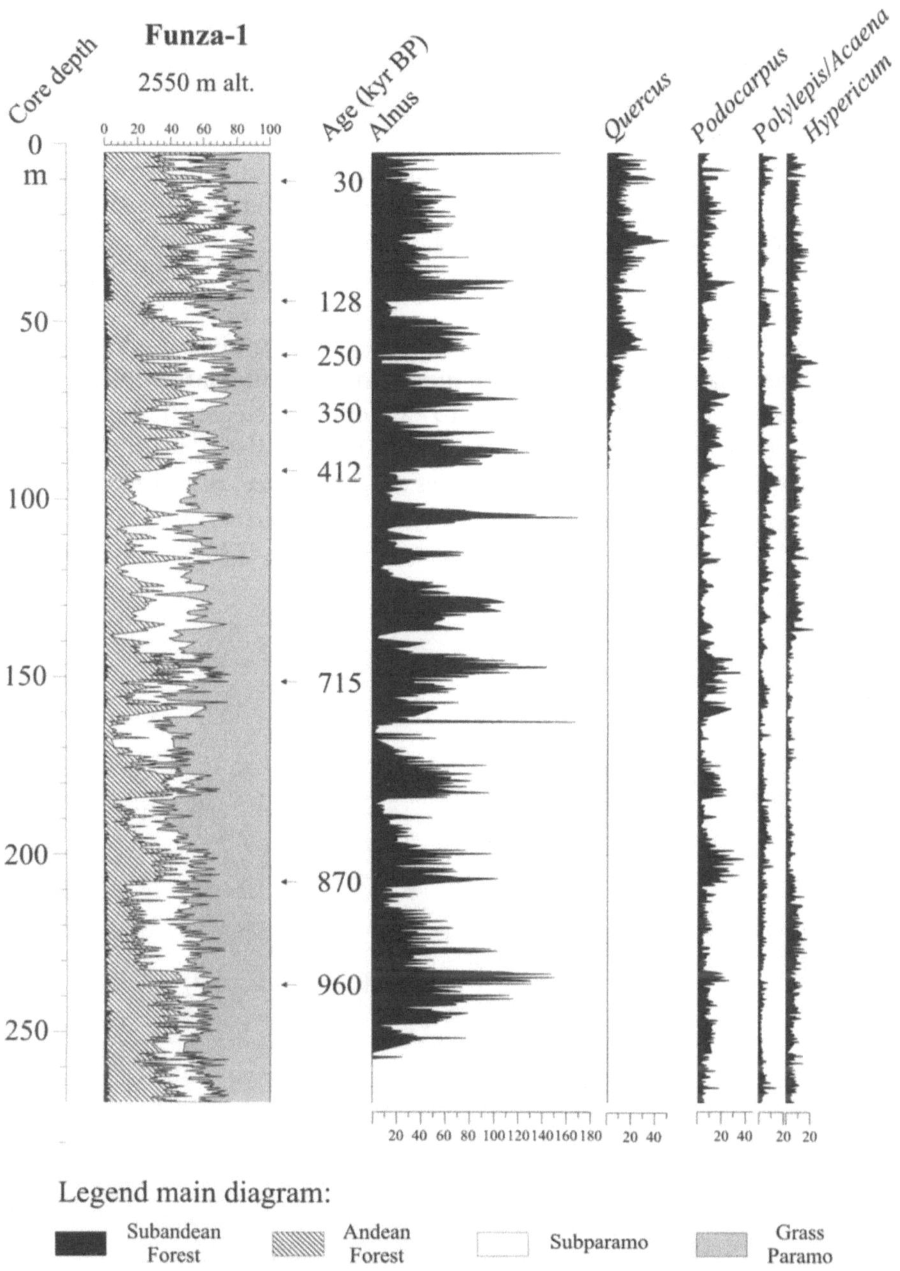

Fig. 2.2. Main pollen diagram of the 357-m-long Funza-1 core, taken from the Bogotá basin at 2,550 m altitude. It shows the individual records of selected trees and shrubs. The age of selected core depths are indicated (after Van't Veer and Hooghiemstra 2000). *Alnus* immigrated about 1.1 million years ago, whereas *Quercus* arrived in the Bogotá basin about 0.47 million years ago (modified after Hooghiemstra 1984)

and *Quercus* around 470 ka BP (Van't Veer and Hooghiemstra 2000). Indeed, fruits of *Alnus* are light and partly wind dispersed, which facilitates rapid migration. *Quercus*, in turn, produces relatively heavy seeds that are animal dispersed (e.g., by squirrels and birds; see Chaps. 13 and 23–36); consequently, a lower migration speed is very plausible.

The different FADs for *Alnus* and *Quercus* might explain the contrasting present-day distribution of both genera in South America: *Alnus* reached as far south as northern Argentina (about 27°S) whereas the southernmost distribution of today's *Quercus* is the Colombian–Ecuadorian border (about 1°N; Fig. 2.3).

Today, the greatest diversity of *Quercus* is found in Mexico where almost 130 different oak species are found, mostly representing large upland forest trees (Rzedowski 1983, Chaps. 1, 8 and 9). Diversity decreases through Central America, to a single species of *Quercus* (*Q. humboldtii*) in the Andes of Colombia (Chap. 11). Neotropical species of *Quercus* are concentrated in mid- to high-elevation forests, although a few species of *Quercus* (e.g., *Q. oleoides*) are found in lowland wet or dry forest of Costa Rica (Chaps. 1 and 10) and

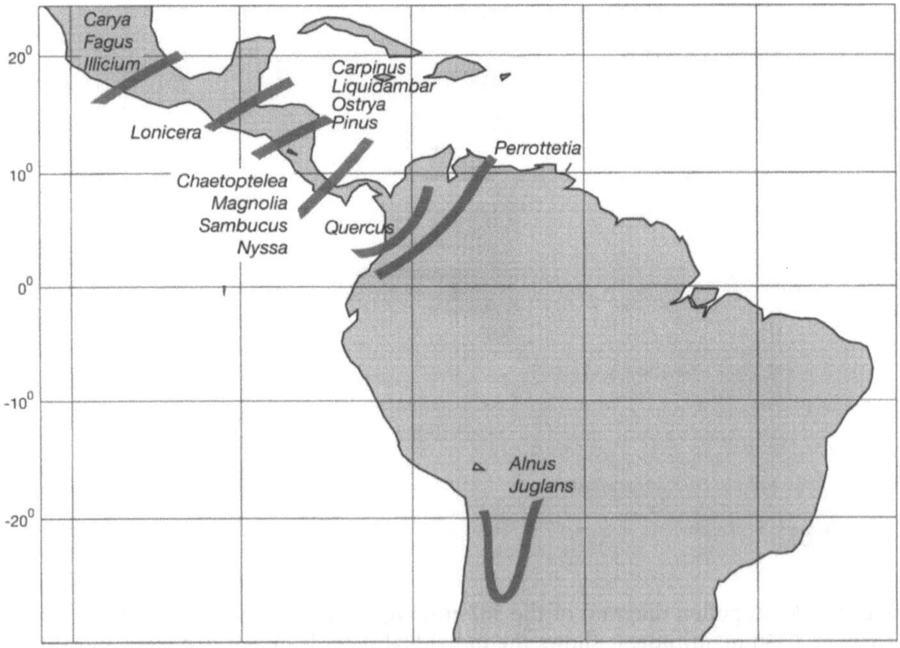

Fig. 2.3. Southward extension of northern hemisphere arboreal taxa in Central America. The southernmost positions of the taxa distribution areas are indicated. Some taxa, such as *Alnus* and *Quercus*, effectively crossed the Panamanian Isthmus and extended their distribution into South America (modified after Webster 1995)

Nicaragua, and in Mexico in a few cases to near sea-level elevations (Hartshorn 2000; Smith et al. 2004, Chap. 1).

In general, *Quercus* species growing at lower elevations were privileged in their ability to pass the Panamanian Isthmus. Genetic analysis might be able to identify which Central American oak species are most closely related to the Colombian *Q. humboldtii* (Chaps. 1 and 11). The genetic reservoir of Mexican oak species of low and mid-elevations also has a high potential for this type of study (Chap. 1).

2.3 Late Pleistocene Records of Neotropical Oak Forest Dynamics

In Colombia, the modern altitudinal range of *Quercus* extends from 1,100 m in the dry Inter-Andean valleys (Cleef et al. 2003) up to the highest humid ecotone forests at 3,200–3,300 m (Cleef and Hooghiemstra 1984; Rangel et al. 2003). Under natural conditions, the largest and possibly most continuous surface of oak forest is found in the Sub-Andean forest belt. During the late Pleistocene, the long river valleys of the Rio Magdalena and the Rio Cauca may have served as routes for easy southward expansion. It is hypothesized that at various places, oak has expanded from the Sub-Andean forest belt to higher elevations in the Andean forest belt, reaching the upper forest line (UFL; A.M. Cleef, personal communication). At several places, for example, where valleys give easy access to Inter-Andean high plains, a close genetic relationship is assumed between low- and high-elevation oak populations. Much of the Sub-Andean forest belt has been cleared for agriculture, and today's last remnants of oak forest are found mostly in the Andean forest belt (Chap. 11).

Quercus is wind pollinated, producing large quantities of pollen grains that can be identified to the generic level only. Therefore, *Quercus* is well represented in pollen records. There are many pollen records in northern South America and Central America that show the dynamic history of oak forest during the late Pleistocene; an overview of Colombian sites is given in Marchant et al. (2001).

The dynamic character of Neotropical oak forest during the last glacial-interglacial cycle is shown by the pollen record of an 18-m-long core, known as 'Fúquene-7C' (Mommersteeg and Hooghiemstra 2006). These sediments were collected at the border of Laguna de Fúquene, located at 2,580 m in the Eastern Cordillera of the Colombian Andes. The chronology of the sediments is based on eight radiocarbon ages (^{14}C years BP; non-calibrated radiocarbon years) in the upper part of the core. In the lower part of the core, the relationship between the record of lake-level oscillations and periods with minimum values for the precession signal of orbital climate forcing was applied (Mom-

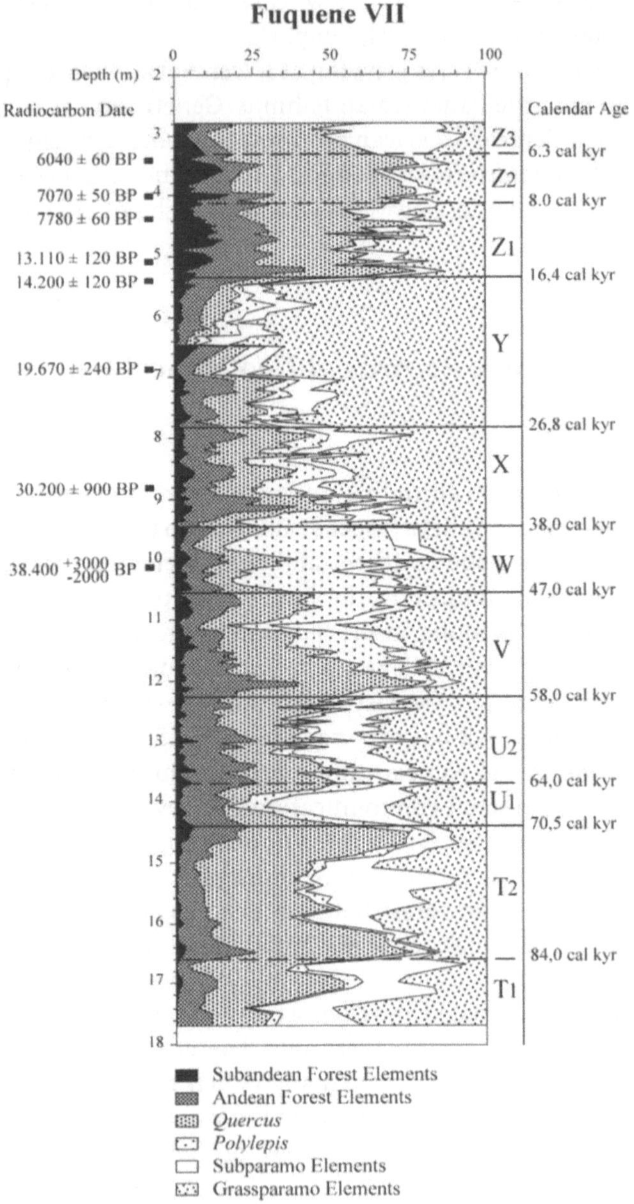

Fig. 2.4. The main pollen diagram of core Fúquene-7C, showing the dynamic vegetation and climate history of the Colombian Andes. Data are given for 5-cm intervals along the core length (about 120-year increments). Laguna de Fúquene is situated at 2,580 m altitude in the present-day Andean forest belt (=upper montane forest belt). The upper forest line (UFL) separates two ecological groups at the right side, and four ecological groups at the left side. The UFL shifted altitudinally from 2,000 m (zone Y) to 3,200 m (zones T2 and Z2). During most of the time period, *Quercus* contributed significantly to the lower and upper montane forests. Temporary replacement by *Polylepis*-dominated forest (zone W) is still insufficiently understood (Mommersteeg and Hooghiemstra 2006)

mersteeg 1998; Mommersteeg and Hooghiemstra 2006). All ages were converted into calendar years before present (cal years BP; calibrated radiocarbon years). Sediments were analyzed at 5-cm increments along the core, providing a document of vegetation change at a 120-year temporal resolution during the last 85,000 cal years BP (Fig. 2.4).

The main pollen record shows changes in the down-core contribution of:

- Sub-Andean forest (=lower montane forest; today's range: 1,000–2,300 m altitude);
- Andean forest (=upper montane forest; today's range: 2,300–3,200 m altitude);
- the proportion of oak (*Quercus*; today's range: 1,100–3,400 m altitude);
- the proportion of *Polylepis* (today mainly as ecotone forest, but also as isolated patches of dwarf forest in the paramo, up to 4,500 m altitude);
- the proportion of subparamo shrub (today's range: 3,200–3,500 m altitude); and
- the proportion of grass-paramo herbs (Poaceae only; today's range: 3,500–4,200 m altitude).

These ecological groups are in competition with each other and, therefore, make up a value of 100 % in the general pollen diagram (van der Hammen 1974). Changes in the dynamic balance cause vegetation belts to migrate in an altitudinal direction. The proportions of different ecological groups are indicative of the altitudinal position of the UFL. For example, in the Colombian Andes the UFL coincides with an arboreal pollen proportion of 40 %, which coincides with the 9.5 °C annual isotherm (Hooghiemstra 1984, 1989). Thus, by inferring the position of the UFL from the pollen record, and applying a temperature gradient of 6 °C per 1,000 m of displacement of the UFL, we can reconstruct the record of temperature change in the past. Within the ecological groups, various taxa show changing contributions. In the scope of this paper, the proportions of two major trees, *Quercus* and *Polylepis*, are plotted separately.

In the following eight subsections, the changing proportions of pollen grains of *Quercus* form the basis to reconstruct past changes in Colombian oak forest. For more precise paleo-environmental reconstructions, and estimates of paleo-temperature and moisture, the reader is referred to Mommersteeg (1998), and Mommersteeg and Hooghiemstra (2006).

1. During the period 88,000–70,500 cal years BP (zones T1, T2), *Quercus* reached 30–55 % representation, evidencing it was the most dominant tree in the ecotone forest around Laguna de Fúquene. The UFL shifted from about 2,700 to 3,200 m.
2. During the period 70,500–64,000 cal years BP (zone U1), arboreal pollen percentages were low and *Quercus* reached values<5 %. The UFL must have been close to the lake, and *Weinmannia* and *Polylepis* dominated in the ecotone forest. The contribution of 'warm forest' elements (*Acalypha*,

Urticaceae-Moraceae) increased, compared to the preceding period. The pollen diagram suggests that during this cold interval, oak became rare in the study area.

3. During the period 64,000–48,000 cal years BP (zone U2), *Quercus* abundance increased again and reached mainly 30–40 % in the record. Oak forest became more abundant, partly replacing *Polylepis* in the ecotone forests.
4. At the start of the period 58,000–47,000 cal years BP (zone V), oak forest was abundant (40–45 %), but the proportion decreased gradually to about 20 % while the proportion of *Polylepis* increased from 10 to 20 %. Also, the proportion of Sub-Andean forest taxa increased, compared to the preceding zone, indicating less extreme climatic conditions. During this time period, the UFL shifted from 3,000–3,200 to 2,800–3,000 m.
5. The period 47,000–35,000 cal years BP (zone W) shows almost equal proportions of 'warm' Sub-Andean forest taxa, *Quercus*, and 'cool' Andean forest taxa, and the contribution of *Polylepis* reached the highest values of the record (up to 60 %). Evidently, *Polylepis* was the most important element of the ecotone forest at that time, and the UFL is estimated at 2,600–2,800 m elevation. We have not observed this setting in the pollen record before. We assume most of the *Polylepis* forest belongs to the zonal belt of Andean forest, and perhaps only a small contribution, as is the case today, comes from small isolated patches of *Polylepis* forest scattered over the paramo. This assumption implies that the inferred paleo-temperatures during this period did not differ much from those of the preceding period (zone V), and this remarkable change in forest composition is unlikely to be driven by temperature change.
6. During the period 35,000–14,500 cal years BP (zones X and Y), the contribution of *Quercus* to the montane forest belts reached proportions similar to those documented before 47,000 cal years BP. At a core depth of 6.5 m, the UFL reached its lowest altitudes (about 2,000 m), reflecting the Last Glacial Maximum (LGM). *Quercus* and *Polylepis* dominated the LGM ecotone forests.
7. In contrast to earlier literature dealing with core Fúquene-2 (van Geel and van der Hammen 1973), the Fúquene-7 record lacks most of the Late Glacial period (about 13,000 to 10,000 cal years BP). A synthesis of pollen evidence from several sites (van der Hammen and Hooghiemstra 1995) shows that, in the period 13–11 ka BP, the temperature rose to about 2 °C lower than that found today. In particular, *Quercus*, *Hedyosmum*, *Myrica* and *Myrsine* responded to this temperature increase. During the 11,000–10,000 cal year BP time interval, a sudden cooling occurred, to a level 4–6 °C lower than that observed today (van der Hammen and Hooghiemstra 1995). Record Fúquene-2 shows that *Quercus* and *Podocarpus* responded least to this cooling, and consequently increased their proportions.

Based on the time control of the Fúquene-7C sediments, we propose the following interpretation: zone Z1 includes part of the temperature increase at the start of the Late Glacial (about 13,000 cal years BP); zone Z1 lacks the remaining part of the Late Glacial, including the cold episode between 11,000 and 10,000 cal years BP (locally called the 'El Abra Stadial'); and zone Z1 is completed with early Holocene sediments.

8. During the Holocene (mainly zones Z2–Z3), the contribution of *Quercus* to the Andean forest in the study area is again large (generally 30–50 %), and proportions are comparable to those of zones T2 (84,000–70,500 cal years BP) and the beginning of zone V (58,000–52,000 cal years BP). Zone Z2 has no signals of human impact, and reflects undisturbed Holocene conditions proper. Zone Z3 shows signals of deforestation reflecting human occupation.

2.4 Discussion

Pollen diagram Fúquene-7C shows a high temporal resolution during most of the Last Glacial–Interglacial Cycle, and belongs to the high-quality documents of past vegetation change and forest dynamics. The composite curves of Sub-Andean and Andean forest elements, including *Quercus* (the three categories at the left in Fig. 2.4), show a large number of cycles with durations of 1,500–3,500 years. This variability seems congruent with the variability of the climate system in the northern Atlantic Ocean, where cycles in the same frequency domain are called Dansgaard-Oeschger Cycles (Mommersteeg 1998; Mommersteeg and Hooghiemstra 2006).

An important observation is that *Quercus*-dominated forest does not show significant periods of stability, but rather a continuous competition with other forest types. The concept of 'climax forest' is based mainly on developments in the floristic composition of early and mid-Holocene forests of Europe and North America, whereby ecosystems are recreated 'from scratch' at the beginning of each warm interval (Roberts 1998). During glacial-interglacial cycles in the tropical mountains, oak migrated altitudinally over, at most, some 1,500 m. Thus, source areas from which to recruit were always at close distance.

There are relatively stable conditions on the slopes of tropical mountains, compared to the higher latitudes where forests have migrated over distances of about 2,000 km. We observe that even under such relatively stable conditions, the floral composition, as well as the proportion of *Quercus* in the montane forest belt, is subject to significant and continuous change. This observation makes it even more important to identify – for all major arboreal taxa of the Neotropics – geographical distribution patterns in combination with ranges of main climatic variables, in order to establish full 'ecological envelopes' (Thompson et al. 2000).

The replacement of *Quercus*-dominated forest by *Polylepis*-dominated forest during the period 55,000–47,000 cal years BP is not yet well understood. Though changing environmental conditions may be inferred, we still do not have strong arguments to infer a significant cooling during this 8,000-year period (Fig. 2.4), even when *Polylepis* forests were part of the zonal upper montane forest. This episode might reflect an example of internal dynamics of Neotropical montane forest: exchanges of taxa without environmental drivers that are clear to science today. More research is needed to better understand forest interior dynamics as a key factor that may be quasi-independent of climate change.

In the Colombian Andes, *Quercus* for the first time formed abundant forests about 330,000 years BP (Van't Veer and Hooghiemstra 2000). Thus, the present interglacial is the third in a sequence with significant zonal oak forests. As the southernmost distribution of *Quercus* is close to the Colombian–Ecuadorian border, we roughly estimate a migration speed of 1,500 km during a 330,000-year period, corresponding to a migration speed of 4–5 km per 1,000 years (i.e., 400–500 m per century). Finally, we contend that this slow migration rate may also be due to the large weight of the oaks' acorns – a seed trait that hampers fast dispersal (Chaps. 13 and 24).

2.5 Conclusions

The genus *Quercus* has a wide northern hemisphere distribution, and occurs with many species in Asia, Europe and North America. During Miocene times, *Quercus* expanded from North America into Central America where it reached high diversity; in Mexico, about 130 different species have been recorded to date (Chap. 1). The closure of the Panamanian Isthmus, mainly between 5 and 3.5 Ma ago, allowed *Quercus* to migrate into South America (Graham 1995; Hooghiemstra and Cleef 1995).

The long continental pollen records from the basin of Bogotá document the FAD of *Quercus* at about 470 ka BP. Accepting the estimated migration speed for the late Quaternary, the actual crossing of the isthmus might have been some 300 ka earlier. In the area of Bogotá, from 470 to 280 ka BP, oak forest was in competition with other types of montane forest in which *Weinmannia* and *Hedyosmum* were abundant; there is no evidence that oak forest had reached significant proportions during this time period of almost 200,000 years. We tentatively estimate oak forest occurred during glacial times between 800 and 2,200 m, and during interglacial times between 1,100 and 3,300 m (coinciding with its present-day distribution). Such altitudinal migrations frequently may have left patches of oak forest in isolation at the lowest and highest elevational intervals (<1,100 and >2,200 m, respectively). Such dynamic history may explain the wide ecological envelope of today's *Quercus* species, but also may have promoted speciation.

Pollen records, reflecting the last ice-age, show episodes during which oak was largely replaced by other forest trees, such as *Polylepis*. A better understanding of the underlying causes of forest dynamics on long time scales is of interest not only for paleo-ecology, but also for conservation and restoration of neotropical montane oak forests.

Acknowledgements I thank Antoine M. Cleef, Maarten Kappelle, and an anonymous reviewer for constructive comment on an earlier version of this manuscript. Jan van Arkel, Vladimir Torres, and Herman Mommersteeg assisted with the preparation of the figures.

References

Cleef AM, Hooghiemstra H (1984) Present vegetation of the area of the high plain of Bogotá. In: Hooghiemstra H (ed) Vegetation and climatic history of the high plain of Bogotá. Diss Bot 79:42–66

Cleef AM, Rangel-Ch JO, Salamanca S (2003) The Andean rain forests of the Parque Los Nevados transect, Cordillera Central, Colombia. In: Van der Hammen T, Dos Santos J (eds) La Cordillera Central Colombiana: transecto Parque los Nevados. Cramer/Borntraeger, Berlin, Studies on Tropical Andean Ecosystems, vol 5, pp 79–141

Fournier GR (1982) Palynostratigraphic analysis of cores from site 493. Proc Deep Sea Drilling Proj Leg 66:661–670

Graham A (1995) Development of affinities between Mexican/Central American and northern South American lowland and lower montane vegetation during the Tertiary. In: Churchill SP, Balslev H, Forero E, Luteyn JL (eds) Biodiversity and conservation of Neotropical montane forests. The New York Botanical Garden, Bronx, NY, pp 11–22

Hartshorn GS (2000) Tropical and subtropical vegetation of Mesoamerica. In: Barbour MG, Billings WD (eds) North American terrestrial vegetation, 2nd edn. Cambridge Univ Press, Cambridge, UK, pp 623–659

Hooghiemstra H (1984) Vegetation and climatic history of the high plain of Bogotá. Cramer, Vaduz, Diss Bot 79

Hooghiemstra H (1989) Quaternary and Upper-Pliocene glaciations and forest development in the tropical Andes: evidence from a long high-resolution pollen record from the sedimentary basin of Bogotá, Colombia. Palaeogeogr Palaeoclimatol Palaeoecol 72:11–26

Hooghiemstra H, Cleef AM (1995) Pleistocene climate change and environmental and generic dynamics in the north Andean montane forest and paramo. In: Churchill SP, Balslev H, Forero E, Luteyn JL (eds) Biodiversity and conservation of Neotropical montane forests. The New York Botanical Garden, Bronx, NY, pp 35–49

Keigwin LD Jr (1978) Pliocene closing of the Isthmus of Panama, based on biostratigraphic evidence from nearby Pacific Ocean and Caribbean Sea cores. Geology 6:630–634

Marchant R, Behling H, Berrio JC, Cleef AM, Duivenvoorden J, Hooghiemstra H, Kuhry P, Melief B, van Geel B, van der Hammen T, Van Reenen G, Wille M (2001) Mid- to late-Holocene pollen-based biome reconstructions for Colombia. Quat Sci Rev 20:1289–1308

Mommersteeg H (1998) Vegetation development and cyclic and abrupt climatic change during the late Quaternary: palynological evidence from the Eastern Cordillera of Colombia. PhD Thesis, University of Amsterdam, Amsterdam

Mommersteeg H, Hooghiemstra H (2006) Centennial-scale vegetation and climate change in the tropical Andes during the last 80 kyr from 18-m Colombian pollen record Fuquene-7. Palaeogeogr Palaeoclimatol Palaeoecol (in press)

Morley RJ (2000) Origin and evolution of tropical rain forests. Wiley, Chichester, UK

Rangel JO, Cleef AM, Salamanca S (2003) The equatorial Inter-Andean and Sub-Andean forests of the Parque Los Nevados transect, Cordillera Central, Colombia. In: Van der Hammen T, Dos Santos J (eds) La Cordillera Central Colombiana: transecto Parque los Nevados. Cramer/Borntraeger, Berlin, Studies on Tropical Andean Ecosystems, vol 5, pp 143–204

Roberts N (1998) The Holocene: an environmental history. Blackwell, Malden, MS

Rzedowski J (1983) Vegetación de México. Editorial Limusa, México DF, Mexico

Smith N, Mori SA, Henderson A, Stevenson DW, Heald SV (eds) (2004) Flowering plants of the Neotropics. Princeton Univ Press, Princeton, NJ

Stehli FG, Webb SD (1985) The great American biotic interchange. Plenum Press, New York

Thompson RS, Anderson KH, Bartlein PJ (2000) Atlas of relations between climatic parameters and distributions of important trees and shrubs in North American hardwoods. USGS, Denver, US Geol Surv Prof Pap 1650-B

Van der Hammen T (1974) The Pleistocene changes of vegetation and climate in tropical South America. J Biogeogr 1:3–26

Van der Hammen H, Hooghiemstra H (1995) The El Abra stadial, a Younger Dryas equivalent in Colombia. Quat Sci Rev 14:841–851

Van Geel B, van der Hammen T (1973) Upper Quaternary vegetational and climatic sequence of the Fuquene area (Eastern Cordillera, Colombia). Palaeogeogr Palaeoclimatol Palaeoecol 4:9–92

Van't Veer R, Hooghiemstra H (2000) Montane forest evolution during the last 650,000 yr in Colombia: a multivariate approach based on pollen record Funza-1. J Quat Sci 15:329–346

Walter H, Straka H (1970) Arealkunde: floristisch-historische Geobotanik. Ulmer, Stuttgart

Webb SD, Rancy A (1996) Late Cenozoic evolution of the Neotropical mammal fauna. In: Jackson JBC, Budd AF, Coates AG (eds) Evolution and environment in tropical America. Univ Chicago Press, Chicago, IL

Webster GL (1995) The panorama of Neotropical cloud forests. In: Churchill SP, Balslev H, Forero E, Luteyn JL (eds) Biodiversity and conservation of Neotropical montane forests. The New York Botanical Garden Press, Bronx, NY, pp 53–77

3 Effects of the Younger Dryas Cooling Event on Late Quaternary Montane Oak Forest in Costa Rica

G.A. Islebe and H. Hooghiemstra

3.1 Introduction

Climate change is expected to profoundly impact global vegetation types. Those changes can be best evaluated if we understand past climate change and its impact on vegetation types. From research by Martin (1964), we know that glacial times had impact on Costa Rican vegetation. Later paleoecological and palynological studies from the Cordillera de Talamanca include Hooghiemstra et al. (1992), Horn and Sanford (1992), Horn (1993), Islebe et al. (1996), Islebe and Hooghiemstra (1997), and Rodgers and Horn (1996). The Younger Dryas cooling event (11,000–10,000 ^{14}C years BP), the last stage of the Pleistocene, is of great interest as there is still an exciting discussion questioning if this event was global, or rather restricted to some regions of the northern hemisphere. To date, the effects of the Younger Dryas cooling have been observed in ice cores (Dansgaard et al. 1989), marine sediments (Kennett 1990), and terrestrial cores from different parts of the world (Peteet 1993, 1995). For Central America and northern South America, a Younger Dryas cooling event has been suggested for Guatemala (Leyden 1995), Costa Rica (Islebe et al. 1995), Colombia (van Geel and van der Hammen 1973; van der Hammen 1978; Kuhry et al. 1993; Hooghiemstra and van der Hammen 1995), Ecuador (Clapperton et al. 1997), and Peru (Thompson et al. 1995). However, the lack of bracketing radiocarbon dates (i.e., dates that delimit events by an upper and lower age boundary) is a problem in many paleorecords (Heine 1993; Van't Veer et al. 2000).

In this chapter, we present data from the La Chonta bog area, located in the Cordillera de Talamanca. Hastenrath (1973) reported several glacier advances, and that at 10,000 years BP the deglaciation process had ended. Our study site is located at 2,310 m altitude and is today surrounded by montane oak forest. We consider this altitude as strategic to understand past vegetation changes. The objective of this chapter is to analyze the impact of

Ecological Studies, Vol. 185
M. Kappelle (Ed.) Ecology and Conservation of Neotropical Montane Oak Forests
© Springer-Verlag Berlin Heidelberg 2006

the Late Glacial–Holocene transitional climatic conditions on montane oak forests.

3.2 Present Vegetation

The La Chonta bog (09°41'N, 83°57'W) is located in Costa Rica's Cordillera de Talamanca at the elevation (2,300 m) where lower and upper montane oak forests meet (Hooghiemstra et al. 1992). The present-day climatic conditions are described by Herrera (1985), who reports a mean annual temperature of 12–15 °C for the bog area, and an annual rainfall of 1,500–2,500 mm. Based on Kappelle (1991), the vegetation has the following altitudinal distribution in the Cordillera de Talamanca: bamboo paramo (3,800–3,300 m), with *Chusquea* as a dominant taxon accompanied by many herbaceous species; subalpine ericacous dwarf forest (3,400–3,100 m), with abundant shrub genera such as *Drimys*, *Hypericum*, *Ilex* and *Myrica*; upper montane myrsinaceous oak forest (3,200–2,400 m), composed of different oak species, *Clusia*, *Nectandra*, *Schefflera* and *Weinmannia*; and lower montane lauraceous oak forest (2,400–1,500 m), composed mainly of *Quercus*, *Alchornea*, *Billia*, *Dendropanax*, *Eugenia*, *Guarea*, *Guatteria*, *Hedyosmum* and other neotropical taxa.

Detailed descriptions of communities and oak forest floristics and ecology are presented in Kappelle (1995, 1996, Chaps. 4 and 10), and Kappelle et al. (1995). Islebe and Kappelle (1994) compared the phytogeographical affinity of subalpine forests of Costa Rica and Guatemala, and reaffirmed the high affinity of Costa Rican vegetation with the northern Andes (Kappelle et al. 1992). At the La Chonta bog, four azonal plant communities occur along a soil moisture gradient. Local vegetation is dominated by *Hypericum*, *Eleocharis*, *Blechnum*, *Puya* and *Eriocaulon*.

3.3 Methods

Sediments of the La Chonta bog were retrieved in sections of 25 cm with a Dachnovsky sampler. Standard acetolysis techniques were used to extract the pollen grains. The fossil pollen was identified with the help of the pollen reference collection of the University of Amsterdam, the morphological descriptions by Hooghiemstra (1984), and the Costa Rica reference collection. *Lycopodium* spores were added to determine pollen concentration. Table 3.1 provides information on all identified pollen and spore taxa. The pollen sum includes arboreal and herbaceous pollen, excluding aquatics and fern spores.

Table 3.1. Pollen taxa from sediment core La Chonta 1, listed according to altitudinal distribution and ecological preferences

Paramo	Subalpine rainforest	Upper montane rainforest	Oak	Lower montane rainforest	Ferns	Aquatics
Caryophyllaceae	Asteraceae	*Alnus*	*Quercus*	*Antidaphne*	Cyatheaceae	Cyperaceae
Cruciferae	Ericaceae	*Drimys*		Rubiaceae	*Cyathea* baculate	*Isoetes*
Gentianaceae	*Hypericum*	*Hedyosmum*			*Salix*	*Hymenophyllum*
Poaceae		*Ilex*		Solanaceae	*Jamesonia*	
Ranunculus		Melastomataceae		*Ulmus*	*Lophosoria*	
Valeriana		*Myrica*		Urticales	*Lycopodium* foveolate	
		Myrtaceae			monolete psilate	
		Rapanea (Myrsine)			monolete verrucate	
		Podocarpus				
		Viburnum				
		Weinmannia				

High-resolution pollen analysis was done at 1-cm intervals between 440 and 430 cm depth, and at 0.5-cm intervals between 430 and 420 cm. Time control was achieved by dating two samples of unidentified organic material at the Utrecht University by means of accelerator mass spectrometry (van der Borg et al. 1987). The following radiocarbon ages were obtained: 9,800±120 ^{14}C years BP (UtC-2,925) at 415 cm depth, and 11,070±130 ^{14}C years BP (UtC-2,927) at 440 cm depth.

3.4 Description of Pollen Zones

Pollen zone 4 (12,000–11,000 ^{14}C years BP)
Poaceae percentages vary in the range 6–30 %, and Asteraceae and Ericaceae in the range 0.6–2 %. *Podocarpus* is present up to 1 %. *Alnus* varies from 11 to 77 %, and *Quercus* reaches up to 55 %. Monolete spores show percentages of 1–6 %. *Isoetes* spores (excluded from pollen sum) are found up to 108 %. Upper montane forest (up to ca. 2,700 m) dominated by *Quercus* was characteristic during this period (see Chap. 2).

Pollen zone 5 (11,000–10,400 ^{14}C years BP)
This zone is characterized by Poaceae with percentages of 20–35 %, somewhat higher than the preceding pollen zone. Percentages for the total of subalpine taxa (Ericaceae, *Ilex*, *Viburnum*, *Drimys*, *Myrica*) are also higher and reach 5 %. Percentages for *Podocarpus* and *Hedyosmum* vary in the range 0.3–1 %. *Alnus* shows percentages up to 18 %, and *Quercus* values of 30–55 %. *Antidaphne*, a parasitic Loranthaceae on oak and some other trees, is present with percentages of 0.5–5 %. Urticales, Umbelliferae, Rubiaceae, Solanaceae and *Salix* have percentages of 0.3–0.7 %. Monolete spores (verrucate and psilate) are much higher than in the preceding zone, with percentages up to 50 %. Cyatheaceae spores show a single peak of nearly 70 %. *Isoetes* percentages are markedly lower, varying in the range 0.5–35 %. Hooghiemstra et al. (1992) described pollen zone 5 of core La Chonta-2 as a zone characterized mainly by relatively high percentages of *Alnus* (up to 75 %) and relatively low percentages of *Quercus* (up to 40 %). However, most samples lacked pollen in that core.

Pollen zone 6 (10,400–9,800 ^{14}C years BP)
In this zone, Poaceace percentages are reduced and vary in the range 5–22 %. Melastomataceae are represented with percentages of 1–3 %. Subalpine taxa show values of <1 %, *Alnus* of 7–94 %, and *Quercus* of 30–70 %. Monolete spores have percentages up to 20 %. *Isoetes* is present with percentages of 10–150 %. Oak-dominated montane forest covered more area during this time period.

3.5 Paleoecology

The La Chonta bog area was located within the paramo belt during the Pleniglacial. During the Last Glacial Maximum (LGM), a cooling of 6–7 °C has been proposed (Islebe and Hooghiemstra 1997), or even of 8–9 °C (Lachniet and Seltzer 1999). The montane oak forests were affected by glacial conditions, and the forest line varied between 2,000 and 2,300 m altitude. During the Late Glacial the upper forest line, which corresponds to the 9–10 °C annual isotherm, was located at 2,700–2,800 m altitude. The bog was surrounded by montane oak forest in that period.

The lowering of the upper montane forest line during the Younger Dryas Chron suggests a cooling event that we named the La Chonta stadial. This lowering of the upper forest line follows the temperature rise at 10,400–9,800 ^{14}C years BP, associated with an increase of taxa from the uppermost montane rainforest (*Ilex*, *Viburnum* and *Drimys*) and from the subalpine forest belt. At present, the subalpine dwarf forest is located at 3,100–3,400 m in the

Cordillera de Talamanca (Islebe and Kappelle 1994). Poaceae pollen indicate a down-slope extension of the paramo vegetation during the Younger Dryas cooling event. Pollen of oak is represented by values less than 50 % in the early stage of the La Chonta stadial, whereas later in the early Holocene the values rise steadily to over 50 %.

The changes recorded at the transition to Holocene conditions show an upward migration of taxa belonging to the montane forest belt. The base of the La Chonta stadial is dated at 11,070±130 ^{14}C years BP, and using a linear interpolation between the two radiocarbon ages, the top has an estimated age of 10,400 ^{14}C years BP. A transition to Holocene conditions is evident at 430–415 cm, and has been dated at 10,400–9,800 ^{14}C years BP. A steady trend in available humidity is also shown by an increase in *Podocarpus*, a montane taxon mainly from the wet Atlantic side of the Cordillera de Talamanca. From Colombian sites, a comparable transitional period has been documented and dated at ca. 10,100–9,600 year BP (van der Hammen and Hooghiemstra 1995). Horn (1993) pointed out that, in the Cordillera de Talamanca, *Isoetes* occurs particularly on lake shores. A drop in lake level would have provided new habitats and brought *Isoetes* populations closer to the core site. Therefore, in our study area, higher percentages of *Isoetes* are most probably related to a lowering of the lake level, resulting in a higher presence of *Isoetes* on the lake shores.

Thus, the La Chonta stadial reflects a moist phase, and the subsequent pollen zone 6 reflects a drier phase. *Isoetes* collected in Colombia are mainly subaquatic (Cleef 1981), and a rise in *Isoetes* in Colombian records indicates higher lake levels (Hooghiemstra 1989). Taxa indicating cool and moist conditions, such as monolete fern spores, decrease in number during the latest stage of the La Chonta stadial. Cyatheaceae peak at the end of the cold event, showing a transitional phase to early Holocene oak forest. We estimate that the upper forest line decreased by ca. 300–400 m during the La Chonta stadial. A temperature drop of 2–3 °C is inferred using the local calculated lapse rate of 0.57 °C/100 m proposed by Kappelle et al. (1995, Chap. 4). The transition to early Holocene conditions is marked by the formation of local *Alnus acuminata* swamp. This phenomenon indicates fast vegetation change, after which the bog became again completely surrounded by oak forest. Between 9,800 and 6,000 ^{14}C years BP, the upper forest line reached the present-day altitude level of 3,300 m.

3.6 Vegetation of the Late Glacial–Holocene Transition

The present data show altitudinal changes of taxa from the oak forest during the Late Glacial–Holocene transition. The vegetation of that period had a mixture of species of montane and subalpine forests; the latter covered a greater area in the Cordillera de Talamanca. So, this represents not simply an altitudinal shift of the oak forest, but rather a rearrangement of taxa within the belt. LGM and Late Glacial vegetation types can be described as dominated by paramo. From the Late Glacial to the early Holocene, oak forest was present up to 2,000–2,400 m altitude. Forest composition may not necessarily have been the same as that of the present day, but *Quercus* was the dominant montane tree species at the time (see Chap. 2). Taxa from the lower montane forest are not that well represented, probably due to the location of the La Chonta coring site, and our data indicate less species diversity during glacial periods, compared to the Holocene. Upper montane forest in Costa Rica today covers an area of 1,037 km^2 (Iremonger et al. 1997), and probably this was reduced to more or less 600 km^2 during the Late Glacial period.

3.7 Regional Younger Dryas

From a lowland site of 300 m altitude in Guatemala, a temperature increase has been suggested for the Late Glacial, promoting a wider distribution of temperate oak forest (Leyden et al. 1994). In the Cordillera de Talamanca, a more humid Late Glacial was associated with an expansion of montane oak forest. Similar conditions were found in Central Mexico (Lozano-Garcia and Ortega-Guerrero 1994). Leyden et al. (1994) have reported a sudden cooling for the Younger Dryas Chron at Lake Quexil in lowland Guatemala (Petén), which is consistent with our Costa Rican data. Results from Panama, however, do not show cooling during the Younger Dryas Chron (Bush et al. 1992). At the Lake La Yaguada site, transition from warm to wet conditions was recorded for the time period 11,000–10,500 years BP (Bush et al. 1992), and Leyden (1995) suggested that monsoonal influence may have guarded against cooler conditions in Panama. The onset of the Holocene in Guatemala and Costa Rica is interpreted as increasing precipitation during the period 10,400–9,800 years BP (Leyden et al. 1994; Islebe et al. 1996).

In the Colombian Andes, a reduction in precipitation, and a decline in temperature of 1–3 °C at 2,550 m altitude has been proposed for the El Abra stadial, the equivalent of the Younger Dryas (Van't Veer et al. 2000). However, the date of the end of the El Abra stadial is still insufficiently documented. The lack of better time control is still a severe problem in our understanding of the Younger Dryas in Central and South America. However, all data available to date indicate the presence of important impacts on vegetation types. Additional high-resolution pollen studies are needed to fully understand past vegetation and climate dynamics.

3.8 Conclusions

Past climate change has profoundly impacted the montane oak forests in Central America, as illustrated by this example from fossil pollen studies carried out in Costa Rica. In fact, montane oak forests were significantly affected by glacial conditions, and the upper forest line shifted over 1,000 m during the Holocene. The forest-bog study site of La Chonta was located within paramo vegetation during the Pleniglacial. Cooling may have ranged from 6 to even 9 °C. In this time period, a sudden cooling event during the last stage of the Pleistocene - known as the Younger Dryas Chron - has been reported for the Costa Rican highlands.

Acknowledgements We thank the late Gerard Noldus, as well as Elly Beglinger, Annemarie Philip, Maarten Kappelle, and Ron van't Veer for support during fieldwork, data analysis, and synthesis.

References

Bush MB, Piperno DR, Colinvaux PA, DeOliveira PE, Krissek LA, Miller MC, Rowe WE (1992) A 14,300 yr paleoecological record profile of a lowland tropical lake in Panama. Ecol Monogr 62:251–275

Clapperton CM, Hall M, Mothes P, Hole MJ, Still JW, Helmens KF, Kuhry P, Gemell AMD (1997) A Younger Dryas icecap in the equatorial Andes. Quat Res 47:13–28

Cleef AM (1981) The vegetation of the paramos of the Colombian Cordillera Oriental. Cramer, Vaduz, Diss Bot 61

Dansgaard W, White JWC, Johnson SJ (1989) The abrupt termination of the Younger Dryas climatic event. Nature 339:532–534

Hastenrath S (1973) On the Pleistocene glaciation of the Cordillera de Talamanca, Costa Rica. Z Gletscherk Glazialgeol 9:105–121

Heine JT (1993) A re-evaluation of the evidence for a Younger Dryas climatic reversal in the tropical Andes. Quat Sci Rev 12:769–779

Herrera W (1985) Clima de Costa Rica. EUNED, San José, Costa Rica

Hooghiemstra H (1984) Vegetation and climatic history of the High Plain of Bogota, Colombia: a continuous record of the last 3.5 million years. Cramer, Vaduz, Diss Bot 79

Hooghiemstra H (1989) Quaternary and Upper Pliocene glaciations and forest development in the tropical Andes: evidence from a long high-resolution pollen record from the Basin of Bogota. Palaeogeogr Palaeoclimatol Palaeoecol 72:11–26

Hooghiemstra H, Van der Hammen T (1995) A Younger Dryas climatic event in Colombia. In: Proc Konink Neder Akad Weten Ser Phys 44:47–48

Hooghiemstra H, Cleef AM, Noldus GW, Kappelle M (1992) Upper Quaternary vegetation dynamics and paleoclimatology of La Chonta bog area (Cordillera de Talamanca, Costa Rica). J Quat Sci 7:205–225

Horn SP (1993) Postglacial vegetation and fire history in the Chirripó Paramo of Costa Rica. Quat Res 40:107–116

Horn SP, Sanford RL (1992) Holocene fires in Costa Rica. Biotropica 24:354–361
Iremonger S, Ravilious C, Quinton T (1997) A statistical analysis of global forest conservation. In: Iremonger S, Ravilious C, Quinton T (eds) A global overview of forest conservation. CIFOR and WCMC, Cambridge, UK
Islebe GA, Hooghiemstra H (1997) Vegetation and climate history in montane Costa Rica since the last glacial. Quat Sci Rev 16:589–604
Islebe GA, Kappelle M (1994) A phytogeographical comparison between subalpine forests of Guatemala and Costa Rica. Feddes Rep 105:73–87
Islebe GA, Hooghiemstra H, Van der Borg K (1995) A cooling event during the Younger Dryas Chron in Costa Rica. Palaeogeogr Palaeoclim Palaeoecol 117:73–80
Islebe GA, Hooghiemstra H, Van't Veer R (1996) Holocene vegetation and water table history in two bogs of the Cordillera de Talamanca, Costa Rica. Vegetatio 124:155–171
Kappelle M (1991) Distribución altitudinal de la vegetación del Parque Nacional Chirripó, Costa Rica. Brenesia 36:1–14
Kappelle M (1995) Ecology of mature and recovering Talamancan montane *Quercus* forests, Costa Rica. University of Amsterdam, Amsterdam
Kappelle M (1996) Los bosques de roble (*Quercus*) de la Cordillera de Talamanca, Costa Rica: biodiversidad, ecología, conservación y desarrollo. Instituto Nacional de Biodiversidad (INBio), Santo Domingo de Heredia
Kappelle M, Cleef AM, Chaverri A (1992) Phytogeography of Talamanca montane *Quercus* forests, Costa Rica. J Biogeogr 19(3):299–315
Kappelle M, Van Uffelen JG, Cleef AM (1995) Altitudinal zonation of montane *Quercus* forests along two transects in the Chirripó National Park, Costa Rica. Vegetatio 119:119–153
Kennett JP (1990) The Younger Dryas cooling event: and introduction. Paleoceanography 5:891–895
Kuhry P, Hooghiemstra H, van der Hammen T, van Geel B (1993) The El Abra stadial in the Eastern Cordillera of Colombia. Quat Sci Rev 12:333–343
Lachniet MS, Seltzer GO (1999) Late Quaternary glaciation of Costa Rica. Geol Soc Am Bull 114:547–558
Leyden BW (1995) Evidence of the Younger Dryas in Central America. Quat Sci Rev 14:833–839
Leyden BW, Brenner M, Hodell DA, Curtis JH (1994) Orbital and internal forcing of climate on the Yucatán Peninsula for the past ca. 36 ka. Palaeogeogr Palaeoclimatol Palaeoecol 109:193–211
Lozano-Garcia S, Ortega-Guerrera B (1994) Palynological and magnetic susceptibility records of Lake Chalco, central Mexico. Palaeogeogr Palaeoclimatol Palaeoecol 109:177–191
Martin PS (1964) Palaeoclimatology and a tropical pollen profile. INQUA, Warsaw, pp 319–321
Peteet DM (1993) Global Younger Dryas? Quat Sci Rev 12:277–355
Peteet DM (1995) Global Younger Dryas, vol 2. Quat Sci Rev 14:811–958
Rodgers JC III, Horn SP (1996) Modern pollen spectra from Costa Rica. Palaeogeogr Palaeoclimatol Palaeoecol 124:53–71
Thompson LG, Mosley-Thompson E, Davis ME, Lin PN, Henderson KA, Cole-Dai J, Bolzan JF, Liu KB (1995) Late Glacial stage and Holocene tropical ice core record from Huascaran, Peru. Science 269:46–50
Van der Borg K, Alderliesten A, Harnton CM, de Jong AF, van Zwol NA (1987) Accelerator mass spectrometry with ^{14}C and ^{10}Be in Utrecht. Nucl Instr Meth B 29:143–145
Van der Hammen T (1978) Stratigraphy and environments of the Upper Quaternary of the El Abra corridor and rock shelters (Colombia). Palaeogeogr Palaeoclimatol Palaeoecol 25:111–162

Van der Hammen T, Hooghiemstra H (1995) The El Abra stadial, a Younger Dryas equivalent in Colombia. Quat Sci Rev 14:841–851

Van Geel B, Van der Hammen T (1973) Upper Quaternary vegetation and climatic sequence of the Fúquene area (Eastern Cordillera, Colombia). Palaeogeogr Palaeoclimatol Palaeoecol 14:9–92

Van't Veer R, Islebe GA, Hooghiemstra H (2000) Climatic change during the Younger Dryas Chron in northern South America: a test of the evidence. Quat Sci Rev 19:1821–1835

4 Altitudinal Zonation of Montane Oak Forests Along Climate and Soil Gradients in Costa Rica

M. KAPPELLE and J.-G. VAN UFFELEN

4.1 Introduction

Altitudinal forest zonation along tropical environmental gradients has fascinated scientists since Alexander von Humboldt's fabulous journey to tropical America at the beginning of the 19th Century (Von Humboldt and Bonpland 1808). Since Cuatrecasas' (1934, 1958) mid-20th Century accounts on altitudinal zonation of Andean forests, numerous scholars have attempted to understand the patterns and causes of forest zonation on wet tropical mountains (Grubb and Whitmore 1966; Whitmore and Burnham 1969; Holdridge et al. 1971; Flenley 1974; Ellenberg 1975; Lauer 1976; Grubb 1977; Tanner 1977; van der Hammen et al. 1983, 1989a; Ohsawa et al. 1985; Gentry 1988; Marrs et al. 1988; Kitayama 1992; Bruijnzeel et al. 1993; Kitayama and Mueller-Dombois 1994a, b; Kappelle and Zamora 1995; Kappelle et al. 1995a; Vitousek et al. 1995a, b; Lieberman et al. 1996; Pendry and Proctor 1997).

Still today, researchers keep wondering about the underlying causes of floristic zonation in the humid tropics, principally paying attention to climatic and edaphic factors governing biological patterns, and the responses of vegetation and individual plant species to those factors (e.g., Bruijnzeel and Veneklaas 1998; Tanner et al. 1998; Vázquez and Givnish 1998; Aiba and Kitayama 1999; Leuschner 2000; Kitayama and Aiba 2002; Ashton 2003; Kappelle 2004).

4.2 Altitudinal Transect Study

4.2.1 Sample Plots

During the last 20 years, the first author and colleagues conducted vascular plant inventories in more than 100 rectangular montane oak forest plots (size: 0.1 ha) at elevations of 1,800–3,400 m above sea level (a.s.l.) along the Pacific

Ecological Studies, Vol. 185
M. Kappelle (Ed.) Ecology and Conservation of Neotropical Montane Oak Forests
© Springer-Verlag Berlin Heidelberg 2006

and Atlantic slopes of Costa Rica's Talamancan mountain range (Kappelle et al. 1989, 1991, 1994, 1995a, b, 2000; Kappelle 1991, 1996; Kappelle and Zamora 1995; Oosterhoorn and Kappelle 2000; Chaps. 10, 15 and 17). Twenty-four of those plots were located in old-growth oak-bamboo forest at 2,000–3,200 m along the slopes of the 3,819-m-high massif of the Chirripó National Park, which is part of the Amistad Biosphere Reserve and has been declared a World Heritage Site, a Center of Plant Diversity, and an Endemic Bird Area (Kappelle and Zamora 1995; Kappelle et al. 1995a; Kappelle 1996). Results from the analysis of these 24 plots are presented here.

Plots of 0.05 ha (20×25 m) were located at altitudinal intervals of 100 m and mapped on topographic sheets (scale: 1:50,000), following van der Hammen et al. (1989b). Twelve plots were located along the Pacific slope, close to the 'Fila Cementerio de la Máquina' trail to Chirripó's summit; another 12 plots were established along the Atlantic slope, near the 'Camino de los Indios' trail in the northern sector of the park.

4.2.2 Climate Measurements

Along the Pacific slope trail, daily courses of air temperature and relative humidity were measured at 1.3 m above the forest floor (breast height), under a closed canopy in lower montane (2,000 m) and upper montane (2,700 m) oak forest. Measurements were done during the 1989 rainy and 1990 dry season, using two Lambrecht thermo-hygrographs installed on a provisional wooden table with a plastic roof and all sides completely open. Additionally, at each altitudinal interval of 100 m, temperatures were measured in top and sub-soils at intervals of 10-cm soil depth, using a digital Consort T 550 thermometer. Averages of hourly recordings (7 days) of air temperature and relative humidity were plotted in diurnal thermo-hygrograms following Ellenberg (1975), and presented in Kappelle et al. (1995a).

4.2.3 Soil Analysis

Following Sevink (1984), soil pits were dug under a closed oak forest canopy at regular altitudinal intervals of 200 m. Soil profiles were described in accordance with FAO (1977). Detailed accounts on soil analysis techniques and full descriptions of soil profiles, humus horizons, soil fauna, texture, organic matter, acidity, cation exchange capacity (CEC), base saturation and dry bulk densities can be found in van Uffelen (1991) and Kappelle et al. (1995a).

4.2.4 Forest Inventory and Community Analysis

In each plot, all terrestrial vascular plant species were collected (Kappelle 1991). Epiphytic vascular plants were collected up to 2.5 m above the forest floor. Vouchers were identified and stored at herbaria in Costa Rica (CR), The Netherlands (ASD, U), and the USA (F, MO, NY, US). Bryophytes and lichens were collected, identified, and stored in Costa Rica (CR) and Germany (B, GOET), and have been dealt with elsewhere (Kappelle and Sipman 1992; Kappelle 1996; Holz et al. 2002, Chaps. 6 and 7).

Species' growth forms were recorded, and aerial crown or shoot cover projection estimated (Braun-Blanquet 1965; Mueller-Dombois and Ellenberg 1974). Average height and cover of forest layers was estimated, and average height of canopy tree layers was measured with a clinometer. Forest structure (tree stem density and relative abundance) was analyzed for stems≥10 cm DBH (diameter at breast height), which were identified, recorded and mapped (Kappelle et al. 1995a).

Using the TWINSPAN software (Jongman et al. 1987), a polythetic divisive classification was undertaken on the data matrix (presence/absence and species-cover data), comprising 24 plots and containing 431 species, each of which occurred in at least two plots. In this way, oak forest communities and their ecological species groups were distinguished and described (Kappelle et al. 1995a).

4.3 Altitudinal Oak Forest Zonation

4.3.1 Plant Species Richness

In Chirripó's 24 oak forest plots, over 3,000 botanical specimens were collected (Kappelle 1991). Identification resulted in 431 species of mainly terrestrial vascular plants: 86 pteridophytes, one gymnosperm, 296 dicots, and 48 monocots (Kappelle et al. 1995a). Table 4.1 presents the most speciose families and genera. Five families were represented by 20 or more species, and four genera were found with 10 or more congeners. Average species richness decreased with increasing altitude, from 73 per plot in lower montane forest to 51 in upper montane forest (Fig. 4.1). Most species-rich was a Pacific plot at 2,200 m a.s.l. (97 spp.), and most species-poor a Pacific plot at 3,000 m (42 spp.). Altitudinal ranges and vernacular names of trees, shrubs, bamboos, climbers and ferns, as well as changes in woody species diversity along the altitudinal gradient have been treated elsewhere (Kappelle et al. 1991; Kappelle and Gómez 1992; Kappelle and Zamora 1995).

Table 4.1. Most diverse families and genera of terrestrial vascular plants found in the montane oak forests of Costa Rica's Chirripó National Park. Numbers of species are given within brackets

Families	Genera
Rubiaceae (26)	*Peperomia* (17)
Piperaceae (21)	*Elaphoglossum* (16)
Asteraceae (20)	*Miconia* (12)
Melastomataceae (20)	*Grammitis* (11)
Polypodiaceae (20)	*Solanum* (9)
Ericaceae (17)	*Ocotea* (8)
Lomariopsidaceae (17)	*Palicourea* (8)
Solanaceae (16)	*Asplenium* (8)
Lauraceae (15)	*Anthurium* (6)
Myrsinaceae (13)	
Araceae (11)	

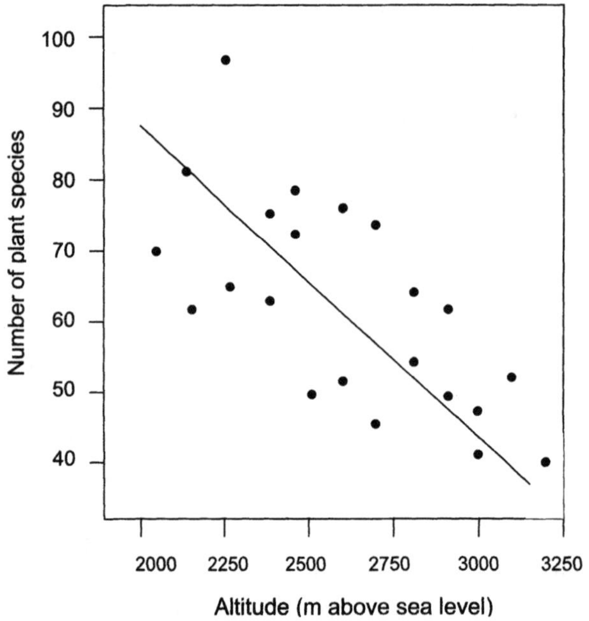

Fig. 4.1. Decrease of vascular plant species richness with altitude in montane oak forest in the Chirripó National Park, Costa Rica. *Dots* represent species numbers at 22 0.05-ha forest plots

4.3.2 Forest Layering

The average height of the canopy and subcanopy layers decreases from lower to upper montane forest, whereas the stature of the shrub, herb, and bryophyte layers remains similar (Table 4.2). This suggests a telescope-like compaction and lowering of forest structure layers at higher elevations (Kap-

Table 4.2. Mean maximum height and cover of forest structure layers for montane old-growth oak forest communities found in the Chirripó National Park, Costa Rica. Roman numbers refer to the communities described in Kappelle et al. (1995a)

Zone Slope Community	LM Pac I	LM Pac II	LM Atl III	LM Atl IV	UM Atl V	UM Atl VI	UM Pac VII	UM Pac-Atl VIII
Mean maximum height (m)								
Canopy tree layer[a]	40	35	35	35	35	35	30	25
Subcanopy tree layer	20	20	15	18	20	16	15	12
Shrub layer[a]	5	5	3	3	5	6	6	6
Herb layer	1	1	0.5	1	1	1	1	1
Bryophyte layer	0.1	0.1	0.1	0.1	0.1	0.1	0.1	0.1
Mean cover (%)								
Canopy tree layer[a]	70	70	90	80	80	80	65	60
Subcanopy tree layer	60	55	60	60	50	45	50	65
Shrub layer[a]	60	75	85	75	40	90	90	70
Herb layer	35	20	85	70	60	65	40	25
Bryophyte layer	10	5	5	12	8	60	35	50

[a] The canopy tree layer corresponds to the *Quercus* tree layer, and the shrub layer to the *Chusquea* bamboo layer as defined by Blaser (1987)

pelle 2004). Average cover values fluctuate between 60 and 90 % for oak-dominated canopy tree layers, and between 45 and 65 % for more open subcanopy tree layers. The shrub layers are dense, and reach high cover percentages due to 6-m-tall *Chusquea* bamboos. The cover of the herb layer has the greatest range (20–85 %), as a consequence of locally abundant species such as *Besleria formosa* and *Hansteinia ventricosa* or Rubiaceae (*Palicourea* and *Psychotria*). The bryophytic ground cover is also very variable, which is probably caused by local differences in light regimes at the forest floor (canopy openings; see also Chap. 15). In general, the bryophyte layer is better developed in upper montane oak forest (Holz et al. 2002, Chap. 7).

4.3.3 Tree Stem Density

Stem densities and relative abundances of tree species (DBH≥10 cm) for 0.015-ha subplots are given in Kappelle et al. (1995a; but see Chap. 10). In general, stem density does not differ significantly between lower and upper montane forest. However, a small increase in average density occurs with increasing altitude (860 stems per ha in lower montane forest vs. 1,180 stems per ha in upper montane forest). Tree density was highest at 3,200 m a.s.l. on the Pacific slope, where *Comarostaphylis arbutoides* (rel. abundance: 76.3 %)

becomes far more abundant than *Quercus costaricensis.* This forest resembles the transition to (non-oak) ericaceous subalpine forest just below the upper forest line (Islebe and Kappelle 1994). At higher elevations (>3,300 m a.s.l.), oak and non-oak forest is replaced by treeless, tropical, alpine, wet bamboo paramo (Kappelle 1991; Kappelle and Horn 2005). Whereas lower montane oak forests show a relatively open and interrupted canopy layer often dominated by emerging oaks, upper montane oak forests show a shorter, more flattened canopy (see also Chap. 10). Another conspicuous feature is the abundance of palms such as *Geonoma hoffmanniana* (Chap. 15) and *Prestoea allenii* in the understorey of the lower montane forests vs. the dense clumps of *Chusquea* bamboos in upper montane forest communities.

4.3.4 Classification of Montane Oak Forest Communities

Using TWINSPAN, eight *Chusquea-Quercus* communities were distinguished: four lower and four upper montane communities (Kappelle et al. 1995a). Montane *Chusquea-Quercus* forests at Chirripó are dominated by *Quercus copeyensis* (now known as Q. *bumelioides*; K.C. Nixon, personal communication), *Q. costaricensis* and *Q. seemannii* in the canopy, and *Chusquea* bamboos in the understorey. Common tree species are *Weinmannia pinnata* and the hemiepiphytic *Clusia stenophylla.* Other wide-ranging trees are *Saurauia veraguasensis, Prunus annularis, Styrax argenteus, Viburnum costaricanum, Ocotea pittieri,* and the large-leaved *Oreopanax capitatus.* The woody climbers *Hydrangea asterolasia* and *Smilax kunthii* occur frequently, as does the vine *Bomarea acutifolia.* Among common herbs figure *Alloplectus ichtyoderma, Begonia udisilvestris,* and the aroid *Anthurium concinnatum* that occurs epiphytically as well as with ground-dwelling habits (Chap. 15). Abundant ferns are *Asplenium serra, Arachniodes denticulata, Elaphoglossum firmum,* and *E. eximium.* Epiphytic species of *Anthurium, Elaphoglossum, Peperomia,* and *Polypodium* inhabit the bases of *Quercus* tree trunks.

Lower Montane Oak Forests

These lauraceous-fagaceous forests occur between (1,800) 2,000 and 2,600 m a.s.l., and are easily recognized because of their abundance of understorey palms (*Geonoma hoffmanniana* (Chap. 15), *Chamaedorea warszewiczii* and *Prestoea allenii*), sometimes accompanied by the bamboo *Aulonemia viscosa* or the cyclanth *Sphaeradenia irazuensis.* Dominant trees are *Quercus copeyensis, Mollinedia pinchotiana, Trichilia havanensis, Ardisia glanduloso-marginata, Tovomitopsis allenii, Billia hippocastanum, Nectandra salicina, Quetzalia (Microtropis) occidentalis, Guarea tonduzii, Alchornea latifolia, Meliosma glabrata, Miconia platyphylla, Lozania mutisiana, Ocotea austinii,* and *O. holdridgeiana*. Important herbaceous taxa (both terrestrial and epiphytic) are *Monstera deliciosa* and *Peperomia palmana*, as well as the ferns *Asplenium harpeodes* and *Vittaria graminifolia*. The climber *Cissus martiniana* is frequently observed.

Upper Montane Oak Forests

These myrsinaceous-fagaceous forests (2,600–3,200 m a.s.l.) are characterized by an understorey that is completely dominated by bamboo (*Chusquea talamancensis*), accompanied by ericads such as *Disterigma humboldtii*, *Cavendishia bracteata*, *Macleania rupestris*, and *Sphyrospermum cordifolium*. In the canopy layer, oak (*Q. costaricensis*) is accompanied by *Schefflera rodriguesiana*. Subcanopy trees include *Rhamnus oreodendron*, *Drymis granadensis*, *Miconia schnellii*, *Zanthoxylum scheryi*, and *Ilex pallida*. The ground cover is made up of herbs (*Maianthemum paniculatum*, *Centropogon costaricae*, and *Peperomia saligna*), ferns (*Blechnum viviparum* and *Elaphoglossum latifolium*), and the terrestrial bromeliad *Vriesea williamsii*. Further details on montane oak forest structure and composition are given in Chap. 10.

4.3.5 Climatic Changes Along Elevations and Between Seasons

Differences between daily courses of air temperature and relative humidity in lower and upper montane forest, and between dry and wet seasons are evident (Kappelle et al. 1995a). Lower relative humidity levels and higher temperatures at noon occurred during the dry season. Temperatures were highest during the dry season in the lower montane forest interior (23.2 °C). The lowest temperature values occurred in upper montane forest (10.8 °C). This feature is confirmed by data recorded over a 43-year period at Villa Mills (3,000 m a.s.l.) at the western border of the Chirripó National Park (Fig. 4.2).

The greatest daily temperature fluctuations were found during the dry season in the forest interior of upper montane forest (12.8–19.6 °C). Relative humidity values oscillated strongly during the dry seasons, and appeared more stable during the wet seasons. Again, the greatest daily fluctuations were recorded in the forest interior of upper montane forest (29 to >95 %). Relative humidity reached values>85 % on almost every recorded day of any season at both altitudes. This is of vital importance to epiphytic bryophytes, which cover trunks and branches in montane oak forests (Holz et al. 2002, Chap. 7).

The diurnal climatic rhythm was less pronounced in upper montane forest. This is a well-known phenomenon in tropical montane forests (Walter 1985).

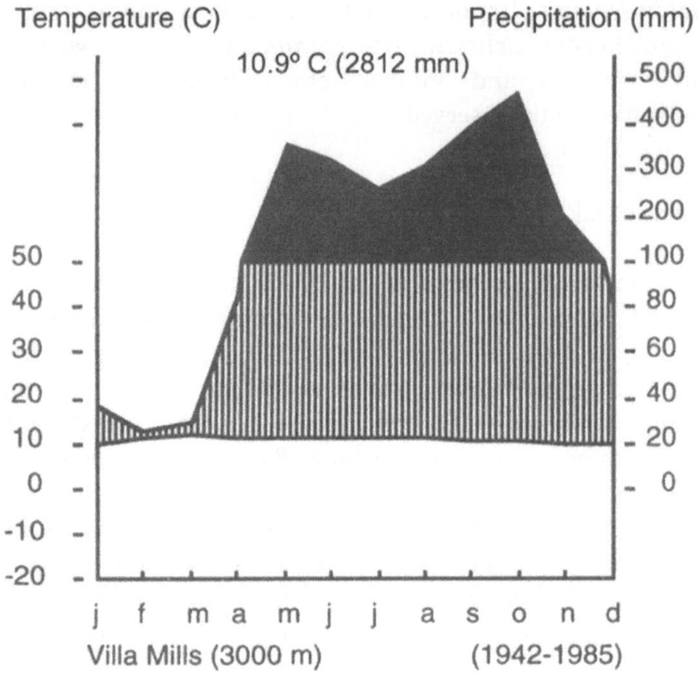

Fig. 4.2. Walter climate diagram from the Villa Mills weather station at 3,000 m a.s.l. in the Talamanca Mountains, west of Chirripó, Costa Rica

Measurements indicate that the average air temperature drops about 4.0 °C with an altitudinal increase of 700 m (3.7 °C for wet season, and 4.2 °C for dry season data). This implies a drop of 0.57 °C per 100-m increase in altitude, a value similar to that calculated for a Venezuelan montane forest (Walter 1985). Similar mean temperature drops can be derived from the subsoil temperature dataset (Kappelle et al. 1995a).

4.3.6 Soil Genesis and Classification

Soils under montane *Quercus* forests occur on moderately steep to very steep positions over both unconsolidated and consolidated substrates, and are developed in residual and colluvial material derived from parent rock (intrusive igneous and volcanic rocks). They have developed on steeply, fluvially dissected terrain, predominantly representing forms of denudational origin influenced by dendritic drainage processes. Locally, sedimentary rocks such as very fine sandstones with calcareous cement are prominent. At gently sloping, imperfectly drained positions at the Atlantic side of the mountain range, thin iron pans have formed in unconsolidated soil material derived from volcanic rock. Organic matter accumulation is higher at the Atlantic side than along the Pacific. At both slopes, andic properties are well-developed at most places, and admixture of volcanic ash in soils is obvious. Soil acidity is very high, with pH values in H_2O varying in the range 4.0–6.5 in the A-horizon. Base saturation is always lower than 15 % for soils formed over basic or acid rock types. Following the FAO (1988) classification system, soil types under closed, old-growth montane oak forest include typic placudand; typic, alic, and acric hapludands; histic, andic, and placic humitropepts; humi-haplic, humi-umbric and humi-mollic andosols; and humi-andic dystric regosol (only at the Atlantic slope; van Uffelen 1991).

4.3.7 Soil Changes Along Elevations and Between Slopes

There are no specific differences in topsoil properties between lower and upper montane oak forests, though Atlantic soils appear to be slightly more clayey than Pacific soils, and their 0-horizons are significantly thicker (Fig. 4.3). In general, the very dark-brown humus profiles are often composed of fine organic material, which is free of litter fragments and may contain

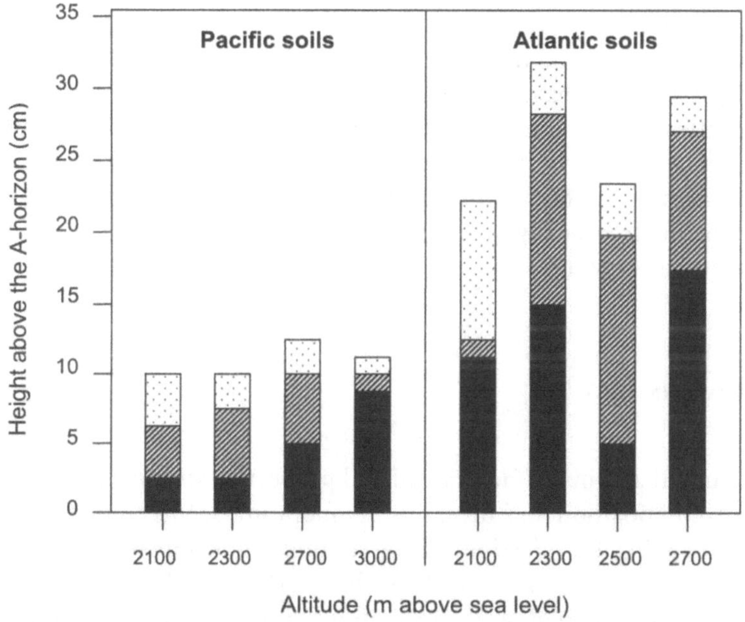

Fig. 4.3. Humus profiles of the 0-horizon of old-growth oak forest soils found along an altitudinal gradient (2,100–3,000 m a.s.l.) at the Pacific and Atlantic slopes of the Chirripó National Park. *Closed bars* Humus horizon, *dashed bars* fermentation horizon, *open bars* litter horizon

some mineral material. On the Atlantic slope, well-decomposed organic horizons are overlaid by less-decomposed horizons, and sometimes by litter layers (Fig. 4.3). On the Pacific slope, the well-decomposed organic horizon is overlaid by a horizon of about equal amounts of more or less fragmented litter and finely divided organic material. This horizon is overlaid by a litter layer. Humus profile thickness ranges from 10–40 cm (Atlantic slope) to 10–20 cm (Pacific slope). In lower parts of humus profiles, thick superficial root mats have been developed.

With increasing elevation along the Pacific slope (2,100–3,200 m a.s.l.), oak forest soils become more yellowish brown in color, with mineral soil material overlaid by layers of more sapric organic material, developed in sandy loams containing less weathered boulders but more fresh gravel and stones. Along this altitudinal gradient, the soil structure changes from very weathered, crumb-like, sub-angular blocky to less weathered and medium granular. At mid-elevation (2,300–2,700 m a.s.l.), dark gray eluvial horizons containing pure quartz grains may occur, and podsolization may take place. At all Pacific elevations, abundant roots traverse the organic soil material, which shows innumerous very fine and fine pores.

Soils along the Atlantic slope at 2,100 m a.s.l. are moderately deep and well-drained, and black to brownish yellow, overlaid by a 20-cm-thick layer of fibric to sapric organic soil material. At 2,700 m, soils are more shallow, imperfectly drained, very dark grayish brown to yellowish brown, with a dark gray, weakly developed, eluvial horizon, overlaid by a 30-cm-thick layer of fibric to sapric soil material. The organic layer thickness clearly peaks at mid-elevation (Fig. 4.3). Podsolization occurs at higher elevation. Along the Atlantic slope, mottles in the higher part of the soil profile are a result of accumulated organic matter (filled-up root spaces). Here, sandy loam is slightly sticky and plastic, and may contain rather high amounts of gravel. Larger roots may abound in the organic layer, and very fine to fine roots are common in the upper mineral horizons.

4.4 Conclusions

We conclude that climate factors and soil properties strongly influence forest structure, composition and diversity. Temperature seems to be the principal factor controlling the distribution of montane oak forest communities on Costa Rica's Chirripó Mountain. This observation is in line with conclusions drawn from studies on other tropical mountains (e.g., van der Hammen et al. 1983, 1989a; Kitayama and Mueller-Dombois 1994a, b; Vázquez and Givnish 1998; Ashton 2003; Kappelle 2004). Amounts and distribution of water vapor, nutrient availability, and light regime also play a major role in determining the forest structure and composition of montane forests on wet tropical mountains such as Chirripó. Observations at this oak-dominated mountain massif support the theory of a close correlation between the lower–upper montane forest ecotone and the diurnal cloud base, as previously documented by Grubb and Stevens (1985) for highland forests in Papua New Guinea. Ashton (2003) adds that the elevation of the diurnal cloud base is set by the relative humidity and rate of cooling of warm lowland air being conducted up slopes as it warms during the morning. This appears to be the case at Chirripó, too.

Climatic changes observed on Cerro Chirripó do not differ much from those found along altitudinal transects in Colombia (van der Hammen et al. 1983, 1989a). On Costa Rican as well as on Colombian neotropical mountains, the diurnal climate is much more pronounced than the yearly cycle. The average temperature in Chirripó's cool-humid montane oak forests depends principally on elevation, as temperature decreases with increasing altitude. A drop of 0.57 °C per 100-m increase in altitude is concordant with values estimated for other tropical mountains (Ohsawa et al. 1985; Walter 1985; Kitayama 1992). Sub-soil temperatures on Chirripó change with elevation, and reflect annual air temperatures.

Differences between hydrological regimes, as expressed in super-humid Atlantic slopes versus wet but seasonally marked Pacific slopes with a clear dry season, also play a crucial role in shaping montane forests in Costa Rica, similarly to other tropical mountains (Grubb 1977; Bruijnzeel et al. 1993; Bruijnzeel and Proctor 1995; Bruijnzeel and Veneklaas 1998). It is well known that average annual rainfall in tropical montane forests is correlated with slope orientation and fluctuates in the range 500–10,000 mm, although yearly precipitation generally shows a range of only 1,000–3,000 mm (Kappelle 2004). Ascending air masses at windward slopes bring increased precipitation to mountain ridges where they cause the formation of condensation belts, especially at mid-elevations. This is particularly the case on Costa Rica's Atlantic slope, which is strongly influenced by trade winds coming in from the Caribbean Sea under influence of the Inter Tropical Convergence Zone (ITCZ; Kappelle 1992). Moreover, the net precipitation or throughfall in these montane cloud forests is significantly enhanced beyond rainfall contribution through direct canopy interception of cloud water (horizontal precipitation), a process also known as cloud stripping (Hölscher et al. 2003, 2005, and Chap. 21). It is therefore not surprising that these magnificent oak forests are particularly rich in epiphytes, which directly obtain water from the perhumid atmosphere (Hölscher et al. 2003, and Chaps. 6, 7, 21 and 29).

Edaphic changes occurring in Costa Rica's montane oak forests appear to be strongly correlated to climate. The yellowish and acid soils on the wetter Atlantic slope are covered with thicker layers of organic material, sometimes even forming peat. Frequently, organic matter becomes more admixed with mineral soil below, and penetrates to greater depth in the soil profile, as has also been noted on Asian mountains (Whitmore and Burnham 1969; Ashton

2003). Such clay-rich soils show crumb structures resembling temperate loams, as Ashton (2003) clearly states. On Chirripó, like on other tropical mountains (Vitousek and Sanford 1986; Tanner et al. 1998; Silver et al. 2001), soils are often waterlogged and suffer from podsolization (van Uffelen 1991), a soil-forming process that causes the leaching of nutrients (lixiviation) from upper soil horizons to lower levels. These nutrient-poor, water-saturated soils may experience an anaerobic environment, associated with impeded root respiration, a reduction in belowground bioactivity, lower decomposition levels, subsequent lower rates of nutrient cycling, and reduced nutrient availability (Vitousek and Sanford 1986; Cuevas and Medina 1988; Tanner et al. 1998; Silver et al. 2001, and Chap. 22). As a result, humus accumulates in top soils (histic horizons, histosols), and nutrients are lost at top and mid soil levels (podsols). In conjunction with this, lowered mineralization rates may lead to larger fine root systems (Chap. 22). All these soil properties appear to correlate strongly with oak forest community distribution (Kappelle et al. 1995a).

The thickness of the humus profile on Chirripó's montane slopes is highest between 2,300 and 2,700 m a.s.l., probably as a consequence of low temperatures, which account for a low degree of soil bioactivity and subsequently slow decomposition processes. With respect to organic carbon levels, soils at Chirripó are similar to those on mountains in New Guinea or Jamaica (Edwards and Grubb 1977, 1982; Tanner 1977). Regarding exchangeable elements (bases), soils at Chirripó are somewhat poorer than their equivalents in Jamaica or Borneo (Tanner 1977; Kitayama 1992), but close to values measured along the La Selva–Barva Volcano altitudinal transect in Costa Rica (Marrs et al. 1988). However, contents of Ca and extractable P resemble those recorded for Mt. Kinabalu (Kitayama 1992).

Acknowledgements Numerous plant taxonomists helped with species identification; their invaluable support is gratefully acknowledged. Major funding was provided by NWO-WOTRO, the University of Amsterdam, and Costa Rica's National University and Biodiversity Institute. Research permission was provided by MINAE.

References

Aiba S, Kitayama K (1999) Structure, composition and species diversity in an altitude-substrate matrix of rain forest tree communities on Mount Kinabalu, Borneo. Plant Ecol 140:139–157

Ashton P (2003) Floristic zonation of tree communities on wet tropical mountains revisited. Perspect Plant Ecol Evol Syst 6(1/2):87–104

Blaser J (1987) Standörtliche und waldkundliche Analyse eines Eichen-Wolkenwaldes (*Quercus* spp.) der Montanstufe in Costa Rica. PhD Thesis, Georg-August Universität, Göttingen

Braun-Blanquet J (1965) Plant sociology: the study of plant communities. Hafner, London

Bruijnzeel LA, Proctor J (1995) Hydrology and biogeochemistry of tropical montane cloud forests: what do we really know? In: Hamilton LS, Juvik JO, Scatena FN (eds) Tropical montane cloud forests. Springer, Berlin Heidelberg New York, Ecological Studies 110, pp 38–78

Bruijnzeel LA, Veneklaas EJ (1998) Climatic conditions and tropical montane forest productivity: the fog has not lifted yet. Ecology 79(1):3–9

Bruijnzeel LA, Waterloo MJ, Proctor J, Kuiters AT, Kotterink B (1993) Hydrological observations in montane rain forests on Gunung Silam, Sabah, Malaysia, with special reference to the „Massenerhebung“ effect. J Ecol 81:145–167

Cuatrecasas J (1934) Observaciones geobotánicas en Colombia. Trab Mus Nac Cienc Nat Ser Bot 27:1–144

Cuatrecasas J (1958) Aspectos de la vegetación natural de Colombia. Rev Acad Col Cienc Exac Fís Nat 10(40):221–268

Cuevas E, Medina E (1988) Nutrient dynamics within Amazonian forests. II. Fine root growth, nutrient availability and leaf litter decomposition. Oecologia 76:222–235

Edwards PJ, Grubb PJ (1977) Studies on mineral cycling in a montane rain forest in New Guinea. I. The distribution of organic matter in the vegetation and soil. J Ecol 65:943–969

Edwards PJ, Grubb PJ (1982) Studies on mineral cycling in a montane rain forest in New Guinea. II. Soil characteristics and the division of mineral elements between the vegetation and soil. J Ecol 70:649–666

Ellenberg H (1975) Vegetationsstufen in perhumiden bis perariden Bereichen der tropischen Anden. Phytocoenology 2:368–387

FAO (1977) Guidelines for soil profile descriptions. UN Food and Agricultural Organization, Rome

FAO (1988) Soil map of the World with revised legend. UN Food and Agricultural Organization, Rome, World Soil Resources Rep 60

Flenley JR (ed) (1974) Altitudinal zonation of forests in Malesia. Trans 3rd Aberdeen-Hull Symp Malesian Ecology, University of Hull, Hull. Dep Miscell Ser 16

Gentry AH (1988) Changes in plant community diversity and floristic composition on environmental and geographical gradients. Ann Missouri Bot Gard 75:1–34

Grubb PJ (1977) Control of forest growth and distribution on wet tropical mountains: with special reference to mineral nutrition. Annu Rev Ecol Syst 8:83–107

Grubb PJ, Stevens PF (1985) The forests of the Fatima Basin and Mt. Kerigomna, Papua New Guinea, with a review of montane and subalpine rain forests in Papuasia. Australian National University, Canberra, Publ BG/5

Grubb PJ, Whitmore TC (1966) A comparison of montane and lowland rain forest in Ecuador. II. The climate and its effects on the distribution and physiognomy of the forests. J Ecol 54:303–333

Holdridge LR, Grenke WC, Hatheway WH, Liang T, Tosi JA (1971) Forest environments in tropical life zones. Pergamon Press, Oxford

Hölscher D, Köhler L, Leuschner C, Kappelle M (2003) Nutrient fluxes in stemflow and throughfall in three successional stages of an upper montane rain forest in Costa Rica. J Trop Ecol 19:557–565

Hölscher D, Köhler L, Leuschner C, Kappelle M (2005) Ecology and use of old-growth and recovering montane oak forest in the Cordillera de Talamanca, Costa Rica. In: Juvik J, Bruijnzeel LA (eds) Mountains in the mist: science for conserving and managing tropical montane cloud forests. University of Hawaii Press, Honolulu (in press)

Holz I, Gradstein SR, Heinrichs J, Kappelle M (2002) Bryophyte diversity, microhabitat differentiation and distribution of life forms in Costa Rican upper montane *Quercus* forest. Bryologist 105(3):334–348

Islebe GA, Kappelle M (1994) A phytogeographical comparison between subalpine forests of Guatemala and Costa Rica. Feddes Rep 105(1/2):73–87

Jongman RHG, Ter Braak CJF, van Tongeren OFR (eds) (1987) Data analysis in community and landscape ecology. Pudoc, Wageningen

Kappelle M (1991) Distribución altitudinal de la vegetación del Parque Nacional Chirripó, Costa Rica. Brenesia 36:1–14

Kappelle M (1992) Structural and floristic differences between wet Atlantic and moist Pacific montane *Myrsine–Quercus* forests in Costa Rica. In: Balslev H, Luteyn JL (eds) Páramo: an Andean ecosystem under human influence. Academic Press, London, pp 61–70

Kappelle M (1996) Los bosques de roble (*Quercus*) de la Cordillera de Talamanca, Costa Rica: biodiversidad, ecología, conservación y desarrollo. Instituto Nacional de Biodiversidad (INBio), Santo Domingo de Heredia, Costa Rica

Kappelle M (2004) Tropical montane forests. In: Burley J, Evans J, Youngquist JA (eds) Encyclopedia of Forest Sciences, vol 4. Elsevier, Oxford, UK, pp 1782–1793

Kappelle M, Gómez LD (1992) Distribution and diversity of montane pteridophytes of the Chirripó National Park, Costa Rica. Brenesia 37:67–77

Kappelle M, Horn SP (eds) (2005) Páramos de Costa Rica. Instituto Nacional de Biodiversidad (INBio), Santo Domingo de Heredia, Costa Rica

Kappelle M, Sipman HJM (1992) Foliose and fruticose lichens of Talamanca montane *Quercus* forests, Costa Rica. Brenesia 37:51–58

Kappelle M, Zamora N (1995) Changes in woody species richness along an altitudinal gradient in Talamancan montane *Quercus* forests, Costa Rica. In: Churchill SP, Balslev H, Forero E, Luteyn JL (eds) Biodiversity and conservation of Neotropical montane forests. New York Botanical Garden Press, Bronx, NY, pp 135–148

Kappelle M, Cleef AM, Chaverri A (1989) Phytosociology of montane *Chusquea-Quercus* forests, Cordillera de Talamanca, Costa Rica. Brenesia 32:73–105

Kappelle M, Zamora N, Flores T (1991) Flora leñosa de la zona alta (2000-3819 m) de la Cordillera de Talamanca, Costa Rica. Brenesia 34:121–144

Kappelle M, van Velzen HP, Wijtzes WH (1994) Plant communities of montane secondary vegetation in the Cordillera de Talamanca, Costa Rica. Phytocoenology 22(4):449–484

Kappelle M, van Uffelen JG, Cleef AM (1995a) Altitudinal zonation of montane *Quercus* forests along two transects in the Chirripó National Park, Costa Rica. Vegetatio 119:119–153

Kappelle M, Kennis PAF, de Vries RAJ (1995b) Changes in diversity along a successional gradient in a Costa Rican upper montane *Quercus* forest. Biodiv Conserv 4:10–34

Kappelle M, van Omme E, Juárez ME (2000) Lista de la flora vascular terrestre de la cuenca superior del Río Savegre, San Gerardo de Dota, Costa Rica. Acta Bot Mex 51:1–38

Kitayama K (1992) An altitudinal transect study of the vegetation of Mount Kinabalu, Borneo. Vegetatio 102:149–171

Kitayama K, Aiba S (2002) Ecosystem structure and productivity of tropical rain forests along altitudinal gradients with contrasting soil P pools on Mount Kinabalu, Borneo. J Ecol 90:37–51

Kitayama K, Mueller-Dombois D (1994a) An altitudinal transect analysis of the windward vegetation on Haleakala, a Hawaiian mountain: (1) climate and soils. Phytocoenology 24:111–133

Kitayama K, Mueller-Dombois D (1994b) An altitudinal transect analysis of the windward vegetation on Haleakala, a Hawaiian mountain: (2) Vegetation zonation. Phytocoenology 24:135-154
Lauer W (1976) Zur Hygrischen Höhenstufung Tropischer Gebirge. In: Sioli H (ed) Neotropische Ökosysteme. Biogeographica 7:169–182
Leuschner C (2000) Are high elevations in tropical mountains arid environments for plants? Ecology 81:1425–1436
Lieberman D, Lieberman M, Peralta R, Hartshorn GS (1996) Tropical forest structure and composition on a large-scale altitudinal gradient in Costa Rica. J Ecol 84:137–152
Marrs RH, Proctor J, Heaney A, Mountford MD (1988) Changes in soil nitrogen – mineralization and nitrification along an altitudinal transect in tropical rainforest in Costa Rica. J Ecol 76:466–482
Mueller-Dombois DR, Ellenberg H (1974) Aims and methods of vegetation ecology. Wiley, New York, NY
Ohsawa M, Nainggolan PHJ, Tanaka N, Anwar C (1985) Altitudinal zonation of forest vegetation on Mount Kerinci, Sumatra: with comparisons to zonation in the temperate region of East Asia. J Trop Ecol 1:193–216
Oosterhoorn M, Kappelle M (2000) Vegetation structure and composition along an interior-edge-exterior gradient in a Costa Rican montane cloud forest. For Ecol Manage 126:291–307
Pendry CA, Proctor J (1997) Altitudinal zonation of rain forest on Bukit Belalong, Brunei: soils, forest structure and floristics. J Trop Ecol 13:221–241
Sevink J (1984) An altitudinal sequence of soils in the Sierra Nevada de Santa Marta. In: Van der Hammen T, Ruiz PM (eds) La Sierra Nevada de Santa Marta (Colombia), transecto Buritaca – La Cumbre. Cramer, Vaduz, Liechtenstein, Studies in Tropical Andean Ecosystems, vol 2, pp 131–138
Silver WL, Marín-Spiotta E, Lugo AE (2001) El Caribe. In: Kappelle M, Brown AD (eds) Bosques nublados del Neotrópico. World Conservation Union (IUCN) – Instituto Nacional de Biodiversidad (INBio), Santo Domingo de Heredia, Costa Rica, pp 155–181
Tanner EVJ (1977) Four montane rain forests of Jamaica: a quantitative characterization of the floristics, the soils and the foliar mineral levels, and a discussion of the interrelations. J Ecol 65:883–918
Tanner EVJ, Vitousek PM, Cuevas E (1998) Experimental investigation of nutrient limitation of forest growth on wet tropical mountains. Ecology 79:10–22
Van der Hammen T, Preciado AP, Pinto P (eds) (1983) La Cordillera Central Colombiana, transecto Parque Los Nevados (introducción y datos iniciales). Cramer, Vaduz, Liechtenstein, Studies in Tropical Andean Ecosystems, vol 1
Van der Hammen T, Diaz S, Álvarez VJ (eds) (1989a) La Cordillera Central Colombiana, transecto Parque Los Nevados (segunda parte). Cramer, Berlin, Studies in Tropical Andean Ecosystems, vol 3
Van der Hammen T, Mueller-Dombois D, Little MA (1989b) Manual of methods for mountain transect studies. International Union of Biological Sciences (IUBS), Paris
Van Uffelen JG (1991) A geological, geomorphological and soil transect study of the Chirripó Massif and adjacent areas, Cordillera de Talamanca, Costa Rica. MSc Thesis, Wageningen Agricultural University, Wageningen, The Netherlands
Vázquez JA, Givnish TJ (1998) Altitudinal gradients in tropical forest compositon, structure, and diversity in the Sierra de Manantlán. J Ecol 86:999–1020
Vitousek PM, Sanford RL (1986) Nutrient cycling in moist tropical forest. Annu Rev Ecol Syst 17:137–167
Vitousek PM, Turner DR, Kitayama K (1995a) Foliar nutrients during long-term soil development in Hawaiian montane rain forest. Ecology 76:712–720

Vitousek PM, Gerrish G, Turner DR, Walker LR, Mueller-Dombois D (1995b) Litterfall and nutrient cycling in four Hawaiian montane rainforests. J Trop Ecol 11:189–203
Von Humboldt A, Bonpland A (1808) Personal narrative of travels to the equinoctial regions of the new continent during the years 1799-1804, vols 1–6. Longman Hurst and Rees, London
Walter H (1985) Vegetation of the earth and ecological systems of the geobiosphere. Springer, Berlin Heidelberg New York
Whitmore TC, Burnham CP (1969) The altitudinal sequence of forests and soils on granite near Kuala Lumpur. Malay Nat J 22:99–118

5 Saprotrophic and Ectomycorrhizal Macrofungi of Costa Rican Oak Forests

G.M. Mueller, R.E. Halling, J. Carranza, M. Mata and J.P. Schmit

5.1 Introduction

5.1.1 Importance of Macrofungi

Macrofungi (e.g., mushrooms, boletes, puffballs, and bracket fungi) are an integral part of all forest systems, since they are intimately involved with such basic processes as nutrient cycling, nutrient uptake, and decomposition of organic matter (see citations in Mueller and Bills 2004). Many trees, including species of *Quercus* (oak), *Alnus* (alder) and *Pinus* (pine), have evolved a highly specialized mutualistic relationship, termed ectomycorrhiza, with certain macrofungi to promote these processes. This relationship is often obligatory for the growth and survival of both parts of the association. Other macrofungi are primary decomposers of cellulose, hemicelluloses, and lignin. Many macrofungi are also important food sources for small mammals, and food sources and egg laying sites for insects and other invertebrates. Additionally, people use macrofungi for food and for their medicinal qualities. In many parts of the world, these fungi make up an important component of the community's diet as well as the local economy, as they are sold to supplement income.

5.1.2 Need for Scientific Knowledge

Information on the species composition, distribution, and host specificity of these fungi that play such fundamental roles remains fragmentary at best, especially for tropical forests (Hawksworth 2001; Mueller and Bills 2004; Hawksworth and Mueller 2005). Knowledge of the fungi involved in ectomycorrhizas and the potential of their host or site specificity is crucial for developing forest management plans and reforestation programs (see citations in

Ecological Studies, Vol. 185
M. Kappelle (Ed.) Ecology and Conservation of Neotropical Montane Oak Forests
© Springer-Verlag Berlin Heidelberg 2006

Mueller and Bills 2004). Likewise, data on the diversity and substrata specificity for tree pathogens and litter-decomposing fungi are crucial for forest management and forest ecology studies (Mueller et al. 2004a). Knowledge on host and substrata specificity of macrofungi in general, and especially for tropical species, is poor. Dramatic shifts in ectomycorrhizal hosts have been postulated by several authors to at least partially explain widely disjunct distributions of some agarics (see citations in Mueller and Halling 1995; Halling 1996, 2001). Additional data are necessary to assess the frequency of such host shifts. Similarly, attention needs to be given to potential substrata specificity of saprotrophic taxa.

Baseline data on species composition in tropical forests also are needed so that potential changes in tropical mycotas due to pollution, global climate changes, forest fragmentation, and/or other factors can be monitored. In Europe, there has been a marked change in species composition of macrofungi during the past 20–30 years, with several previously common fungi now no longer encountered and a number of other species placed on several countries' Red Lists (see Pegler et al. 1993). Without baseline data, these observed changes would not have been detected.

Additionally, data on tropical fungal diversity and species composition are necessary for understanding the evolutionary history of fungi and the organisms with which they are intimately associated. For example, the mode of formation and composition of the Central American flora and fauna following the closure of the Panama gap has been discussed many times (e.g., Raven and Axelrod 1975; Stehli and Webb 1985; Kappelle et al. 1992; papers being published as part of the 2004 Missouri Botanical Garden Fall Symposium, and Chap. 2).

Little has been published on the development of fungal communities in relation to the Great American Interchange (see Mueller and Halling 1995; Halling 2001; Halling and Mueller 2002, 2005). According to current estimates (e.g., Hawksworth 2001; Schmit et al. 2005; Hawksworth and Mueller 2005), fungi are the second largest (next to insects) and least known group of eukaryotic organisms, with less than 5 % of the hypothesized 1.5 million species described. However, since the model used to make these predictions is based almost exclusively on the knowledge of the European and United States mycotas, information on the mycotas from other regions, especially the tropics, is necessary to test the hypothesis (May 1991). Evidence of strong host specificity and narrowly distributed species in tropical countries would be congruent with this estimate. Conversely, a preponderance of fungal species that grow in both temperate and tropical regions would cast doubt on the model. Rigorous data on the diversity and species composition of tropical fungi are essential for addressing these issues (Schmit et al. 2005).

5.1.3 Macrofungal Research in Costa Rica

In this chapter, we summarize macrofungal biodiversity data that we have compiled during nearly 20 years of work in Costa Rica. Costa Rica is a logical choice to undertake studies aimed at understanding the diversity, species composition, and evolution of tropical fungi. Costa Rica ranks as one of the top 20 countries in biodiversity of plants and animals. Within a relatively small area, there are numerous ecosystems ranging from mangrove forests on the coasts to alpine vegetation in the páramo. Due to the relatively small number of mycorrhizal hosts in the country, (no native species of Pinaceae, 12 species of *Quercus*, *Alnus acuminata*, and *Comarostaphylis arbutoides*), Costa Rican forests are appropriate systems to analyze host and substrata specificity, and to investigate whether fungi that form ectomycorrhizas shift hosts during migration.

Costa Rican *Quercus*-dominated forests are also excellent laboratories to study the Great American Interchange, as their dominant canopy-tree species of oak is of temperate origin, whereas their understorey is principally composed of tropical elements (Kappelle 1996). Much general data are available on the ecology and species composition of plants and animals (Janzen 1983, www.inbio.ac.cr). The knowledge of Costa Rica's mycotas is growing through our work and that of others (Mueller and Mata 2001, and references listed on www.nybg.org/bsci/res/hall and www.ots.ac.cr/cn/binabitrop.shtml). Importantly, Costa Rica was until recently the only Central American country to have an active systematic mycology program, thus allowing for the collaboration essential for long-term year-round studies. Our work is being done in conjunction with the National Biodiversity Inventory of Costa Rica. The National Inventory covers fungi (macrofungi, microfungi and lichens), several arthropod groups, plants, and nematodes, so our data are part of a large dataset that facilitates comparisons across taxonomic groups. Additionally, a major goal of the National Biodiversity Inventory of Costa Rica is to make biodiversity data available to diverse public sources, so that it can be widely used.

The Costa Rican Fungal Inventory, of which our study is a part, is generating the most complete dataset on fungal diversity of any large region outside of a few countries in Western Europe. In addition to our survey of macrofungi, there are ongoing inventories of microfungi (Huhndorf and Umaña, unpublished data), lichenized fungi (Lücking et al. 2004), and slime molds (Schnittler and Stephenson 2000). Combining results from these studies will enable us to directly test current diversity estimates, and to predict fungal diversity in Central America and the neotropics.

5.2 Methods

5.2.1 Macrofungal Sampling

Collecting and documenting taxa of macrofungi are very time-consuming activities, so logistic constraints dictated the number of sites and frequency of sampling that could be undertaken during the project (Lodge et al. 2004). Macrofungi are identifiable in the field only by the characters found in the basidiomata (macroscopic spore-producing structures, i.e., the mushrooms, brackets, etc.), but the periodicity of ephemeral basidiomata production of most macrofungi is impossible to predict, except to say that sufficient moisture/rainfall availability is an overriding requirement. Because of the unlikelihood of being able to find all, or even most, of the macrofungal species represented in a given area in a single visit, repeated trips to selected collecting sites were undertaken. Sampling for additional years and at additional sites will undoubtedly uncover additional species, and refine our knowledge of species distributions and community composition (Schmit et al. 1999; Mueller et al. 2004b; Schmit et al. 2005; Hawksworth and Mueller 2005).

5.2.2 Information Sources and Data Analysis

Information on Costa Rican macrofungi is available from a number of sources. A large number of journal articles describing new taxa, or describing distribution patterns of Costa Rican fungi have been published, as well as several field guides on Costa Rican fungi (Mata 1999; Mata et al. 2003; Halling and Mueller 2005). A new website, www.ots.ac.cr/cn/binabitrop.shtml, created by the Organization for Tropical Studies (OTS), provides a searchable database of systematic papers published on Costa Rican biota, including fungi (BINABITROP-OTS, Bibliografía Nacional en Biología Tropical, www.ots.ac.cr/cn/binabitrop.shtml). The reader should refer to this website, and the list of publications on www.nybg.org/bsci/res/hall for citations of Costa Rican macrofungal publications, as we do not provide a literature review in this paper, due to space limitations. Data on many of the specimens that form the basis for this chapter are available at www.nybg.org/bsci/res/hall and http://atta.inbio.ac.cr/attaing/atta03.html. We analyze our data on fleshy macrofungi (Agaricales sensu Singer 1986; Euagarics and Russulales, sensu Monclavo et al. 2002), and tough and woody polyporoid fungi (Hawksworth et al. 1995) separately, as they each show distinct patterns of diversity and levels of endemism. Fleshy macrofungi are further separated in this chapter into those that putatively form ectomycorrhizas with species of *Quercus*, and those that are putatively saprotrophic in Costa Rican oak-dominated forests.

5.3 Results

5.3.1 Polyporid Fungi

The woody or tough macrofungi (e.g., bracket, coral, resupinate, and tooth fungi) are a heterogeneous group scattered among a number of orders (e.g., woody macrofungi with poroid hymenophores are now classified in Poriales, Hymenochaetales, Bondarzewiales, Fistulinales, Ganodermatales, and Herimembers; Hawksworth et al. 1995). Ryvarden (1991) reports 132 genera of pore fungi; 91 of these genera have a tropical distribution. Extensive studies have been carried out in Costa Rica and to date, 72 genera and 231 species have been reported from this country (Carranza 1996; Carranza and Ryvarden 1998). However, some areas have not been well sampled (e.g., the Osa and Caribbean La Amistad (Talamanca) conservation areas), so it is likely that new or unreported species for the country await discovery. Other groups of woody or tough macrofungi, such as the diverse resupinate „corticioids", have been examined only recently (K.H. Larsson, unpublished data), and numerous new species and records in these groups undoubtedly remain to be discovered, too.

Fewer species of polyporoid fungi are found in *Quercus*-dominated forests than in other Costa Rican forest types. Polypore fungi found in oak forest are adapted to significant daily fluctuations in temperature and to high humidity levels throughout the year. Polypore species richness in Costa Rica is greatest from sea level up to 900 m where tree diversity is high and more favorable environmental conditions are found throughout the year. Most of the polypore genera occurring in Costa Rican *Quercus*-dominated forests are cosmopolitan; common genera include *Ganoderma, Bjerkandera, Coltricia, Coriolopsis, Cyclomyces, Daedalea, Fistulina, Fomes, Fuscocerrena, Laetiporus, Perenniporia, Phellinus, Polyporus, Tyromyces,* and *Trametes.* Some species, such as *Fistulina hepatica,* are restricted to *Quercus* and *Alnus* trees. The genus *Phellinus* seems to be well adapted to decaying oak wood, and is commonly found in these forests. It represents the most species-rich genus of polyporoid fungi in Costa Rica. A list of polypore fungi commonly encountered in Costa Rican *Quercus*-dominated forests is provided in Table 5.1.

Table 5.1. List of polypore fungi commonly encountered in Costa Rican *Quercus*-dominated forests

Species
Anomoporia myceliosa (Peck) Pouzar
Bjerkandera adusta (Willd.: Fr.) P. Karst
Cerrena unicolor (Bull.: Fr.) Murrill
Daedalea quercina L.: Fr.
Fuscocerrena portoricensis (Spreng.: Fr.) Ryvarden
Ganoderma applanatum (Pers.) Pat.s.l.
Ischnoderma resinosum (Fr.) P. Karst.
Laetiporus sulphureus (Bull.: Fr.) Murrill
Oxyporus latemarginatus (Durieu & Mont.) Donk
Phellinus fastuosus (Lév.) Ryvarden
Phellinus ferrugineo-velutinus (Henn.) Ryvarden
Phellinus portoricensis (Overh.) O. Fidalgo
Phellinus sarcites (Fr.) Ryvarden
Phellinus umbrinellus (Bres.) Ryvarden
Phellinus undulatus (Murrill) Ryvarden
Phylloporia pectinata (Klotzsch) Ryvarden
Polyporus dictyopus Mont.
Polyporus tricholoma Mont.
Trametes versicolor (L.: Fr.) Pilát
Trichaptum biforme (Fr.) Ryvarden
Trichaptum sector (Ehrenb.: Fr.) Kreisel
Tyromyces cerifluus (Berk. & M.A. Curtis) Murrill

5.3.2 Fleshy Macrofungi

General Aspects

The Agaricales (mushrooms and boletes) sensu Singer (1986), corresponding to euagaric, bolete and russuloid clades sensu Monclavo et al. (2002), is the second largest order of Basidiomycetes (the order containing the rusts has fewer families and genera but more reported species). Singer (1986) recognized 18 families, 230 genera, and slightly over 5,000 species in the Agaricales. There are undoubtedly many more genera and species of Agaricales, based on the rate of new taxa being described (see Hawksworth 2001; Hawksworth and Mueller 2005). We have collected undescribed taxa (a number of which are published, plus many others waiting further work) in each of our trips to Costa Rican and Colombian *Quercus* forests (see citations listed on www.ots.ac.cr/cn/binabitrop.shtml, and www.nybg.org/bsci/res/hall). Approximately 10 % of 223 representative species illustrated on our website www.nybg.org/bsci/res/hall are unpublished. We have not seen a decline in the rate at which novel taxa are being found.

Mueller and Halling (1995) reported a high degree of endemism in the genera that they surveyed, 28–100 % of the species being unknown outside of montane Costa Rica and/or Colombia. We have identified nearly 400 species from Costa Rican *Quercus*-dominated montane forests (NYBG, FMNH, and INBio databases, plus Halling and Mueller's website since 1997). Many species in genera such as *Agaricus*, *Cortinarius*, *Inocybe*, *Marasmius* sensu lato, *Mycena*, *Psathyrella*, and *Russula* remain unidentified, and we estimate that there are approximately 600 agaric species in these forests, with an additional 400–500 species occurring in other forest types of Costa Rica. These estimates are based on several observations, including (1) there is little species overlap between different *Quercus*-dominated forests in Costa Rica, and (2) the species discussed by Ovrebo (1996) from La Selva, and those we have encountered in lowland and mid-elevation forests in the Osa Penninsula and Arenal region show very little overlap with the species that we report from *Quercus*-dominated montane forests.

Roughly half of the 400 identified agarics from Costa Rican montane *Quercus*-dominated forests are ectomycorrhizal, the other half being putatively saprotophic. However, whereas nearly 1/3 of the ectomycorrhizal species are putative, neotropical montane *Quercus* forest endemics, less than 10 % of the reported saprotrophs, are restricted to such forests. Thus, although the reported species richness of the two ecological guilds is similar, biogeographic patterns among the species vary greatly. Halling and Mueller (2002) discuss biogeographic patterns of montane neotropical Agaricales.

Ectomycorrhizal Species

Ectomycorrhizal fungi are limited in Costa Rica to forests dominated by oaks, with the exception of some species associated with *Alnus acuminata* and *Comarostaphylis arbutoides*, and those growing with planted pines and *Eucalyptus*. Southern neotropical oak-associated ectomycorrhizal agarics and boletes exhibit the following distribution patterns: north temperate origin with a distribution into Costa Rica; north temperate origin with a distribution southward through Costa Rica into southern Colombia; neotropical montane endemics; and local endemics with restricted distributions. Table 5.2 lists the ectomycorrhizal fungi commonly encountered in Costa Rican oak-dominated forests.

A high percentage of ectomycorrhizal species are putative endemics of Central and South American oak-dominated forests. Most of the non-endemic ectomycorrhizal species appear to have migrated from eastern North America with their trophic partners. Although there are some very commonly encountered ectomycorrhizal species in these forests, a number of species have rarely been observed and/or are restricted to one or two sites. Table 5.3 lists many of the putative endemic neotropical montane *Quercus*-dominated forest agarics, boletes, and russuloids. The percentage of putative endemics

Table 5.2. List of ectomycorrhizal fungal species commonly encountered in Costa Rican *Quercus*-dominated forests

Species
Amanita brunneolocularis Tulloss, Ovrebo & Halling
Boletus auriporus Peck
Boletus quercophilus Halling & G.M. Mueller
Calostoma cinnabarinum Desvaux
Cantharellus cibarius Fr.
Cantharellus ignicolor R.H. Petersen
Cortinarius violaceus (Fr.) Gray
Craterellus boyacensis Singer
Gyroporus castaneus (Bull.: Fr.) Quél.
Hydnum repandum L.: Fr.
Hygrocybe cantharellus Schw.) Fr.
Hygrocybe conica (Fr.) P. Kumm.
Hygrocybe laeta (Fr.) P. Kumm.
Hygrocybe miniata (Fr.) P. Kumm.
Laccaria amethystina Cooke
Laccaria gomezii G.M. Mueller & Singer
Laccaria major G.M. Mueller et al. nom. prov.
Laccaria ohiensis (Mont.) Singer
Lactarius chrysorheus Fr.
Lactarius deceptivus Peck
Lactarius rubidus (Hesler & Smith) Methven
Lactarius gerardii Peck var. *gerardii* Peck
Lactarius indigo (Schwein.) Fr.
Lactarius plinthofragilis Methven & Halling, nom. prov.
Lactarius rimosellus Peck
Leccinum andinum Halling
Leccinum monticola Halling & G.M. Mueller
Leccinum talamancae Halling, L.D. Gómez & Lannoy
Leotia lubrica Fr.
Phylloporus phaeoxanthus Singer & L.D. Gómez
Pulveroboletus ravenelii (Berk. & M.A. Curtis) Murrill
Russula burlinghamiae Singer
Russula compacta Frost in Peck
Russula nigricans sensu lato
Strobilomyces spp.
Tricholoma caligatum (Viv.) Ricken
Tylopilus bulbosus Halling & G.M. Mueller
Tylopilus cartagoense Wolfe & Bougher
Tylopilus obscurus Halling

Table 5.3. Partial list of putative neotropical *Quercus*-dominated forest ectomycorrhizal macrofungal endemics

Species
Amanita arocheae Tulloss, Ovrebo & Halling
Amanita brunneolocularis Tulloss, Ovrebo & Halling
Amanita colombiana Tulloss, Ovrebo & Halling
Amanita conara Tulloss & Halling
Amanita costaricensis Tulloss, Halling, G.M. Mueller & Singer
Amanita flavoconia G.F. Atk. var. *inquinata* Tulloss, Ovrebo & Halling
Amanita fuligineodisca Tulloss, Ovrebo & Halling
Amanita garabitoana Tulloss, Halling & G.M. Mueller
Amanita sororcula Tulloss, Ovrebo & Halling
Amanita talamancae nom. prov.
Amanita xylinivolva Tulloss, Ovrebo & Halling
Boletus flavoniger Halling, G.M. Mueller & L.D. Gómez
Boletus lychnipes Halling & G.M. Mueller
Boletus neoregius Halling & G.M. Mueller
Boletus quercophilus Halling & G.M. Mueller
Cantharellus atrolilacinus Eyssartier, Buyck & Halling
Chalciporus chontae Halling & M. Mata
Cortinarius „chaconii" nom. prov.
Cortinarius „rubicolor" nom. prov.
Cortinarius aureopigmentatus Ammirati et al., nom. prov.
Cortinarius grandibasalis Ammirati et al., nom. prov.
Cortinarius quercoarmillatus nom. prov.
Cortinarius savegrensis Ammirati et al., nom. prov.
Craterellus boyacensis Singer
Laccaria gomezii G.M. Mueller & Singer
Laccaria major G.M. Mueller et al., nom. prov.
Lactarius costaricensis Singer
Leccinum andinum Halling
Leccinum monticola Halling & G.M. Mueller
Leccinum neotropicalis Halling
Leccinum tablense Halling & G.M. Mueller
Leccinum talamancae Halling, L.D. Gómez & Lannoy
Phylloporus aurantiacus Halling & G.M. Mueller
Phylloporus centroamericanus Singer & L.D. Gómez
Phylloporus phaeoxanthus Singer & L.D. Gómez
Phylloporus phaeoxanthus var. *simplex* Singer & L.D. Gómez
Rozites colombiana Halling & Ovrebo
Russula „atromarginata" nom. prov.
Russula cartaginis Buyck & Halling
Russula quercophila Buyck & Halling
Tylopilus alkalixanthus Halling & Amtoft
Tylopilus bulbosus Halling & G.M. Mueller
Tylopilus cartagoense Wolfe and Bougher
Tylopilus gomezii Singer
Tylopilus obscurus Halling
Tylopilus pseudoobscurus nom. prov.

does not seem to be evenly distributed throughout agaric genera and families. For example, we have reported a high number of new bolete and *Amanita* species from southern neotropical oak forests, whereas the number of unique species in *Russula* and *Lactarius* is currently relatively low. Questions regarding the actual distribution of the putative endemics remain. For example, several species previously known only from southern neotropical oak forests (Costa Rica and Colombia) have been found in oak-dominated forests of Guatemala, e.g., *Amanita garabitoana*, *Laccaria gomezii*, and *L. major* (G.M. Mueller, O. Morales, and R. Cáceras, unpublished data).

Saprotrophic Species

In general, saprotrophic agarics are more widely distributed than species that form ectomycorrhizas. However, fewer data are currently available on the diversity, distribution, and species composition of saprotrophic agarics in neotropical oak-dominated forests, because of lack of resources for identification. Less work has been done on these fungi worldwide than on ectomycorrhizal fungi. Our preliminary data indicate that north temperate, tropical, and endemic elements occur in neotropical montane forests. Whereas many

Table 5.4. Partial list of putative neotropical *Quercus*-dominated forest saprotrophic agaric endemics

Species
Clitopilus griseobrunneus T.J. Baroni & Halling
Gymnopus lodgeae (Singer) J.L. Mata
Gymnopus omphaloides (Berk.) Halling
Gymnopus macropus Halling
Lentinula aciculospora J.L. Mata & R.H. Petersen
Marasmius perlongispermus Singer
Marasmius tropalis nom. prov.
Phaeocollybia talamancensis E. Horak & Halling, nom. prov.
Phaeocollybia ambigua E. Horak & Halling
Phaeocollybia caudata E. Horak & Halling
Phaeocollybia oligoporpa Singer
Phaeocollybia pseudolugubris Bandala & E. Horak
Phaeocollybia quercetorum Singer
Phaeocollybia singularis E. Horak & Halling
Rhodocybe mellea T.J. Baroni & Ovrebo
Rhodocybe umbrosa T.J. Baroni & Halling
Rhodocollybia pandipes Halling & J.L. Mata
Rhodocollybia popayanica (Halling) Halling
Rhodocollybia tablensis J.L. Mata
Rhodocollybia turpis (Halling) Halling
Tricholomopsis humboldtii Singer, Ovrebo & Halling
Tulostoma matae Calonge & Carranza

saprotrophic species occur in both anectotrophic and ectotrophic forests, species composition of saprotrophic agarics varies between oak-dominated and other types of tropical forests in Costa Rica. Observed distribution patterns for saprotrophic macrofungi include amphi-Atlantic, neotropical, pantropical, Cordilleran, and Gondwanan endemics. Species of *Mycena* sensu lato, *Marasmius* sensu lato, *Marasmiellus*, *Hygropus*, *Gymnopus*, *Rhodocollybia*, *Hypholoma*, *Galerina*, *Pleurotus*, *Crepidotus*, *Coprinus* sensu lato, *Phaeocollybia*, and *Psathyrella* are commonly encountered in these forests. Table 5.4 lists some putative endemic saprotrophic agarics found in neotropical montane oak forests.

5.4 Conclusions

The macrofungi of Costa Rican *Quercus*-dominated forests are taxonomically rich, with over 500 agaric and polypore species reported and another 300–400 species estimated to occur in these forests. Most of these species have distinct distribution patterns, and except for some of the polypore species, cosmopolitan species are rare. Our data to date are congruent with the hypothesis of high global species numbers, in that many of the fungi display some level of host specificity and are relatively narrowly distributed.

Wood-inhabiting polypore fungi are the best-known group of macrofungi in these forests. Although most polypore species are widely distributed, the species composition documented from tropical oak-dominated forests is distinct from that of other Costa Rican forest types. The diversity of polypore fungi is lower in oak-dominated montane forests than in most other forest types in Costa Rica, as these forests lack many tropical elements. Many of the polypore fungi in these forests show nearly cosmopolitan distributions.

Roughly half of the 400 identified agarics from Costa Rican montane *Quercus*-dominated forests are ectomycorrhizal, the other half being putatively saprotophic. However, whereas nearly 1/3 of the ectomycorrhizal species are putative, neotropical montane *Quercus* forest endemics, less than 10 % of the reported saprotrophs, are restricted to such forests. Thus, although the reported species richness of the two ecological guilds is similar, biogeographic patterns among the species vary greatly. Our data from *Quercus*-dominated forest suggest that ectomycorrhizal fungi migrated from the north with their trophic partners (oaks), as most of the species collected either show a range extension from eastern North America or are new species, with their putative sister taxon found in eastern North America. We have seen little indication of long-distance dispersal, as few taxa previously known only to areas outside of the Americas have been found. Observed distribution patterns for saprotrophic macrofungi include amphi-Atlantic, neotropical, pantropical, Cordilleran, and Gondwanan endemics.

The results presented in this chapter were obtained through an unprecedented collaboration between multiple institutions in Costa Rica and the USA. Combining the complementary strengths of each of the partner institutions enabled us to obtain, interpret, and disseminate data in here-to-fore impossible ways. We hope that this study is not only generating the most completed dataset on fungi from the tropics, but that it will also serve as a model for how to undertake biotic inventories of megadiverse countries (Mueller and Mata 2001).

Acknowledgements This project has been supported through a number of grants to the authors. G.M. Mueller and R.E. Halling gratefully acknowledge the support of the National Science Foundation for its support through three grants – The Agaricales of Costa Rican *Quercus* Forests (NSF and USAID, grant number DEB-9300798) and Macrofungi of Costa Rica (NSF grant numbers DEB-9972018 and DEB-9972027). G.M. Mueller received funding from the MacArthur Foundation to initiate this work. The World Bank, NORAD (Norwegian Aid for Developing Countries Organization), and The Netherlands government are acknowledged for their support of the National Biodiversity Inventory of Costa Rica. Many people contributed to this project through invaluable help in the field and herbarium. This list includes the parataxonomists and technicians at INBio, Mitzi Campos and students at USJ, Betty Strack, Sabine Huhndorf and Robert Lücking at Field Museum, and the many mycologists who accompanied us in the field. We wish to especially mention a debt of gratitude to Luis Diego Gómez for initiating the study of Costa Rican agarics and boletes through his work, and by introducing Rolf Singer to the incredible diversity of Costa Rican fungi. Luis Diego was the one who first took G.M. Mueller into the Costa Rica oak-dominated forests in 1986, and then in 1991 introduced us to the incredible montane oak forests near San Gerardo de Dota, where much of the data forming the basis of this chapter were collected.

References

Carranza J (1996) Distribution of pore fungi (Aphyllophorales: Basidiomycotina) in the biotic units of Costa Rica. Rev Biol Trop 44 Suppl 4:103–109

Carranza J, Ryvarden L (1998) Additional list of pore fungi of Costa Rica. Mycotaxon 69:377–390

Halling RE (1996) Boletaceae (Agaricales): latitudinal biodiversity and biological interactions in Costa Rica and Columbia. In: Carranza J, Mueller GM (eds) Fungi of Costa Rica: selected studies on biodiversity and ecology. Rev Biol Trop 44 Suppl 4: 111–114

Halling RE (2001) Ectomycorrhizae: co-evolution, significance and biogeography. Ann Missouri Bot Gard 88:5–13

Halling RE, Mueller GM (1997) Macrofungi of Costa Rica. http://www.nybg.org/bsci/res/hall

Halling RE, Mueller GM (2002) Agarics and boletes of Neotropical oakwoods. In: Watling R, Frankland JC, Ainsworth AM, Isaac S, Robinson CH (eds) Tropical mycology: macromycetes, vol 1. CABI, Wallingford, Oxon, UK, pp 1–10

Halling RE, Mueller GM (2005) Common mushrooms of the Talamanca Mountains, Costa Rica. The New York Botanical Garden Press, Bronx, NY

Hawksworth DL (2001) The magnitude of fungal diversity: the 1.5 million species estimate revisited. Mycol Res 105:1422–1432

Hawksworth DL, Mueller GM (2005) Fungal communities: their diversity and distribution. In: Dighton J, Oudemans P, White J (eds) The fungal community: its organization and role in the ecosystem, 3rd edn. Dekker, New York pp 27–37

Hawksworth DL, Kirk PM, Sutton BC, Pegler, DN (1995) Ainsworth and Bisby's dictionary of fungi, 8th edn. CABI, Wallingford, Oxon, UK

Janzen DH (ed) (1983) Costa Rican natural history. Univ Chicago Press, Chicago, IL

Kappelle M (1996) Los bosques de roble (*Quercus*) de la Cordillera de Talamanca, Costa Rica: biodiversidad, ecología, conservacíon y desarrollo. Instituto Nacional de Biodiversidad INBio), Santo Domingo de Heredia, Costa Rica

Kappelle M, Cleef AM, Chaverri A (1992) Phytogeography of Talamanca montane *Quercus* forest, Costa Rica. J Biogeogr 19:299–315

Lodge JD, Ammirati JF, O'Dell TE, Mueller GM, Huhndorf SM, Wang CJ, Stokland JN, Schmit JP, Ryvarden L, Leacock PR, Mata M, Umaña L, Wu QX, Czederpiltz DL (2004) Terrestrial and lignicolous macrofungi. In: Mueller GM, Bills GF, Foster MS (eds) Biodiversity of fungi: inventory and monitoring methods. Elsevier/Academic Press, San Diego, CA, pp 128–172

Lücking R, Sipman HJM, Umaña Tenorio L (2004) Ticolichen – the Costa Rican lichen biodiversity inventory as a model for lichen inventories in the tropics. In: Poster Abstr Vol 5th IAL Symp Lichens in Focus, August 2004, Tartu, Estonia

Mata M (1999) Macrohongos de Costa Rica, vol 1. Instituto Nacional de Biodiversidad (INBio), Santa Domingo de Heredia, Costa Rica

Mata M, Halling RE, Mueller GM (2003) Macrohongos de Costa Rica, vol 2. Instituto Nacional de Biodiversidad (INBio), Santa Domingo de Heredia, Costa Rica

May R (1991) A fondness for fungi. Nature 352:475–476

Moncalvo J M, Vilgalys R, Redhead SA, Johnson JE, James TY, Aime MC, Hofstetter V, Verduin SJW, Larsson E, Baroni TJ, Thorn RG, Jacobsson S, Clemencon H, Miller OK Jr (2002) One hundred and seventeen clades of euagarics. Mol Phylogen Evol 23:357–400

Mueller GM, Bills GF (2004) Introduction. In: Mueller GM, Bills G, Foster MS (eds) Biodiversity of fungi: inventory and monitoring methods. Elsevier/Academic Press, San Diego, CA, pp 1–4

Mueller GM, Halling RE (1995) Evidence for high biodiversity of Agaricales (Fungi) in neotropical montane *Quercus* forests. In: Churchill SP, Balslev H, Forero E, Luteyn JL (eds) Biodiversity and conservation of neotropical montane forests. The New York Botanical Garden Press, Bronx, NY, pp 303–312

Mueller GM, Mata M (2001) Costa Rican national fungal inventory: a large scale collaborative project. Inoculum 52:1–4

Mueller GM, Bills GF, Foster MS (eds) (2004a) Biodiversity of fungi: inventory and monitoring methods. Elsevier/Academic Press, San Diego, CA

Mueller GM, Schmit JP, Huhndorf SM, Ryvarden L, O'Dell TE, Lodge DJ , Leacock PR, Mata MM, Umaña L, Wu QX, Czederpiltz D (2004b) Recommended protocols for sampling macrofungi. In: Mueller GM, Bills GF, Foster MS (eds) Biodiversity of fungi: inventory and monitoring methods. Elsevier/Academic Press, San Diego, CA, pp 168–172

Ovrebo CL (1996) The agaric flora (Agaricales) of La Selva Biological Station, Costa Rica. Rev Biol Trop 44 Suppl 4:39–57

Pegler DN, Boddy L, Ing B, Kirk PM (1993) Fungi of Europe: investigation, recording and conservation. The Royal Botanical Gardens, Kew, UK

Raven PH, Axelrod, DI. (1975) History of the flora and fauna of Latin America. Am Sci 63:420–429

Ryvarden L (1991) Genera of Polypores: nomenclature and taxonomy. Synopsis Fungorum 5, Fungiflora, Norway
Schmit JP, Murphy JF, Mueller GM (1999) Macrofungal diversity in a temperate oak forest: a test of species richness estimators. Can J Bot 77:1014–1027
Schmit JP, Mueller GM, Leacock PR, Mata JL, Wu QX, Huang YQ (2005) Assessment of tree species richness as a surrogate for macrofungal species richness. Biol Conserv 121:99–110
Schnittler M, Stephenson SL (2000) Myxomycete biodiversity in four different forest types in Costa Rica. Mycology 92:626–637
Singer R (1986) The Agaricales in modern taxonomy, 4th edn. Koeltz, Koenigstein, Germany
Stehli FG, Webb SD (eds) (1985) The great American biotic interchange. Plenum Press, New York

6 Diversity and Biogeography of Lichens in Neotropical Montane Oak Forests

H.J.M. Sipman

6.1 Introduction

Oaks, being slow-growing hardwood trees, are a very suitable substrate for lichens. This is shown by observations in the temperate zone of the northern hemisphere, where *Quercus* species are often dominant trees in the natural forest. In Great Britain, for instance, over 300 lichen species, or 22 % of the total lichen flora, occur as epiphytes on oak (Rose 1974). More recently, the oak forests of the Mediterranean have received considerable attention (Fos 1998; Atienza 1999; Alvarez Andrés and Carballal Durán 2000; Zedda 2002b). These authors report 102–331 species on oak in the investigated forests. Lichen inventories were used for phytogeographical considerations (Barreno et al. 1992) and for monitoring of habitat modification (Zedda 2002a). Knops et al. (1997) studied the effect of lichens on nutrient cycling in oak wood, and Wolseley and Pryor (1999) developed a monitoring system by means of the lichen growth on twigs. Rose (1974) treats lichen community patterns on different parts of oak trees.

The oak forests of the neotropical mountains often show abundant lichen growth, seemingly no less than in the temperate zone. The montane environment, with often high precipitation, frequent fog and moderate temperatures, is very suitable for lichens. Crown twigs may carry loads of the yellowish, bushy beard lichen (*Usnea* spp.), whereas older branches are usually covered with whitish patches of leafy lichens belonging to the families Parmeliaceae and Physciaceae, in particular the genera *Hypotrachyna*, *Parmotrema* and *Heterodermia*. In more shady situations, large individuals of the genera *Lobaria* and *Sticta* are conspicuous, and most of the bark not covered by these lichens – or bryophytes (Chap. 7) – tends to be covered by greyish crustose lichens.

Nevertheless, the lichen flora of these forests is little known. This can largely be explained by insufficient general knowledge of the neotropical

Ecological Studies, Vol. 185
M. Kappelle (Ed.) Ecology and Conservation of Neotropical Montane Oak Forests
© Springer-Verlag Berlin Heidelberg 2006

lichen flora, which hampers or prevents reliable identifications. This applies even to conspicuous genera such as *Lobaria*, *Sticta* and *Usnea*. These genera are evidently represented by dozens of species, but currently only few can be reliably identified. Recent decades have seen a considerable increase in taxonomic knowledge of tropical lichens, but still many, predominantly crustose, genera lack a modern treatment or key. Therefore, it is no surprise that few people have ever studied the lichen flora of neotropical oak forests in any detail. Kappelle included epiphytic macrolichens while investigating the regeneration of oak forests in the Cordillera de Talamanca, Costa Rica (Kappelle and Sipman 1992). Holz (2003, Chap. 7) extended his detailed investigations of the epiphytic flora of primary and secondary oak forests in the same area to macrolichens. Phytogeographical relations of the lichen flora of Chiapas oak forests were treated by Sipman (1996) on a poster. Otherwise, information on the lichen flora is restricted to herbarium collections. This is difficult to retrieve, because the pertinent herbarium specimens often remain unpublished, or are published scattered in taxonomic literature. Moreover, label information on forest type and carrying tree is usually restricted. The high tree diversity in the neotropical mountains usually prohibits an adequate characterisation of the lichen habitat.

Consequently, the preparation of a representative list of epiphytic lichens on neotropical oaks is beyond the scope of this chapter. Rather, a list of epiphytic lichens from oak forests is presented, which includes also species growing on epiphytic substrates other than *Quercus* – on bark, decorticated wood or leaves (Appendix 6.1). It is compiled from the literature indicated above, augmented with information from the lichen herbarium database of the Botanical Museum in Berlin. This includes data from oak woods in five countries which, however, is in no case based on any intensive inventory of such forests. Consequently, the list is very incomplete and does not reflect well the actual distribution of the species. The distributional information is included only to give some estimate of the frequency of recording.

6.2 Floristic Composition

The dominant genera of the list are *Cladonia*, *Heterodermia*, *Hypotrachyna*, *Leptogium*, *Parmotrema*, *Ramalina* and *Sticta*. All belong to the morphotype macrolichens: *Cladonia* and *Ramalina* are fruticose, the others foliose. Genera with crustose morphotype are less well represented, and all records identified to genus only belong here. This dominance of macrolichens certainly reflects the investigations by Kappelle and Sipman (1992) and Holz (2003), which were restricted to macrolichens. Still, the lichen flora of montane oak woods seems often dominated by foliose and fruticose lichens. Another cause might be the advanced taxonomic knowledge of these groups. For four of the listed genera (*Cladonia*, *Heterodermia*, *Hypotrachyna*, *Parmotrema*), modern revisions are available, and for two more (*Leptogium*, *Ramalina*), partial or unpublished revisions.

The good representation of recently revised groups and the poor representation of crustose lichens are an indication that the list is far from complete for the neotropical oak forests. Another indication is that many widespread species are represented only once in the list, in particular crustose lichens from substrates requiring special attention like decorticated wood and living leaves. Therefore, it can be inferred that the actual epiphytic lichen flora of the neotropical oak forests is much larger than the 464 listed species, and probably closer to 1,000 species.

This value, compared with those for European oak forests presented above, suggests that the tropical oak forests have a richer epiphytic lichen flora. This assumption is supported by observations in other tropical forests, where species numbers observed on individual trees far exceed those from temperate areas. Individual oaks in Britain have usually no more than 30 species, with a maximum of 52 (Rose 1974, p. 266), and a temperate tree with 76 species is considered extremely rich (Jarman and Kantvilas 1995). In tropical (lowland) forest, Komposch and Hafellner (2000) found a range of 45–84, with a mean of 65 species per individual tree. The highest number ever found was 173 on a tree in tropical montane forest in New Guinea (Aptroot 1997). Additional research is needed to assess whether oak forests are richer than other neotropical forests.

6.3 Phytogeographical Considerations

The present information suggests that 85 % of the epiphytic lichen species are widespread, occurring all over the neotropics (Table 6.1) or (66 %) all over the tropics. This dominance of widespread species is common among lichens. It may have been intensified by the fact that widespread species are more likely to have been described and identified.

A small number of neotropical species are of special phytogeographical interest because they seem to have their limit in the oak forests. The most remarkable is a group of nine species distributed predominantly in the northern temperate zone, which seem to reach their southern limit in the neotropical oak forest. This includes four widespread temperate lichens, *Cetrelia cetrarioides*, *Lobaria pulmonaria*, *Mycoblastus sanguinarius* and *Pyrrhospora elabens*, and five North American taxa, *Bryoria furcellata*, *Fuscopannaria leucosticta*, *Parmotrema stuppeum*, *Pseudevernia consocians* and *P. intensa*. Most of them reach Costa Rica; only *Cetrelia cetrarioides*, *Fuscopannaria leucosticta* and *Pseudevernia intensa* were not found south of Chiapas. Four species have a restricted range in the northern part of the oak forest region, and

Table 6.1. Main lichen distribution types and their frequency. Most lichens have a wide distribution

Main distribution type	Frequency
Widespread in the tropics and beyond	290 species=66 %
Widespread in the neotropics	127 species=19 %
Restricted to the oak forest region	17 species=4 %
Southern limit in the oak forest region	Nine species=2 %
Western limit the oak forest region	Four species=1 %
Northern limit in the oak forest region	16 species=4 %
Distribution type unknown	29 species
Total	464 species

extend into the Caribbean: *Cladonia botryocarpa, C. caribaea, C. pulverulenta* and *Parmotrema viridiflavum*. A further 16 species have their main distribution further south in the neotropics, mainly in the northern Andes, and reach their northern limit in the oak forest region, usually at high elevations in Costa Rica: *Anzia leucobates, A. masonii, A. parasitica, Erioderma gloriosum, E. granulosum, E. laminisorediatum, E. marcellii, Everniastrum fragile, Hypotrachyna caraccensis, H. halei, H. monilifera, H. partita, Lecania sulphureofusca, Megalospora admixta, Oropogon lorobic* and *Parmotrema virescens*.

To our current knowledge, only 17 species are restricted to the neotropical oak forest region. This includes six inconspicuous, crustose species which are easily overlooked (*Acanthothecis subclavulifera, Chrismofulvea omalia, Graphis stygioarachnoides, Phaeographina strigops, Thelotrema conveniens, Thelotrema occlusum*). The remainder are macrolichens whose distributions seem fairly well established, suggesting the presence of two diversity centres. Most of them, 11 species, are essentially restricted to Mexico: *Anzia* cf. *masonii, Graphis stygioarachnoides, Phaeographina strigops, Cladonia jaliscana, Everniastrum neocirrhatum, Oropogon caespitosus, O. mexicanus, Parmotrema chiapense, P. eurysacum, P. mesogenes* and *P. moreliense*. These were found mainly in Chiapas, a few further north, and two have extensions into Guatemala and El Salvador. Five species seem restricted to the southern part of the neotropical oak forest region, Colombia, Panama and Costa Rica: the crustose species *Chrismofulvea omalia, Thelotrema conveniens* and *T. occlusum*, and the macrolichens *Hypotrachyna protoboliviana* and *Sticta ferax*. One species, *Acanthothecis subclavulifera*, deviates from this pattern, in that it is rather widely distributed in the oak forest region from Costa Rica to Chiapas.

6.4 Conclusions

The neotropical montane oak forests harbour a rich lichen flora, which is currently very little known. A preliminary list of 464 species encountered in such forests suggests that the species richness is at least comparable with that of temperate oak forests, but probably the actual diversity is at least twice as large. Most of the lichens have a wide distribution in the neotropics, often also in the other tropical regions, sometimes even with extensions into the temperate zones. They are apparently able to colonize various tree species under tropical conditions, and have no strong affinity with oaks in the neotropics. Only 17 species are, as far as is currently known, restricted to the neotropical oak countries – 11 to Mexico, five to the region Colombia–Costa Rica, and one from Costa Rica to Chiapas. Information about their host trees is scanty, and there is no evidence that they are bound to oak or oak forest. Noteworthy are nine species occurring mainly in the northern temperate zone, which have their southern limit in the oak forests of Chiapas or Costa Rica, four predominantly Caribbean species which reach the area, and 16 species which have their northern limit in the area. They are the only evidence for a bridge function of the area for lichens.

Acknowledgements Knowledge of the neotropical oak forest lichens owes much to ecologists paying attention to 'lesser' organisms in their study plots. Therefore, I acknowledge gratefully I. Holz and M. Kappelle. Persons too numerous to list most kindly enabled the gathering of further data by supporting my fieldwork. A special mention deserve J. Aguirre (Bogotá, Colombia), M. Escobar (Medellín, Colombia), R. Gradstein (Göttingen, Germany), P. Maas (Utrecht, The Netherlands), H. Valencia (Bogotá, Colombia), L. Umaña (Santo Domingo, Costa Rica), R. Veloso (Popayán, Colombia), B. ter Welle (Zeist, The Netherlands) and J. Wolf (at that time at San Cristobal de las Casas, Mexico). I am very grateful to all of them.

Appendix 6.1. Epiphytic lichens observed in neotropical montane oak forests (COL=Colombia, 224 species; CR=Costa Rica, 164 species; GUA=Guatemala, three species; MEX=México, 213 species, SAL=El Salvador, 112 species)

Species
Acanthothecis subclavulifera Staiger & Kalb (**MEX**)
Anisomeridium sp. (**COL**)
Anzia americana Yoshim. & Sharp (**COL; CR**), *A. leucobates* (Nyl.) Müll.Arg. (**CR**), *A. masonii* Yoshim. (**CR**), *A.* cf. *masonii* (**MEX**), *A. parasitica* (Fée) Zahlbr. (**CR**)
Arthonia cinnabarina (DC.) Wallr. (**COL; SAL; MEX**), *A. palmulacea* (Müll.Arg.) R.Sant. (**COL**)
Arthopyrenia sp. (**CR**)
Arthothelium sp. (**CR**)
Aspidothelium cinerascens Vain. (**COL**)
Astrothelium cf. *gigasporum* R.C.Harris (**MEX**)
Auriculora byssomorpha (Nyl.) Kalb (**COL; CR**)
Bacidia fragilis Vezda (**COL**), *B. sublecanorina* (Nyl.) Zahlbr. (**COL**)
Bacidina apiahica (Müll.Arg.) Vezda (**COL**)
Bacidiopsora squamulosula (Nyl.) Kalb (**COL; CR**)
Bapalmuia palmularis (Müll.Arg.) Sérus. (**COL; SAL**)
Bathelium cf. *madreporiforme* (Eschw.) Trevis. (**MEX**)
Brigantiaea leucoxantha (Spreng.) R.Sant. & Hafellner (**COL; MEX**)
Bryoria furcellata (Fr.) Brodo & D.Hawksw. (**CR**)
Bulbothrix ventricosa (Hale & Kurok.) Hale (**SAL**)
Bunodophoron melanocarpum (Sw.) Wedin (**COL; CR; SAL; MEX**)
Byssoloma chlorinum (Vain.) Zahlbr. (**MEX**), *B. fadenii* Vezda (**SAL**), *B. leucoblepharum* (Nyl.) Vain. (**COL**), *B. subdiscordans* (Nyl.) P.James (**COL**), *B. tricholomum* (Mont.) Zahlbr. (**COL**)
Calenia conspersa (Stirt.) R.Sant. (**COL**)
Calicium abietinum Pers. (**COL**), *C. glaucellum* Ach. (**COL**), *C. hyperelloides* Nyl. (**COL**), *C. tricolor* F.Wils. (**COL; MEX**)
Calopadia puiggarii (Müll.Arg.) Vezda (**COL**)
Caloplaca brebissonii (Fée) J.Sant. ex Hafellner & Poelt (**SAL; MEX**)
Candelaria concolor (Dicks.) Stein (**MEX**), *C. fibrosa* (Fr.) Müll.Arg. (**MEX**)
Candelariella cf. *reflexa* (Nyl.) Lettau (**MEX**), *C.* cf. *xanthostigma* (Ach.) Lettau (**MEX**)
Canomaculina neotropica (Kurok.) Elix (**MEX**), *C. subsumpta* (Nyl.) Elix (**COL; MEX**), *C. subtinctoria* (Zahlbr.) Elix (**MEX**)
Canoparmelia caroliniana (Nyl.) Elix & Hale (**COL; SAL; MEX**), *C.* cf. *carneopruinata* (Zahlbr.) Elix & Hale (**MEX**), *C. eruptens* (Kurok.) Elix & Hale (**SAL**), *C. texana* (Tuck.) Elix & Hale (**SAL; MEX**)
Catinaria sp. (**SAL**)
Cetrelia cetrarioides (Delise ex Duby) W.L.Culb. & C.F.Culb. (**MEX**)
Chaenotheca brunneola (Ach.) Müll.Arg. (**COL**), *C. chlorella* (Ach.) Müll.Arg. (**COL**), *C. chrysocephala* (Turn. ex Ach.) Th.Fr. (**CR; MEX**), *C. olivaceorufa* Vain. (**COL**), *C. trichialis* (Ach.) Th.Fr. (**CR**)
Chaenothecopsis pusilla (Ach.) A.Schmidt (**COL**)
Chrismofulvea omalia Marbach (**CR**)
Chroodiscus sp. (**SAL**)
Chrysothrix candelaris (L.) Laundon (**COL; CR; MEX**)
Cladia aggregata (Sw.) Nyl. (**COL; MEX**)

Appendix 6.1. (*Continued*)

Species
Cladonia bacillaris Nyl. (**COL**), *C. botryocarpa* G.Merr. (**MEX**), *C. capitata* (Michx.) Spreng. (**COL**), *C. caribaea* Stenroos (**MEX**), *C. cartilaginea* Müll.Arg. (**CR**), *C. ceratophylla* (Sw.) Spreng. (**COL; CR; SAL; MEX**), *C. corniculata* Ahti & Kashiw. (**CR**), *C. didyma* (Fée) Vain. (**COL; CR; MEX**), *C. granulosa* (Vain.) Ahti (**COL; CR**), *C. grayi* G.Merr. ex Sandst. (**COL**), *C. jaliscana* Ahti & Guzman-Dávalos (**MEX**), *C. lepidophora* Ahti & Kashiw.? (**CR**), *C. ochrochlora* Flörke (**COL; CR**), *C. pulverulenta* (L.Scriba) Ahti (**CR**), *C. ramulosa* (With.) J.R.Laundon (**COL**), *C. rappii* A.Evans (**COL**), *C. squamosa* (Scop.) Hoffm. var. *squamosa* (**COL; CR; MEX**), *C. squamosa* var. *subsquamosa* (Nyl. ex Leight.) Vain. (**CR**), *C. subradiata* (Vain.) Sandst. (**CR; SAL**), *C. subsquamosa* Kremp. (**COL; CR**), *C. verruculosa* (Vain.) Ahti (**COL**)
Coccocarpia domingensis Vain. (**COL; CR**), *C. erythroxyli* (Spreng.) Swinscow & Krog (**CR; MEX**), *C. palmicola* (Spreng.) Arv. & D.J.Galloway (**COL; CR; MEX**), *C. pellita* (Ach.) Müll.Arg. (**COL; MEX**), *C. stellata* Tuck. (**COL**)
Coenogonium fallaciosum (Müll.Arg.) Kalb & Lücking (**SAL**), *C. linkii* Ehrenb. (**MEX**), *C. moniliforme* Tuck. (**MEX**), *C. subluteum* (Rehm) Kalb & Lücking (**COL**)
Collema glaucophthalmum Nyl. var. *glaucophthalmum* (**COL; MEX**)
Cresponea leprieurii (Mont.) Egea & Torrente (**CR**)
Cryptothecia rubrocincta (Ehrenb.: Fr.) Thor (**COL; CR; SAL; MEX**)
Dichosporidium sp. (**SAL**)
Dictyonema glabratum (Spreng.) D.Hawksw. (**CR; MEX**), *D. sericeum* (Sw.) Berk. (**MEX**)
Diorygma monophorum (Nyl.) Kalb, Staiger & Elix (**COL; SAL**)
Echinoplaca leucotrichoides (Vain.) R.Sant. (**COL**), *E. lucernifera* Kalb & Vezda (**CR**), *E. pellicula* (Müll.Arg.) R.Sant. (**COL**)
Erioderma gloriosum P.M.Jørg. & Arv. (**CR**), *E. granulosum* P.M.Jørg. & Arv. (**COL; CR**), *E. laminisorediatum* P.M.Jørg. & Arv. (**CR**), *E. leylandii* (Taylor) Müll.Arg. (**MEX**), *E. marcellii* P.M.Jørg. & Arv. (**COL; CR**), *E. mollissimum* (Samp.) Du Rietz (**COL; CR; MEX**), *E. wrightii* Tuck. (**CR**)
Everniastrum cirrhatum (Fr.) Hale ex Sipman (**COL; CR; SAL; MEX**), *E. fragile* Sipman (**COL; CR**), *E. lipidiferum* (Hale & M.Wirth) Hale ex Sipman (**SAL; MEX**), *E. neocirrhatum* (Hale & M.Wirth) Hale ex Sipman (**MEX**), *E. nigrociliatum* (Bouly de Lesd.) Hale ex Sipman (**SAL; MEX**), *E. sorocheilum* (Vain.) Hale ex Sipm. (**CR; MEX**), *E. vexans* (Zahlbr.) Hale (**COL; CR; SAL; MEX**)
Fellhanera bouteillei (Desm.) Vezda (**COL**), *F. dominicana* (Vain.) Vezda (**COL**), *F.* cf. *longispora* Lücking (**COL**), *F. stanhopeae* (Müll.Arg.) Lücking (**COL**)
Fissurina dumastii Fée (**CR**), *F. triticea* (Nyl.) Staiger (**MEX**)
Flavoparmelia caperata (L.) Hale (**MEX**)
Flavopunctelia flaventior (Stirt.) Hale (**COL; MEX**). *F. praesignis* (Nyl.) Hale (**MEX**), *F. soredica* (Nyl.) Hale (**MEX**)
Fuscopannaria leucosticta (Tuck.) P.M.Jørg. (**MEX**)
Gassicurtia rufofuscescens (Vain.) Marbach (**MEX**)
Graphina elongata (Vain.) Zahlbr. (**MEX**), *G.* cf. *nuda* H.Magn. (**COL**)
Graphis acharii Fée (**COL; MEX**), *G. anguilliformis* Tayl. (**MEX**), *G. longula* Kremp. (**COL**), *G. macella* Kremp. (**MEX**), *G. proserpens* Vain. (**COL**), *G. stygioarachnoides* M.Wirth & Hale (**MEX**)
Gyalidea epiphylla Vezda (**SAL**)
Haematomma africanum (J.Steiner) C.W.Dodge (**SAL**), *H. collatum* (Stirton) C.W.Dodge (**COL**), *H. rufidulum* (Fée) A.Massal. (**MEX**)
Hemithecium rufopallidum (Vain.) Staiger (**MEX**)

Appendix 6.1. (*Continued*)

Species
Heterodermia albicans (Pers.) Swinscow & Krog **(MEX)**. *H. antillarum* (Vain.) Swinscow & Krog **(MEX)**, *H. barbifera* (Nyl.) K.P.Singh **(SAL; MEX)**, *H. casarettiana* (A.Massal.) Trevis. **(CR; SAL; MEX)**, *H. circinalis* (Zahlbr.) W.A.Weber **(COL; CR)**, *H. corallophora* (Tayl.) Vain. **(COL)**, *H. crocea* R.C.Harris **(MEX)**, *H. echinata* (Tayl.) W.L.Culb. **(MEX)**, *H. galactophylla* (Tuck.) W.L.Culb. **(COL; SAL; MEX)**, *H. isidiophora* (Vain.) Awasthi **(COL)**, *H. lamelligera* (Tayl.) Follmann & Redon **(CR; MEX)**, *H. leucomela* (L.) Poelt **(COL; CR; SAL; MEX)**, *H. leucomela* ssp. *boryi* (Fée) Swinscow & Krog **(MEX)**, *H. lutescens* (Kurok.) Follmann & Redon **(COL; CR; MEX)**, *H. magellanica* (Zahlbr.) Swinscow & Krog **(CR; MEX)**, *H. obscurata* (Nyl.) Trevis. **(CR; MEX)**, *H. squamulosa* (Degel.) W.L.Culb. **(COL; CR; MEX)**, *H. subcomosa* (Nyl.) Elix **(SAL)**, *H. tropica* (Kurok.) Sipman **(SAL; MEX)**, *H. verrucifera* (Kurok.) W.A.Weber **(MEX)**, *H. vulgaris* (Vain.) Follmann & Redon **(COL; CR)**
Hypocenomyce scalaris (Ach.) Choisy **(CR)**
Hypotrachyna bogotensis (Vain.) Hale **(COL; CR; SAL; MEX)**, *H. caraccensis* (Tayl.) Hale **(CR)**, *H. chlorina* (Müll.Arg.) Hale **(COL; CR)**, *H. citrellla* (Kurok.) Hale **(CR)**, *H. consimilis* (Vain.) Hale **(MEX)**, *H. costaricensis* (Nyl.) Hale **(COL; CR; MEX)**, *H. croceopustulata* (Kurok.) Hale **(COL; CR; MEX)**, *H. dactylifera* (Vain.) Hale **(COL; SAL; MEX)**, *H. degelii* (Hale) Hale **(CR)**, *H. densirhizinata* (Kurok.) Hale **(COL; CR; MEX)**, *H. ducalis* (Jatta) Hale **(CR)**, *H. enderythraea* (Zahlbr.) Hale **(CR)**, *H. endochlora* (Leight.) Hale **(COL)**, *H. ensifolia* (Kurok.) Hale **(CR)**, *H. exsplendens* (Hale) Hale **(MEX)**, *H. gondylophora* (Hale) Hale **(CR)**, *H. halei* ad int. **(CR)**, *H. imbricatula* (Zahlbr.) Hale **(COL; CR; SAL; MEX)**, *H. isidiocera* (Nyl.) Hale **(MEX)**, *H. laevigata* (Smith) Hale **(COL; CR; SAL; MEX)**, *H. longiloba* (H.Magn.) Hale **(COL; CR)**, *H. microblasta* (Vain.) DR. **(COL; CR; MEX)**, *H. monilifera* (Kurok.) Hale **(CR)**, *H. norlopezii* ad int. **(CR)**, *H. osseoalba* (Vain.) Park & Hale **(COL; MEX)**, *H. partita* Hale **(CR)**, *H. physcioides* (Nyl.) Hale **(COL; CR; MEX)**, *H. prolongata* (Kurok.) Hale **(COL; CR; MEX)**, *H. protoboliviana* (Hale) Hale **(CR)**, *H. pulvinata* (Fée) Hale **(COL; CR; MEX)**, *H. reducens* (Nyl.) Hale **(COL; CR)**, *H. rockii* (Zahlbr.) Hale **(COL; CR; MEX)**, *H. sinuosa* (Smith) Hale **(CR)**, *H. sublaevigata* (Nyl.) Hale **(SAL)**
Imshaugia venezolana (Elix) Hale **(CR; MEX)**
Lecanactis epileuca (Nyl.) Tehler **(COL; MEX)**
Lecania sulphureofusca (Fée) Müll.Arg. **(COL)**
Lecanora arthothelinella Lumbsch **(COL)**, *L. caesiorubella* Ach. **(CR; MEX)**, *L. flavidomarginata* Bouly de Lesd. **(SAL)**, *L. pseudoargentata* Lumbsch **(COL)**
Lecidella sp. **(MEX)**
Lepraria sp. **(CR)**
Leprocaulon arbuscula (Nyl.) Nyl. **(CR; SAL)**
Leptogium adpressum Nyl. **(MEX)**, *L. andinum* P.M.Jørg. **(COL; CR)**, *L. azureum* (Ach.) Mont. **(MEX)**, *L. burgessii* (L.) Mont. **(COL; CR; SAL; MEX)**, *L. cochleatum* (Dicks.) P.M.Jørg. & P.James **(COL; CR; MEX)**, *L. coralloideum* (Meyen & Flot.) Vain. **(COL; CR; MEX)**, *L. cyanescens* (Ach.) Körb. **(CR; MEX)**, *L. diaphanum* (Sw.) Mont. **(COL; CR)**, *L. furfuraceum* (Harm.) Sierk **(MEX)**, *L. hibernicum* P.M.Jørg. **(MEX)**, *L. hypotrachynum* Müll.Arg. **(MEX)**, *L. laceroides* Bouly de Lesd. **(COL; CR)**, *L. mandonii* P.M.Jørg. **(CR)**, *L. olivaceum* (Hook.) Zahlbr. **(COL; CR; SAL; MEX)**, *L. papillosum* (Bouly de Lesd.) C.W.Dodge **(COL; CR)**, *L. phyllocarpum* (Pers.) Mont. **(COL; CR; MEX)**, *L. vesiculosum* (Sw.) Malme **(COL; MEX)**
Leucodecton sp. **(MEX)**

Appendix 6.1. (*Continued*)

Species
Lobaria crenulata (Hook. in Kunth) Trevis. (**CR**), *L. denudata* (Tayl.) Yoshim. (**COL**), *L. dissecta* (Nyl.) Vain. (**COL; MEX**), *L. pallida* (Hook. in Kunth) Trevis. (**COL; CR; MEX**), *L. pulmonaria* (L.) Hoffm. (**CR; MEX**), *L. subdissecta* (Nyl.) Vain. (**COL; CR**), *L. subexornata* Yoshim. (**COL; CR; MEX**), *L. submultiseriata* ad int. (**CR**), *L. tenuis* Vain. (**COL; MEX**)
Lopezaria versicolor (Fée) Kalb & Hafellner (**COL; CR; SAL; MEX**)
Malcolmiella piperis (Spreng.) Kalb & Lücking (**COL; MEX**)
Mazosia ocellata (Nyl.) R.C.Harris (**MEX**), *M. phyllosema* (Nyl.) Zahlbr. (**COL**)
Megalaria sp. (**SAL**)
Megaloblastenia marginiflexa (Hook. & Tayl.) Sipman var. *dimota* (Malme) Sipman (**SAL**)
Megalospora admixta (Nyl.) Sipman (**COL; CR**), *M. sulphurata* Meyen & Flot. var. *nigricans* (Müll.Arg.) Riddle (**COL**, Tolim (**SAL; MEX**), *M. tuberculosa* (Fée) Sipman (**COL; CR; SAL; MEX**)
Melaspilea diplasiospora (Nyl.) Müll.Arg. (**COL; CR**)
Mycoblastus sanguinarius (L.) Norm. (**CR**)
Mycomicrothelia captiosa (Kremp.) D.Hawksw. (**COL**), *M. punctata* Aptroot (**CR**)
Mycoporum sparsellum Nyl. (**COL**)
Myriotrema bahianum (Ach.) Hale (**CR** Puntarenas; **SAL; MEX**), *M. hartii* (Müll.Arg.) Hale (**COL; SAL; MEX**), *M. insigne* (Zahlbr.) Hale (**CR**), *M. protocetraricum* (Hale) Hale (**COL**), *M. urceolare* (Ach.) Hale (**CR**)
Nephroma helveticum Ach. (**CR**)
Normandina pulchella (Borrer) Nyl. (**COL; CR; SAL; MEX**)
Ocellularia calvescens (Fée) Müll.Arg. (**CR**), *O. cavata* (Ach.) Müll.Arg. (**COL; SAL; MEX**), *O. comparabilis* (Kremp.) Müll.Arg. (**SAL**), *O. domingensis* (Fée) Müll.Arg. (**SAL; MEX**), *O. interpositum* (Nyl.) Hale (**SAL**), *O. leucomelanum* (Nyl.) Hale (**SAL**), *O. maxima* (Hale) Hale (**COL; SAL**), *O. perforata* (Leight.) MA (**CR; MEX**), *O. rhodostroma* (Mont.) Zahlbr. (**CR**), *O. tenuis* (Hale) Hale (**MEX**)
Ochrolechia isidiata (Malme) Verseghy (**MEX**), *O. mexicana* Vain. (**MEX**), *O. pallescens* (L.) Massal. (**CR**)
Opegrapha filicina Müll.Arg. (**COL**)
Oropogon bicolor Essl. (**CR**), *O. caespitosus* Essl. (**MEX**), *O. diffractaica* Essl.? (**MEX**), *O. formosanus* Asah. (**COL; MEX**), *O. granulosus* Essl. (**MEX**), *O. lorobic* Essl. (**CR**), *O. loxensis* (Fée) Th.Fr. (**CR; MEX**), *O. mexicanus* Essl. (**MEX**), *O. sperlingii* Essl. (**COL; CR**)
Pannaria conoplea (Ach.) Bory (**COL; CR; MEX**), *P. mosenii* C.W.Dodge (**CR; SAL**), *P. prolificans* Vain. (**SAL**), *P. rubiginosa* (Ach.) Bory (**COL; MEX**), *P. stylophora* Vain. (**COL**), *P. tavaresiana* P.M.Jørg. (**SAL; MEX**)
Parmeliella miradorensis Vain. (**COL; MEX**), *P. pannosa* (Sw.) Müll.Arg. (**COL; CR; GUA** Alta Verapaz; **SAL; MEX**)
Parmelinella wallichiana (Tayl.) Elix & Hale (**SAL**)
Parmelinopsis horrescens (Tayl.) Elix & Hale (**COL; SAL; MEX**), *P. minarum* (Vain.) Elix & Hale (**SAL**), *P. spumosa* (Asah.) Elix & Hale (**CR; SAL**), *P. subfatiscens* (Kurok.) Elix & Hale (**SAL**)

Appendix 6.1. (*Continued*)

Species
Parmotrema arnoldii (Du Rietz) Hale (COL; CR; MEX), *P. austrosinense* (Zahlbr.) Hale (COL; SAL), *P. chiapense* (Hale) Hale (SAL), *P. conformatum* (Vain.) Hale (CR), *P. crinitum* (Ach.) Choisy (COL; SAL; MEX), *P. dilatatum* (Vain.) Hale (COL), *P. dominicanum* (Vain.) Hale (SAL), *P. eciliatum* (Nyl.) Hale (SAL; MEX), *P. endosulphureum* (Hillm.) Hale (SAL), *P. eurysacum* (Hue) Hale (MEX), *P. hababianum* (Gyeln.) Hale (SAL; MEX), *P. latissimum* (Fée) Hale (MEX), *P. leucosemotheta* (Hue) Hale (MEX), *P. madagascariaceum* (Hue) Hale (COL), *P. mellissii* (C.W.Dodge) Hale (CR; SAL; MEX), *P. mesogenes* (Nyl.) Hale (SAL), *P. moreliense* (Bouly de Lesd.) W.LCulb. & C.F.Culb. (MEX), *P. rampoddense* (Nyl.) Hale (COL; CR; SAL), *P. robustum* (Degel.) Hale (COL; CR; SAL; MEX), *P. sancti-angelii* (Lynge) Hale (SAL), *P. stuppeum* (Tayl.) Hale (MEX), *P. subrugatum* (Kremp.) Hale (SAL; MEX), *P. tinctorum* (Nyl.) Hale (GUA Alta Verapaz; SAL; MEX), *P. ultralucens* (Krog) Hale (MEX), *P. virescens* Hale (COL), *P. viridiflavum* (Hale) Hale (COL; MEX), *P. xanthinum* (Müll.Arg.) Hale (SAL; MEX)
Peltigera austroamericana Zahlbr. (CR; MEX), *P. collina* (Ach.) Schrad. (MEX), *P. dolichorhiza* (Nyl.) Nyl. (COL; CR), *P. pulverulenta* (Tayl.) Nyl. (COL)
Pertusaria sp. (SAL)
Phaeographina strigops M.Wirth & Hale (MEX)
Phaeographis dendritica (Ach.) Müll.Arg. (COL; MEX), *P. exaltata* (Mont. & v.d.Bosch) Müll.Arg. (CR; MEX), *P. intricans* (Nyl.) Staiger (COL; SAL), *P. scalpturata* (Ach.) Staiger (COL)
Phaeophyscia endococcinodes (Poelt) Essl. (MEX), *P. hispidula* (Ach.) Moberg (COL; MEX)
Phlyctella andensis (Nyl.) Nyl. (COL; SAL)
Phlyctidea boliviensis (Nyl.) Müll.Arg. (COL)
Phyllopsora buettneri (Müll.Arg.) Zahlbr. (CR; SAL), *P. chlorophaea* (Müll.Arg.) Zahlbr. (MEX), *P. corallina* (Eschw.) Müll.Arg.? (MEX), *P. cuyabensis* (Malme) Zahlbr. (MEX), *P. furfuracea* (Pers.) Zahlbr. (MEX)
Physcia atrostriata Moberg (MEX), *P. integrata* Nyl. (MEX), *P. lacinulata* Müll.Arg. (MEX), *P. lopezii* Moberg (CR), *P. sorediosa* (Vain.) Lynge (MEX)
Piccolia conspersa (Fée) Hafellner (COL)
Platythecium allosporellum (Nyl.) Staiger (MEX), *P. grammitis* (Fée) Staiger (SAL; MEX)
Polychidium dendriscum (Nyl.) Henssen (CR)
Polymeridium catapastum (Nyl.) R.C.Harris (SAL), *P. proponens* (Nyl.) R.C.Harris (MEX)
Porina barvica Lücking (COL), *P. distans* Vezda (MEX), *P. epiphylla* (Fée) Fée (COL; SAL), *P. exasperatula* Vain. (COL), *P. fulvella* Müll.Arg. (COL), *P. heterospora* (Fink) R.C.Harris (MEX), *P. leptosperma* Müll.Arg. (COL; SAL), *P. mastoidea* (Ach.) Müll.Arg. (SAL; MEX), *P. nitidula* Müll.Arg. (COL), *P. nucula* Ach. (CR), *P. octomera* (Müll.Arg.) Schilling (SAL), *P. pseudofulvella* Sérus. (COL), *P. rubentior* (Stirt.) Müll.Arg. (SAL), *P. rufula* (Kremp.) Vain. (COL; SAL), *P. umbilicata* (Müll.Arg.) Vezda (COL)
Protoparmelia sp. (SAL)
Pseudevernia consocians (Vain.) Hale & W.L.Culb. (CR; MEX), *P. intensa* (Nyl.) Hale & W.L.Culb. (MEX)
Pseudocyphellaria aurata (Ach.) Vain. (COL; CR; GUA Alta Verapaz; MEX), *P. clathrata* (De Not.) Malme (MEX), *P. crocata* (L.) Vain. (COL; CR; MEX), *P. intricata* (Del.) Vain. (SAL)
Pseudoparmelia cubensis (Nyl.) Elix & Nash (SAL)

Appendix 6.1. (*Continued*)

Species
Punctelia reddenda (Stirt.) Krog (**MEX**), *P. rudecta* (Ach.) Krog (**COL; MEX**), *P. semansiana* (W.L.Culb. & C.F.Culb.) Krog (**MEX**), *P. subrudecta* (Nyl.) Krog (**COL; MEX**)
Pyrenula anomala (Ach.) Vain. (**SAL**), *P. aspistea* (Ach.) Ach. (**CR**), *P. astroidea* (Fée) R.C.Harris (**COL**), *P. dermatodes* (Borr.) Schaer. (**SAL**), *P. martinicana* (Vain.) R.C.Harris (**CR**)
Pyrrhospora elabens (Fr.) Hafellner (**CR**), *P. russula* (Ach.) Haf. in Kalb & Haf. (**SAL**)
Pyxine eschweileri (Tuck.) Vain. (**MEX**), *P. obscurascens* Malme (**SAL**), *P. rhodesiaca* Vain. (**SAL**)
Ramalina anceps Nyl. (**MEX**), *R. andina* V.Marcano & A.Morales (**COL**), *R. asahinae* W.L.Culb. & C.F.Culb. (**COL**), *R. aspera* Räsänen (**MEX**), *R. bogotensis* Nyl. (**COL**), *R. camptospora* Nyl. (**SAL**), *R. canaguensis* V.Marcano & A.Morales (**COL**), *R. canaguensis* var. *colombiana* ad int. (**COL**), *R. celastri* (Spreng.) Krog & Swinsc. (**COL**), *R. chiguarensis* V.Marcano & A.Morales (**COL**), *R. chilensis* Bert. (**COL**), *R. cochlearis* Zahlbr. (**COL; MEX**), *R. cumanensis* Fée (**COL**), *R. dendriscoides* Nyl. (**COL**), *R. leptosperma* Nyl. (**SAL**), *R. reducta* Krog & Swinscow (**COL**), *R. rigida* (Pers.) Ach. (**CR**), *R. subcalicaris* (Nyl.) Kashiw. (**CR**), *R. tenaensis* V.Marcano & A.Morales (**COL**), *R. tenuissima* V.Marcano & A.Morales (**COL**), *R. usnea* (L.) R.Howe (**COL**), *R. vareschii* V.Marcano & A.Morales (**COL**)
Reimnitzia santensis (Tuck.) Kalb (**SAL**)
Relicina abstrusa (Vain.) Hale (**SAL; MEX**)
Rimelia cetrata (Ach.) Hale & A.Fletcher (**COL; CR; MEX**), *R. commensurata* (Hale) Hale & Fletcher (**COL**), *R. reticulata* (Tayl.) Hale & A.Fletcher (**COL; CR; SAL; MEX**), *R. simulans* (Hale) Hale & A.Fletcher (**MEX**), *R. subisidiosa* (Müll.Arg.) Hale & A.Fletcher (**COL; SAL**)
Rinodina neglecta Aptroot (**COL**)
Sarcographa cinchonarum Fée (**SAL**), *S. tricosa* (Ach.) Müll.Arg. (**MEX**)
Sclerophora sanguinea (Tibell) Tibell (**COL**)
Siphula decumbens Nyl. (**CR**)
Sticta cf. *damaecornis* (Sw.) Ach. (**COL; CR**), *S.* cf. *dufourii* Delise (**COL**), *S. ferax* Müll.Arg. (**CR**), *S.* cf. *ferax* (**MEX**), *S.* cf. *filicinella* Nyl. (**CR**), *S. fulginosa* (Dicks.) Ach. (**COL; CR**), *S. granatensis* Nyl. (**COL**), *S. gyalocarpa* (Nyl.) Trevis. (**CR**), *S.* cf. *humboldtii* Hook. (**CR**), *S.* cf. *laciniata* Ach. (**COL; CR**), *S. lenormandii* (Nyl.) Zahlbr. (**COL**), *S. orizabana* Nyl. (**COL**), *S. peltigerella* (Nyl.) Trevis. (**COL**), *S. subscrobiculata* (Nyl.) Gyeln. (**COL**), *S. tomentosa* (Sw.) Ach. (**COL; CR**), *S. tomentosa* var. *dilatata* Nyl. (**COL**), *S. weigelii* (Ach.) Vain. (**SAL; MEX**), *S.* cf. *weigelii* (**COL; MEX**)
Stirtonia sp. (**MEX**)
Strigula concreta (Fée) R.Sant. (**COL**), *S. nitidula* Mont. (**COL; MEX**), *S. obducta* (Müll.Arg.) R.C.Harris (**SAL**), *S. platypoda* (Müll.Arg.) R.C.Harris (**COL**), *S. smaragdula* Fr. (**COL**)
Syncesia farinacea (Fée) Tehler (**CR; SAL; MEX**), *S. psaroleuca* (Nyl.) Tehler (**CR; MEX**)
Tapellaria epiphylla (Müll.Arg.) R.Sant. (**COL**), *T. nana* (Fée) R.Sant. (**COL; MEX**)
Teloschistes exilis (Michx.) Vain. (**MEX**), *T. flavicans* (Sw.) Norm. (**COL; CR; SAL; MEX**)
Tephromela atra (Huds.) Hafellner (**MEX**)
Thalloloma cinnabarinum (Fée) Staiger (**SAL**)
Thelotrema conveniens Nyl. (**COL**), *T. lepadinum* (Ach.) Ach. (**COL; CR; SAL; MEX**), *T. occlusum* Nyl. (**COL**), *T. spondaicum* (Nyl.) Hale (**SAL**), *T.* cf. *subtile* Tuck. (**MEX**), *T. stylothecium* Vain. (**CR; SAL; MEX**), *T. tenue* Hale (**CR**)

Appendix 6.1. (*Continued*)

Species
Tricharia farinosa R.Sant. **(COL)**, *T. subalbostrigosa* Lücking **(COL)**, *T.* cf. *vainioi* R.Sant. **(COL)**
Trichothelium bipindense F.Schill. **(COL)**, *T. epiphyllum* Müll.Arg. **(COL)**
Trypethelium nitidiusculum (Nyl.) R.C.Harris **(MEX)**
Trypethelium ochroleucum (Eschw.) Nyl. (SAL)
Tylophoron protrudens Nyl. **(COL; MEX)**
Usnea ceratina Ach. **(COL)**, *U. rubicunda* Stirt. **(COL; CR)**

References

Alvarez Andrés J, Carballal Durán R (2000) Flora liquénica sobre *Quercus robur* L. en Galicia (NW España). Cryptogamie Mycologie 21(2):103–117

Aptroot A (1997) Lichen biodiversity in Papua New Guinea, with the report of 173 species on one tree. Bibl Lichenol 68:203–213

Atienza V (1999) Hongos liquenizados epifitos de los bosques con galler (*Quercus faginea* Lam.) al norte de la Comunidad Valenciana. But Soc Micol Valenciana 4/5:5–24

Barreno E, Sanz MJ, Atienza V, Muñoz A (1992) Biogeografía y ecología comparadas de líquenes epífitos de alcornocales ibéricos y sardos. In: Actes Simp Int Cryptogamia, 1988, Botánica Pius Font i Quer, vol 1, pp 179–185

Fos S (1998) Líquenes epífitos de los alcornocales ibéricos: correlaciones bioclimáticas, anatómicas y densimétricas con el corcho de reproducción. Servicio Editorial de la Universidad del País Vasco, Bilbao, Guinean A, no 4

Holz I (2003) Diversity and ecology of bryophytes and macrolichens in primary and secondary montane *Quercus* forests, Cordillera de Talamanca, Costa Rica. PhD Dissertation, Göttingen University, Göttingen

Jarman SJ, Kantvilas G (1995) Epiphytes on an old Huon pine tree (*Lagarostrobus franklinii*) in Tasmanian rainforest. NZ J Bot 33:65–78

Kappelle M, Sipman HJM (1992) Foliose and fruticose lichens of Talamanca montane *Quercus* forests, Costa Rica. Brenesia 37:51–58

Knops JMH, Nash TH III, Schlesinger WH (1997) The influence of epiphytic lichens on the nutrient cycling of a blue oak woodland. USDA Forest Service Gen Tech Rep PSW-GTR 160:75–82

Komposch H, Hafellner J (2000) Diversity and vertical distribution of lichens in a Venezuelan tropical lowland rain forest. Selbyana 21(1/2):11–24

Rose F (1974) The epiphytes of oak. In: Morris MG, Perring FH (eds) The British oak. Classey, Faringdon, pp 250–273

Sipman HJM (1996) The lichen flora of the Chiapas oak/pine forests, tropical or northern-temperate? In: Abstr Vol Int Senckenberg Conf Global Biodiversity Research in Europe, 9–13 December 1996, Frankfurt, p 73

Wolseley PA, Pryor KV (1999) The potential of epiphytic twig communities on *Quercus petraea* in a welsh woodland site (Tycanol) for evaluating environmental changes. Lichenology 31(1):41–61

Zedda L (2002a) Development of a hemeroby scale for oak forests in Sardinia (Italy) based on changes in the epiphytic lichen flora. In: Llimona X, Lumbsch HT, Ott S (eds) Progress and problems in lichenology at the turn of the millennium. Cramer, Berlin, Bibl Lichenol 82:257–268

Zedda L (2002b) The epiphytic lichens on *Quercus* in Sardinia (Italy) and their value as ecological indicators. Englera 24:1–457

7 Epiphytic Communities of Bryophytes and Macrolichens in a Costa Rican Montane Oak Forest

I. Holz

7.1 Introduction

Because of their complexity and variety of microhabitats, lowland and montane tropical rain forests are the habitat of many bryophytes and lichens. Most of the bryophytes and lichens of tropical rain forests are epiphytes. Even though they are often small and inconspicuous, bryophytes and lichens are an important component of tropical forest ecosystems, especially montane ones, in terms of ecosystem functioning, biomass and biodiversity (Pócs 1980, 1982; Veneklaas and Van Ek 1990; Veneklaas et al. 1990; Hofstede et al. 1993; Wolf 1993; Clark et al. 1998a, b, Chap. 6).

Whereas increasing attention has been paid to the taxonomy and diversity of tropical bryophytes and lichens, little is known about their ecology and the impacts of deforestation on these communities. Relevant aspects are degradation of biomass, loss of species diversity, and change in microclimate associated with forest destruction and fragmentation. Deforestation is generally considered to have a deleterious effect on the bryophyte flora of the primary forest, and may lead to a considerable loss of species. Therefore, analyses of epiphytic cryptogam communities should be considered a research priority for conserving biodiversity and ecosystem functions.

The present book chapter summarizes recent research on the cryptogamic vegetation of the upper montane oak forests of the Cordillera de Talamanca, Costa Rica (Holz 2003). Aspects dealt with in this chapter are the diversity and biogeography of bryophytes, the distribution patterns of life forms and species in microhabitats and along ecological gradients, the host preference and community composition of epiphytic bryophytes and macrolichens, and the secondary succession of the epiphytic cryptogam vegetation.

Ecological Studies, Vol. 185
M. Kappelle (Ed.) Ecology and Conservation of Neotropical Montane Oak Forests
© Springer-Verlag Berlin Heidelberg 2006

7.2 Study Area

The Costa Rican oak forest vegetation under study is situated in the Cordillera de Talamanca, the geological backbone of southern Central America.

The study area has a Cf climate, according to the Köppen Climate System. In general, a short dry season and a long wet season can be distinguished. At 3,000 m a.s.l. (meteorological station Villa Mills; a.s.l., above sea level), the dry season starts in December and ends in April. Climatic conditions in the Cordillera de Talamanca are very diverse, due to the region's large size, its geographic location, which includes Pacific and Caribbean watersheds, marked altitudinal differences, and the irregular and abrupt topography.

The evergreen high-elevation tropical oak forests occur in the upper montane forest belt sensu Grubb (1974), or montane belt sensu Holdridge (1967), situated between the lower montane forest belt, which has its upper limit at about 2,100 m a.s.l., and the subalpine dwarf forest belt at 3,000–3,100 m a.s.l. In general, the *Quercus* forests under study comprise stands over 40 m tall, and consist of about five vegetation layers: (1) a uniform canopy layer, generally made up of only *Quercus* trees (mainly *Quercus copeyensis* and *Q. costaricensis*), sometimes intermingled with a few other tree species; (2) a diverse 10–25 m tall subcanopy layer, covering 30–50 % of the surface; (3) a shrub layer; (4) a herb layer; and (5) a bryophyte layer. Detailed information on vegetation, structure, and physiognomy of the forests is given in Chaps. 4, 10 and 17 (see also references in these chapters).

7.3 Primary Forest

7.3.1 Species Richness and Biogeography

Montane oak forests in the Cordillera de Talamanca show a high diversity of bryophyte and macrolichen species, and great diversification of microhabitats. In all, 251 bryophyte species (128 hepatics, one hornwort, 122 mosses) were found in a recent inventory of the oak forests of Cordillera de Talamanca (Holz and Gradstein 2005b). *Lejeuneaceae*, *Plagiochilaceae* and *Lepidoziaceae* were the most important liverwort families in terms of number of species; *Dicranaceae*, *Neckeraceae*, *Meteoriaceae* and *Orthotrichaceae* were the most species-rich families of mosses. For details on lichens inhabiting neotropical montane oak forests, see Chap. 6.

In fact, 93 % of all species found have a tropical distribution. Showing a value of 74 %, the neotropical species are most important, whereas 27 % of the species are tropical Andean-centered (montane and alpine) and 7 % are restricted to the northern Andes (north of northern Peru). Represented by 21 species (8 %), the Central American 'endemics' are an important element in the flora of the oak forests. Only 4 % of the species are temperate, and 3 % are cosmopolitan in distribution. The bryophyte floras of different microhabitats within the oak forest show a phytogeographical make-up similar to that of the total oak forest bryophyte flora. However, the analysis shows that the temperate species are found only in forest floor habitats. The cosmopolitan species are also mainly restricted to the forest floor.

A comparison of the phytogeographical make-up of the bryophyte flora with that of vascular plant genera of the oak forests (Kappelle et al. 1992) shows that the bryophyte flora is more tropical in character than is the case for the vascular plant flora. The latter has more temperate and amphi-pacific taxa. We hypothesize that differences in migration rates and speciation between vascular and bryophyte species have played only a minor role in this respect, and suggest that an analysis of the phytogeography of the vascular flora at species level might confirm the trends found in the bryophytes.

7.3.2 Microhabitats and Life Forms

Microhabitats inventoried in 6 ha of primary forest at 'Los Robles' forests reserve included soil, rocks, logs, shrubs, living leaves, tree bases, trunks, branches, and twigs in the tree crowns (Holz et al. 2002). Tree bases (cf. 33 % of all species), rotting logs (34 %) and soil (34 %) are the richest habitats for bryophytes, followed by shrubs in the understorey (32 %), trunks (30 %), rocks and stones (19 %), branches of the inner canopy (17 %), twigs of the outer canopy (7 %), and leaves in the understorey (7 %). Canopy leaves were devoid of bryophytes. Canopy habitats (twigs, branches and upper portions of trunks) had less than half the number of species (73) as the forest understorey (all other habitats; 184 species). The contribution of forest floor habitats to total bryophyte species richness is much higher than in forests at lower elevations. In all, 25 % of the species occurred both in the canopy and the understorey. Species richness of hepatics and mosses was equal on logs, soil and stones, but epiphytic habitats were generally richer in hepatics.

Similarities in species composition show a strong relationship between *forest floor habitats* (including the tree base), in contrast to *epiphytic habitats.* Tree bases are a transition zone between the species of the forest floor and those growing on trunks of large trees. The bryophytes growing on living leaves (the *phyllosphere*) form a distinct group, with little relation to those of other epiphytic microhabitats.

Eight types of bryophyte life forms were recognized: cushions, feathers, mats, pendants, tails, treelets, turfs and wefts (Richards 1984; Bates 1998). Distributions of species and life forms in different forest microhabitats are correlated with humidity and light regimes, and show the importance of the pronounced dry season in the oak forests of the Cordillera de Talamanca, especially for epiphytic bryophytes.

7.3.3 Host Preference, Vertical Distribution and Community Composition

Vertical gradients in the distribution of bryophyte and lichen species on their host trees in the tropics were also demonstrated by Cornelissen and ter Steege (1989), Montfoort and Ek (1990), Wolf (1993), Gradstein et al. (2001b), Holz et al. (2002), and Acebey et al. (2003). However, a detailed analysis of the vertical distribution of cryptogamic epiphytes on trees in upper montane tropical rain forests is still lacking.

Most studies have been limited to the tree base and the lower part of the trunk, and have neglected the richness of the canopy flora. However, the epiphytic vegetation of the tree base is often heterogeneous, and more similar to that of the surrounding terrestrial vegetation than to the trunk and canopy flora (Hietz and Hietz-Seifert 1995; Clement et al. 2001; Holz et al. 2002). This limits the usefulness of the tree base flora and communities as an indicator of epiphytic diversity, or in predicting that of the canopy.

Host specificity or host preference of cryptogamic epiphytes in the tropics is widely considered to be of minor importance (e.g., Pócs 1982; Richards 1984). However, Cornelissen and ter Steege (1989) demonstrated that host specificity may indeed occur in tropical lowland forests, and Smith (1982) suggested that host preference is common among tropical bryophytes, except in very humid montane forests. Quantitative data to support this suggestion are still lacking, especially for montane rain forests and canopy species.

In addition to single species, community composition and community changes along ecological gradients may provide important information on the ecology of ecosystems. In tropical forests, conservation concepts based on single species (indicator species) may be inadequate to predict the vulnerability of this ecosystem, due to the complex niche diversification of these forests. However, beside the studies conducted by Wolf (1993) in Colombia, there is hardly any information on community composition of epiphytic bryophytes and lichens in tropical montane forests, especially with respect to the canopy.

A study of bryophytes and macrolichens on standing mature *Quercus copeyensis* and *Q. costaricensis* trees, the dominating tree species of oak forests in the Cordillera de Talamanca, provided deeper insights into community composition and distribution of epiphytic species. Ten trees (five for each of two host oak tree species) were sampled from the base up to the twigs of the outer canopy, using a single rope climbing technique. Coverage of corticolous bryophyte and macrolichen species was estimated and compared using detrended correspondence analysis (DCA, Hill and Gauch 1980) and nonmetric multi-response permutation procedure (MRPP, Mielke 1984).

Of the 153 taxa found on the ten host trees (Table 7.1), 57 were lichens, 56 hepatics, and 40 mosses. In addition to the vertical gradient, the two host tree species themselves proved to be the most important factor in community differentiation of epiphytic cryptogams, as indicated by DCA (Fig. 7.1). Many of the species are preferentially associated, or restricted to either *Quercus copeyensis* or *Q. costaricensis*. Also, non-metric MRPP confirmed significant differences in community composition of individual height zones on different host trees (Holz 2003).

Species richness per plot (~600 cm²) was highly variable, with a mean of 9.7 species (4.7 hepatics, 2.7 lichens, and 2.3 mosses) and a high standard deviation of 3.5 species. There was no difference between the average number

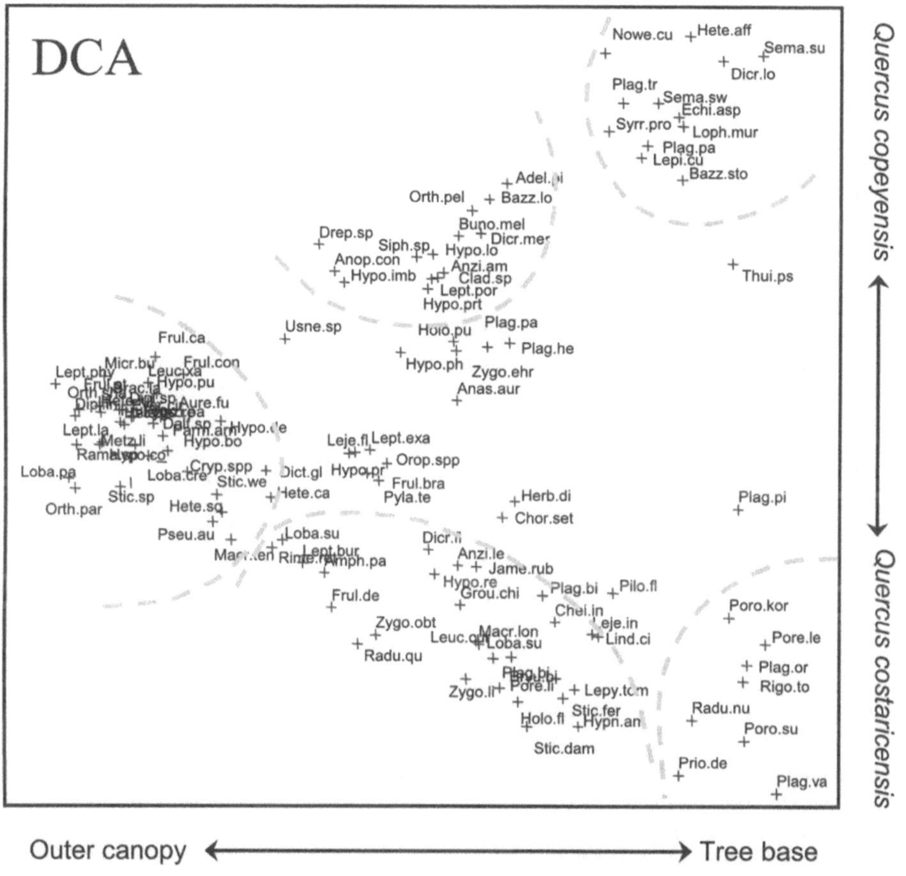

Fig. 7.1. Ordination of species in the epiphyte species space using detrended correspondence analysis (DCA) and Beals smoothing (Hill and Gauch 1980; Beals 1984; McCune 1994). X-axis: axis 1, y-axis: axis 2. *Hatched lines* indicate main ecological species groups. For explanation of acronyms, see Table 7.1

Table 7.1. Epiphytic bryophytes and macrolichens found on ten *Quercus copeyensis* and *Q. costaricensis* trees (five of each host tree species) in a primary upper montane oak forest of Cordillera de Talamanca, Costa Rica

Taxon	Acronym
Bryophytes	
Hepatics	
Adelanthus decipiens (Hook.) Mitt.	Adel.dec
Adelanthus pittieri (Steph.) Grolle	Adel.pit
Amphilejeunea patellifera (Spruce) R.M.Schust.	Amph.pat
Anastrophyllum auritum (Lehm.) Steph.	Anas.aur
Anoplolejeunea conferta (C.F.W.Meissn.) A.Evans	Anop.con
Aureolejeunea fulva R.M.Schust.	Aure.ful
Bazzania longistipula (Lindenb.) Trevis.	Bazz.lon
Bazzania stolonifera (Sw.) Trevis.	Bazz.sto
Brachiolejeunea laxifolia (Taylor) Schiffner	Brac.lax
Cheilolejeunea inflexa Hampe ex Lehm. & Lindenb.	Chei.inf
Diplasiolejeunea involuta S. Winkl.	Dipl.inv
Diplasiolejeunea replicata (Spruce) Steph.	Dipl.rep
Diplasiolejeunea spec. A	Dipl.spA
Drepanolejeunea spp.	Drep.spp
Echinocolea asperrima (Spruce) R.M.Schust.	Echi.asp
Frullania brasiliensis Raddi	Frul.bra
Frullania caulisequa (Nees) Nees	Frul.cau
Frullania convoluta Lindenb. & Hampe	Frul.con
Frullania ecklonii (Spreng.) Spreng.	Frul.eck
Frullania stenostipa Spruce	Frul.ste
Frullanoides densifolia Raddi ssp. *densifolia*	Frul.den
Harpalejeunea stricta (Lindenb. & Gottsche) Steph.	Harp.str
Herbertus divergens (Steph.) Herzog	Herb.div
Iwatsukia jishibae (Steph.) N.Kitag.	Iwat.jis
Jamesoniella rubricaulis (Nees) Grolle	Jame.rub
Lejeunea flava (Sw.) Nees	Leje.fla
Lejeunea intricata J.B.Jack & Steph.	Leje.int
Lejeunea laetevirens Nees & Mont.	Leje.lae
Lepidozia cupressina (Sw.) Lindenb.	Lepi.cup
Leptoscyphus amphibolius (Nees) Grolle	Lept.amp
Leptoscyphus porphyrius (Nees) Grolle	Lept.por
Leucolejeunea xanthocarpa (Lehm. & Lindenb.) A.Evans	Leuc.xan
Lindigianthus cipaconeus (Gottsche) Kruijt & Gradst.	Lind.cip
Lophocolea muricata (Lehm.) Nees	Loph.mur
Metzgeria liebmanniana Lindenb. & Gottsche	Metz.lie
Microlejeunea bullata (Tayl.) Steph.	Micr.bul
Nowellia curvifolia (Dicks.) Mitt.	Nowe.cur
Omphalanthus filiformis (Sw.) Nees	Omph.fil
Plagiochila bicuspidata Gottsche	Plag.bic
Plagiochila bifaria (Sw.) Lindenb.	Plag.bif
Plagiochila cf. *vagae* (sect. Contiguae)	Plag.vag
Plagiochila heterophylla Lindenb. ex Lehm.	Plag.het
Plagiochila oresitropha Spruce	Plag.ore
Plagiochila pachyloma Tayl.	Plag.pac

Table 7.1. (*Continued*)

Taxon	Acronym
Plagiochila papillifolia Steph.	Plag.pap
Plagiochila patzschkei Steph.	Plag.pat
Plagiochila pittieri Steph.	Plag.pit
Plagiochila retrorsa Gottsche	Plag.ret
Plagiochila stricta Lindenb.	Plag.str
Plagiochila trichostoma Gottsche	Plag.tri
Porella leiboldii (Lehm.) Trevis.	Pore.lei
Porella liebmanniana (Lindenb. & Gottsche) Trevis.	Pore.lie
Radula nudicaulis Steph.	Radu.nud
Radula quadrata Gottsche	Radu.qua
Telaranea nematodes (Austin) M.Howe	Tela.nem
Trichocolea floccosa Herzog & Hatch.	Tric.flo
Mosses	
Brachymenium systylium (Müll.Hal.) A. Jaeger	Brac.sys
Braunia squarrulosa (Hampe) Müll.Hal.	Brau.squ
Bryum billarderi Schwägr.	Bryu.bil
Chorisodontium setaceum (E.B.Bartram) E.B.Bartram	Chor.set
Cryphaea spp.	Cryp.spp
Daltonia spp.	Dalt.spp
Dicranodontium longisetum (Hook.) R.S.Williams	Dicr.lon
Dicranodontium meridionale E.B.Bartram	Dicr.mer
Dicranum flagellare Hedw.	Dicr.fla
Groutiella chimborazensis (Spruce ex Mitt.) Florsch.	Grou.chi
Heterophyllium affine (Hook.) M.Fleisch.	Hete.aff
Holomitrium flexuosum Mitt.	Holo.fle
Holomitrium pulchellum Mitt.	Holo.pul
Hypnum amabile (Mitt.) Hampe	Hypn.ama
Leptodontium exasperatum Cardot	Lept.exa
Leptodontium flexifolium (Dicks.) Hampe	Lept.fle
Lepyrodon tomentosus (Hook.) Mitt.	Lepy.tom
Leucobryum antillarum Schimp. ex Besch.	Leuc.ant
Leucodon curvirostris Hampe	Leuc.cur
Macrocoma tenuis subsp. *sullivantii* (Müll.Hal.) Vitt	Macr.ten
Macromitrium longifolium (Hook.) Brid.	Macr.lon
Mittenothamnium reptans (Hedw.) Cardot	Mitt.rep
Orthodontium pellucens (Hook.) B.S.G.	Orth.pel
Orthotrichum pariatum Mitt.	Orth.par
Orthotrichum sharpii H.Rob.	Orth.sha
Pilotrichella flexilis (Hedw.) Ångström	Pilo.fle
Porotrichodendron superbum (Taylor) Broth.	Poro.sup
Porotrichum korthalsianum (Dozy & Molk.) Mitt.	Poro.kor
Prionodon densus (Sw. ex Hedw.) Müll.Hal.	Prio.den
Pylaisiadelpha tenuirostris (Sull.) W.R.Buck	Pyla.ten
Renauldia mexicana (Mitt.) H.A.Crum	Rena.mex
Rigodium toxarion (Schwägr.) A.Jaeger	Rigo.tox
Sematophyllum subsimplex (Hedw.) Mitt.	Sema.sub
Sematophyllum swartzii (Schwägr.) W.H.Welch & H.A.Crum	Sema.swa

Table 7.1. (*Continued*)

Taxon	Acronym
Syrrhopodon prolifer Schwägr.	Syrr.pro
Thuidium pseudoprotensum (Müll.Hal.) Mitt.	Thui.pse
Zygodon ehrenbergii Müll.Hal.	Zygo.ehr
Zygodon liebmannii Schimp.	Zygo.lie
Zygodon obtusifolius Hook.	Zygo.obt
Zygodon reinwardtii (Hornsch.) A.Braun	Zygo.rei
Macrolichens	
Anzia americana Yoshim. & Sharp	Anzi.ame
Anzia leucobates (Nyl.) Müll.Arg.	Anzi.leu
Anzia masonii Yoshim.	Anzi.mas
Anzia parasitica (Fée) Zahlbr.	Anzi.par
Bunodophoron melanocarpum (Sw.) Wedin	Buno.mel
Cladonia spp.	Clad.spp
Coccocarpia erythroxyli (Spreng.) Swinscow & Krog	Cocc.ery
Dictyonema glabratum (Spreng.) D.L.Hawksw.	Dict.gla
Erioderma mollissimum (Samp.) DR.	Erio.mol
Everniastrum cirrhatum (E.Fr.) Hale ex Sipman	Ever.cir
Heterodermia casarettiana (Massal.) Trevis.	Hete.cas
Heterodermia leucomela (Fée) Swinsc. & Krog	Hete.leu
Heterodermia lutescens (Koruk.) Follm. & Redon	Hete.lut
Heterodermia obscurata (Nyl.) Trev.	Hete.obs
Heterodermia squamulosa (Degel.) W.Culb.	Hete.squ
Heterodermia vulgaris (Vain.) Follmann & Redon	Hete.vul
Hypotrachyna bogotensis (Vain.) Hale	Hypo.bog
Hypotrachyna costaricensis (Nyl.) Hale	Hypo.cos
Hypotrachyna densirhizinata (Kurok.) Hale	Hypo.den
Hypotrachyna ducalis (Jatta) Hale	Hypo.duc
Hypotrachyna ensifolia (Kurok.) Hale	Hypo.ens
Hypotrachyna imbricatula (Zahlbr.) Hale	Hypo.imb
Hypotrachyna longiloba (H.Magn.) Hale	Hypo.lon
Hypotrachyna monilifera (Kurok.) Hale	Hypo.mon
Hypotrachyna physcioides (Nyl.) Hale	Hypo.phy
Hypotrachyna prolongata (Kurok.) Hale	Hypo.pro
Hypotrachyna protoboliviana (Hale) Hale	Hypo.prt
Hypotrachyna pulvinata (Fée) Hale	Hypo.pul
Hypotrachyna reducens (Nyl.) Hale	Hypo.red
Hypotrachyna rockii (Zahlbr.) Hale	Hypo.roc
Leptogium burgessii (L.) Mont.	Lept.bur
Leptogium coralloideum (Mey. & Flot.) Vain.	Lept.cor
Leptogium laceroides Bouly de Lesd.	Lept.lac
Leptogium phyllocarpum (Pers.) Mont.	Lept.phy
Lobaria crenulata (Hook.) Trev.	Loba.cre
Lobaria pallida (Hook.) Trevis.	Loba.pal
Lobaria pulmonaria (L.) Hoffm.	Loba.pul
Lobaria subdissecta (Nyl.) Vain.	Loba.sud
Lobaria subexornata Yoshim.	Loba.sub
Nephroma helveticum Ach.	Neph.hel

Table 7.1. (*Continued*)

Taxon	Acronym
Oropogon loxensis (Fée) Th.Fr.	Orop.lox
Oropogon spp.	Orop.spp
Pannaria spec. A	Pann.spA
Parmeliopsis venezuelana (Hale) DePriest & Hale	Parm.ven
Parmotrema arnoldii (DR.) Hale	Parm.arn
Physcia lopezii Moberg	Phys.lop
Pseudocyphellaria aurata (Ach.) Vain.	Pseu.aur
Pseudocyphellaria crocata (L.) Vain.	Pseu.cro
Ramalina spec. A	Rama.spA
Rimedia cetrata (Ach.) Hale & A.Fletcher	Rime.ret
Siphula spec. A	Siph.spA
Sticta damaecornis agg.	Stic.dam
Sticta ferax Müll. Arg.	Stic.fer
Sticta spp.	Stic.spp
Sticta weigelii (Isert) Ach.	Stic.wei
Teloschistes spec. A	Telo.spA
Usnea spp.	Usne.spp

of species per plot on the two host tree species, but species richness generally increased with height in the tree. This was also true for the richness of lichens, whereas richness of mosses generally decreased. There was no general trend with increasing height for hepatics.

A perennial life form is the predominant ecological strategy of epiphytic bryophytes on tree bases and the lower parts of trunks. As a mechanism of adaptation to promote interspecific competition for space and light, many species on the tree base tend to grow in pure patches due to their growth form and vegetative reproduction (e.g., *Bazzania* spp., *Rigodium toxarium*, *Thuidium* spp., *Plagiochila* spp.). This is also the case in many lichens growing on the trunk and in the inner canopy (e.g., *Hypotrachyna* spp., *Lobaria* spp.). In the outer canopy, community structure and ecological strategies of species are very different. Many species are restricted to this height zone, and are early-successional 'sun epiphytes' or pioneers occurring also on twigs in the understorey (Cornelissen and ter Steege 1989). Average species richness per plot and species frequency are higher in the outer canopy than on the tree base and tree trunk, and beta diversity is low. Similar observations were reported by McCune et al. (2000) in an old-growth conifer forest in western Washington, and is apparently a general characteristic of the twig community. Outer canopy twigs are a relatively young habitat, and light and humidity conditions in this habitat are more extreme than in the understorey.

Species of the outer canopy community are generally characterized by small stature, low cover, and copious production of diaspores promoting fast establishment (Cornelissen and ter Steege 1989; van Leerdam et al. 1990). These are 'r' strategists, whereas those of the tree base and trunk are 'K' strategists (MacArthur and Wilson 1967). The principles of dispersal and life strategies of the rain forest bryophytes and lichens remain poorly understood (Schuster 1988; Gradstein 1992), however, and require long-term monitoring observations on succession and colonization.

7.3.4 Factors Controlling the Microhabitat Differentiation

In contrast to most upper montane forests, especially true cloud forests, in which it is possible to distinguish as many different epiphytic habitats on a single tree as can be done in lowland forests (Pócs 1982), the microhabitats of the oak forest in Cordillera de Talamanca show remarkably distinct bryophyte and macrolichen synusiae, and a clear differentiation between tree bases, trunks, branches and twigs in terms of species assemblages. Distributions of species, life forms, and epiphytic cryptogam communities in different forest microhabitats reflect the vertical variation of humidity and light regimes in these oak forests. In addition, they mirror the influence of a pronounced dry season, and specific structural characters (tree height, stem and branch diameter, stratification, host tree species). Furthermore, bark pH, water capacity (Köhler 2002, Chap. 21), and bark hardness differ significantly among host tree species, and may well explain the observed host–epiphyte relations.

Comparing species distribution on the two studied host trees, it can be recognized that most species occupy the same height zone on different host trees. However, many of these species show a broader height spectrum on the one tree species than on the other tree species (e.g., many of the species restricted to the outer canopy of *Q. copeyensis* are common in the outer canopy, the inner canopy, or even the upper trunk of *Q. costaricensis*). It seems that substrate factors (bark chemistry and/or bark physics) are more important for the distribution of these species than are microclimatic factors, including light conditions. Physiological and transplanting experiments might help to resolve the factors responsible for stratification with height.

7.4 Recovering Forests

7.4.1 General Aspects

Secondary forest communities are widely distributed and are increasingly becoming the most important repository of biodiversity in tropical uplands (Brown and Lugo 1990; Chazdon 1994; Holl and Kappelle 1999; Helmer 2000).

Studies on the recovery of cryptogamic communities in secondary tropical forests are very few, and focus only on lowland, submontane or lower montane rain forests (e.g., Sillet et al. 1995; Costa 1999; Acebey et al. 2003), none on upper montane ones.

Canopy trees of *Quercus copeyensis* were sampled in upper montane primary (PF), early secondary (ESF) and late secondary oak forests (LSF) of the Cordillera de Talamanca, Costa Rica, with the aim of gaining insight into patterns and processes of epiphyte succession and recovery of diversity in secondary forest following forest clearing (Holz and Gradstein 2005a).

7.4.2 Species Diversity

In total, 168 epiphytic bryophyte and macrolichen species (60 lichens, 67 hepatics, 41 mosses) were found in 437 plots (of ca. 600 cm²) located on 15 trees in PF, ESF and LSF. Figure 7.2 shows species-accumulation curves of randomly pooled plots from the three forest types. Total species richness was remarkably similar for all three forest types, with highest numbers found in LSF and lowest in PF. Although total number of species in PF is relatively low compared to that of the two secondary forest types, PF has the highest number of species exclusive to one forest type (46 % of all species in PF; 27 % of all species found in the three forest types).

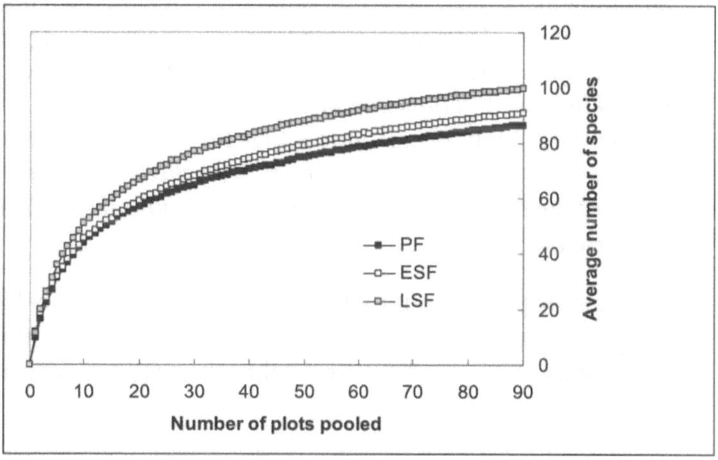

Fig. 7.2. Species-accumulation curves (rarefaction) of cryptogamic epiphyte plots taken from *Quercus copeyensis* canopy trees in primary forest (*PF*), early secondary forest (*ESF*), and late secondary forest (*LSF*)

7.4.3 Indicator Species

Epiphytic cryptogams are of great value as ecological indicator species in tropical forest ecosystems (Hietz 1999; Gradstein et al. 2001a). Floristic changes due to deforestation may be large, depending on the amount and type of damage inflicted upon the forest. Clearcutting results in the immediate loss of epiphytic cryptogams, and selective logging will modify forest structure and microclimate. After secondary forest regeneration in clearcut areas or on plantations (and agroforest ecosystems), at least part of the species may reestablish. The resulting distribution patterns of cryptogamic epiphytes and their communities are diverse, reflecting the microclimatic and substrate conditions in their secondary microhabitat, and the progress and speed of succession.

Ecological species groups and indicator species of forest types and height zones were determined using ordination of species by DCA after Beals smoothing (Beals 1984; Hill and Gauch 1980), and calculation of indicator values following the methodology outlined in Dufrene and Legendre (1997). Species with highest calculated indicator values for the three forest types are the following:

1. Species with highest indicator values for PF: *Leptodontium exasperatum*, *Frullania brasiliensis*, *Plagiochila heterophylla*, *Zygodon ehrenbergii*, *Dicranodontium meridionale*, *Hypotrachyna imbricatula*, *Bunodophoron melanocarpum*, *Herbertus divergens*, *Hypotrachyna physcioides* and *Holomitrium pulchellum*.
2. Species with highest indicator values for ESF: *Microlejeunea bullata*, *Daltonia longifolia*, *Metzgeria liebmanniana*, *Metzgeria agnewii*, *Brachiolejeunea laxifolia*, *Heterodermia leucomela*, *Diplasiolejeunea replicata*, *Frullania ecklonii* and *Plagiochila bicuspidata*.
3. Species with highest indicator values for LSF: *Lejeunea intricata*, *Zygodon reinwardtii*, *Plagiochila patzschkei*, *Aptychella proligera*, *Metzgeria* sp. A, *Hypotrachyna costaricensis*, *Porotrichum mutabile*, *Frullania stenostipa* and *Lejeunea flava*.

It should be pointed out that these results are valid only for the forest types investigated, and that the indicator species listed above may be common in other habitats, too, or on host trees other than *Quercus copeyensis*.

7.4.4 Recovery of Cryptogamic Epiphyte Communities

Although species richness is high in the secondary forests (both ESF and LSF) studied here, the rate of floristic recovery, expressed by floristic similarity to the primary forest, is relatively slow. Similarity in species composition in secondary forests compared to primary forests increases with age, but even after

40 years of forest succession more than one third (46 %) of primary forest species of cryptogams were not found in the secondary forest. By contrast, 40 % (68 species) of all species recorded were restricted to secondary forests, which demonstrates the important contribution of secondary forests to total species diversity in the Talamancan oak forests.

Kappelle et al. (1996) estimated about 85 years as the minimum time needed for structural recovery of upper montane oak forests following clearing. This estimation was based on the development of basal area of trees and canopy height using linear regressions. As the oldest secondary forest included in the calculation was less than 35 years old, the estimation is not yet sufficiently validated, and it remains largely hypothetical if all characteristics of the different microhabitats of the forest will recover within such a time period. We suggest that at least 100 years is needed for the complete recovery of floristic and community composition, and possibly centuries if recovery follows non-linear trends. Predicting how similar the non-vascular epiphyte vegetation of the mature secondary forest will be compared to the original primary forest remains difficult, and requires more work on the reproductive biology of the species (local epiphyte propagule supply, fragments from which species regenerate), their physiological ecology and competition for resources. Future sampling of cryptogamic epiphyte communities in over 40-year-old secondary forests will be needed in order to better understand long-term trends in secondary succession in the montane oak forests of Costa Rica.

7.5 Conclusions

Upper montane oak forests in Cordillera de Talamanca show a high diversity of bryophyte and lichen species, and a great diversification of microhabitats. Similarities in species composition show a strong relationship between forest floor habitats (including tree bases), in contrast to epiphytic habitats.

The bryophyte flora of the oak forests is dominated by neotropical species. Andean-centered species are a conspicuous element, reflecting the close geographical connection between the montane bryophyte floras of Costa Rica and South America. A high percentage of Central American endemics is found in the oak forest bryophyte flora.

Host preference and vertical gradients on host trees play an important role in the differentiation of epiphytic cryptogam communities in these forests. Different life strategies of epiphytic bryophytes and lichens are found in the canopy ('r' strategists) and on the tree base and trunk ('K' strategists). Distributions of species, life forms, and epiphytic cryptogam communities in different forest microhabitats reflect the vertical variation of humidity and light regimes in these oak forests. In addition, they show the impact of a pronounced dry season and of structural characters of the forest. Furthermore, bark pH, water capacity, and bark hardness differ significantly among host tree species, and may well explain host–epiphyte relations.

Total species richness in secondary and primary Talamancan montane oak forests was very similar, showing that primary forests are not necessarily more diverse than secondary forests. Similarity in species composition in secondary forests compared to primary forest increases with age, but even after 40 years more than one third of the primary forest species have not colonized the secondary forest. By contrast, 40 % of all species found in the studied forest types are restricted to secondary forests alone. In the succession of cryptogamic epiphyte communities toward a mature secondary forest, the diversity in microsites due to tree growth is of utmost importance. The high number of species found only in the primary forest indicates that a long period will be needed for the reestablishment of microhabitats, and colonization by species adapted to different niches.

It may thus be recommended that, in order to maintain high biodiversity, management practices should be adopted to maintain all successional stages present in the forest landscape to preserve the diversity of non-vascular epiphytes.

Acknowledgements I would like to thank Bruce Allen, William R. Buck, Riclef Grolle, Dick Harries, Jochen Heinrichs, Maria I. Morales Z., Denise Pinheiro da Costa, Ronald Pursell, William D. Reese, M. Elena Reiner-Drehwald, Alfons Schäfer-Verwimp, Harrie J.M. Sipman, Jiri Váña and Kohsaku Yamada for help with species identifications. Thanks are also due to Maarten Kappelle, Nelson Zamora and Armando Soto, Instituto Nacional de Biodiversidad (INBio), for logistic support during fieldwork in Costa Rica, and to Lars Köhler (University of Göttingen) for assistance in the field. The friendly hospitality of the Chacón, Monge and Serrano families in the Los Santos region is very much appreciated. Jürgen Franzaring (University of Hohenheim), Rob Gradstein (University of Göttingen) and Martin Schnittler (University of Greifswald) are thanked for providing helpful comments on earlier versions of the manuscript.

References

Acebey A, Gradstein SR, Krömer T (2003) Species richness and habitat diversification of bryophytes in submontane rain forests and fallows of Bolivia. J Trop Ecol 19:9–18

Bates JW (1998) Is 'life-form' a useful concept in bryophyte ecology? Oikos 82:223–237

Beals EW (1984) Bray-Curtis ordination: an effective strategy for analysis of multivariate ecological data. Adv Ecol Res 14:1–55

Brown S, Lugo AE (1990) Tropical secondary forests. J Trop Ecol 6:1–32

Chazdon RL (1994) The primary importance of secondary forests in the tropics. Tropinet 5:1

Clark KL, Nadkarni NM, Schaefer D, Gholz HL (1998a) Atmospheric deposition and net retention of ions by the canopy in a tropical montane forest, Monteverde, Costa Rica. J Trop Ecol 14:27–45

Clark KL, Nadkarni NM, Schaefer D, Gholz HL (1998b) Cloud water and precipitation chemistry in a tropical montane forest, Monteverde, Costa Rica. Atmos Environ 32(9):1595–1603

Clement JP, Moffet MW, Shaw DC, Lara A, Alarcon D, Larrain O (2001) Crown structure and biodiversity in *Fitzroya cupressoides*, the giant conifers of Alerce Andino National Park, Chile. Selbyana 22:76–88

Cornelissen JHC, ter Steege H (1989) Distribution and ecology of epiphytic bryophytes and lichens in dry evergreen forest of Guyana. J Trop Ecol 5:29–35

Costa D (1999) Epiphytic bryophyte diversity in primary and secondary lowland rainforests in south-eastern Brazil. The Bryologist 102:320–326

Dufrene M, Legendre P (1997) Species assemblages and indicator species: the need for a flexible asymmetrical approach. Ecol Monogr 67:345–366

Gradstein SR (1992) The vanishing tropical rain forest as an environment for the bryophytes and lichens. In: Bates JW, Farmer A (eds) Bryophytes and lichens in a changing Environment. Oxford Science Publ, Oxford, UK, pp 232–256

Gradstein SR, Churchill SP, Salazar-NA (2001a) Guide to the bryophytes of tropical America. Mem NY Bot Gard 86:1–590

Gradstein SR, Griffin D, Morales MI, Nadkarni NM (2001b) Diversity and habitat differentiation of mosses and liverworts in the cloud forest of Monteverde, Costa Rica. Caldasia 23:203–212

Grubb PJ (1974) Factors controlling the distribution of forest-types on tropical mountains: new facts and new perspective. In: Flenley JR (ed) Altitudinal zonation in Malesia. University of Hull, Misc Ser 16:13–45

Helmer EH (2000) The landscape ecology of tropical secondary forest in montane Costa Rica. Ecosystems 3:98–114

Hietz P (1999) Diversity and conservation of epiphytes in a changing environment. Pure Appl Chem 70:2114 (www.iupac.org/symposia/proceedings/phuket97/hietz.html)

Hietz P, Hietz-Seifert U (1995) Structure and ecology of epiphyte communities of a cloud forest in central Veracruz, Mexico. J Veg Sci 6:719–728

Hill MO, Gauch HG (1980) Detrended correspondence analysis: an improved ordination technique. Vegetatio 42:47–58

Hofstede RGM, Wolf JHD, Benzing DH (1993) Epiphytic biomass and nutrient status of a Colombian upper montane rain forests. Selbyana 14:37–45

Holdrige LR (1967) Life zone ecology, rev edn. Tropical Science Center, San José, Costa Rica

Holl KD, Kappelle M (1999) Tropical forest recovery and restoration. Trends Ecol Evol 14:378–379

Holz I (2003) Diversity and ecology of bryophytes and macrolichens in primary and secondary montane *Quercus* forests, Cordillera de Talamanca, Costa Rica. PhD Thesis, University of Göttingen, Göttingen

Holz I, Gradstein SR (2005a) Cryptogamic epiphytes in primary and recovering upper montane oak forests of Costa Rica – species richness, community composition and ecology. Plant Ecol 178:89–109

Holz I, Gradstein SR (2005b) Phytogeography of the bryophyte floras of oak forests and páramo of the Cordillera de Talamanca, Costa Rica. J Biogeogr 32:1591–1609

Holz I, Gradstein SR, Heinrichs J, Kappelle M (2002) Bryophyte diversity, microhabitat differentiation and distribution of life forms in Costa Rican upper montane *Quercus* forest. The Bryologist 105:334–348

Kappelle M, Cleef AM, Chaverri A (1992) Phytogeography of Talamanca montane *Quercus* forests, Costa Rica. J Biogeogr 19(3):299–315

Kappelle M, Geuze T, Leal ME, Cleef AM (1996) Successional age and forest structure in a Costa Rican upper montane *Quercus* forest. J Trop Ecol 12:681–698

Köhler L (2002) Die Bedeutung der Epiphyten im ökosystemaren Wasser- und Nährstoffumsatz verschiedener Altersstadien eines Bergregenwaldes in Costa Rica. PhD Thesis, University of Göttingen, Göttingen

MacArthur RH, Wilson EO (1967) The theory of island biogeography. Princeton University, Princeton, NJ

McCune B (1994) Improving community analysis with the Beals smoothing function. Ecoscience 1:82–86

McCune B, Rosentreter R, Ponzetti JM, Shaw DC (2000) Epiphyte habitats in an old conifer forest in Western Washington, USA. The Bryologist 103:417–427

Mielke PW Jr (1984) Meteorological applications of permutation techniques based on distance functions. In: Krishnaiah PR, Sen PK (eds) Handbook of Statistics, vol 4. Elsevier, Amsterdam, pp 813–830

Montfoort D, Ek R (1990) Vertical distribution and ecology of epiphytic bryophytes and lichens in a lowland rain forest in French Guyana. MSc Thesis, Utrecht University, Utrecht

Pócs T (1980) The epiphytic biomass and its effect on the waterbalance of two rain forest types in the Uluguru mountains. Acta Bot Acad Sci Hung 26:143–167

Pócs T (1982) Tropical forest bryophytes. In: Smith AJE (ed) Bryophyte ecology. Chapman and Hall, London, pp 59–104

Richards PW (1984) The ecology of tropical forest bryophytes. In: Schuster RM (ed) New Manual of Bryology, vol 2. The Hattori Botanical Laboratory, Nichinan, Miyazaki, Japan, pp 1233–1270

Schuster RM (1988) Ecology, reproduction biology and dispersal of Hepaticae in the Tropics. J Hattori Bot Lab 64:237–269

Sillet SC, Gradstein SR, Griffin D III (1995) Bryophyte diversity of *Ficus* tree crowns from cloud forest and pasture in Costa Rica. The Bryologist 98:251–260

Smith AJE (1982) Epiphytes and epiliths. In: Smith AJE (ed) Bryophyte ecology. Chapman and Hall, London, pp 191–227

Van Leerdam A, Zagt RJ, Veneklaas EJ (1990) The distribution of epiphyte growth-forms in the canopy of Colombian cloud-forest. Vegetatio 87:59–71

Veneklaas E, Ek R (1990) Rainfall interception in two tropical montane rain forests, Colombia. Hydrol Proc 4:311–326

Veneklaas E, Zagt R, Van Leerdam A, Ek R, Broekhoven G, Genderen M (1990) Hydrological properties of the epiphyte mass of a montane tropical rain forest, Colombia. Vegetatio 89:183–192

Wolf JHD (1993) Ecology of epiphytes and epiphyte communities in montane rain forests, Colombia. PhD Thesis, University of Amsterdam, Amsterdam

Part III
Stand Structure and Composition

8 Composition and Structure of Humid Montane Oak Forests at Different Sites in Central and Eastern Mexico

I. Luna-Vega, O. Alcántara-Ayala, C.A. Ruiz-Jiménez, and R. Contreras-Medina

8.1 Humid Montane Oak Forests in Mexico

Mexican humid montane oak forests are highly diverse, exhibit a large biological heterogeneity, and are characterized by a remarkable intermingling of taxa frequently found in northern and southern biotas, which are mixed with endemic taxa (Miranda and Sharp 1950, Chap. 9). Structural complexity of these forests follows an elevational and latitudinal gradient; this complexity decreases toward high elevations and latitudes, and varies from slope to slope, depending on sun exposure, soil type, wind regime and various micro-environmental features. They are present at an elevational range of 600–3,200 m, but are best developed at 1,000–2,500 m, in areas with high precipitation, often at sites with characteristic, frequent cloudiness.

Some areas with the vegetation discussed herein include localities in eastern and central Mexico; the climate in the eastern localities is temperate, with rains usually produced by the prevailing winds from the northeast, causing temperatures in the upper zone of the escarpment to be relatively cool, mainly during the winter season. In the central part of Mexico, the climate is not affected by these winds, and an alternation of wet and dry seasons is less evident. In all cases, these forests are developed in gorge environments, in places with a rough topography.

Many of the species found in this type of vegetation are threatened and/or endemic, and have been included in risk categories laid down by the Mexican government in the official document Norma Oficial Mexicana (NOM) 059 (SEMARNAT 2002), some also in the CITES species list (CITES 2003) and in the IUCN Red List (IUCN 2003). Indeed, there is an urgent challenge to preserve and study these ecosystems, since human settlements are steadily growing and contributing to the large-scale loss of these forests. Due to their varied climatic characteristics, they have been used for coffee, bean and corn plantations, among other crops, and also for animal husbandry.

Ecological Studies, Vol. 185
M. Kappelle (Ed.) Ecology and Conservation of Neotropical Montane Oak Forests
© Springer-Verlag Berlin Heidelberg 2006

8.2 Study Area

In the present study, floristic composition and vegetation structure were investigated in four different patches of humid montane oak forests in central and eastern Mexico. These were located in four Mexican states, Hidalgo (Molocotlán and Lolotla), Veracruz (Teocelo-Ixhuacán), and Morelos-México (Ocuilan), and two different physiographic and floristic provinces (Sierra Madre Oriental and Transmexican Volcanic Belt, both included in the Mesoamerican Mountain region of Rzedowski 1978). The forests are located at different altitudinal and latitudinal ranges (Table 8.1), and have different floristic composition and vegetation structure.

8.3 Localities and Sampled Sites

8.3.1 Selection of Localities and Floristic Composition

Based on previous floristic assessments in central and eastern Mexico (Luna et al. 1988, 1989; Mayorga et al. 1998; Escutia 2004), four localities were selected where mature upper montane oak forest patches were present. In all cases, the canopy trees are represented by species of present-day holarctic distribution, and the understory trees by a mixture of species of present-day tropical and holarctic distribution.

Two sites in each locality were sampled, and were named as follows: Lolotla (LT), Hidalgo; Molocotlán (ML), Hidalgo; Ocuilan (OC), México-Morelos; and Teocelo-Ixhuacán (IX), Veracruz. These sites were selected based on conservation criteria. Some topographic and physiographic data for these localities are provided in Table 8.1.

8.3.2 Vegetation Sampling

Vegetation sampling campaigns were conducted from August 2003 to January 2004. The method used was a modification of Gentry's technique (Gentry 1995); in each locality, we sampled a total area of 0.2 ha, corresponding to two sample sites that included 10 rectangles of 50×2 m each. Our modification of Gentry's technique consisted of an increment from 2.5 to 3.18 cm in diameter at breast height (DBH) for the inclusion criteria.

Total height was calculated and recorded by using a clinometer for all large woody stems. Crown cover diameter was calculated for all tree individuals. Crown cover diameter of each individual was calculated on the basis of two

Table 8.1. Topographic and physiographic data of eight Mexican humid montane oak forest sites

Site	Mexican states	Location	Altitude (m)	Latitude N	Longitude W	Aspect
Lolotla 1 (LT1)	Hidalgo	1.17 km ENE from Lolotla	1,255	20°51'21"	98°40'46"	N
Lolotla 2 (LT2)	Hidalgo	2.87 km ENE from Lolotla	1,425	20°51'12"	98°41'29"	S
Molocotlán 1 (ML1)	Hidalgo	1.21 km ESE from Molocotlán	1,480	20°44'46"	98°42'23"	W
Molocotlán 2 (ML2)	Hidalgo	0.89 km E from Molocotlán	1,540	20°44'58"	98°42'33"	SW
Teocelo-Ixhuacán 2 (IX 2)	Veracruz	0.88 km NW from Ixhuacán de los Reyes	2,048	19°21'40"	97°07'22"	NE
Teocelo-Ixhuacán 1 (IX1)	Veracruz	1.87 km NE from Ixhuacán de los Reyes	2,095	19°22'03"	97°06'16"	S–SW
Ocuilan 2 (OC2)	México	1.27 km NNE from Tlaltizapan	2,350	18°59'13"	99°20'23"	SW
Ocuilan 1 (OC1)	México	0.36 km S from Tlaltizapan	2,430	18°58'21"	99°20'34"	N

Table 8.2. Vegetation structure and diversity parameters of eight 0.1-ha plots in four Mexican humid montane oak forests. Site abbreviations are given in the main text

Plot site	LT1	LT2	ML1	ML2	IX1	IX2	OC1	OC2
Max. canopy height (m)	34.64	34.66	39.02	37.50	26.90	21.51	27.00	27.66
# individuals per ha	1800	1610	1430	1540	2150	650	740	1720
Basal area (m^2/ha)	29.50	25.34	23.50	52.30	29.30	21.36	50.08	28.04
Richness	23	22	16	16	20	6	12	18
Crown cover (%)	282.21	214.40	163.21	346.71	266.29	169.50	209.23	188.69
Maximum DBH (m)	0.66	0.78	0.60	1.05	0.54	0.44	0.78	0.47

perpendicular crown diameters projected onto the forest floor. Diameter at breast height values were assessed for all recorded trees.

Plant specimens were identified, in some cases with the help of specialists, and voucher specimens were stored at the Herbario de la Facultad de Ciencias (FCME), UNAM.

8.4 Composition and Structure Analyses

A complete list of the species, and analyses of the floristic composition of each locality can be found in Luna et al. (1988, 1989), Mayorga et al. (1998) and Escutia (2004). The forests studied herein have a high canopy (21–39 m), high density, generally with low basal area values, foliage cover values generally over 100 %, and DBH values generally less than 100 cm. Species richness in these forests is variable, from six to 23 species (Table 8.2). Table 8.3 summarizes the quantitative parameters of the eight study sites.

8.4.1 Lolotla (LT)

The forests at LT1 and LT2 are dense and have low basal area values (Table 8.2). Structurally, the two more important species of LT1 are included in the NOM-059 (SEMARNAT 2002); this site includes a higher number of species in some risk category in this official document.

8.4.2 Molocotlán (ML)

The forest in ML1 is dense and has low basal area values. The quantitative structure of this forest reflects its location in areas with high sun exposure. The forest in ML2 is also dense, and shows average basal area values. At this site, the largest amount of trees with high DBH values was found (Table 8.2). The forest is located on slopes protected from sun exposure and strong winds. Based on its composition and structure, this forest represents a typical temperate forest of the Sierra Madre Oriental.

8.4.3 Teocelo-Ixhuacán (IX)

The forest at IX1 is dense, with low basal area values. Site IX2 is an open forest, and oak species demonstrate high structural values (Table 8.2). At this site, six trees are structurally important, five of which are oaks.

Table 8.3. Quantitative forest structure at eight sample sites in four Mexican humid montane oak forests. The three most important species at each locality are highlighted in bold for each parameter (*D* density, *DR* relative density, *BA* basal area, *BAR* relative basal area, *F* frequency, *FR* relative frequency, *C* tree crown cover, *CR* relative crown cover, *RIV* relative importance value)

Species	D (ind ha^{-1})	DR (%)	BA (m^2 ha^{-1})	BAR (%)	F	FR (%)	C (m^2 ha^{-1})	CR (%)	RIV
Lolotla 1 (LT1)									
Ostrya virginiana (Mill.) K. Koch	**530**	29.44	**5.10**	17.32	**100**	13.70	**7,217.60**	25.57	**86.04**
Carpinus caroliniana Walt.	**360**	20.00	**4.30**	14.70	**80**	10.96	**5,200.10**	18.43	**64.08**
Quercus sartorii Liebm.	**200**	11.11	**6.90**	23.33	**70**	9.59	**3,911.60**	13.86	**57.89**
Quercus germana Schltdl. & Cham.	70	3.89	4.70	15.92	50	6.85	1,587.30	5.62	32.28
Inga huastecana M. Sousa	190	10.56	0.70	2.30	60	8.22	2,378.70	8.43	29.51
Other species (18)	450	25.00	7.80	26.43	370	50.68	7,926.02	28.09	130.20
Total	1,800	100.0	29.5	100.0	730	100.0	28,221.32	100.0	400.0
Lolotla 2 (LT2)									
Beilschmiedia mexicana (Mez) Kosterm.	**540**	33.54	2.20	8.86	**100**	13.33	**4,488.40**	20.97	**76.70**
Quercus germana Schltdl. & Cham.	90	5.59	**8.30**	32.88	**60**	8.00	**3,196.90**	14.94	**61.41**
Liquidambar macrophylla Oerst.	40	2.48	**7.50**	29.48	40	5.33	**2,662.40**	12.44	**49.74**
Nectandra salicifolia (Kunth) Nees	**210**	13.04	**2.60**	10.15	**60**	8.00	1,967.30	9.19	40.38
Turpinia occidentalis (Sw.) G. Don	**140**	8.69	1.00	4.10	**70**	9.33	2,123.20	9.92	32.05
Other species (17)	590	36.66	3.74	14.53	420	56.01	6,966.11	32.54	139.72
Total	1,610	100.0	25.34	100.0	750	100.0	21,404.31	100.0	400.0

Table 8.3. (*Continued*)

Species	D (ind ha^{-1})	DR (%)	BA (m^2 ha^{-1})	BAR (%)	F	FR (%)	C (m^2 ha^{-1})	CR (%)	RIV
Molocotlán 1 (ML1)									
Quercus affinis Scheid.	**180**	12.59	**7.59**	32.30	**90**	15.25	**5,521.98**	33.83	**93.97**
Carya ovata K. Koch.	**560**	39.16	**3.26**	13.86	**80**	13.56	**3,597.24**	22.04	**88.62**
Quercus germana Schltdl. & Cham.	70	4.90	2.81	11.94	40	6.78	**1,701.94**	10.43	**34.04**
Lyonia squamulosa M. Martens & Galeotti	**230**	16.08	0.34	1.44	**70**	11.86	598.81	3.67	33.06
Pinus patula Schiede & Deppe ex Schltdl. & Cham.	20	1.40	**4.22**	17.94	20	3.39	952.30	5.83	28.57
Other species (11)	370	25.87	5.28	22.52	290	49.16	3,949.27	24.20	121.74
Total	1,430	100.0	23.5	100.0	590	100.0	16,321.54	100.0	400.0
Molocotlán 2 (ML2)									
Carpinus caroliniana Walt.	**460**	29.87	**6.60**	12.64	**90**	12.86	**8,320.10**	24.0	**79.36**
Quercus affinis Scheid.	50	3.25	**17.20**	32.83	50	7.14	**6,719.50**	19.38	**62.60**
Ostrya virginiana (Mill.) K. Koch	**360**	23.38	4.50	8.64	**80**	11.43	**5,359.10**	15.46	**58.90**
Liquidambar macrophylla Oerst.	**150**	9.74	**10.40**	19.89	**90**	12.86	4,937.20	14.24	56.73
Clethra mexicana Greenm.	90	5.84	3.30	6.22	70	10.00	1,868.40	5.39	27.45
Other species (11)	430	27.92	10.30	19.78	320	45.71	7,466.81	21.53	114.96
Total	1,540	100.0	52.3	100.0	700	100.0	34,671.11	100.0	400.0
Ixhuacán 1 (IX1)									
Quercus ocoteaefolia Liebm.	**640**	29.77	**19.57**	66.79	**100**	11.36	**12,390.76**	46.53	**154.45**
Clethra macrophylla M. Martens & Galeotti	**270**	12.56	**1.56**	5.32	**90**	10.23	**1,685.77**	6.33	**34.44**
Vaccinium leucanthum Schltdl.	**240**	11.16	**1.56**	5.31	**80**	9.09	1,581.90	5.94	**31.51**
Ostrya virginiana (Mill.) K. Koch	130	6.05	1.46	4.98	70	7.95	**2,722.81**	10.22	29.20
Gaultheria sp.	150	6.98	0.54	1.85	**80**	9.09	1,071.28	4.02	21.94
Other species (15)	720	33.48	4.61	15.75	460	52.28	7,176.71	26.96	128.46
Total	2,150	100.0	29.30	100.0	880	100.0	26,629.23	100.0	400.0

Ixhuacán 2 (IX2)									
Quercus sapotiifolia Liebm.	**220**	33.85	**8.00**	33.72	**80**	26.67	**6,691.70**	39.48	**133.71**
Quercus affinis Scheid.	**180**	27.69	**4.60**	22.77	**80**	26.67	**4,072.40**	24.02	**101.15**
Quercus ocoteaefolia Liebm.	**120**	18.46	3.20	15.90	**50**	16.67	**2,804.70**	16.55	**67.57**
Quercus aff. *pinnativenulosa* C.H. Mull.	50	7.69	**4.30**	21.04	40	13.33	1,888.60	11.14	53.20
Clethra macrophylla M. Martens & Galeotti	60	9.23	1.16	5.93	40	13.33	1,160.70	6.85	35.34
Quercus x laurina Humb. & Bonpl.	20	3.08	0.10	0.64	10	3.33	332.89	1.96	9.03
Total	650	100.0	21.36	100.0	300	100.0	16,950.99	100.0	400.0
Ocuilan 1 (OC1)									
Quercus laurina Humb. & Bonpl.	**310**	41.89	**29.10**	58.11	**100**	26.32	**10,148.80**	48.50	**174.83**
Cleyera integrifolia (Benth.) Choisy	**110**	14.86	**4.30**	8.57	**70**	18.42	**2,450.00**	11.71	**53.57**
Symplocos prionopylla Hemsl.	70	9.46	**6.60**	13.23	**50**	13.16	1,855.80	8.87	**44.72**
Styrax ramirezii Greenm.	**100**	13.51	1.50	3.00	**50**	13.16	1,490.30	7.12	36.79
Carpinus caroliniana Walt.	50	6.76	4.00	8.02	20	5.26	**2,816.30**	13.46	33.50
Other species (7)	100	13.52	4.58	9.07	90	23.68	2,162.14	10.34	56.59
Total	740	100.0	50.08	100.0	380	100.0	20,923.34	100.0	400.0
Ocuilan 2 (OC2)									
Quercus laurina Humb. & Bonpl.	**280**	16.28	**10.70**	38.14	**90**	12.33	**4,636.00**	24.57	**91.32**
Zinowiewia coccinea Lundell	**450**	26.16	**4.10**	14.77	70	9.59	**4,085.50**	21.65	**72.18**
Cleyera integrifolia (Benth.) Choisy	160	9.30	**5.00**	17.82	**80**	10.96	**2,607.80**	13.82	**51.90**
Quercus candicans Née	80	4.65	3.70	13.09	50	6.85	1,511.50	8.01	32.60
Styrax ramirezii Greenm.	**170**	9.88	0.70	2.49	**80**	10.96	1,333.70	7.07	30.40
Other species (13)	580	33.73	3.84	13.69	360	49.31	4,695.05	24.88	121.60
Total	1,720	100.0	28.04	100.0	730	100.0	18,869.55	100.0	400.0

8.4.4 Ocuilan (OC)

Site OC1 is an open forest, with average basal area values. *Quercus laurina* Humb. & Bonpl. has high structural values, in comparison with the other tree species (Table 8.2). The forest in OC2 is dense, with low basal area values.

8.4.5 Comparison of Localities

The structural characteristics of the IX2 forest at Teocelo-Ixhuacán are exceptional, compared to those of the other forests; it has the lowest values in terms of canopy height (21.2 m), density (650 ind/ha), basal area (21.36 m^2/ha), richness (six species), and maximum DBH (0.44 m; Table 8.2). However, this forest contains a higher number of species of *Quercus* (five), and structurally the oaks possess the highest relative importance values among all the forests studied (364.65; Table 8.4). The structural characteristics of this site are the result of management by neighboring human communities; settlers selectively extract some wood, mainly from *Pinus* (furniture) and *Quercus* (fuelwood). The same use pattern (albeit less intense) is found in the Lolotla communities; these two examples show that human activities are reflected in the forest structure.

Figure 8.1 presents all tree stems at each site, considering the height of the trees grouped in eight different classes and DBH in 11 classes. From these graphs, it becomes clear that only ML1 and ML2 include individuals with more than 35 m height. In general, the higher classes are poorly represented in our sample, and the lower classes include a high proportion of tree stems, demonstrating that the forest is disturbed. In ML2, taller trees (35–40 m) are present with high DBH values (almost 110 cm); although in IX2 trees with lower DBH values (0–19 cm) and less tall (almost 20 m) are present, this site was found to be the most diverse in oak species (five).

Only *Quercus ocoteaefolia* Liebm. is present in IX1, with a high relative importance value (154.45), and in LT2 only *Q. germana* Schltdl. & Cham. is found, with a relative importance value of 61.41. At the remaining sites, the number of species of *Quercus* varies in the range 2–5, and in all cases, oaks are structurally important elements of the forest (Table 8.4). These forests have variable relative values (density, basal area, frequency, and foliage cover), which together contribute with high percentages of relative importance values (Table 8.2).

The studied forests have different relative structural values; IX2 shows the highest values, since five species of oaks at this locality have the highest relative importance index. The other sampled forests do not exceed a relative density value of 44 %, relative basal area of 67 %, relative frequency of 29 %, and relative crown cover of 51 % (Table 8.2). The conservation status of these

Table 8.4. Structural contribution of oaks (*Quercus*: *Q.*) in relation to the relative variables at each sample site in Mexican humid montane oak forest sites

Site	Number of oak species	Other tree species	Relative density	Relative basal area	Relative frequency	Relative crown cover	Relative importance index	Oak species
Lolotla 1 (LT1)	2	21	15.00	39.25	16.44	19.44	90.17	*Q. sartorii*, *Q. germana*
Lolotla 2 (LT2)	1	21	5.59	32.88	8.00	14.94	61.41	*Q. germana*
Molocotlán 1 (ML1)	4	12	20.99	54.51	28.80	50.99	155.29	*Q. affinis*, *Q. germana*, *Q. sartorii*, *Q. glabrescens*
Molocotlán 2 (ML2)	3	13	7.80	40.67	14.29	26.32	89.06	*Q. affinis*, *Q. sartorii*, *Q. eugenifolia*
Teocelo-Ixhuacán 1 (IX1)	1	19	29.77	66.79	11.36	46.53	154.45	*Q. ocoteaefolia*
Teocelo-Ixhuacán 2 (IX2)	5	1	90.77	94.08	86.67	93.15	364.65	*Q. sapotiifolia*, *Q. affinis*, *Q. ocoteaefolia*, *Q. aff. pinnativenulosa*, *Q. x laurina*
Ocuilan 1 (OC1)	2	10	43.24	58.22	28.95	48.88	179.30	*Q. laurina*, *Q. obtusata*
Ocuilan 2 (OC2)	3	15	22.09	51.84	20.55	33.27	127.74	*Q. laurina*, *Q. candicans*, *Q. aff. acutifolia*

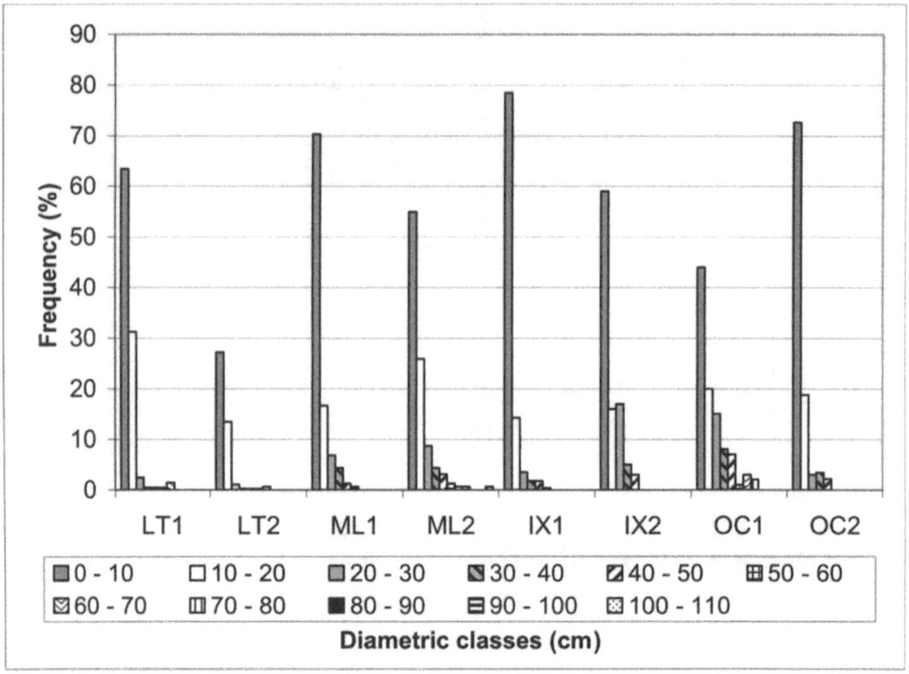

a

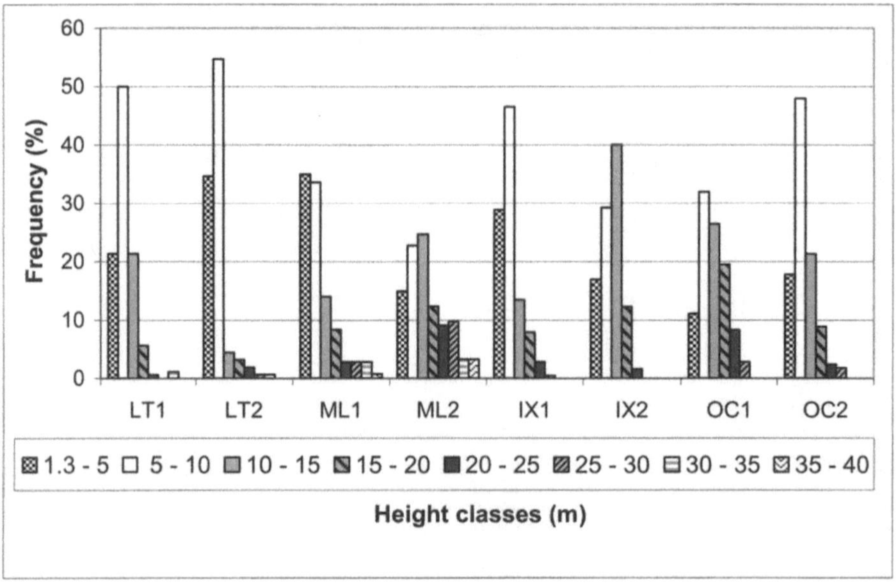

b

Fig. 8.1a, b. Frequency distributions of trees per stem diameter class (a) and per height class (b), as found at eight sample sites in eastern and central Mexico forests does not significantly change the structural contribution of the oaks to these forests (Table 8.4).

8.5 Conclusions

The forests of the eastern part of Mexico are characteristically dominated by *Liquidambar macrophylla* Oerst., several species of *Quercus* (i.e., *Q. affinis* Scheid., *Q. aristata* Hook. & Arn., *Q. crassipes* Humb. & Bonpl., *Q. conspersa* Benth., *Q. eugeniifolia* Liebm., *Q. germana* Schltdl. & Cham., *Q. glaucescens* Humb. & Bonpl., *Q. laurina* Humb. & Bonpl., *Q. sartorii* Liebm., and *Q. salicifolia* Née), and pines (*Pinus greggii* Engelm. ex Parl., *P. patula* Schiede & Deppe ex Schltdl. & Cham., *P. montezumae* Lamb., *P. tenuifolia* Salisb., and *P. pseudostrobus* Lindl.), among others. In the central part of Mexico (Ocuilan), there is a total absence of *L. macrophylla* Oerst., and the forest is mainly dominated by *Carpinus caroliniana* Walt., *Pinus leiophylla* Schltdl. & Cham., and oaks (*Quercus candicans* Née, *Q. crassifolia* Humb. & Bonpl., *Q. laurina* Humb. & Bonpl., and *Q. rugosa* Née). The forests of the eastern part of Mexico (Sierra Madre Oriental) are more diverse than those located at Ocuilan (Transmexican Volcanic Belt). These forests have variable relative values (density, basal area, frequency and foliage cover), and the oak species together contribute with high percentages of relative importance values (Table 8.2). Many other species, mainly from the intermediate and upper tree layers, are also structurally important to these humid oak forests.

Over the last decades, the humid montane oak forests of the central and eastern parts of Mexico, as for many other Mexican vegetation types, have severely deteriorated. Indeed, their vegetation structure and composition have been modified by anthropogenic influence, mainly as a result of forest extraction and livestock grazing. As in many other Mexican humid montane oak forests, selective logging of some canopy elements, i.e., species of *Quercus*, may favor some heliophyte species, such as species of Ericaceae and *Pinus*.

It is important to conserve these forests, as they contain a high proportion of threatened and/or endemic taxa. In view of their restricted distribution, and populations composed of only few individuals, many more of these species must be studied to be included in the CITES list and the IUCN Red List. Some species that inhabit the central and eastern Mexican forests are already listed in risk categories of the Norma Oficial Mexicana NOM-059 (SEMARNAT 2002), for example, *Acer negundo* var. *mexicanum* (DC.) Standl. & Steyerm., *Aporocactus flagelliformis* (L.) Lem., *Carpinus caroliniana* Walt., *Ceratozamia mexicana* Brongn., *Cupressus lusitanica* Mill., *Diospyros riojae* Gómez Pompa, *Juglans pyriformis* Liebm., *Magnolia schiedeana* Schltdl., and *Ostrya virginiana* (Mill.) K. Koch. Three species that are structurally important in these humid oak forests (*Carpinus caroliniana* Walt., *Ostrya virginiana*

(Mill.) K. Koch, and *Cyathea mexicana* Schltdl. & Cham.) are also generally represented in the temperate forests of the Sierra Madre Oriental and the Transmexican Volcanic Belt.

We conclude that anthropogenic disturbance has strongly changed the forest structure and floristic composition of the localities in central and eastern Mexico studied here, mainly by agriculture, timber extraction, animal husbandry, and firewood extraction.

Acknowledgements Assistance in the field provided by Ana Quintos, Dafne Saavedra, Elizabeth Olivos, Maribel Paniagua, Sandra Córdoba, Alberto González, Hamlet Santa Anna, Armando Ponce, Jorge Escutia, and Rogelio Aguilar is gratefully appreciated. The oaks species were identified by Susana Valencia. This research was supported by project PAPIIT IN206202 of the DGAPA (UNAM).

References

CITES (2003) CITES species list. Convention on International Trade in Endangered Species of Wild Fauna and Flora (http://www.cites.org)

Escutia J (2004) Análisis estructural del bosque mesófilo de montaña de Monte Grande de Lolotla, Hidalgo, México. BSc Thesis, Facultad de Ciencias, Universidad Nacional Autónoma de México (UNAM), Mexico

Gentry A (1995) Patterns of diversity and floristic composition in Neotropical montane forests. In: Churchill SP, Balsev H, Forero E, Luteryn JL (eds) Biodiversity and conservation of Neotropical montane forests. New York Botanical Garden, Bronx, NY, pp 103–126

IUCN (2003) Red list of threatened species. World Conservation Union (http://www.redlist.org/serch/details.php?species=36160)

Luna I, Almeida L, Villers L, Lorenzo L (1988) Reconocimiento florístico y consideraciones fitogeográficas del bosque mesófilo de montaña de Teocelo, Veracruz. Bol Soc Bot Mex 48:35–63

Luna I, Almeida L, Llorente J (1989) Florística y aspectos fitogeográficos del bosque mesófilo de montaña de las cañadas de Ocuilan, estados de Morelos y México. An Inst Biol UNAM Ser Bot 59:63–87

Mayorga R, Luna I, Alcántara O (1998) Florística del bosque mesófilo de montaña de Molocotlán, Molango-Xochicoatlán, Hidalgo, México. Bol Soc Bot Mex 63:101–119

Miranda F, Sharp AJ (1950) Characteristics of the vegetation in certain temperate regions of eastern Mexico. Ecology 31:313–333

Rzedowski J (1978) Vegetación de México. Limusa, Mexico

SEMARNAT (2002) Norma Oficial Mexicana (NOM) 059-ECOL-2001, Protección ambiental, especies nativas de México y de flora y fauna silvestres, categorías de riesgo y especificaciones para su inclusión, exclusión o cambio-lista de especies en riesgo. Secretaría de Medio Ambiente y Recursos Naturales, Diario Oficial de la Federación, 6 March 2001, Mexico, pp 1–80

9 Oak Forests of the Hyper-Humid Region of La Chinantla, Northern Oaxaca Range, Mexico

J.A. Meave, A. Rincón, and M.A. Romero-Romero

9.1 Introduction

The southern part of Mexico lies within the tropical region of North America. This geographical consideration, however, contrasts with the fact that much of its existing or potential vegetation has a temperate character (Rzedowski 1978). This is due to the presence of large mountain ranges or sierras, which largely characterize the Mexican landscape (de Czerna 1989). The geo-climatic history of these mountains appears to be responsible for the extreme diversification of *Quercus* and *Pinus*, two typical holarctic genera (Perry 1991; Nixon 1993; Valencia-A and Nixon 2004; Chap. 1). Only in Central Mexico, there are 45 oak species, suggesting that this is one of the major diversification centres of this genus (Valencia-A 2004) and of other plant groups.

Despite this large biodiversity and the fascination it causes among ecologists and evolutionary biologists, highland forests of Mexico are disappearing very rapidly. For one, areas originally covered by oak forests have been preferred for agricultural development owing to their benign climate and good soils (Challenger 1998, and see Chap. 8). In contrast to this trend, La Chinantla, located in the northern part of Oaxaca State, is one of the few regions in Mexico where large, undisturbed tracts of oak forests still remain.

In this chapter, we synthesize the existing literature for the oak forests of La Chinantla. We combine floristic information with quantitative descriptions derived from vegetation sampling at some localities. Descriptions are given not only for oak-dominated forests, but also for some other communities, mostly various kinds of cloud forests, where oaks form part of the forest structure.

Ecological Studies, Vol. 185
M. Kappelle (Ed.) Ecology and Conservation of Neotropical Montane Oak Forests
© Springer-Verlag Berlin Heidelberg 2006

9.2 La Chinantla Region

La Chinantla is a culturally defined region in the northern part of Oaxaca State, southern Mexico. Despite the narrow scope given by Schultes (1941) to this geographical name, La Chinantla sensu lato (17°22'–18°12'N and 95°43'–96°58'W) roughly corresponds to the geographical distribution of the Chinantec ethnic group. This region comprises a heterogeneous array of very diverse landscapes (Martin 1993). La Chinantla is considered as one of the most complex regions of Oaxaca, and even of the entire country (Martin and Madrid 1992; Rodrigo-Álvarez 1994).

In this region, there is a sharp transition from the Gulf of Mexico Coastal Plain to the adjacent Northern Oaxaca Range (NOR), an imposing and extremely complex mass of mountains raising from almost sea level to around 3,200 m, with over 120 mountain tops above 2,500 m. The abruptness of the altitudinal gradient and the ruggedness of the terrain are illustrated by the horizontal distance between the town of Valle Nacional (ca. 65 m above sea level, a.s.l.) and the top of the Humo Chico mountain (ca. 3,200 m), which is a little less than 30 km, and by the frequency distribution of slope inclinations: 17.3 % of slopes show values of 0–6°, 38.3 % values of 6–18°, 43.3 % values of 18–45°, and 1 % are steeper than 45° (Ortíz-Pérez et al. 2004).

The geomorphologic complexity of the NOR has been pointed out repeatedly (de Czerna 1989; Centeno-García 2004). According to Ferrusquía-Villafranca (1993), this region forms part of the Sierra Madre del Sur (Southern Mother Range) morphotectonic province, specifically of the Oaxaca-Puebla Uplands subprovince. The oldest rocks are of Late Palaeozoic to Early Mesozoic age (Centeno-García 2004). Uplifting of the NOR, which begun around 14 Ma ago, was caused by the activity of the Oaxaca fault, which today runs along the western margin of the sierra (Centeno-García 2004). Substrate instability near mountain summits commonly results in massive landslides.

Regional soils are also poorly studied. In general, they are shallow, strongly influenced by water erosion, and may be generally classified as lithosols (leptosols; Alfaro 2004). In some places, high organic matter, N and P contents have been detected (van der Wal 1998), but areas of infertile soils, classified as oxisols, have been reported at lower elevations (van der Wal 1996).

Climatic conditions at La Chinantla are still poorly understood, due to the scarcity of meteorological records. In summer months (May–October), trade winds bring large amounts of moisture into the southern parts of the country (Trejo 2004). Despite the high amount of rain falling on the Gulf of Mexico Coastal Plain, air masses moving further inland still carry much moisture until they reach the high peaks of the NOR, where adiabatic cooling during rising results in substantial condensation and precipitation. In late summer, there is additional rain because of the influence of tropical cyclones. All these phenomena result in very high levels of precipitation, particularly at mid elevation, making La Chinantla the rainiest region of Mexico and warranting its recognition as hyper-humid. This can be illustrated by climatic data from selected localities: at Tuxtepec (10 m a.s.l.) and Valle Nacional (65 m), two lowland localities, annual precipitation and mean annual temperature are 2,304 mm and 24.9 °C, and 3,590 mm and 24.8 °C respectively; at Vista Hermosa (1,450 m a.s.l.) and Yaxila (1,730 m), two stations typical of intermediate elevations, annual rainfall is >5,000 mm (5,800 and 5,499 mm respectively, with temperatures of 16.5 and 16.3 °C respectively); finally, at the even higher location of Humo Chico (3,240 m a.s.l.), both annual precipitation (3,616 mm) and mean annual temperature (8.7 °C) are much lower than those at the two intermediate stations.

During the period of winter drought observed in most of Mexico, La Chinantla receives some rain brought by the *nortes*, which are cold, humidity-loaded winds coming from the northern latitudes. In addition, the condensation of fog on the leaves and branches of plants makes a substantial contribution of water to the system, particularly during the dry season (Vogelmann 1973). Rainfall is commonly larger than evapotranspiration (Trejo 2004), which results in an excess of water which is drained by numerous creeks and rivers, eventually forming the Papaloapan River and discharging into the Gulf of Mexico.

As a consequence of the decrease in air temperature with increasing elevation, at La Chinantla there is a gradual transition from a hot climate (mean annual temperature above 22 °C), through a semi-hot (18–22 °C) and temperate (12-18°C) climate, to a cold climate (5–12 °C) at elevations around 3,000 m. Along this marked climatic gradient, a complex series of plant communities are more or less organized in altitudinal belts, including forests ranging from lowland rain and evergreen forests, through several types of montane rain (cloud) forests, to oak and pine forests (Torres-Colín 2004). The oak forests examined here are located at elevations above 200 m a.s.l., on the windward side of the mountains. Large tracts of drier, very tall oak forests occurring on the leeward slopes are excluded from this chapter.

9.3 Floristic Survey and Vegetation Sampling

Most information presented here derives from a floristic survey conducted in the higher parts of La Chinantla in the time period 1993–1997. The floristic information was supplemented with quantitative data used for descriptions of forest structure. To this end, we used Gentry's (1982) sampling method, based on ten 50×2 m rectangles which are placed parallel to each other (in our case, at random distances of 10–20 m). Along these transects, all trees, shrubs and lianas with diameter at breast height (DBH)≥2.5 cm were sampled. Girths were measured at breast height in order to calculate DBH values, and vouchers of each plant were collected for their determination by specialists at the National Herbarium of Mexico (MEXU), where the first set of vouchers is kept. The main advantages of this method is the rapidity with which structural data are obtained, particularly in forest stands located on very rugged terrain, and the possibility of performing direct comparisons with many studies based on the same sampling methods (e.g. Boyle 1996; Phillips and Miller 2002). Our descriptions of oak forests at La Chinantla incorporated some data produced by Boyle (1996) in the same region. For the structural descriptions, the contribution of individual species to forest structure was evaluated in terms of the importance value (IV) of each species, calculated by adding the relative values of their frequencies, densities, and basal areas (Curtis and McIntosh 1951; Barbour et al. 1999).

9.4 Altitudinal Distributions of Oak Species at La Chinantla

Oak forests of the humid slopes of the NOR harbour only six species. These are largely segregated along the altitudinal gradient of this mountain range (Fig. 9.1). Only the lower altitudinal distributions of *Q. glaucescens* and *Q. elliptica* (700–1,400 m a.s.l. in our data) are almost identical. *Quercus* aff. *eugeniifolia* has an intermediate altitudinal position (1,600–2,000 m). All other species occur in the highest parts of the sierra only. *Q. corrugata* also displays a large altitudinal range (1,900–2,500 m). By contrast, the highest occurrences shown by two oak species have the narrowest belts: *Q. ocoteifolia* (2,400–2,700 m) and *Q. macdougallii* (2,600–3,000 m). Only some individuals of this latter species, and a few more belonging to the genus *Pinus* reach this altitude, marking the timber line. Around 3,000 m, mountain ridges are cov-

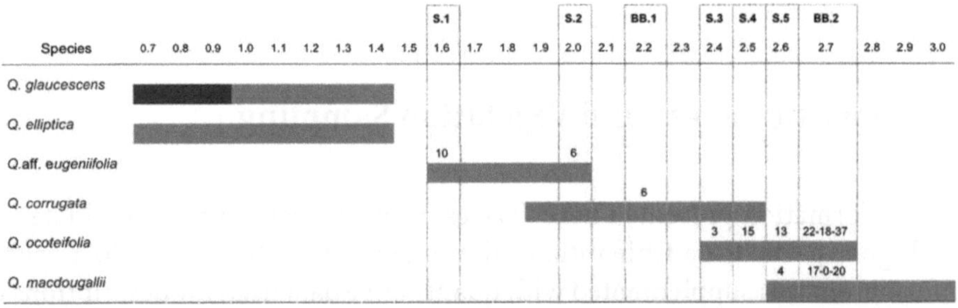

Fig. 9.1. Altitudinal distribution of six oak species found at La Chinantla, Oaxaca State, southern Mexico (elevations are given in meters×1,000). *Horizontal bars* represent species ranges based on collected specimens and vegetation sampling. Numbers above the bars indicate densities in 0.1-ha samples at the five sites in the present study (*S.1–S.5*), and two sites (*BB.1* and *BB.2*) described by Boyle (1996)

ered by Ericaceae, whose physiognomy somewhat resembles that of *páramo* communities (Cleef et al. 1984), or by bunch grassland communities similar to those described for the Volcanic Transversal Axis (Rzedowski 1978).

9.5 Higher-Elevation Oak Forests at the Watershed Divide

The most humid oak forests of the windward slopes of the sierra are by and large restricted to the highest elevations. We have structural data for two sites classified as oak forest (site 4 at 2,500 m and site 5 at 2,560 m a.s.l.). Site 4 is located very close to the watershed divide, slightly to the south, and thus is the only site on the leeward side of the mountains. Data are available for another three sites which do not strictly qualify as oak forests but where oaks are present: site 1 (1,650 m a.s.l.), site 2 (2,010 m), and site 3 (2,430 m). In the following descriptions, a decreasing altitudinal sequence is followed.

At site 5, dominance is associated uniquely to *Quercus ocoteifolia* (Table 9.1). This species accounts for more than 40 % of the stand's basal area. *Q. macdougallii* is also present, but shows a subordinate position, and in fact accounts for a smaller contribution to community structure than is the case for several typical cloud forest species. The strong dominance of *Q. ocoteifolia* coincides with the relatively low richness recorded here (22 species). Leaves of *Q. ocoteifolia* are microphyllous (mean leaf area=19.9 cm^2), and have a leaf mass per unit area of 126.8 g m^{-2}, whereas *Q. macdougallii* also has microphyllous leaves (mean leaf area=9.4 cm^2), a slightly larger leaf mass per unit area (146.0 g m^{-2}), and also a stomatal density of 471 mm^{-2} (Velázquez-Rosas et al. 2002).

The forest sampled at site 4 is similar to that at site 5. *Q. ocoteifolia*, the sole species of oak occurring here, accounts for ca. 45 % of total basal area (Table 9.1). Even though all of the accompanying species (another 28) belong to typical cloud forests taxa, the physiognomy of this forest warrants its classification as oak forest. Leaf characteristics reflect the protected conditions occurring at this site: *Q. ocoteifolia* has larger (notophyllous) leaves than those at site 5 (mean leaf area=22.7 cm^2), a specific leaf weight of 139.4 g m^{-2}, and a stomatal density of 490 mm^{-2} (Velázquez-Rosas 1997).

The forest at site 3 represents a particular community type very common on the ridges. This is a much denser forest, where structural dominance is based on *Vaccinium consanguineum* and *Weinmannia tuerckheimii* (Table 9.1). According to its importance value (IV), the only oak species occurring here, *Q. ocoteifolia*, ranked no. 13. Total species richness at this site was 37. We do not have data on leaf morphology and anatomy for this oak species at this site.

The plant community occurring at site 2 is a type of cloud forest which has been referred to as „elfin forest". Dominance is associated to *Zinowiewia* sp.,

Table 9.1. Importance values of the three most important species recorded at each site; if not included among the former, data are shown for all oak species present. Absolute values are given for stems with DBH ≥2.5 cm in 0.1 ha. *IV* = Importance value

	Species	Basal area (cm^2)	Density (ind)	Frequency (%)	Relative basal area (%)	Relative density (%)	Relative frequency (%)	IV (%)
Site 5 (2,560 m)								
1	*Quercus ocoteifolia* Liebm.	31,346.17	13	12	43.29	8.07	8.00	59.36
2	*Clethra galeottiana* Briquet	5,972.25	21	18	8.25	13.04	12.00	33.29
3	*Persea chamissonis* vel aff. Mez	7,616.74	17	15	10.52	10.56	10.00	31.08
9	*Quercus macdougallii* Martínez	8,478.12	4	4	11.71	2.48	2.67	16.86
	Remaining 18 species	18,997.23	106	101	26.24	65.84	67.33	159.41
	Totals	72,410.51 (7.24 m^2)	161	150	100	100	100	300
Site 4 (2,500 m)								
1	*Quercus ocoteifolia* Liebm.	31,173.14	15	14	44.1	4.89	5.56	54.5
2	*Cornus disciflora* Moc. et Sessé ex DC.	10,530.99	28	22	14.9	9.12	8.73	32.7
3	*Styrax glabrescens* Benth.	6,856.42	36	28	9.69	11.7	11.1	32.5
	Remaining 26 species	22,168.04	228	188	31.34	74.27	74.60	180.21
	Totals	70,728.58 (7.07 m^2)	307	252	100	100	100	300
Site 3 (2,430 m)								
1	*Vaccinium consanguineum* Klotzsch	9,129.73	165	77	18.51	19.83	15.13	53.47
2	*Weinmannia tuerckheimii* Engl.	9,362.25	168	57	18.98	20.19	11.20	50.37
3	*Viburnum acutifolium* Benth.	4,214.08	84	55	8.54	10.10	10.81	29.45
13	*Quercus ocoteifolia* Liebm.	2,286.44	3	3	4.64	0.36	0.59	5.59
	Remaining 33 species	27,321.79	264	234	39.49	80.24	79.86	198.95
	Totals	49,319.32 (4.93 m^2)	832	509	100	100	100	300

Site 2 (2,010 m)								
1	*Zinowiewia* sp.	5,815.41	102	61	13.13	23.72	17.89	54.74
2	*Clethra conzattiana* L.M.González	2,509.25	51	41	5.66	11.86	12.02	29.55
3	*Myrsine juergensenii* (Mez) Lundell	3,591.54	44	36	8.11	10.23	10.56	28.90
6	*Quercus* aff. *eugeniifolia* Liebm.	7,556.04	6	6	17.06	1.40	1.76	20.21
	Remaining 39 species	24,831.65	227.0	197.0	56.05	52.79	57.77	166.61
	Totals	44,303.89 (4.43 m^2)	430	341	100	100	100	300
Site 1 (1,650 m)								
1	*Cyrilla racemiflora* L.	25,631.65	17	16	36.44	5.17	5.46	47.07
2	*Ticodendron incognitum* Gómez-Laur. et L.D. Gómez	1,998.05	35	32	2.84	10.64	10.92	24.40
3	*Pinus chiapensis* (Mart.) Andersen	14,348.30	3	3	20.40	0.91	1.02	22.34
13	*Quercus* aff. *eugeniifolia* Liebm.	1,033.95	10	8	1.47	3.04	2.73	7.24
	Remaining 49 species	24,326.82	412	317	49.33	49.52	62.28	161.12
	Totals	70,333.74 (7.03 m^2)	329	293	100	100	100	300

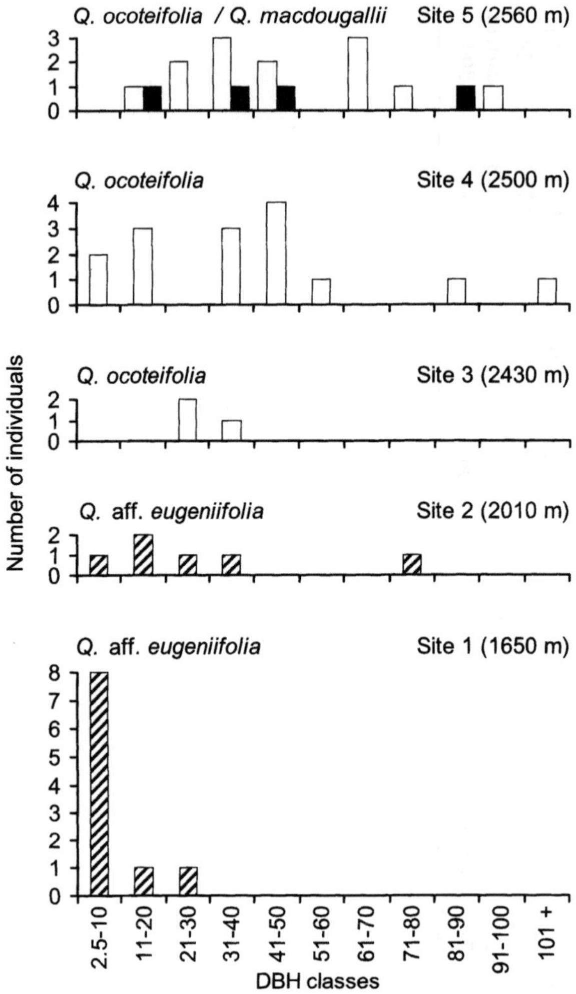

Fig. 9.2. Frequency distributions of stem diameter (*DBH*) classes for the most abundant oaks found during vegetation sampling in the La Chinantla region, Oaxaca State, southern Mexico

a species represented by more than 100 individuals in the sampling area (Table 9.1). Although the single oak species encountered here – *Quercus* aff. *eugeniifolia* – ranked no. 6 in terms of IV, it made the largest contribution to basal area. In addition, total richness (43 species) increased noticeably above that recorded in the other forests described above. At site 2, mean leaf area of *Quercus* aff. *eugeniifolia* was 19.5 cm^2, leaf mass per unit area was 170.5 g m^{-2}, and stomatal density 379 mm^{-2} (Velázquez-Rosas et al. 2002).

The forest sampled at site 1 represents an uncommon community dominated by *Cyrilla racemiflora*, a very rare species in the region (Gallardo et al. 1998). Relative contribution of *C. racemiflora* to forest structure was 36.5 % of total basal area, but only 5.2 % of total density (Table 9.1). These values reflect the massive size of this species. Only *Quercus* aff. *eugeniifolia* was present here, and ranked no. 13 in terms of IV. The greatest richness of all sites we sampled was recorded at this site (53 species). Again, no data are available on leaf morphology or anatomy for individuals of *Quercus* aff. *eugeniifolia* occurring here.

Based on the structural data presented above, it is difficult to draw firm conclusions about the dynamics of these forests. However, diameter structures of oak species may provide some indication of the regeneration of these populations. Figure 9.2 shows the frequency distributions of DBH classes of *Q.* aff. *eugeniifolia* (sites 1 and 2), *Q. ocoteifolia* (sites 3, 4 and 5), and *Q. macdougallii* (site 5). Despite the very limited number of trees used for this analysis, it is noteworthy that classes of smaller diameters are present in all cases, indicating that regeneration does indeed take place.

9.6 Lower-Elevation Oak Forests

At elevations below the cloud forest belt, there are tracts of oak forests which constitute systems completely different from those located at higher altitudes. Although we lack structural information for these forests, the study conducted by van der Wal (1996) provides interesting ecological information on this topic. *Quercus glaucescens* is undoubtedly the dominant species in these communities, although *Q. elliptica* is also present. In fact, these are the only forests of La Chinantla with absolute canopy monodominance.

These lowland oak forests are restricted to a particular soil type classified as truncated oxisol (van der Wal 1996). Chinantec people refer to it as „dry and hard“, and consider it as being unsuitable for agriculture. Consequently, they have put into practice a procedure known as „ringing“, which consists of killing trees by removing a ring of bark around the tree's trunk. Trees die standing, which causes large changes in light and temperature regimes underneath. The plant communities resulting from the secondary succession triggered by this ringing technique have a more complex structure than that of the original oak forests. During this process, oaks face difficulty to establish, although their seedlings and saplings may be found sporadically in secondary vegetation stands (Romero-Romero et al. 2000). Ultimately, the oak forests are replaced gradually by the expansion of the tropical rain forest.

9.7 Discussion

The humid oak forests of La Chinantla harbour only a small proportion (8.6 %) of the entire oak diversity of Oaxaca. With its 70 species (48 % of Mexico's total), this state boasts the largest richness of oaks in the country (Valencia-A 2004). The proportion of the country's oak diversity (161 species; Valencia-A 2004) represented at La Chinantla is even smaller (3.7 %). These results imply that these very high regional and national oak richness levels may be achieved only through high spatial species turnover (β-diversity sensu Whittaker 1972). Considering that only 8 % of Oaxaca is covered by oak forests, and a further 17 % by mixed pine-oak forests (i.e. a total of 25 %; Palacio-Prieto et al. 2000), oak β-diversity across the state must be very large, which is possible given the environmental complexity of the Oaxacan territory. In this context, it is interesting that two species encountered in our survey, namely *Q.* aff. *eugeniifolia* and *Q. macdougallii*, are Oaxaca endemics (Valencia-A and Nixon 2004).

The observed altitudinal distribution of oak species at La Chinantla is in agreement with altitudinal ranges reported previously for these species, based on multiple collections from various regions (Valencia-A and Nixon 2004). *Q. glaucescens* shows the only important discrepancy, as the previously reported range corresponded to a typical lowland species (250–650 m a.s.l.), whereas we collected specimens of this species much higher, at 1,400 m.

The structural information from La Chinantla forests allows us to compare with other regions. For example, the oak forests studied by Kappelle (1996, Chap. 10) in the Cordillera de Talamanca (Costa Rica) generally have taller trees (35–38 m) than those of the present study, in which the tallest trees reached 30 m but usually were much shorter. By contrast, density at our site 5 (1,610 ind ha^{-1}) is similar to those reported for equivalent trees in Talamanca (1,840, 1,979 and 1,820 ind ha^{-1}). When density is calculated for a 10-cm DBH cut-off, however, all values for Talamanca (range: 510–700 ind ha^{-1}) are smaller than those from La Chinantla (range: 770–1,560 ind ha^{-1}, including data of Boyle 1996). Such discrepancies become larger towards lower elevations at La Chinantla, where oaks are present but not dominant. At elevations of 2,000–2,500 m in the Sierra de Manantlán (Jalisco, W Mexico), Vázquez-G and Givnish (1998) found higher densities (range: 2,500–3,000 ind ha^{-1}) in forests dominated by *Q. castanea* or *Q. laurina*.

Differences in basal area between La Chinantla and other regions are smaller. Again, values for Talamanca (range: 57.5–64.7 m^2 ha^{-1}) are lower than those documented at La Chinantla (range: 44.3–97.7 m^2 ha^{-1}). Interestingly, our two oak forests at La Chinantla (sites 4 and 5) are among those showing the largest basal area, with the exception of the forest dominated by *Cyrilla racemiflora* in our study (site 1, 70.3 m^2 ha^{-1}), and a site investigated by Boyle which had an unusually high value (97.7 m^2 ha^{-1}). Regarding this variable, oak forests of Manantlán (range: 44–52 m^2 ha^{-1}) lie below those of La Chinantla (Vázquez-G and Givnish 1998).

With respect to floristic richness, when only high-elevation oak forests are compared, large similarities emerge between these three regions. The values for La Chinantla are 22 and 29 species, those for Talamanca are 18, 20 and 21 (Kappelle 1996), and the majority of those reported for Sierra de Manantlán are in the range 17–20.

Interestingly, oak densities appear to be generally lower at La Chinantla than at Talamanca. Kappelle (1996, Chap. 10) reported densities of 470, 630 and 820 ind ha^{-1} for *Q. copeyensis*, and of 120, 20 and 10 ind ha^{-1} for *Q. costaricensis*, whereas in our oak forest communities, the common *Q. ocoteifolia* had densities of 130 and 150 ind ha^{-1}, although Boyle (1996) recorded values of 220, 180 and 370 ind ha^{-1} for this species. Densities for all other species were more in the range observed by Kappelle (1996, Chap. 10) in Costa Rica.

9.8 Conclusions

Considering these structural similarities and the fact that oak forests of La Chinantla share many species with equivalent communities in Costa Rica (Lawton and Dryer 1980; Kappelle 1996), we contend that La Chinantla forms part of a single Mesoamerican biogeographical region of oak humid forests. Moreover, in view of the antiquity and the boreo-tropical origin of some taxa occurring at La Chinantla (Wendt 1993; Meave et al. 1997; Gallardo et al. 1998), as well as the southward migration of the holarctic genus *Quercus* into Central America (Chap. 2), it is likely that La Chinantla forests represent a centre from which homologous communities could develop at latitudes further south.

Acknowledgements We are grateful to Susana Valencia of the Herbarium FCME, Faculty of Sciences, National Autonomous University of Mexico, for her assistance in the determination of oak species collected at La Chinantla. This study received financial support from the Mexican National Commission of Biodiversity (CONABIO) through project P069.

References

Alfaro GS (2004) Suelos. In: García-Mendoza A, Ordóñez MJ, Briones-Salas M (eds) Biodiversidad de Oaxaca. Instituto de Biología, Universidad Nacional Autónoma de México (UNAM), Fondo Oaxaqueño para la Conservación de la Naturaleza and WWF, México DF, pp 55–65

Barbour MG, Burk JH, Pitts WD, Gilliam FS, Schwartz MW (1999) Terrestrial plant ecology, 3rd edn. Benjamin/Cummings, Menlo Park

Boyle BL (1996) Changes on altitudinal and latitudinal gradients in Neotropical montane forests. PhD Dissertation, Washington University, St Louis, MO

Centeno-García E (2004) Configuración geológica del estado. In: García-Mendoza A, Ordóñez MJ, Briones-Salas M (eds) Biodiversidad de Oaxaca. Instituto de Biología, Universidad Nacional Autónoma de México (UNAM), Fondo Oaxaqueño para la Conservación de la Naturaleza y WWF, México DF, pp 29–42

Challenger A (1998) Utilización y conservación de los ecosistemas terrestres de México: pasado, presente y futuro. CONABIO, Universidad Nacional Autónoma de México (UNAM), and Agrupación Sierra Madre, México DF

Cleef AM, Rangel O, van der Hammen T, Jaramillo R (1984) La vegetación de las selvas del transecto Buritaca. In: van der Hammen T, Ruiz PM (eds) La Sierra Nevada de Santa Marta (Colombia), transecto Buritaca-La Cumbre. Cramer, Vaduz, Studies in Tropical Andean Ecosystems, vol 2

Curtis JT, McIntosh RP (1951) An upland forest continuum in the prairie-forest border region of Wisconsin. Ecology 32:476–498

De Czerna Z (1989) An outline of the geology of Mexico. In: Bally AW, Palmer AR (eds) The geology of North America: an overview. Geological Society of America, Boulder, pp 233–264

Ferrusquía-Villafranca I (1993) Geology of Mexico: a synopsis. In: Ramamoorthy TP, Bye R, Lot A, Fa J (eds) Biological diversity of Mexico: origins and distributions. Oxford Univ Press, New York, pp 3–107

Gallardo C, Meave J, Rincón A (1998) Plantas leñosas raras de bosque mesófilo de montaña. IV. *Cyrilla racemiflora* L. (Cyrillaceae). Bol Soc Bot Méx 62:183–186

Gentry AH (1982) Patterns of Neotropical plant species diversity. Evol Biol 15:1–84

Kappelle M (1996) Los bosques de roble (*Quercus*) de la Cordillera de Talamanca, Costa Rica: biodiversidad, ecología, conservación y desarrollo. Instituto Nacional de Biodiversidad (INBio), Santo Domingo de Heredia, Costa Rica

Lawton R, Dryer V (1980) The vegetation of Monteverde cloud forest reserve. Brenesia 18:101–116

Martin GJ (1993) Ecological classification among the Chinantec and Mixe of Oaxaca, Mexico. Etnoecológica 1:17–33

Martin GJ, Madrid S (1992) Ethnobotany, distribution, and conservation status of *Ticodendron incognitum* in northern Oaxaca, Mexico. J Ethnobiol 12:227–231

Meave J, Gallardo C, Rincón A (1997) Plantas leñosas raras del bosque mesófilo de montaña. II. *Ticodendron incognitum* Gómez-Laurito & Gómez P. (Ticodendraceae). Bol Soc Bot Méx 59:149–152

Nixon KC (1993) The genus *Quercus* in Mexico. In: Ramamoorthy TP, Bye R, Lot A, Fa J (eds) Biological diversity of Mexico: origins and distributions. Oxford Univ Press, New York, pp 447–458

Ortiz-Pérez MA, Hernández-Santana JR, Figueroa Mah-Eng JM (2004) Reconocimiento fisiográfico y geomorfológico. In: García-Mendoza A, Ordóñez MJ, Briones-Salas M (eds) Biodiversidad de Oaxaca. Instituto de Biología, Universidad Nacional Autónoma de México (UNAM), Fondo Oaxaqueño para la Conservación de la Naturaleza and WWF, México DF, pp 43–54

Palacio-Prieto JL, Bocco G, Velázquez A, Mas JF, Takaki-Takaki F, Victoria A, Luna-González L, Gómez-Rodríguez G, López-García J, Palma-Muñoz M, Trejo-Vázquez I, Peralta-Higuera A, Prado-Molina J, Rodríguez-Aguilar A, Mayorga-Saucedo R, González-Medrano F (2000) La condición actual de los recursos forestales en México: resultados del Inventario Forestal Nacional 2000. Bol Invest Geogr 43:183–203

Perry JP (1991) The pines of Mexico and Central America. Timber Press, Portland, OR

Phillips O, Miller JS (2002) Global patterns of plant diversity: Alwyn H. Gentry's forest transect data set. Missouri Botanical Garden, St Louis, MO

Rodrigo-Álvarez L (1994) Geografía general del estado de Oaxaca. Carteles Editores, Oaxaca, Mexico

Romero-Romero MA, Castillo S, Meave J, van der Wal H (2000) Análisis florístico de la vegetación secundaria derivada de la selva húmeda de montaña de Santa Cruz Tepetotutla (Oaxaca), México. Bol Soc Bot Méx 67:89–106

Rzedowski J (1978) Vegetación de México. Limusa, Mexico DF, Mexico

Schultes RE (1941) The meaning and usage of the Mexican place-name „Chinantla". Bot Mus Leaf Harvard Univ 9:101–116

Torres-Colín R (2004) Tipos de vegetación. In: García-Mendoza A, Ordóñez MJ, Briones-Salas M (eds) Biodiversidad de Oaxaca. Instituto de Biología, Universidad Nacional Autónoma de México (UNAM), Fondo Oaxaqueño para la Conservación de la Naturaleza and WWF, México DF, Mexico, pp 105–117

Trejo I (2004) Clima. In: García-Mendoza A, Ordóñez MJ, Briones-Salas M (eds) Biodiversidad de Oaxaca. Instituto de Biología, Universidad Nacional Autónoma de México (UNAM), Fondo Oaxaqueño para la Conservación de la Naturaleza y WWF, México DF, pp 67–85

Valencia-A S (2004) Diversidad del género *Quercus* (Fagaceae) en México. Bol Soc Bot Méx 75:33–53

Valencia-A S, Nixon KC (2004) Encinos. In: García-Mendoza AJ, Ordóñez MJ, Briones-Salas M (eds) Biodiversidad de Oaxaca. Instituto de Biología, Universidad Nacional Autónoma de México (UNAM), Fondo Oaxaqueño para la Conservación de la Naturaleza and WWF, México DF, Mexico, pp 219–225

van der Wal H (1996) Modificación de la vegetación y el suelo por los chinantecos de Santiago Tlatepusco, Oaxaca, México. Etnoecológica 3:37–57

van der Wal H (1998) Chinantec shifting cultivation and secondary vegetation: a case-study on secondary vegetation resulting from indigenous shifting cultivation in the Chinantla, Mexico. BOS Foundation, Wageningen

Vázquez-G JA, Givnish TJ (1998) Altitudinal gradients in tropical forest composition, structure and diversity in the Sierra de Manantlán. J Ecol 86:999–1020

Velázquez-Rosas N (1997) Características foliares de los árboles de bosques húmedos de montaña en la región de La Chinantla, Sierra Norte de Oaxaca. BSc Thesis, Universidad Nacional Autónoma de México (UNAM), México DF, Mexico

Velázquez-Rosas N, Meave J, Vázquez-Santana S (2002) Elevational variation of leaf traits in montane rain forest tree species at La Chinantla, southern Mexico. Biotropica 34:534–546

Vogelmann HW (1973) Fog precipitation in the cloud forests of eastern Mexico. BioScience 23:96–100

Wendt T (1993) Composition, floristic affinities, and origins of the canopy tree flora of the Mexican Atlantic slope rain forests. In: Ramamoorthy TP, Bye R, Lot A, Fa J (eds) Biological diversity of Mexico: origins and distributions. Oxford Univ Press, New York, pp 595–680

Whittaker RH (1972) Evolution and measurement of species diversity. Taxon 21:213–251

10 Structure and Composition of Costa Rican Montane Oak Forests

M. Kappelle

10.1 Introduction

Montane forests in the humid tropics differ significantly from tropical lowland forests (Richards 1952; Grubb and Whitmore 1966; Churchill et al. 1995; Hamilton et al. 1995; Kappelle 2004). The diurnal presence of clouds and mist is often the most remarkable characteristic of these forests (Stadtmüller 1987). The specific atmospheric humidity regime and strong diurnal temperature oscillations are probably the main environmental causes generating such a different structure and composition in tropical highland forest systems, compared to tropical lowland rainforests (Bruijnzeel and Veneklaas 1998).

A peculiar forest type frequently found in tropical and subtropical highland regions is the oak/beach-bamboo forest. Mature phases of this forest type generally have a canopy layer dominated by 30- to 50-m-tall fagaceous species, and an understorey characterized by 3- to 6-m-tall woody bamboos (Kappelle 1996). Such forests occur in the Americas as well as in Asia. Examples are beach forest (*Fagus*) with *Sasa* bamboo in Japan (Nakashizuka 1988), *Nothofagus* forest with *Chusquea* bamboo in southern South America (Veblen et al. 1981), *Nothofagus* forest with *Nastus* bamboo in Papua New Guinea (van Valkenburg and Ketner 1994), *Castanopsis* forest and *Lithocarpus* forest in Kalimantan and Sumatra (Ohsawa et al. 1985; Kitayama 1992), *Colombobalanus* (formerly known as *Trigonobalanus*) forest in Colombia (van der Hammen and Cleef 1983, Chaps. 1 and 11), and oak forest (*Quercus*) often with *Arundinaria* bamboo in the Himalayas (Saxena and Singh 1982), on Kalimantan and Java (Werner 1986), or with *Chusquea*, *Aulonemia* and *Rhipidocladum* bamboos in tropical Mexico, Central America and Colombia (Lozano and Torres 1974; Soderstrom et al. 1988; Pohl 1991; Widmer 1993; Kappelle 1996; Kappelle and Brown 2001; Chaps. 1, 10 and 11). Figure 10.1 shows the distribution of oak in Costa Rica.

Ecological Studies, Vol. 185
M. Kappelle (Ed.) Ecology and Conservation of Neotropical Montane Oak Forests
© Springer-Verlag Berlin Heidelberg 2006

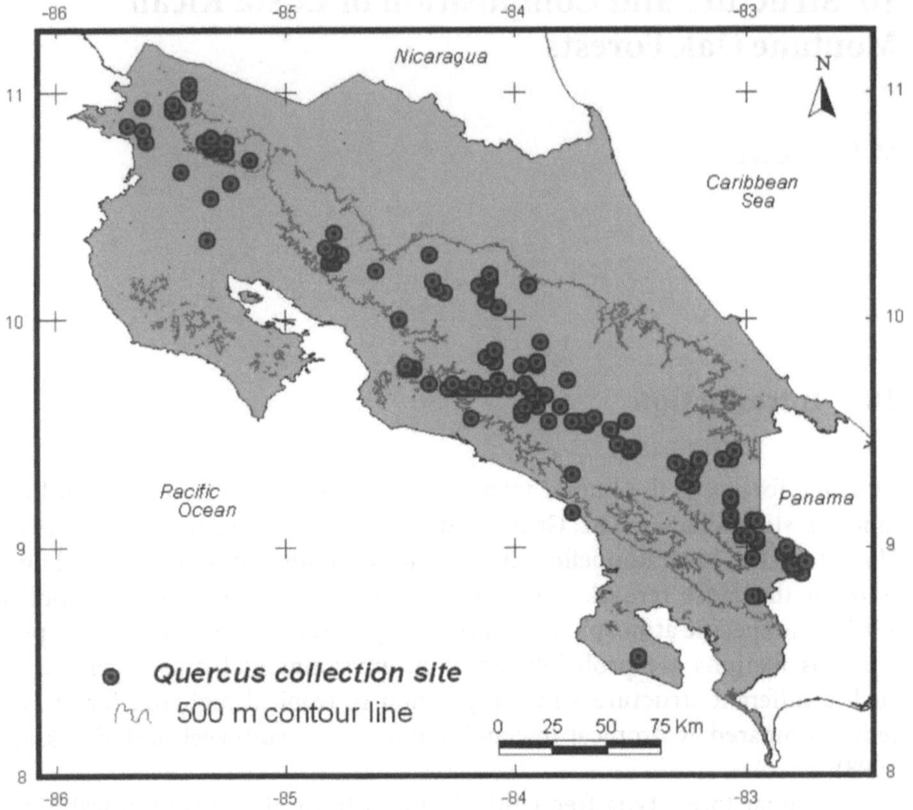

Fig. 10.1. Location of 168 Costa Rican collection sites of *Quercus* specimens stored at INBio's herbarium (INB). The 500-m contour line is drawn to show that most collections are from mid and high elevations. Only *Q. oleoides* has been collected below 500 m elevation, in the drier northern Pacific lowlands of Guanacaste. The collection site on the Osa Peninsula in the southern Pacific region corresponds to a cloud forest at the summit of a >700 m high hill where *Q. rapurahuensis* and *Q. insignis* were found (Kappelle et al. 2003). *Q. costaricencis* and *Q. corrugata* have been included in IUCN's Red List

The oak forests of upland Costa Rica are a good example of these tropical montane fagaceous-bamboo forests. They differ in many aspects from oak forests in temperate lowland North America (Hammitt and Barnes 1989) and Mediterranean Europe (Romane and Terradas 1992; Roda et al. 1999). This chapter presents a characterization of their distribution, structure, composition and diversity, and serves as an introduction to other chapters in this book on oak forest paleoecology (Chap. 2), non-vascular plants and lichens (Chaps. 6 and 7), population dynamics (Chaps. 15, 18, 23, 24, 25, 26 and 27), ecosystem functioning (Chaps. 21 and 22), and conservation and sustainable use (Chaps. 30, 31, 32 and 33).

10.2 Geographic Forest Distribution

Montane oak forests in Costa Rica are principally found along Pacific slopes at altitudes of 1,500–3,400 m, and along Atlantic slopes at 1,800–3,100 m elevation (see also Chap. 4). Most montane oak forest stands are concentrated in Costa Rica's Talamanca Range, though small, dispersed patches of oak forest stands occur in the volcanic mountain chains to the northwest (Kappelle 1996), including the Monteverde Cloud Forest Preserve (Nadkarni and Wheelwright 2000). Occasionally, highland oak trees may appear in patchy distribu-

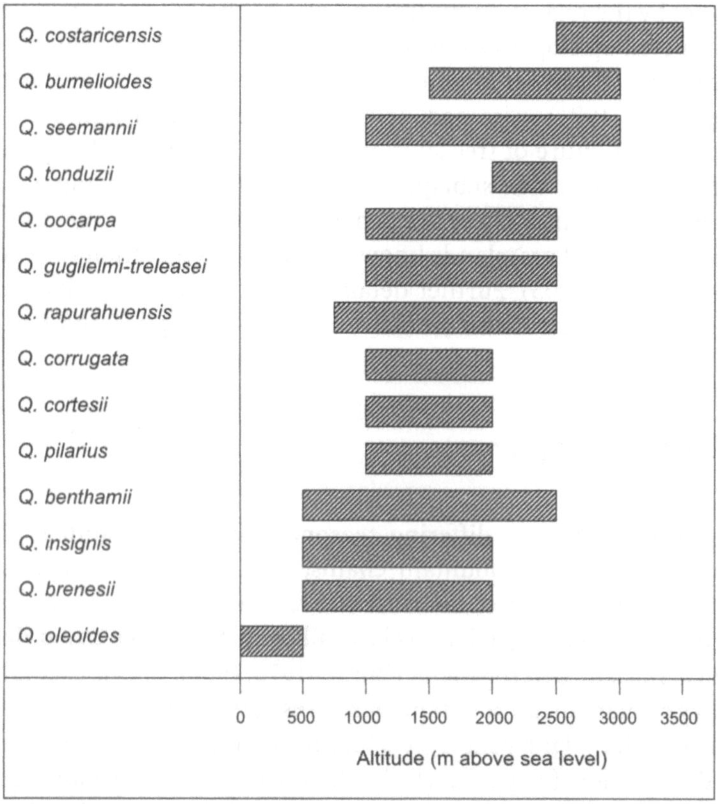

Fig. 10.2. Altitudinal distribution of 14 oak (*Quercus*) species occurring in Costa Rica. Distributions are in accordance with Burger (1977), Kappelle (1987, 1996), and reviews of herbarium specimens at CR and INB. Following Burger (1977), *Q. eugeniaefolia* and *Q. sapotaefolia* have been included in *Q. seemannii*. However, *Q. bumelioides*, which Burger (1977) also classified under *Q. seemannii*, has been treated here as a separate species, as recommended by N. Zamora at INB (personal communication; see www.inbio.ac.cr). *Q. bumelioides* is synonymous with *Q. copeyensis* (K.C. Nixon, personal communication). Previously, *Q. benthamii* and *Q. cortesii* had not been reported for Costa Rica (Burger 1977)

tions at lower elevations. One species of Costa Rican oak, *Quercus oleoides*, is even restricted to dry lowland forests in Costa Rica's northwestern, Pacific Guanacaste region. Only four of 14 *Quercus* species known from Costa Rica reach elevations below 1,000 m a.s.l. (Fig. 10.2).

As on many other tropical mountains (e.g., Aiba and Kitayama 1999; Ashton 2003), Talamanca's highland oak forests are zoned in sequential altitudinal belts: the upper montane oak forests (2,200–3,400 m), and the lower montane oak forests (1,500–2,400 m). Upper montane oak forests generally have a higher cloud and mist incidence (persistence) than is the case for lower montane oak forests. In fact, according to Grubb and Stevens (1985), there is a close correlation between the lower–upper montane forest ecotone and the diurnal cloud base. In the tropics, the elevation of the diurnal cloud base is generally set by the relative humidity and rate of cooling of warm lowland air being conducted up slopes as it warms during the morning (Ashton 2003). In Costa Rica, below the lower montane oak forest belt, a premontane belt occurs (Holdridge et al. 1971) – immediately above the lowland rainforest zone – dominated by a mixture of tree species including Lauraceae (Kappelle 2004). At higher elevations, the subalpine (3,100–3,500 m) and alpine (3,300–3,819 m) belts are found. These are generally dominated by cold and humid, low-stature scrub and grasslands known as paramo vegetation (Körner 1999; Kappelle and Horn 2005). Further details on altitudinal gradients and elevational zonation in Costa Rican montane oak forests are given in Chap. 4.

10.3 Plant Geography

In Costa Rica's highlands, differing seasonal patterns of rainfall, superimposed on discontinuous mountain chains, rich mineral volcanic soils, the nearness of large species-rich continental areas, a past history as an archipelago, and the influence of glaciations have all contributed to a dynamic system of high local floristic heterogeneity (Burger 1975, 1980). This is exemplified by the country's Talamancan montane oak forests, in which almost 75 % of 253 censused vascular plant genera (excluding orchids and bromeliads) has a tropical distribution (Kappelle et al. 1992), whereas the remaining 25 % is made up of temperate (17 %) and cosmopolitan (8 %) genera (Fig. 10.3).

Important temperate plant genera include holarctic *Alnus*, *Arenaria*, *Cornus*, *Myrica*, *Quercus*, *Prunus*, *Rhamnus*, *Ribes*, *Rubus*, *Vaccinium* and *Viburnum*, and austral-Antarctic *Acaena*, *Drymis*, *Escallonia*, *Fuchsia*, *Gaiadendron*, *Gaultheria*, *Pernettya*, *Podocarpus* and *Weinmannia*. Within the tropical component, the neotropical element is best represented and contributes to almost half of all recorded genera (46 %). Some characteristic neotropical tree genera are *Billia*, *Brunellia*, *Freziera*, *Guatteria* and *Mollinedia*. The tropical afro-American element is very poorly represented (3 %, *Guarea*, *Lippia*,

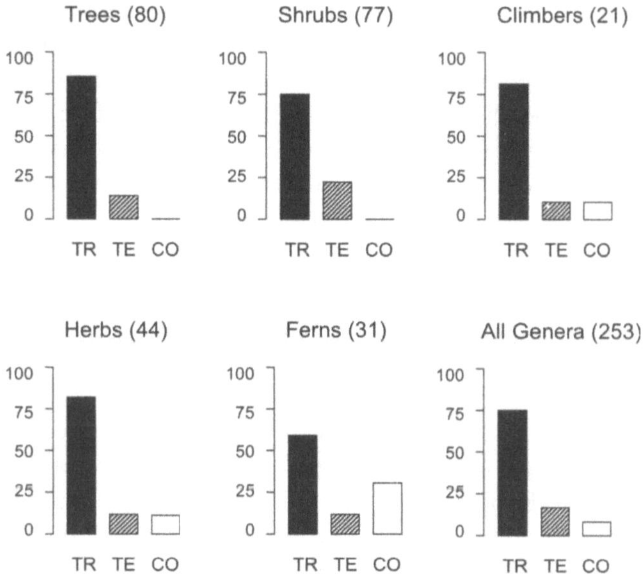

Fig. 10.3. Biogeographical distribution of 253 terrestrial vascular plant genera per growth form in upper montane *Quercus* forests in Costa Rica. *Closed bars* Tropical genera (*TR*), *dashed bars* temperate genera (*TE*), *open bars* cosmopolitan genera (*CO*). Y-axis values are percentages of the total number of genera per growth form. *Numbers within brackets* indicate the number of genera per growth form

Trichilia, *Urera*). Most of the 80 tree genera observed are neotropical, tropical Asian-American or pantropical in distribution. *Clethra*, *Cleyera*, *Hedyosmum*, *Magnolia*, *Meliosma*, *Persea*, *Phoebe*, *Styrax*, *Symplocos* and *Turpinia* are of tropical Asian-American origin. The only cosmopolitan tree genus that has been found is *Ilex*.

Shrubs amount to 77 genera and are principally neotropical (over 60 %), pantropical or northern/southern temperate. Neotropical montane shrub genera are mostly Andean-centered, and originated as a result of very active speciation or even explosive evolution as a consequence of tropical Andean orogenesis (Gentry 1982, 1985). Herb genera (44 genera in total) are basically neotropical, pantropical and wide-temperate. Climbers (21) are principally neotropical or pantropical. Ferns (31) show mainly cosmopolitan, pantropical or neotropical distributions. Most cosmopolitan genera are herbs (14 %) or ferns (29 %).

A comparative, phytogeographical analysis demonstrates a great floristic affinity of Costa Rican montane oak forests with equivalents in the Colombian Andes (Cordillera Oriental, Chap. 11), and lower levels of similarity with Mexican mesophyllous montane oak forests, such as found in the transversal Neovolcanic mountain range and surroundings (Kappelle et al. 1992, Chaps. 8 and 9). The greater affinity with Colombia may be due to climatic similarities between Costa Rica's Talamanca mountains and the Colombian Andes, which both display humid to per-humid conditions. The Mexican Neovolcanic mountain belt is much drier, favoring a set of drought-resistant upland plant genera of northern origin not known in Costa Rica (e.g., *Liquidambar* and

Pinus). Similarly, some moisture-loving plant genera of neotropical or austral-Antarctic origin that are well spread in the Colombian Andes seem to have migrated northward to Costa Rica's Talamanca range, but have not been able to reach the drier Mexican montane oak forests. More in-depth, regional-scale biogeographic studies are needed to help reveal the precise floristic – as well as faunistic, for that matter – affinities and dissimilarities among related biota of neotropical montane oak forests.

10.4 Forest Structure and Physiognomy

Tropical montane oak forests demonstrate a clear vertical structure with a number of horizontal layers, similar to the stratification of temperate oak forests (Kappelle 2004). In mature old-growth stands in Costa Rica, the dominant canopy oaks are normally 25–40 m tall, though some giant, emergent individuals may reach heights of 50–60 m. It has been hypothesized that maximum tree height is principally limited by water transport constraints, leaf water stress, and the resulting reductions in leaf photosynthesis (Koch et al. 2004). Table 10.1 presents some stand structure and diversity data for old-growth oak forest (OGF) at 2,900–3,000 m a.s.l. in Costa Rica.

Immediately below the upper oak forest line at altitudes of 3,000–3,200 m where subalpine forests commence (Islebe and Kappelle 1994), *Q. costaricensis* trees become lower in stature (<25 m) and more stunted (Chap. 4). Here, they may form twisted branches with densely packed, small-sized leaves (Holdridge et al. 1971; Kappelle and Leal 1996). Oak branches and twigs are often thickly covered with epiphytic aroids, bromeliads, ericads, orchids, ferns and parasitic loranths, alternated with pending, atmospheric moisture-capturing mosses, hepatics and lichens (Holdridge et al. 1971; Kappelle et al. 1989; Chaps. 6, 7 and 21). Mature oaks lack real buttresses, but expanded bases do occur on some larger trees (Holdridge et al. 1971).

Stem densities in numbers per ha range from 5,000–8,400 for stems>1 cm DBH (diameter at breast height), to 700–1,000 for stems>5 cm DBH, and 455–510 for stems>10 cm DBH (Blaser 1987; Jiménez et al. 1988). Occasionally, the DBH of giant oaks may reach values over 120 cm. Values of basal area are among the highest found in tropical forests: 50–53 m^2 per ha for stems>1 cm DBH, 48–51 m^2 per ha for stems>10 cm DBH, and 32–37 m^2 per ha for stems>50 cm DBH (Blaser 1987; Jiménez et al. 1988). *Q. costaricensis* and *Q. copeyensis* (now known as *Q. bumelioides*) alone may account for up to 90 % of both density and basal area for stems>50 cm DBH, and thousands of juveniles (seedlings, saplings) may fit into a single hectare (Chap. 18).

Family importance values (FIV), which include measures of relative dominance, density and diversity (Mori et al. 1983), were measured for stems>3.0 cm DBH in a 0.1-ha plot of old-growth, mature oak forest. Highest

Table 10.1. Stand structure and diversity data for three 0.1-ha plots in montane old-growth oak forest (OGF) at ~2,950 m a.s.l. in Costa Rica. Values are based on plot data presented in Kappelle et al. (1995a, 1996). Similar data for successional oak forest are presented in Chap. 17

Variable[a]	Plot 1	Plot 2	Plot 3	Mean+1 SE
Canopy height (m)	35	38	35	36.5+2.1
Number of stems ≥3 cm DBH per plot	184	197	182	189.5+7.5
Number of stems ≥10 cm DBH per plot	69	51	70	60.5+9.5
Number of stems per diameter class				
Stems 3–5 cm DBH per plot	58	81	65	73+8
Stems 5–10 cm DBH per plot	57	65	47	56+9
Stems 10–20 cm DBH per plot	33	27	39	33+6
Stems 20–40 cm DBH per plot	20	13	20	16.5+3.5
Stems 40–80 cm DBH per plot	14	8	8	8+0
Stems ≥80 cm DBH per plot	2	3	3	3+0
Stem density (stems ≥3 cm DBH per ha)	1,840	1,970	1,820	1,895+75
Basal area for stems ≥3 cm DBH ($m^2 ha^{-1}$)	64.7	57.5	58.7	58.1+0.6
Species richness per plot (terr. vasc.)[b]	62	79	68	73.5+5.5
Species richness per plot (trees only)	18	21	20	20.5+0.5
Shannon-Wiener's index (terr. vasc.)	3.70	4.45	4.38	4.18+0.24
Shannon-Wiener's index (trees only)	3.27	3.18	3.48	3.33+0.15
Reciprocal Simpson's index (terr. vasc.)	6.30	9.97	11.31	10.64+0.67
Reciprocal Simpson's index (trees only)	6.14	4.87	7.98	6.43+1.56
Species density (terr. vasc.)	20.7	26.3	22.7	24.5+1.8
Evenness or equitability index (terr. vasc.)	0.62	0.71	0.72	0.72+0.01

[a] Shannon-Wiener's index, reciprocal Simpson's index, species density, and evenness index were measured following procedures presented in Magurran (1988)

[b] Terr. vasc., all terrestrial vascular plant species

FIV values were recorded in Fagaceae (122), followed by Myrsinaceae (30), Cunoniaceae (22), Styracaceae (18), Araliaceae (16), Lauraceae (15), and Theaceae (11) (Kappelle et al. 1996).

10.5 Plant Diversity

Costa Rican montane oak forests are extraordinarily rich in vascular plant species. For information on the diversity of non-vascular plant species, fungi and lichens, I refer to Chaps. 5, 6, 7 and 21. Epiphytic vascular species are particularly abundant, with at least 100 orchid and 25 bromeliad species (Kappelle 1996). As some 1,000 native orchid species are known to reside in Costa Rica (N. Zamora, personal communication), we may assume that – given the size of the country and the extent of intact montane oak forest – many more orchid species than the 100 recorded grow in the high canopies of these oak forests.

To date, a total of 1,300 vascular plant species has been recorded for old-growth and recovering Talamancan montane oak forest (2,000–3,400 m elevation; Atlantic and Pacific slopes). For species lists, the reader is referred to Kappelle et al. (1991, 2000), Kappelle and Gómez (1992), van Velzen et al. (1993), Kappelle (1996), Kappelle and van Omme (1997), and MNCR (2001). Almost 500 of these species are woody, and include hundreds of trees and shrubs as well as a few woody climbers such as *Hydrangea* and hemi-epiphytes such as *Clusia* (Kappelle and Zamora 1995). Angiosperms account for some 1,000 species, and are distributed between 750 species of dicots (Magnoliopsida) and 250 species of monocots (Liliopsida). Only three species are native gymnosperms (conifers), belonging to Podocarpaceae. Over 300 species are pteridophytes, including at least 250 ferns, 35 clubmosses (Lycopodiaceae, Selaginellaceae), one quillwort (*Isoetes*), and one horsetail (*Equisetum*).

Most speciose angiosperm plant families are Asteraceae (>60 species), Ericaceae (>30), Lauraceae (>35), Melastomataceae (>35), Myrsinaceae (>20), Piperaceae (>40), Poaceae (>20), Rosaceae (>20), Rubiaceae (>50), and Solanaceae (>30). The most diverse fern families are Adiantaceae (>20), Grammitidaceae (>40), Hymenophyllaceae (>25), Lomariopsidaceae (>35), and Polypodiaceae (>35). Tree ferns account for at least 14 species, spread over Cyatheaceae (11), Dicksoniaceae (3), and Lophosoriaceae (1). Extremely rich epiphytic vascular genera include the tongue fern *Elaphoglossum* (>30 species), the small, sclerophyllous dicot herb *Peperomia* (>25), and the orchid *Maxillaria* (>20). The most speciose terrestrial vascular plant genus is the shrub *Miconia* (>20 species). Diverse vascular genera with at least 15 species are the epiphytes *Anthurium* (see also Chap. 15), *Asplenium*, *Begonia*, *Epidendrum* and *Huperzia*, the bamboo *Chusquea*, the shrubs *Piper*, *Psychotria* (including *Cephaelis*) and *Solanum*, the climber *Passiflora*, and the lauraceous tree *Ocotea*, an important fruit tree for the Resplendant Quetzal (Chap. 25). Other, less diverse but still rich genera with over ten species include the ground-rooted tree *Ficus*, the hemi-epiphytic tree *Oreopanax*, the dwarf palm *Chamaedorea*, and the shrubs *Palicourea* and *Rubus* (Kappelle and Zamora 1995; Kappelle 1996; MNCR 2001).

Alpha diversity was measured for terrestrial vascular plants in three separate 0.1-ha mature old-growth oak forest plots, using different diversity indices (Magurran 1988; Table 10.1). Species richness varied in the range 62–79 species per plot, species density was 20.7–26.3, Shannon-Wiener's index 3.70–4.45, Simpson's reciprocal index 6.30–11.31, and the equitability index – a measure of evenness – showed rounded values of 0.62–0.72 (Kappelle et al. 1995a; Table 10.1).

10.6 Floristic Composition

The 20- to 40-m-tall canopy layer of old-growth stands of Talamanca's montane oak forests is almost exclusively dominated by the genus *Quercus*. At elevations over 2,000 m, endemic *Q. copeyensis*, endemic *Q. costaricensis* and wide-ranging *Q. seemannii* dominate, each within its specific altitudinal range (Burger 1977; Kappelle et al. 1989, 1991). *Q. oocarpa* and *Q. rapurahuensis* are also frequently observed, albeit in lower numbers, mainly at mid elevation (2,000–2,400 m) along less wet Pacific slopes. Other important canopy tree genera are *Magnolia*, *Podocarpus*, *Prumnopitys*, *Schefflera* and *Weinmannia* (Kappelle et al. 1995b; MNCR 2001). *Clusia* may occasionally occur as an (hemi)epiphytic tree on canopy branches of oak. Parasitic Loranthaceae, including Viscaceae, such as *Dendrophthora*, *Phoradendron* and *Struthantus*, share oak – and other species' – branches and twigs with epiphytic non-parasitic vascular genera in the Araceae, Begoniaceae, Bromeliaceae, Cyclanthaceae, Ericaceae, Gesneriaceae, Orchidaceae, Piperaceae and ferns.

The 5- to 20-m-tall subcanopy layer of mature oak forest is composed of a complex mixture of tree species. They include genera such as *Abatia*, *Aiouea*, *Alchornea*, *Alfaroa*, *Alnus*, *Ardisia*, *Billia*, *Brunellia*, *Buddleja*, *Cinnamomum* (including *Phoebe*), *Clethra*, *Cleyera*, *Clusia*, *Comarostaphylis*, *Cornus*, *Croton*, *Dendropanax*, *Drimys*, *Escallonia*, *Eugenia*, *Freziera*, *Guatteria*, *Guarea*, *Hedyosmum*, *Ilex*, *Inga*, *Ladenbergia*, *Lippia*, *Lozania*, *Meliosma*, *Monnina*, *Myrcianthes*, *Myrsine*, *Nectandra*, *Ocotea*, *Oreopanax*, *Panopsis*, *Parathesis*, *Persea*, *Picramnia*, *Prunus*, *Quetzalia* (synonymous with *Microtropis*), *Rhamnus*, *Rondeletia*, *Roupala*, *Salix*, *Sapium*, *Saurauia*, *Styrax*, *Symplocos*, *Ticodendron*, *Trichilia*, *Turpinia*, *Ulmus*, *Vaccinium*, *Viburnum* and *Zanthoxylum*. Often, these species are accompanied by young trees of *Magnolia*, *Quercus*, *Podocarpus*, *Schefflera* and *Weinmannia*, waiting for a tree fall to continue their journey to the higher canopy (Kappelle et al. 1989, 1991, 1995a).

The 1- to 5-m-high understorey layer is dominated largely by bamboo species of the genus *Chusquea* and, to a lesser extent, *Aulonemia*. Most common are *Chusquea longifolia*, *C. talamancensis* and *C. tomentosa*. Bamboos are often associated with dwarf palms (*Chamaedorea*, *Geonoma*), cyclanths (*Asplundia*, *Sphaeradenia*) and treeferns (*Alsophila*, *Cnemidaria*, *Culcita*, *Cyathea*, *Dicksonia*, *Lophosoria* and *Sphaeropteris*); see also Kappelle et al. (1989, 1995b). In this layer, shrubs in the Ericaceae, Melastomataceae, Rubiaceae and Solanaceae are also common. Climbers include *Bomarea*, *Cissus*, *Cyclanthera*, *Cynanchum*, *Dioscorea*, *Iresine*, *Hydrangea*, *Passiflora*, *Sechium*, *Smilax* and *Tropaeolum* (Kappelle 1996).

Ground-dwelling vascular plant species shorter than 1 m, and often recorded in forest tree fall gaps and at forest edges, include a number of herbs in the Acanthaceae, Apiaceae, Asteraceae, Brassicaceae, Campanulaceae, Caryophyllaceae, Commelinaceae, Convallariaceae, Cyperaceae, Gentianaceae, Geraniaceae, Gesneriaceae, Gunneraceae, Heliconiaceae, Iridaceae, Juncaceae, Lamiaceae, Malvaceae, Onagraceae, Oxalidaceae, Phytolaccaceae, Piperaceae, Plantaginaceae, Rosaceae, Scrophulariaceae, Urticaceae, Valerianaceae and Violaceae (Kappelle 1996).

10.7 Conclusions

The current chapter provides some insight into the structure, composition and diversity of Costa Rica's montane oak forest. It is meant to set the stage on this particularly rich and voluminous forest, in order to better understand its spatial and temporal patterns and processes, and above all, its functioning as an ecosystem. In subsequent chapters (Chaps. 15, 17, 23, 24, 26 and 30), these themes will be dealt with by the author, co-authors and colleagues who have studied the magnificent Talamancan montane oak forest environment over the last two decades.

Acknowledgements I am very grateful to friends, colleagues and students who supported my research in Costa Rica's montane oak forest over the last 20 years. I especially want to thank Antoine M. Cleef at the University of Amsterdam (UvA), and Luis Poveda, Nelson Zamora, and the late Adelaida Chaverri (1947–2003) at Costa Rica's National University (UNA) and National Biodiversity Institute (INBio). Marco Castro prepared Fig. 10.1. Major funding was provided by UvA, UNA, INBio and NWO-WOTRO. Research permission was granted by MINAE.

References

Aiba S, Kitayama K (1999) Structure, composition and species diversity in an altitude-substrate matrix of rain forest tree communities on Mount Kinabalu, Borneo. Plant Ecol 140:139–157

Ashton P (2003) Floristic zonation of tree communities on wet tropical mountains revisited. Perspect Plant Ecol Evol Syst 6(1/2):87–104

Blaser J (1987) Standörtliche und waldkundliche Analyse eines Eichen-Wolkenwaldes (*Quercus* spp.) der Montanstufe in Costa Rica. PhD Thesis, Georg-August Universität, Göttingen

Bruijnzeel LA, Veneklaas EJ (1998) Climatic conditions and tropical montane forest productivity: the fog has not lifted yet. Ecology 79(1):3–9

Burger WC (1975) The species concept in *Quercus*. Taxon 24:45–50

Burger WC (1977) Fagaceae. In: Burger WC (ed) Flora costaricensis. Field Bot Ser 40:59–80

Burger WC (1980) Why are there so many kinds of flowering plants in Costa Rica? Brenesia 17:371–388

Churchill SP, Balslev H, Forero E, Luteyn JL (eds) (1995) Biodiversity and conservation of Neotropical montane forests. New York Botanical Garden Press, Bronx, NY

Gentry AH (1982) Neotropical floristic diversity: phytogeographical connections between Central and South Amercia: Pleistocene climatic fluctuations, or an accident of the Andean orogeny? Ann Missouri Bot Gard 69:557–593

Gentry AH (1985) Contrasting phytogeographic patterns of upland and lowland Panamanian plants. In: D'Arcy WG, Correa MD (eds) The botany and natural history of Panama. Missouri Bot Gard Press, St Louis, MO

Grubb PJ, Stevens PF (1985) The forests of the Fatima Basin and Mount Kerigomna, Papua New Guinea. Research School of Pacific Studies, Australian National University, Canberra

Grubb PJ, Whitmore TC (1966) A comparison of montane and lowland rain forest in Ecuador. II. The climate and its effects on the distribution and physiognomy of the forests. J Ecol 54:303–333

Hamilton LS, Juvik JO, Scatena FN (eds) (1995) Tropical montane cloud forests. Springer, Berlin Heidelberg New York, Ecological Studies, vol 110

Hammitt WE, Barnes BV (1989) Composition and structure of an old-growth oak-hickory forest in southern Michigan over 20 years. In: Rink G, Budelsky CA (eds) Proc 7th USDA Forest Service Central Hardwood Conf, St Paul, MN. Gen Tech Rep NC-132, pp 247–253

Holdridge LR, Grenke WC, Hatheway WH, Liang T, Tosi JA Jr (1971) Forest environments in tropical life zones: a pilot study. Pergamon Press, Oxford, UK

Islebe GA, Kappelle M (1994) A phytogeographical comparison between subalpine forests of Guatemala and Costa Rica. Feddes Rep 105:73–87

Jiménez W, Chaverri A, Miranda R, Rojas I (1988) Aproximaciones silviculturales al manejo de un robledal (*Quercus* spp.) en San Gerardo de Dota, Costa Rica. Turrialba 38(3):208–214

Kappelle M (1987) A phytosociological analysis of oak forests in the western Talamanca Range, Costa Rica. MSc Thesis, University of Amsterdam, Amsterdam, The Netherlands

Kappelle M (1996) Los bosques de roble (*Quercus*) de la Cordillera de Talamanca, Costa Rica: biodiversidad, ecología, conservación y desarrollo. Instituto Nacional de Biodiversidad (INBio), Santo Domingo de Heredia, Costa Rica

Kappelle M (2004) Tropical montane forests. In: Burley J, Evans J, Youngquist JA (eds) Encyclopedia of Forest Sciences, vol 4. Elsevier, Oxford, UK, pp 1782–1793

Kappelle M, Brown AD (eds) (2001) Bosques nublados del Neotrópico. Instituto Nacional de Biodiversidad (INBio), Santo Domingo de Heredia, Costa Rica

Kappelle M, Gómez LD (1992) Distribution and diversity of montane pteridophytes of the Chirripó National Park, Costa Rica. Brenesia 37:67–77

Kappelle M, Horn SP (eds) (2005) Páramos de Costa Rica. Instituto Nacional de Biodiversidad (INBio), Santo Domingo de Heredia, Costa Rica

Kappelle M, Leal ME (1996) Changes in leaf morphology and foliar nutrient status along a successional gradient in a Costa Rican upper montane *Quercus* forest. Biotropica 28(2):331–344

Kappelle M, van Omme E (1997) Lista de las plantas de los bosques nubosos subalpinos de la Cordillera de Talamanca en Costa Rica. Brenesia 47/48:55–71

Kappelle M, Zamora N (1995) Changes in woody species richness along an altitudinal gradient in Talamancan montane *Quercus* forests, Costa Rica. In: Churchill SP, Balslev H, Forero E, Luteyn JL (eds) Biodiversity and conservation of Neotropical montane forests. New York Botanical Garden Press, Bronx, NY, pp 135–148

Kappelle M, Cleef AM, Chaverri A (1989) Phytosociology of montane *Chusquea-Quercus* forests, Cordillera de Talamanca, Costa Rica. Brenesia 32:73–105

Kappelle M, Zamora N, Flores T (1991) Flora leñosa de la zona alta (2000-3819 m) de la Cordillera de Talamanca, Costa Rica. Brenesia 34:121–144

Kappelle M, Cleef AM, Chaverri A (1992) Phytogeography of Talamanca montane *Quercus* forests, Costa Rica. J Biogeogr 19(3):299–315

Kappelle M, Kennis PAF, de Vries RAJ (1995a) Changes in diversity along a successional gradient in a Costa Rican upper montane *Quercus* forest. Biodiv Conserv 4:10–34

Kappelle M, van Uffelen JG, Cleef AM (1995b) Altitudinal zonation of montane *Quercus* forests along two transects in the Chirripó National Park, Costa Rica. Vegetatio 119:119–153

Kappelle M, Geuze T, Leal ME, Cleef AM (1996) Successional age and forest structure in a Costa Rican upper montane *Quercus* forest. J Trop Ecol 12:681–698

Kappelle M, van Omme E, Juárez ME (2000) Lista de la flora vascular terrestre de la cuenca superior del Río Savegre, San Gerardo de Dota, Costa Rica. Acta Bot Mex 51:1–38

Kappelle M, Castro M, Acevedo H, González L, Monge H (2003) Ecosystems of the Osa Conservation Area (ACOSA), Costa Rica. Instituto Nacional de Biodiversidad (INBio), Santo Domingo de Heredia, Costa Rica

Kitayama K (1992) An altitudinal transect study of the vegetation of Mount Kinabalu, Borneo. Vegetatio 102:149–171

Koch GW, Sillet SC, Jennings GM, Davis SD (2004) The limits to tree height. Nature 428:851–854

Körner C (1999) Alpine plant life. Springer, Berlin Heidelberg New York

Lozano G, Torres JH (1974) Aspectos generales sobre la distribución, sistemática fitosociológica y clasificación ecológica de los bosques de robles (*Quercus*) en Colombia. Ecol Trop 1(2):45–79

Magurran AE (1988) Ecological diversity and its measurement. Croom Helm, London

MNCR (2001) Caracterización de la vegetación de la cuenca del Río Savegre. Proyecto Araucaria, Museo Nacional de Costa Rica (MNCR) and Instituto Nacional de Biodiversidad (INBio), San José, Costa Rica

Mori SA, Boom BM, de Carvallo AM, dos Santos TS (1983) Southern Bahian moist forest. Bot Rev 49:155–232

Nadkarni N, Wheelwright N (eds) (2000) Monteverde: ecology and conservation of a tropical cloud forest. Oxford Univ Press, Oxford

Nakashizuka T (1988) Regeneration of beech (*Fagus crenata*) after simultaneous death of undergrowing dwarf bamboo (*Sasa kurilensis*). Ecol Res 3:21–35

Ohsawa M, Nainggolan PHJ, Tanaka N, Anwar C (1985) Altitudinal zonation of forest vegetation on Mount Kerinci, Sumatra: with comparisons to zonation in the temperate region of east Asia. J Trop Ecol 1:193–216

Pohl RW (1991) Blooming history of the Costa Rican bamboos. Rev Biol Trop 39(1):111–124

Richards PW (1952) The tropical rain forest. Cambridge Univ Press, Cambridge, UK

Roda F, Retana J, Gracia CA, Bellot J (eds) (1999) Ecology of Mediterranean evergreen oak forests. Springer, Berlin Heidelberg New York, Ecological Studies, vol 137

Romane F, Terradas J (eds) (1992) *Quercus ilex* ecosystems: function, dynamics and management. Springer, Berlin Heidelberg New York, Advances in Vegetation Science, vol 13

Saxena AK, Singh JS (1982) A phytosociological analysis of woody species in forest communities of a part of Kumaun Himalaya. Vegetatio 50:3–22

Soderstrom TR, Judziewicz EJ, Clark LG (1988) Distribution patterns of neotropical bamboos. In: Heyer WR, Vanzolini PE (eds) Proc Worksh Neotropical Distribution Patterns, Academia Brasileira de Ciências, Rio de Janeiro, pp 121–157

Stadtmüller T (1987) Cloud forests in the humid tropics: a bibliographic review. United Nations University (UNU), Tokyo, Japan

Van der Hammen T, Cleef AM (1983) *Trigonobalanus* and the tropical amphi-pacific element in the North Andean forest. J Biogeogr 10:437–440

Van Valkenburg JLCH, Ketner P (1994) Vegetation changes following human disturbance of mid-montane forest in the Wau area, Papua New Guinea. J Trop Ecol 10:41–54

Van Velzen, HP, Wijtzes WH, Kappelle M (1993) Lista de especies de la vegetación secundaria del piso montano pacífico, Cordillera de Talamanca, Costa Rica. Brenesia 39/40:147–161

Veblen TT, Donoso C, Schlegel FM, Escobar B (1981) Forest dynamics in south-central Chile. J Biogeogr 8:211–247

Werner WL (1986) A comparison between two tropical montane ecosystems in Asia: Pidurutalagala (Ceylon/Sri Lanka) and Pangrango-Gede (Java). Mount Res Dev 6:335–344

Widmer Y (1993) Bamboo and gaps in the oak forests of the Cordillera de Talamanca, Costa Rica. Verh Gesell Ökol 22:329–332

11 Structure and Composition of Colombian Montane Oak Forests

M.T. Pulido, J. Cavelier, and S.P. Cortés-S

11.1 Biogeography

Quercus is a young immigrant in Colombia. The palynological record shows oak pollen in sediments dating from 250,000 (van der Hammen and González 1963) to 340,000 years BP (Hooghiemstra and Sarmiento 1991; Hooghiemstra and Ran 1994; Chap. 2). The relatively recent migration of this genus from North and Central America into South America correlates with the uplifting of the Andes during the Pliocene, and the formation of the Isthmus of Panama. During the migration from northern higher to lower neotropical latitudes, the genus appears with a decreasing number of species. Indeed, *Quercus* has 150 species in Mexico (Rzedowski 1978; see Chaps. 1, 8 and 9), 12–17 in Costa Rica (Müller 1942; Burger 1975; see Chap. 10), ten in Panamá (Müller 1942), and only one (*Q. humboldtii*) in Colombia (Müller 1942; Chap. 1). The most likely ancestors of *Q. humboldtii* are *Q. benthami* and *Q. costaricencis* (Müller 1942). In Colombia, *Quercus humboldtii* shows a wide altitudinal distribution, ranging from 1,100 to 3,200 m a.s.l. (above sea level), and a latitudinal range of 8°N (Cerro Tacarcuna, Darién-Chocó) to 1°N (Pasto Airport, Nariño; Fig. 11.1). There are no records of *Quercus* in either Ecuador or Venezuela.

11.2 Taxonomy

Although Müller (1942) accepted only *Q. humboldtii* for Colombia, the number of *Quercus* species in Colombia has always been a controversial issue, with seven "species" identified: *Quercus humboldtii* Bonpl., *Q. tolimensis* Bonpl., *Q. lindenii* A.DC., *Erythrobalanus duqueana* Schwartz, *Q. colombiana* Cuatrec., *Q. boyacensis* Cuatrec., and *Q. almaguerensis* Bonpl. (Table 11.1). The uncertainty about the number of *Quercus* species in Colombia may be related to the

Ecological Studies, Vol. 185
M. Kappelle (Ed.) Ecology and Conservation of Neotropical Montane Oak Forests
© Springer-Verlag Berlin Heidelberg 2006

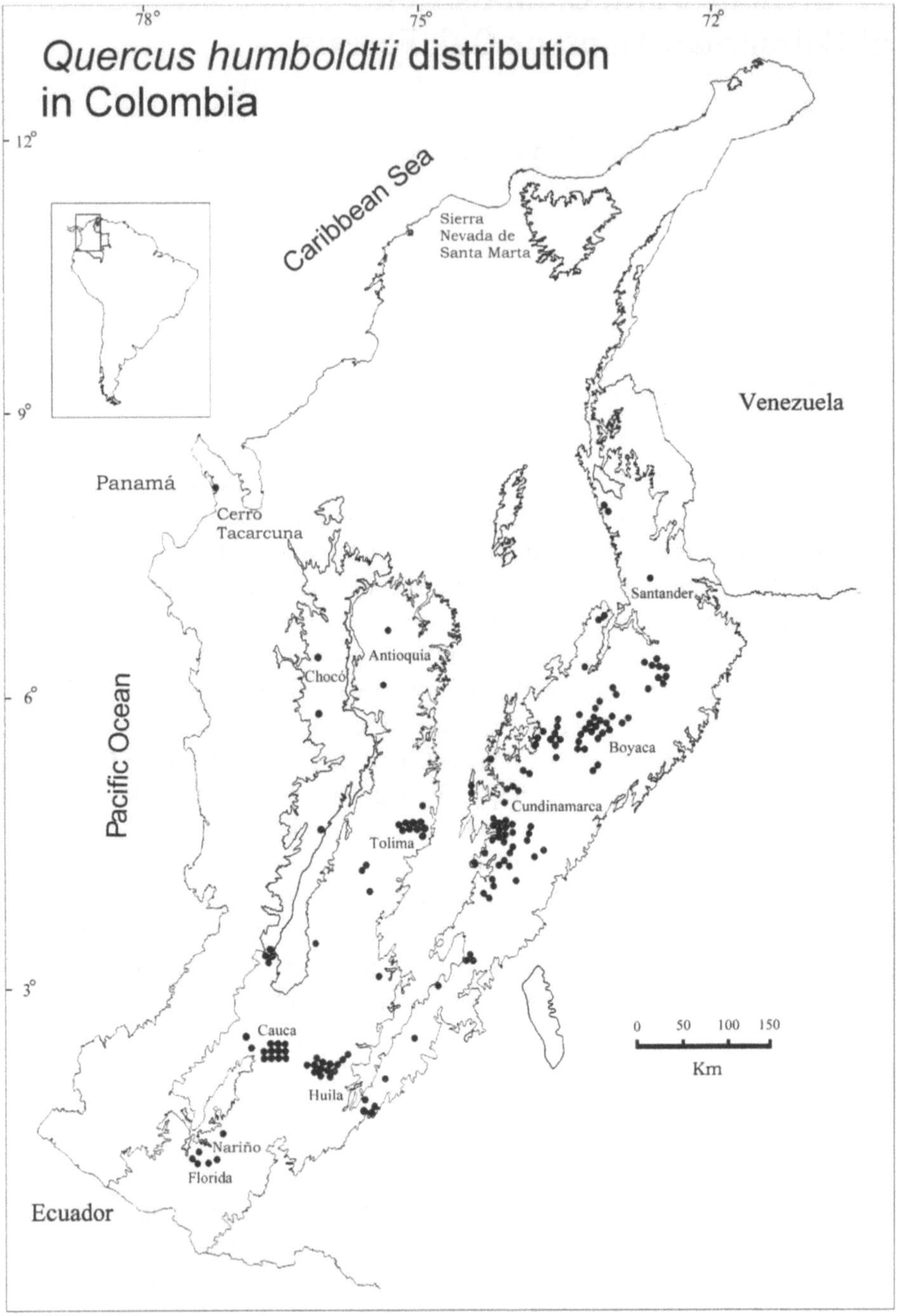

Fig. 11.1. Geographical distribution of *Quercus humboldtii* in Colombia. Each data point represents the collection site of a herbarium specimen. The *continuous line* represents the 1,000-m depth contour

Table 11.1. Fagaceae species recorded in Colombia. Type locality is given precisely as it appeared in the original publication (except for punctuation). *Dto.* Departamento (political division

Species	Type locality
Quercus boyacensis CUATRECASAS 1944 (Cuatrecasas 1944)	Dto. Boyacá, cordillera oriental, quebrada de Susacón, 3,300–3,100 m
Q. colombiana CUATRECASAS 1944 (Cuatrecasas 1944)	Dto. Boyacá, bosques de Arcabuco, 2,600–2,700 m
Erytrobalanus duqueana SCHWARTZ 1937 (Schwartz 1937)	Dto. Valle, Cordillera central, Rio Nima cerca de Palmira, 1,800–2,400 m
Q. humboldtii BONPL. 1809 (Humboldt and Bonpland 1809)	Regno Bogotensi inter vicum Ascensionis et la Vega de San Lorenzo
Q. almaguerensis BONPL. 1809 (Humboldt and Bonpland 1809)	Novogranatensium Andibus, juxta urbem Almaguer
Q. lindeni A. DC. 1864 (De Candolle 1864)	In Novae Granadae provincia Tunja
Q. tolimensis BONPL. 1809 (Humboldt and Bonpland 1809)	Montis Quindiu

genus' high hybridization potential (Burger 1975) and/or high morphological variability. The majority of the vouchers stored at Colombia's National Herbarium (COL) are identified only down to genus level or as *Quercus humboldtii.*

11.3 Morphological Variability

Pulido (1996) recently analyzed herbarium specimens of *Quercus* to further study the geographic variations of this genus in Colombia. Eight leaf and five fruit characteristics were measured in each of the 162 herbarium specimens stored at COL (Table 11.2). The specimens included the holotype of *Q. boyacensis* and *Q. duqueana*, the isotype of *Q. colombiana*, and the topotype of *Q. humboldtii* and *Q. lindenii.*

On each voucher, leaf and fruit variables were measured at most five times at a precision of 1.0 mm. Floristic characteristics were not measured, given the few fertile samples in the collection. Because it was not possible to measure fruits in all specimens, two databases were created: (1) an H matrix with 162 specimens and eight leaf variables; (2) an F matrix with 47 specimens, with eight leaf variables and five fruit variables. Each matrix was analyzed by means of a principal component analysis (PCA) using the NTSYS 2.1 software package. The numerical values of each variable of the basic data matrix (H and F matrices) were standardized by subtracting the mean value of the variable per specimen from the average for this variable in all specimens studied,

Table 11.2. Morphological characters of *Quercus* measured on botanical vouchers stored at COL. Most characters are quantitatively continuous, only three (*) being quantitatively discontinuous. In addition, the eigenvectors in the first, second and third principal components (PC1, PC2, and PC2) are given for matrices H and F. Bold values represent the most important characters that explain patterns of variation as evidenced by the PCA ordination of morphological data obtained from botanical vouchers

	H matrix			F matrix		
Characters	PC1	PC2	PC3	PC1	PC2	PC3
Leaves						
Leaf length (L> in mm)	**0.907**	0.227	0.039	**0.776**	0.269	0.180
Leaf width (A> in mm)	0.746	0.548	–0.018	0.674	0.202	0.145
Base width (BW in mm)	–0.263	**0.796**	–0.175	–0.452	–0.581	–0.036
Apex width (AW in mm)	–0.589	0.529	0.052	–0.670	–0.052	0.009
Drip–tip length (AP in mm)	0.106	–0.304	**–0.884**	0.029	–0.446	**0.743**
Secondary veins (SV* in #)	0.646	–0.167	0.244	0.488	0.337	0.493
Petiole length (PL in mm)	0.574	0.260	–0.291	0.694	0.043	–0.262
Leaf indument (LI* in %)	–0.412	0.250	–0.160	–0.448	–0.377	–0.223
Fruits						
Acorn diameter (AD in mm)				**0.734**	–0.396	–0.063
Acorn length (AL in mm)				0.621	–0.551	–0.104
Cupule's length (CL in mm)				0.652	–0.499	–0.346
Scaly rows (SR* in #)				0.342	–0.300	–0.143
Acorn tip length (LAP in mm)				–0.099	**–0.702**	0.434

and then dividing by the standard deviation for this variable. PCA was performed on correlation matrices between calculated variables by using Pearson's rank correlation coefficient.

In the ordination of the H matrix for leaf variables, 65 % of the total variation was explained by the first three components (34 % PC1, 18.9 % PC2, 12.3 % PC3), with stronger contribution of leaf length, base width and drip–tip length (Fig. 11.2, Table 11.2). In the ordination of the F matrix for leaf and fruit variables, 58 % of the total variation was explained by the first three components (31.6 % PC1, 16.9 % PC2, 10.0 % PC3), with stronger contribution of leaf length, acorn diameter, length of acorn tip, and drip–tip length (Fig. 11.2). Specimens were distributed over a continuous gradient within the space between the first two principal components (Fig. 11.2).

Specimens were grouped according to the region of origin. The H matrix showed that many specimens from the Boyacá region in the northeast sector of the Colombian Andes occupied the left side of PC1, whereas many specimens of the Cauca region in the south were clustered on the right side of PC1. This apparent separation was masked by specimens from Huila, Nariño, and other regions, located in the center of the PCA graph. The Boyacá–Cauca sep-

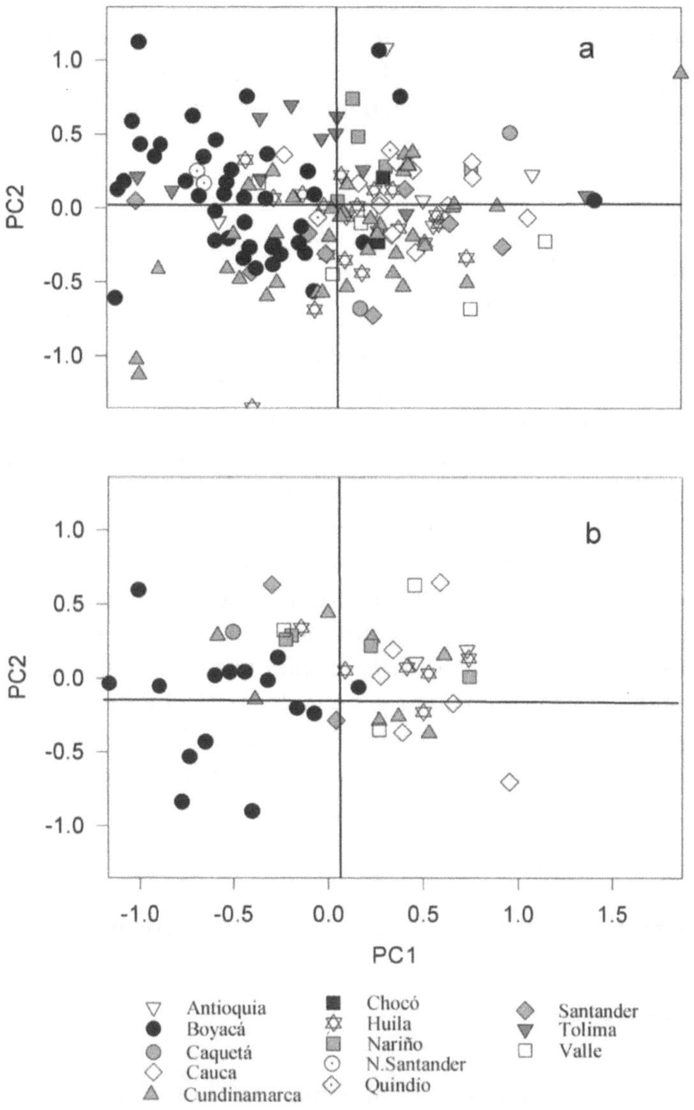

Fig. 11.2a, b. First two components from principal component analysis for the matrices H and F. **a** Matrix H: ordination of 162 specimens of *Quercus*, including eight foliar variables. **b** Matrix F: ordination of 47 specimens of *Quercus*, including eight foliar and five fruit variables. Each *symbol* represents the collection site of a herbarium specimen

aration was explained mainly by differences in leaf length, with Boyacá specimens having shorter leaves (Table 11.2). Similarly, the F matrix showed that the Boyacá specimens occupied the left side of PC1, whereas specimens from other regions are on the right side (Fig. 11.2). In general, the Boyacá specimens had shorter leaves and smaller fruits (smaller acorn diameter). In short, the ordination analysis showed that there is a continuous morphological variation in both leaves and fruits of *Quercus*, supporting the idea that in Colombia only one single oak species (*Q. humboldtii*) exists. Although some regional differentiation does occur, this phenomenon is not strong enough to suggest a true separation at the species level.

11.4 Molecular Variability

Some molecular studies have provided initial information on the genetic diversity of *Q. humboldtii*. Fernandez et al. (2000) found three hypervariable microsatellite loci useful to study gene flow between populations and genetic diversity in *Q. humboldtii*. Cavelier et al. (1993) found low genetic variation using four primers to DNA amplification. Samples for these analyses included dry specimens at COL, and fresh samples from the type localities of *Q. boyacensis*, *Q. lindenii* and *Q. colombiana* as well as samples of the "red" and "white" oaks in the Villa de Leyva region (Boyacá). A total of 29 bands were found, the majority with high frequencies (>0.8). These results were compared with ten samples of *Poulsenia armata* from different American countries, using the same primers. The genetic diversity of *P. armata* – with a very large geographic range – was greater than that in *Q. humboldtii*. In addition, samples of the apparently different "red" and "white" oaks had the largest value for Jaccard's similarity coefficient.

11.5 Floristic Composition and Phytosociology

11.5.1 Composition

The floristic composition of the Colombian oak forests is known from a few studies carried out mostly along the Eastern Cordillera (Table 11.3). In the present chapter, the information reported by ten authors (see Table 11.3) was compared and compiled to obtain general data about the floristic composition of oak forest in Colombia. The floristic richness of Colombian oak forest includes 577 species, 332 genera, and 127 families of vascular plants. The richness of Colombian oak forest is lower than that of similar forests in Costa Rica where up 1,095 species, 419 genera, and 145 families have been reported (Kappelle 1996; Chaps. 4 and 10). The largest families are the same for both Colombia and Costa Rica, including Asteraceae, Orchidaceae, Melastomataceae, Rubiaceae, and Rosaceae (Table 11.4). The largest genera in Colombian oak forest are *Miconia* (17 spp.), *Weinmannia* (nine spp.), *Piper* (nine spp.), *Poly-*

Table 11.3. Floristic studies conducted in Colombian montane oak forests

Author	Department	Locality	Altitude (m)	Families (#)	Genera (#)	Species (#)
Becerra (1989)	Boyacá	Duitama, sector La Sierra	2,200–2,900	n.a.[a]	n.a.	33*
Cuatrecasas (1934)**	Tolima	Municipio de Ibague, sector la Suiza	2,500	37	44	50
Devia and Arenas (1997)**	Cundinamarca, Boyacá and Santander		1,500 and 3,200	24	29	31*
Galvis (1994)	Boyacá	Santuario Nacional de Fauna y Flora de Iguaque	2,700–2,900	94	242	478
Lozano and Torres (1965)**	Cundinamarca	Municipio de Bojacá	2,500–2,700	75	86	258
Lozano et al. (1979)	Cundinamarca	Municipio de Viotá, sector La Vieja	1,200–1,560	34	n.a.	56
Lozano et al. (1979)	Nariño	Municipio de La Florida	2,300	41	n.a.	111
Lozano et al. (1979)	Santander	Municipio de Onzaga, Vereda Chaguaca	2,640	52	n.a.	148
Marín (1996)**	Boyacá	Santuario Nacional de Fauna y Flora de Iguaque	2,740–2,900	27	34	53
Ramírez (1999)**, Diazgranados (1999)**	Cundinamarca	Soacha-Bojacá, Parque Natural Chicaque	2,320	34	57	90
Rangel and Lozano (1989)	Huila and Cauca	Valle del río Magdalena – volcán del Puracé	1,800–2,700	61	91	118
Romero (1966)**	Boyacá	Duitama, sector La Sierra	2,200–2,900	n.a.	n.a.	46*
Torres-Novoa (1997)**	Boyacá	Santuario Nacional de Fauna	2,700	31	38	46*
Van der Hammen and Gonzalez (1963)**	Cundinamarca, Boyacá and Santander	y Flora de Iguaque	2,000–3,500	83	120	148
Velez and Fresneda (1992)**	Antioquia	Medellín, Quebrada Piedras Blancas	2,200–2,600	n.a.	n.a.	132
Zerning and Betancur (1994)	Boyacá	Santuario Nacional de Fauna y Flora de Iguaque	2,700–2,900	80	203	399

[a] N.a., information not available; *, canopy trees only; **, studies analyzed for us with respect to floristic richness of Colombian oak forest

Table 11.4. Most diverse plant families, with species and genus richness, in Colombian montane oak forests

Families	Species (#)	Genera (#)
Asteraceae	50	34
Orchidaceae	30	19
Melastomataceae	29	9
Rubiaceae	27	18
Rosaceae	20	7
Piperaceae	17	2
Ericaceae	15	8
Solanaceae	14	6
Bromeliaceae	14	5
Poaceae	12	10
Lauraceae	12	8
Polypodiaceae	12	3
Myrsinaceae	11	5
Euphorbiaceae	10	7

podium (nine spp.), *Rubus* (nine spp.), *Tillandsia* (eight spp.), *Peperomia* (eight spp.), *Palicourea* (eight spp.), *Ficus* (seven spp.), *Brunellia* (six spp.), *Pleurothallis* (six spp.), *Solanum* (six spp.), *Anthurium* (five spp.), *Schefflera* (five spp.), *Ageratina* (five spp.), *Myrsine* (five spp.), and *Passiflora* (five spp.). The dominant growth forms are trees (40 %), shrubs (20 %), and herbs (20 %), followed by epiphytes (9 %), climbers (6 %), and scandent plants (2 %).

In addition to vascular plants, macrofungi of Colombian oak forests have been studied in Antioquia, Cundinamarca, Nariño and Cauca.

11.5.2 Phytosociology

Floristic descriptions of oak forests in Colombia have shown the occurrence of dominant *Quercus* trees, accompanied by a variable set of canopy, subcanopy and understory woody elements, some of them unique to the sites for which plant associations and alliances have been described. In some rare cases, *Quercus* forests were in fact dominated by *Colombobalanus*, a related and physiognomically similar tree, also in the Fagaceae.

The first phytosociological description of a *Quercus* forest was made by Cuatrecasas (1934). He named this association *Quercetum tolimensis* (J. Cuatrecasas 1934), based on a description of an oak forest in the area of La Suiza (2,500 m a.s.l.), between Pereira and Armenia. The upper stratum of this association is dominated by *Quercus tolimensis*, with trees more than 40 m in height. Other trees included *Clethra* sp., *Clusia* spp. and *Saurauia excelsa*. Smaller trees and shrubs included *Tibouchina lepidota*, *Miconia mutisiana*, *Cavendishia obtusa*, and *Inga* sp. In the understory grow *Bomarea* sp. and *Mutisia grandiflora*.

Lozano and Torres (1965, 1974), and Lozano et al. (1979) suggested the name *Quercetum humboldtii* after studying an oak forest in the Eastern Cordillera at the locality of San Antonio, Municipio de Bojacá, Departamento de Cundinamarca. This forest was dominated by *Quercus*, followed by *Palicourea anacardifolia*, *Viburnum pichinchense*, *Oreopanax floribundum*, and *Maytenus laxiflora*, unique to the site. Characteristic species included *Miconia theaezans*, *Palicourea crocea*, *Saurauia anolaimensis* and *Cytharexylum* sp. Among shrubs appear *Cestrum parvifolium* and *Berberis glauca*, the latter also a species unique to this site. In the understory grow *Smilax floribunda*, *Mikania lehmannii*, *Tillandsia tetrantha* and *T. suescana*.

Rangel and Lozano (1986, 1989) described two associations (*Hedyosmo-Quercetum humboldtii* and *Alfaroo-Quercetum humboldtii*) and one alliance (*Monotropo-Quercion humboldtii*). The *Hedyosmo-Quercetum humboldtii* association was described based on a site at the Reserva Merenberg (2,400–2,500 m a.s.l.) near the Municipio de La Plata, Huila. Trees more than 25 m tall were included, besides *Quercus*, *Brunellia putumayensis*, *Weinmannia glabra*, *Ocotea karsteniana*, *Miconia floribunda*, *Prunus integrifolia*, and *Miconia pedicellata*. Among the shrubs, the most common species were *Palicourea cuatrecasasii*, *Hedyosmum racemosum*, and *Calyptranthes* aff. *bipennis*. Other species included the treefern *Lophosoria quadripinnata*, and the shrubs *Viburnum lasiophyllum* and *Ardisia* cf. *sapoana*. In the understory occurred *Mollinedia* cf. *latifolia* and *Besleria reticulate*, and the epiphytic *Tillandsia biflora*. The association *Alfaroo-Quercetum humboldtii* was described based on a site at Serranía de las Minas (1,850–2,300 m), Municipio de la Argentina (La Plata Vieja), Huila, also in the Central Cordillera (Rangel and Lozano 1986, 1989). In this association, trees included *Weinmannia sorbifolia*, *Cinchona officinalis* and *Alfaroa* spp. Subcanopy trees and shrubs included *Cybianthus cuatrecasasii*, *Palicourea* aff. *abbreviata*, *Schefflera decagyna*, and the small palm *Geonoma margyraffia*. Some unique understory species were *Mandevilla fendleri* and *Dictyostega orobanchioides*, as well as the ephiphytic *Stelis lentiginosa*, *Peperomia hartwegiana* and *Grammitis serrulata*, and the climber *Mikania* aff. *stuebelii*.

The alliance described by Rangel and Lozano (1989) from a site in the sub-Andean belt between 1,800 and 2,600 m was named *Monotropo-Quercion humboldtii*. The canopy of this alliance was dominated by *Quercus humboldtii*. Other trees included *Billia columbiana*, *Rapanea ferruginea*, *Myrsine guianensis*, *Clethra fagifolia*, *Clusia multiflora* and *Inga codonantha*. In the subcanopy, characteristic species were *Cyathea caracasana*, *Conomorpha pastensis*, and *Solanum lepidotum*, whereas *Monotropa uniflora* occurred in the understory.

11.6 Conclusions

The current dataset evidences that the morphological variability observed in leaves and fruits of Colombian oak is continuous. It reflects the presence of only one single species of oak in this country – *Quercus humboldtii*. Current molecular data support this hypothesis. However, more detailed molecular analysis should be done in order to draw firmer conclusions. Similarly, more specific research should be conducted to find out if floristic richness of oak forest in Colombia is really as low as past, limited studies report.

References

Becerra JE (1989) Estructura y crecimiento de un bosque secundario de roble (*Quercus humboldtii*). Universidad Distrital Francisco José de Caldas, Bogotá, Rev Col For 3:1–64

Burger WC (1975) The species concept in *Quercus*. Taxon 24:45–50

Cavelier J, Aide TM, Lozano G, Pulido MT, Rivera E (1993) Especiación del género *Quercus* (robles) en Colombia: un siglo y medio de incertidumbre. Fondo FEN, Bogotá, Colombia

Cuatrecasas, J (1934) Observaciones geobotánicas en Colombia. Trab Mus Nac Cienc Nat Ser Bot 27(45/48):1–222

Cuatrecasas J (1944) Notas a la flora de Colombia. VI. Rev Acad Col Cienc Ex Fis Nat 6(21):32–67

De Candolle A (1864) Prodromus systematis naturalis regni vegetabilis. V16. Sect Post 320

Devia CA, Arenas H (1997) Evaluación de status ecosistémico y de manejo de los bosques de fagáceas (*Quercus humboldtii* y *Trigonobalanus excelsa*) en el norte de la cordillera oriental (Cundinamarca, Santander y Boyacá). In: Cárdenas F (ed) Desarrollo sostenible en los Andes de Colombia. IDEADE – Universidad Javeriana, Bogotá, pp 63–77

Diazgranados CM (1999) Estructura de la vegetación del parque natural Chicaque. Bachelor Thesis, Universidad Javeriana, Bogotá

Fernandez JF, Sork VL, Gallego G, López J, Bohorques A, Tohme J (2000) Croo-amplification of microsatellite loci in a neotropical *Quercus* species and standardization of DNA extraction from mature leaves dried in silica gel. Plant Mol Biol Rep 18:397a–397e

Galvis M (1994) Inventario de las plantas fanerógamas en el Santuario de flora y fauna de Iguaque, Boyacá. In Cavelier J, Uribe A (eds) Resúmenes Simp Nacl Diversidad Biológica Conservación y Manejo de los Ecosistemas de Montaña en Colombia, Universidad de los Andes, Santafé de Bogotá, p 44

Hooghiemstra H, Ran ETH (1994) Late Pliocene-Pleistocene high resolution pollen sequence of Colombia: an overview of climatic change. Quat Int 21:63–80

Hooghiemstra H, Sarmiento G (1991) Long continental pollen record from a tropical intermontane basin: Late Pliocene and Pleistocene history from a 540-meter core. Episodes 14(2):107–115

Humboldt FA, Bonpland A (1809) Plantes Équinoxiales, vol 2. Schoell, Paris, pp 1–191

Kappelle M (1996) Los bosques de roble (*Quercus*) de la Cordillera de Talamanca, Costa Rica: biodiversidad, ecología, conservación y desarrollo. Instituto Nacional de Biodiversidad (INBio), Santo Domingo de Heredia

Lozano G, Torres JH (1965) Estudio fitosociológico de un bosque de robles (*Quercus humboldtii* H. & B.) de la Merced (Cundinamarca). Instituto de Ciencias Naturales (ICN), Universidad Nacional de Colombia, Bogotá

Lozano G, Torres JH (1974) Aspectos generales sobre la distribución, sistemática fitosociológica y clasificación ecológica de los bosques de robles (*Quercus*) en Colombia. Ecol Trop 2:1–79

Lozano G, Diaz S, Torres JH (1979) Inventario florístico de algunos bosques de robles (*Quercus*) en Colombia, COLCIENCIAS. Instituto de Ciencias Naturales (ICN), Universidad Nacional de Colombia, Bogotá

Marín-C CA (1996) Flora y vegetación del santuario de flora y fauna de Iguaque. Bachelor Thesis, Departamento de Biología, Universidad Nacional de Colombia, Bogotá

Müller CH (1942) The Central American species of *Quercus*. USDA Misc Publ 477:1–216

Pulido MT (1996) Variación morfológica y biogeografía del género *Quercus* in Colombia. Bachelor Thesis, Facultad de Ciencias, Universidad de los Andes, Bogotá

Ramírez-H W (1999) Composición florística y diversidad alfa de la vegetación del parque Chicaque, San Antonio del Tequendama, Cundinamarca. Bachelor Thesis, Facultad de Ciencias Naturales, Universidad Javeriana, Bogotá

Rangel-Ch JO, Lozano GL (1986) Un perfil de vegetación entre La Plata (Huila) y el Volcán Puracé. Caldasia 14(68/70):53–547

Rangel-Ch JO, Lozano GL (1989) La vegetación selvática y boscosa del Valle de la Plata entre el río Magdalena y el parque Natural Puracé. In: Herrera LF, Drenan R, Uribe C (eds) Cacicazgos prehispánicos del Valle de la Plata, vol 1. El contexto medioambiental de la ocupación humana. Univ Pittsburg Mem Lat Am Archaeol 2:95–118

Romero-A E (1966) Algunos aspectos ecológicos y silvícolas de los bosques de robles (*Quercus humboldtii*) de "La Sierra" Boyacá – Colombia. MSc Thesis, Universidad Distrital Francisco José de Caldas, Bogotá

Rzedowski J (1978) Vegetación de México. Limusa, México DF, Mexico

Schwartz O (1937) Fagaceae. Notizbl Bot Gart Berlin-Dahlem 13:495–496

Torres-Novoa ND (1997) Estructura y composición floristica y crecimiento inicial de un bosque secundario de roble en el Santuario de Flora y Fauna de Iguaque (Boyacá). Tesis de Grado Universidad Distrital, Facultad de Ingeniería Forestal, Bogotá

Van der Hammen T, Gonzalez E (1963) Historia del clima y vegetación del Pleistoceno Superior y del Holoceno de la Sabana de Bogotá. Bol Geol 40(1/3):189–266

Velez-S G, Fresneda E (1992) Diversidad florística en las comunidades de robledal y rastrojo alto en la cuenca de la quebrada Piedras Blancas, Antioquia. Rev Fac Nacl Agron 45(2):3–25

Zerning K, Betancur J (1994) Flora de Iguaque. In: Cavelier J, Uribe A (eds) Resúmenes Simp Nacl Diversidad Biológica, Conservación y Manejo de los Ecosistemas de Montaña en Colombia. Universidad de los Andes, Bogotá, p 93

Part IV
Population Dynamics

12 Regeneration and Population Dynamics of *Quercus rugosa* at the Ajusco Volcano, Mexico

C. Bonfil

12.1 Introduction

Mexico has the largest oak species diversity in the western hemisphere, with 150–200 species, most of which are located in the main mountain ranges of the country, where they form either pure oak stands or mixed pine-oak forests (Chap. 1). However, oaks are found in a broad array of climatic conditions and thus are important elements of many different vegetation types (González Rivera 1993; Nixon 1993; Chap. 1). The systematics of Mexican oaks is still incomplete, and ecological knowledge of oak species is only starting to develop. Basic ecological research on most species is lacking, and only a few studies address the regeneration and management, from an ecological perspective, of a handful of species (Quintana Ascencio et al. 1992; Muhler-Using 1994; Eckelman 1995; Moreno-Gómez et al. 1995; Zavala and García Moya 1997; Bonfil and Soberón 1999; López-Barrera and González Espinoza 2000; Peña and Bonfil 2003; Alfonso-Corrado 2004; Chaps. 13, 14, 16, 18, 19 and 28). However, this information is particularly relevant to design sound management and restoration programs of oak forests, as they maintain a high biodiversity and are among the most disturbed vegetation types of Mexico (Rzedowski 1981; Challenger 1998; Chaps. 8, 9, 13, 14, 16 and 28).

Oaks have been eliminated from most plains and low hills, and many present-day oak forests were heavily cut during the first half of the 20th century, in order to supply charcoal – the main domestic fuel – to a growing population (Rzedowski 1981; Bonfil 1991), similarly to some parts of Costa Rica (Chap. 31). At present, oak wood is still the favourite domestic fuel in many rural villages, and cattle is raised in the understorey of numerous oak forests. As a result, regeneration problems arise from a variety of causes, such as overgrazing, frequent forest fires (set to stimulate grass growth), and changes in microhabitat conditions (see also Chap. 16).

Ecological Studies, Vol. 185
M. Kappelle (Ed.) Ecology and Conservation of Neotropical Montane Oak Forests
© Springer-Verlag Berlin Heidelberg 2006

In this chapter, I present the results of studies addressing the seedling and population dynamics of *Quercus rugosa* at the Ajusco Volcano, south of Mexico City, in order to identify those processes critical to regeneration and population growth. These results are then discussed in terms of their application to the design of restoration-oriented management programs.

Q. rugosa is a white oak (section *Quercus*) which is widely distributed in Mexico. It is found in the main mountain ranges of Mexico, at altitudes of 1,800–2,900 m, where it is frequently intermingled with other oak or pine species, although it may form pure stands, too (González-Villarreal 1986; González-Rivera 1993). A tree may reach up to 30 m; its leaves are thick and rigid, partially shed in October–November, when seeds are dispersed.

12.2 The Ajusco Volcano

The Ajusco Volcano, south of Mexico City, is one of the main mountains of the basin of Mexico. Various oak species establish there, forming either oak forests (below 2,700 m a.s.l.) or pine-oak forests (2,800–2,900 m). Fir (*Abies religiosa*) and pine (mainly *Pinus hartwegii*) are the dominant species at higher altitudes (Benítez 1986). The north-facing piedmont of the Ajusco, as well as portions of the valley south of Mexico City, experienced several lava flows around 2,000 years ago (Cordova et al. 1994). As a result, lower elevation lands were covered by basaltic rock, whereas at the piedmont (around 2,600 m) small hills rose with older, well-developed soil in a matrix of basaltic rock. As succession proceeded, a xerophytic shrub developed in the basaltic rock bed (which holds only a thin (<5 cm) soil layer), and at the piedmont this shrubland was slowly colonized by oaks, spreading from forest patches uncovered by the lava flow. Therefore, *Q. castanea* and *Q. rugosa* became the dominant elements of a vast xerophytic shrubland, the former species being the dominant tree at lower altitudes.

In 1988, part of this area was disturbed by a temporary human settlement. Both vegetation and rock were removed, as the latter was used as building material. Subsequently, the area was protected and a restoration program started in 1990, with the main goal of re-establishing the original vegetation. The studies reported here were carried out in this protected zone, the Parque Ecológico de la Ciudad de México. Mean annual temperature is 15 °C, and precipitation during the rainy season (May–October) accounts for 80 % of the annual total of 1,000 mm.

12.3 Seedling Dynamics

Seed predation, seed germination, and seedling survival and growth are processes crucial for regeneration. To determine the main characteristics of favourable sites for oak regeneration (Harper 1977; Crow 1992), these processes were studied in three main environments present in the study area: (1) oak forest with deep soils; (2) vegetation at the forest border or in the transition zone between the edge and the oak-shrubland; and (3) disturbed shrubland. A brief summary of the results obtained in studies conducted in the area is given below (Bonfil 1995; Bonfil and Soberón 1999; Bonfil et al. 2000).

Seed consumers, mainly small mammals, ate a large proportion of acorns at all three sites (cf. Chap. 13). Acorns were more easily detected if found in large clusters (e.g. 25 seeds), rather than in smaller groups (e.g. five seeds) or as single seeds. The increased consumption of clumped acorns is well known (Price and Jenkins 1986; Quintana-Ascencio et al. 1992). Higher predation rates of seed clusters occur both in the forest interior and at the forest border than was the case for the disturbed shrubland, probably because small mammals are more easily found by predators and therefore more vulnerable in the latter (see also Chap. 26).

Despite the high seed predation, some seeds remain in the ground for more than 35 days, which is enough time for a seed to germinate if suitable conditions are met (Robledo-Jiménez 1997). Microsites with high humidity and partial shade provide such conditions, as acorns placed in completely open sites dry out in a few days (Bonfil and Soberón 1999). The dark rock, thin soil layer, and sparse vegetation of the disturbed shrubland prevents seed germination and seedling establishment (Table 12.1), which implies that past colo-

Table 12.1. Seed germination, seedling establishment and seedling survival (mean values ± SE) of *Quercus rugosa*, and seedling survival of *Q. laurina* at three sites in the Parque Ecológico de la Ciudad de México, Ajusco. Germination and establishment values are from Bonfil and Soberón (1999), and based on data collected from July 1991 to August 1992

	Quercus rugosa Germination (%)	Establishment (%)	Survival (%)	*Q. laurina* Survival (%)
Disturbed shrubland	46.8 (11.5)[a*]	0.9 (0.9)[a]	34	17
Forest border	73.0 (06.5)[b]	20.6 (4.5)[b]	77	75
Forest interior	92.2 (03.0)[b]	15.4 (4.5)[b]	49	42

* Different letters in the same column indicate significant differences between sites, according to Tukey's HSD test ($P<0.05$)

nization of the basaltic rock bed by oaks must have been a very slow process which needed, as a prerequisite, the development of a protective layer of vegetation and litter providing adequate conditions for oak establishment.

It is very likely that seedling recruitment is episodic. Although trees of *Q. rugosa* produce seeds almost every year (seed fall peaks during October–November), seed production is variable, including mast years and low production periods. However, seeds germinate soon after they are shed, and the fragile seedlings must survive a dry period of variable length (up to 6 months, until the rainy season arrives in May–June). Therefore, only in those years with higher than average winter rains (i.e. a reduced dry season) are conditions met for abundant seedling establishment. In average years, recently germinated seedlings die during their first dry season, and acorns of *Q. rugosa* do not remain viable in the soil long enough to germinate at the onset of the rainy season.

This pattern of seasonal mortality continues during the first years after seedling establishment. When nursery-grown seedlings of two oak species (*Q. rugosa* and *Q. laurina*) were planted at the three sites mentioned above, survival was highest at the forest border, intermediate in the forest interior, and lowest in the disturbed shrubland (Table 12.1), with no significant differences between species. Thus, the semi-open canopy found at the forest border produces light and moisture conditions conducive to seedling establishment and survival. Other studies with oak species in seasonally dry environments have found that medium to low light levels result in higher survival of oak seedlings, as they are protected from direct solar radiation and desiccation (Quintana-Ascencio et al. 1992; Thadani and Ashton 1995; Retana et al. 1999; Chap. 18). In the forest interior, although soil moisture is relatively high, low light levels preclude seedling growth, and after a period of 2–3 years, most seedlings die from rot, herbivory, or coverage by the abundant litter which falls during autumn. As a result, accumulation of advanced reproduction, necessary for the natural regeneration of most oak forests, does not take place under the forest canopy (Larsen and Johnson 1998).

Desiccation during the dry season is the main cause of seedling death in the disturbed shrubland. The porosity of the basaltic rock results in low water availability, which causes high seedling mortality. Thus, in these conditions seedling persistence depends on the presence of a protective shade cover, provided by an established tree or shrub acting as a nurse plant. The facilitating effect of established shrubs for oak seedling survival at sites with periodic water deficiency has been reported (Muick 1991; Callaway 1992).

Dependence on nurse plants is shared by the two oak species present in the shrubland (*Q. rugosa* and *Q. castanea*), as well as by other dominant plant species of this community (such as *Senecio praecox*; Rodríguez de la Vega 2003). However, the period of dependence may be brief, as shown by the fact that survival of 2-year-old nursery-grown seedlings of *Q. rugosa* planted at the site was independent of their association with a nurse plant, whereas that of 1-year-old seedlings increased substantially if placed under the canopy of a shrub (Bonfil et al. 2000). However, in a harsh dry season, this partial shade is not effective and seedling mortality increases even in more protected environments, such as at the forest border (Bonfil and Soberón 1999). Therefore, initial seedling survival in the disturbed shrubland is closely linked to water availability. Herbivory by rabbits and rodents (which is conspicuous only during the dry months) is also an important cause of seedling death. Even though seedlings resprout after aerial biomass loss (Bonfil 1998a), the proportion of seedlings resprouting is higher during the rainy season, and repeated herbivory may result in seedling death.

Despite their initial requirement for shade, seedlings of *Q. rugosa* are relatively shade-intolerant, as shown by their low survival rates and lack of growth under the forest canopy, and reported for other oak species, too (Takenaka 1986; Crow 1992; Quintana-Ascencio et al. 1992; Chaps. 14 and 16). By contrast, after two growing seasons, seedlings surviving in the shrubland attain larger sizes than those found at other, more shaded sites (Table 12.2). Overall, the above results show that regeneration of *Q. rugosa* takes place mainly in the semi-open environment of the forest border, whose conditions permit the accumulation of seedlings from several cohorts. However, further growth depends on the specific light environment, and therefore only those individuals in relatively open microsites grow into saplings and small trees, as reported for other broad-leaved trees (Quintana-Ascencio et al. 2004; Chaps. 14 and 16). Thus, seedlings and saplings receiving low direct sunlight may survive several years, resprouting repeatedly, without any increase in height. Only if a disturbance opens the canopy will they be released from this suppressed condition.

Table 12.2. Shoot height, basal diameter and crown area of *Quercus rugosa* seedlings at the end of the first (November 1991) and second (November 1992) growing season. Data are from Bonfil and Soberón (1999)

	Height (cm)		Diameter (cm)		Area (cm²)	
	Nov. 1991	Nov. 1992	Nov. 1991	Nov. 1992	Nov. 1991	Nov. 1992
Disturbed shrubland	12.3[a]*	15.0[a]	0.21[a]	0.39[a]	58.3[a]	125.6[a]
Forest border	11.7[a]	15.4[a]	0.20[a]	0.28[b]	49.4[a,b]	88.9[a]
Forest interior	10.0[b]	9.9[b]	0.17[b]	0.21[c]	40.6[b]	35.6[b]

* Different letters in the same column indicate significant differences between sites, according to Tukey's HSD test ($P<0.05$)

12.4 Population Dynamics

As regeneration occurs mainly at the forest border, the dynamics of the population depend on demographic processes occurring at this spot. In a demographic study carried out during 1991–1994 (Bonfil 1998b), a Lefkovitch population projection matrix (in which individuals were classified into seven size categories) was built, and the methods explained in Caswell (1989) were used to estimate the finite population growth rate (λ). Results showed that the population is close to equilibrium and grows very slowly under present conditions (λ value close to 1). This is the outcome of several demographic processes. Those related to seedling recruitment and growth have been explained in the preceding section: seed predation and the limited availability of safe sites for seedling establishment restrict seedling recruitment; the latter depends partially also on prevailing weather conditions (i.e. amount and distribution of precipitation).

Once seedlings have established, most of them remain in relatively shaded conditions which prevent growth, and therefore saplings accumulate in the understorey. Accordingly, analysis of the structure of the population shows that the highest proportion of individuals corresponds to the sapling stage (35 %), whereas seedlings account for only 8 %; the three adult and two juvenile categories comprise 57 %. Transition probabilities (which reflect the proportion of individuals growing into the next size category) are low, consistent with the low individual growth rates of juveniles and adult trees characteristic of many oak species. Although trees grow slowly in the semi-open conditions of the forest border (a 6-year period was needed to detect increments in trunk diameter), significant growth may not occur where tree density and competition for light are higher (i.e. in the forest interior).

If the conditions present during the study period were maintained for a longer time period, the oak population would remain stable and population size would increase very slowly. This demographic behaviour is determined both by the characteristic growth pattern of the species and by the restrictive conditions found in the disturbed shrubland; it is feasible that at higher-quality sites (i.e. deep, rich soils and higher water availability), population growth rates would be higher.

On the long run, succession will proceed and the oaks will slowly recolonize the contiguous shrubland. As individuals at the present forest border grow, the canopy will close and regeneration will proceed towards the expanding outer borders. Simulations for matrix population models may be useful to explore the potential effect of different restoration-oriented management practices by focussing on those techniques which increase population growth rates.

12.5 Conclusions

Several important implications for the reintroduction of oaks at the disturbed site emerge from these results. Planting acorns is not practical, as adequate burial of acorns and high-density sowing are not feasible in this substrate, and acorn desiccation and predation will limit seedling establishment. Seedling planting may prove to be more successful, especially if suitable microsites are carefully chosen. Seedlings should be placed under the canopy of shrubs which do not loose all their leaves during the dry season. Two-year old seedlings will have higher survival rates, and mortality may be further lowered if they are wire mesh-fenced during the dry season, to avoid herbivory and shoot loss from small mammals. Also, in case of a long and harsh dry season, they should be watered.

These protective measures are probably needed for only 2–3 years after seedling planting. Subsequently, shade from nurse plants should be removed, as saplings and young trees attain higher growth rates under more open conditions. Overall, the tree population may recover over a period of 30–50 years, when transplanted trees reproduce steadily and the population is able to develop without further assistance from forest managers.

Acknowledgments The author thanks V. Peña and H. Rodríguez de la Vega for their assistance with fieldwork and data analysis. Financial and academic support for fieldwork was provided by J. Soberón. R. Evans made useful suggestions which improved the text. This study was carried out with partial support from the Consejo Nacional de Ciencia y Tecnología (CONACyT) through a PhD studentship to the author.

References

Alfonso-Corrado C (2004) Ecología, manejo y conservación de *Quercus potosina* y *Q. eduardii* (Fagaceae) en Sierra Fría, Aguascalientes. PhD Thesis, Universidad Nacional Autónoma de México (UNAM), México DF, Mexico

Benítez-B G (1986) Árboles y flores del Ajusco. Instituto de Ecología y Museo de Historia Natural de la Ciudad de México, México DF, Mexico

Bonfil C (1991) Los encinos y la Ciudad. Oikos, Boletín del Centro de Ecología, Universidad Nacional Autónoma de México (UNAM), México DF, Mexico

Bonfil C (1995) Establecimiento, sobrevivencia y crecimiento de plántulas de dos especies de encinos en el Ajusco, D.F. In: Marroquín J (ed) Mems III Sem Nacional Utilización de Encinos, Facultad de Ciencias Forestales, Universidad Autónoma de Nuevo León. Rep Cien Espec 15, pp 350–365

Bonfil C (1998a) The effects of seed size, cotyledon reserves and herbivory on seedling survival and growth in *Quercus rugosa* and *Q. laurina* (Fagaceae). Am J Bot 85:79–87

Bonfil C (1998b) Dinámica poblacional y regeneración de *Quercus rugosa*: implicaciones para la restauración de bosques de encinos. PhD Thesis, Instituto de Ecología, Universidad Nacional Autónoma de México (UNAM), México DF, Mexico

Bonfil C, Soberón J (1999) *Quercus rugosa* seedling dynamics as related to its re-introduction in a disturbed Mexican landscape. Appl Veg Sci 2:189–200

Bonfil C, Rodríguez de la Vega H, Peña Ramírez V (2000) Evaluación del efecto de las plantas nodrizas en el establecimiento de una plantación de *Quercus*. Rev Cien For Méx 25:59–73

Callaway RM (1992) Effect of shrubs on recruitment of *Quercus douglasii* and *Quercus lobata* in California. Ecology 73:2118–2128

Caswell H (1989) Matrix population models. Sinauer, Sunderland, MS

Challenger A (1998) Utilización y conservación de los ecosistemas terrestres de México: pasado, presente y futuro. Comisión Nacional para el Conocimiento y Uso de la Biodiversidad (CONABIO), Instituto de Biología, Universidad Nacional Autónoma de México (UNAM), México DF, Mexico

Cordova C, Martín del Pozo AL, López-Camacho J (1994) Paleolandforms and volcanic impact on the environment of prehistoric Cuicuilco, southern Mexico City. J Archeol Sci 21:585–596

Crow TR (1992) Population dynamics and growth patterns for a cohort of northern red oak (*Quercus rubra*) seedlings. Oecologia 91:192–200

Eckelmann CM (1995) Regeneración y dinámica natural de un bosque de pino-encino en la Sierra Madre Oriental, en el noreste de México. In: Marroquín J (ed) Mem III Sem Nacional Utilización de Encinos, Facultad de Ciencias Forestales, Universidad Autónoma de Nuevo León. Rep Cien Espec 15, pp 199–212

González Rivera R (1993) La diversidad de los encinos mexicanos. Rev Soc Mex Hist Nat XLIV:125–142

González-Villarreal LM (1986) Contribución al conocimiento del género *Quercus* (Fagaceae) en el Estado de Jalisco. Instituto de Botánica, Universidad de Guadalajara, Guadalajara, Mexico

Harper J (1977) Population biology of plants. Academic Press, New York

Larsen DR, Johnson PS (1998) Linking the ecology of natural oak regeneration to silviculture. For Ecol Manage 106:1–7

López-Barrera F, González Espinoza M (2000) Influence of litter on emergence and early growth of *Quercus rugosa*: a laboratory study. New For 21:59–70

Moreno-Gómez S, Olvera-Vargas M, Figueroa-Rangel BL (1995) Sistemas silvícolas para encinares en Cerro Grande, Sierra de Manantlán, Jalisco. In: Marroquín J (ed) Mem III Sem Nacional Utilización de Encinos, Facultad de Ciencias Forestales, Universidad Autónoma de Nuevo León. Rep Cien Espec 15, pp 301–319

Muhler-Using B (1994) Contribuciones al conocimiento de los bosques de encino y pino-encino en el noreste de México. Facultad de Ciencias Forestales, Universidad Autónoma de Nuevo León, Rep Cien Espec 14

Muick PC (1991) Effects of shade on blue oak and coast live oak regeneration in California annual grasslands. USDA For Serv Gen Tech Rep PSW-126:21–24

Nixon KC (1993) The genus *Quercus* in México. In: Ramamoorthy TP, Bye R, Fa JE (eds) Biological diversity of Mexico. Oxford Univ Press, Oxford, UK

Peña Ramírez V, Bonfil C (2003) Efecto del fuego en la estructura poblacional y la regeneración de dos especies de encinos (*Quercus liebmanii* Oersted y *Quercus magnoliifolia* Née) en la región de La Montaña, Guerrero. Bol Soc Bot Mex 72:5–20

Price MV, Jenkins SH (1986) Rodents as seed consumers and dispersers. In: Murray DR (ed) Seed dispersal. Academic Press, New York, pp 191–235

Quintana-Ascencio PF, González-Espinosa M, Ramírez-Marcial N (1992) Acorn removal, seedling survivorship, and seedling growth of *Quercus crispipilis* in successional forests of the highlands of Chiapas, Mexico. Bull Torrey Bot Club 119:6–18

Quintana-Ascencio PF, Ramírez-Marcial N, González-Espinosa M, Martínez-Icó M (2004) Sapling survival and growth of coniferous and broad-leaved trees in successional highland habitats in Mexico. Appl Veg Sci 7:81–88

Retana J, Espelta JM, Gracia M, Riba M (1999) Seedling recruitment. In: Roda F, Retana J, Gracia CA, Bellot J (eds) Ecology of Mediterranean evergreen oak forests. Springer, Berlin Heidelberg New York, Ecological Studies 137, pp 89–103

Robledo-Jiménez A (1997) Germinación y crecimiento de plántulas de cuatro especies de encinos del Ajusco, D.F.: efecto del tamaño de semilla. BSc Thesis, Facultad de Estudios Superiores Zaragoza, Universidad Nacional Autónoma de México, Mexico

Rodríguez de la Vega H (2003) Estructura poblacional y distribución espacial de *Senecio praecox* en el Ajusco Medio, D.F.: implicaciones para su reintroducción en sitios perturbados. BSc Thesis, Facultad de Ciencias, Universidad Nacional Autónoma de México (UNAM), México DF, Mexico

Rzedowski J (1981) Vegetación de México. Limusa, México DF, Mexico

Takenaka A (1986) Comparative ecophysiology of two representative *Quercus* species appearing in different stages of succession. Ecol Res 1:129–140

Thadani R, Ashton PMS (1995) Regeneration of banj oak (*Quercus leucotrichopora* A. Camus) in the central Himalaya. For Ecol Manage 78:217–224

Zavala-Chávez F, García-Moya E (1997) Plántulas y rebrotes en la regeneración de encinos en la sierra de Pachuca, Hidalgo. Agrociencia 31:323–329

13 Ecology of Acorn Dispersal by Small Mammals in Montane Forests of Chiapas, Mexico

F. López-Barrera and R.H. Manson

13.1 Introduction

The highlands of central and eastern Mexico are the major centre of diversity (60–75 species) for the genus *Quercus* (Nixon 1993; Chap. 1). Oaks are canopy dominants in many forests in the mountains of Mexico, and provide a wide range of biological resources for insect, mammal and bird species (Quintana-Ascencio et al. 1992; Tovar-Sanchez et al. 2003; Chaps. 14, 16 and 20). The occurrence of temporal fruiting synchronicity in oak populations (usually termed masting or mast seeding) has cascading effects in acorn consumers. Understanding such mast-dependent ecological chain reactions is important for predicting, managing and conserving montane forests. In more northern temperate forests, small mammal populations rely on acorns as an important food source (Wolff 1996), and acorns rely on small mammals as potential seed dispersers (Price and Jenkins 1986; Steele and Smallwood 2002). However, there is little information available on the extent to which oaks rely on dispersal in Neotropical montane forests.

Most oak species suffer reproductive failure under their own canopies (Crow 1988; Lorimer et al. 1994; Figueroa-Rangel and Olvera-Vargas 2000; Chap. 28), relying instead on the dispersal of acorns to forest edges or clearings where establishment success is increased (López-Barrera 2003, see Chaps. 14 and 16). In the Chiapas Highlands, the traditional land use of slash-and-burn agriculture, existing since pre-Columbian times, has resulted in forest mosaics comprised of small clearings (0.5–2 ha; pastures, cornfields and shrublands), surrounded by fragments of secondary forest, evergreen cloud forest, oak forest, and pine–oak forest with various levels of disturbance (see Chap. 16), and resulting in the creation of different forest edge types varying in sharpness, such as hard and soft edges (Fig. 13.1).

Oak regeneration in this landscape of small-scale, scattered forest disturbances (Gonzalez-Espinosa et al. 1991; Chap. 16) is primarily due to oak resprouting ability and the activity of acorn dispersers. This landscape con-

Ecological Studies, Vol. 185
M. Kappelle (Ed.) Ecology and Conservation of Neotropical Montane Oak Forests
© Springer-Verlag Berlin Heidelberg 2006

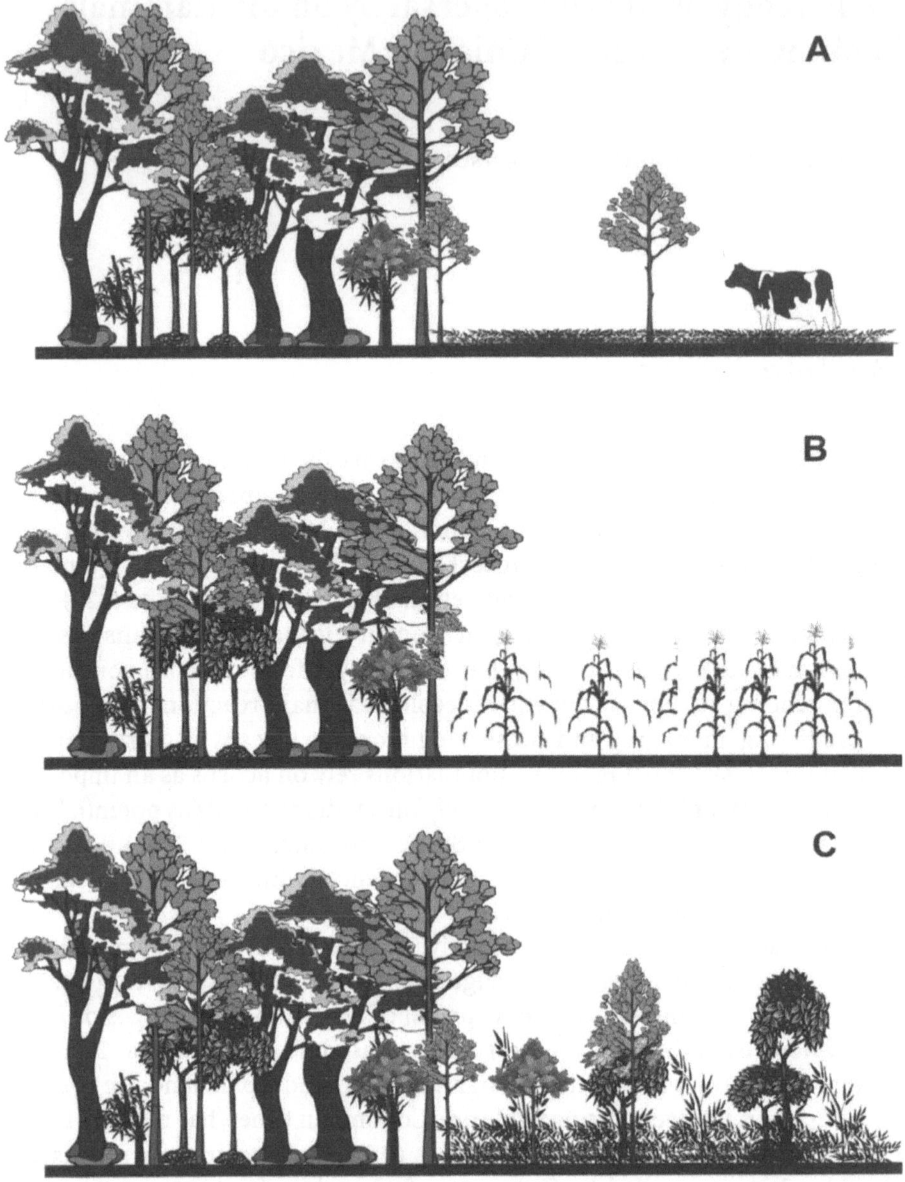

Fig. 13.1A–C. Common non-forested habitats adjacent to the pine–oak forest of the Chiapas Highlands, including **A** permanent pastures, **B** cornfields, and **C** abandoned clearing with scattered presence of shrubs and tree saplings. These land uses adjacent to forests represent a gradient of vegetation cover and microclimate for oak dispersal and establishment, and create borders with different vegetation contrasts ranging from **A** hard, **B** intermediate, and **C** soft edges (López-Barrera 2003)

trasts with those found in temperate and lowland tropical forests, where large and relatively well-preserved forest fragments have been isolated and surrounded by highly transformed, semi-permanent types of cover. In such landscapes, the activity of birds as long-distance oak dispersers may be more important than that of small-distance dispersers such as small mammals (Guevara and Laborde 1993; Gomez 2003). Determining how acorns are dispersed into openings by small mammals, and which factors modulate this interaction is key to understanding Neotropical montane forest regeneration.

This chapter considers changes in the role of small mammals as acorn consumers and/or dispersers as a function of mast-seeding years, and how this interaction may be modulated by different patterns of forest fragmentation (Fig. 13.2). The data we present were obtained from studies conducted in six experimental plots in Rancho Merced Bazom (2,020–2,560 m), located in the central highlands of Chiapas (1,500–2,700 m) in southern Mexico (Chap. 16). The canopy of the forest at this study site is dominated by oaks (*Quercus laurina*, *Q. crassifolia* and *Q. rugosa*) and, to a lesser extent, by pines (for a detailed description of the study area, see González-Espinosa et al. 1991; Chap. 16).

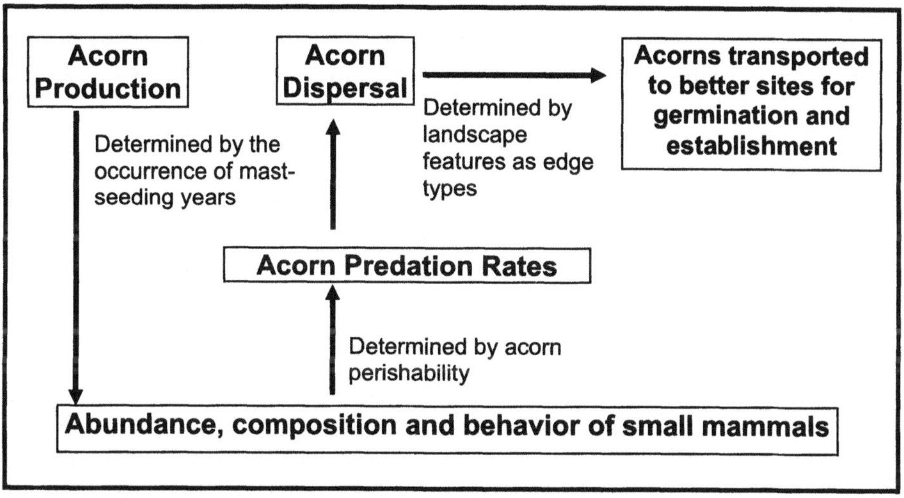

Fig. 13.2. Schematic representation of the ecological processes and interactions reviewed in this chapter. *Boxes* represent the processes studied during two consecutive years in six experimental plots in Rancho Merced Bazom, in the Chiapas Highlands (López-Barrera 2003)

13.2 The Role of Mast Seeding in Oak Dispersal and Recruitment

In oaks, seed production varies considerably from year to year, with pulses of acorn production (mast years) occurring at 2–6 year intervals (Sork 1993; Crawley and Long 1995). There are no long-term studies documenting acorn production patterns in Neotropical montane forests, but observations in our study area suggest that acorn production is not synchronised among oak species and fluctuates at 2–5 year intervals. In temperate forests, the relationship between masting and population fluctuations of small mammals is well documented. Mast seeding produces a large pulse of food resources which allow populations of acorn predators and/or dispersers such as *Peromyscus* spp. to increase into the following year (Wolff 1996; Margaletic et al. 2002; Schnurr et al. 2002; Chap. 26). However, decreased acorn production in years following a masting event is unable to sustain high-seed predator populations, resulting in population crashes (McShea 2000) and increases in the probability of predation satiation during the subsequent mast-seeding year, when food production exceeds the number of seeds which can be consumed (Janzen 1971).

In the highlands of Chiapas, our studies have shown that mast seeding increases small mammal populations in the subsequent years (Fig. 13.3), and this pattern strongly affects the rate of removal of *Q. laurina* acorns. During a mast-seeding year, acorn removal rates were lower compared to the subsequent non-masting year (75% within 213 days vs. 85% removal within

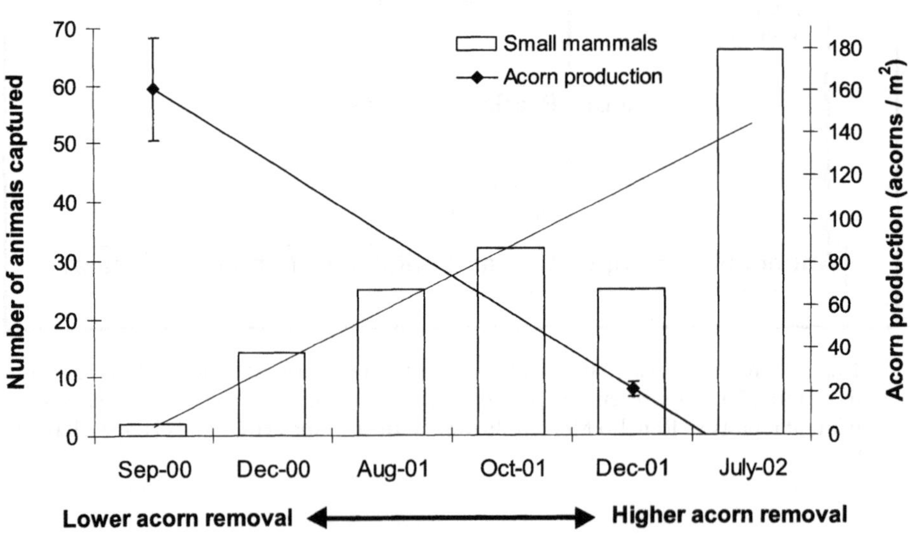

Fig. 13.3. The number of rodents captured (per date, 486 trap nights) versus acorn production from September 2000 to July 2002 (López-Barrera 2003)

138 days, respectively). This time delay in the functional response of small mammal populations means that the two key years for acorn survival are the mast-seeding year (when seed survival is the highest) and the post-masting year (when seed survival is the lowest), compared to the other years with intermediate levels of seed predation (Fig. 13.3).

In addition to fluctuations in the abundance of small mammals, during years of high seed production, animals store seeds far in excess of what they could ever recover and eat, thus allowing many dormant dispersed seeds to germinate and establish (Jansen and Forget 2001). Higher seedling recruitment as a consequence of mast-seeding years has been documented in studies in temperate forests (Jensen 1982; Crawley and Long 1995; Yu et al. 2003), and in tropical forests (Guariguata and Saenz 2002; Chap. 18).

13.3 Forest Fragmentation Effects on Patterns of Acorn Removal and Dispersal by Rodents

13.3.1 Acorn Removal Rates and Edge Effects

We studied the relationship between acorn removal and small mammal abundance in the Chiapas Highlands during two consecutive years (masting vs. non-masting), while simultaneously controlling for edge type (hard or soft), and distance from the edge (both into forest patches and neighbouring grasslands; Fig. 13.1). At sites with hard edges, rates of acorn dispersal and removal were higher within the forest edge and forest interior, compared to adjacent grasslands. By contrast, acorn removal at sites with soft edges was similar at all points along the forest–edge–grassland gradient (López-Barrera 2003), suggesting that, as the structural similarity of the ground cover increases between two adjacent habitats, the permeability of edges to small mammal acorn consumers also increases. Higher rates of seed removal were related to increased vegetation cover and abundance of *Peromyscus* spp. (Fig. 13.4), these relationships coinciding with those reported in others studies in temperate forests (Ostfeld et al. 1999; Kollmann and Buschor 2002). Although in the permanent pastures (Fig. 13.1) there is higher acorn germination and seedling establishment than is the case for the forest interior, there was lower acorn removal and movement of *Peromyscus* spp. into grasslands at sites with hard edges (only two animals were recorded vs. 29 at sites with soft edges), which means that oak colonisation into these habitats is suppressed by low acorn availability (López-Barrera 2003).

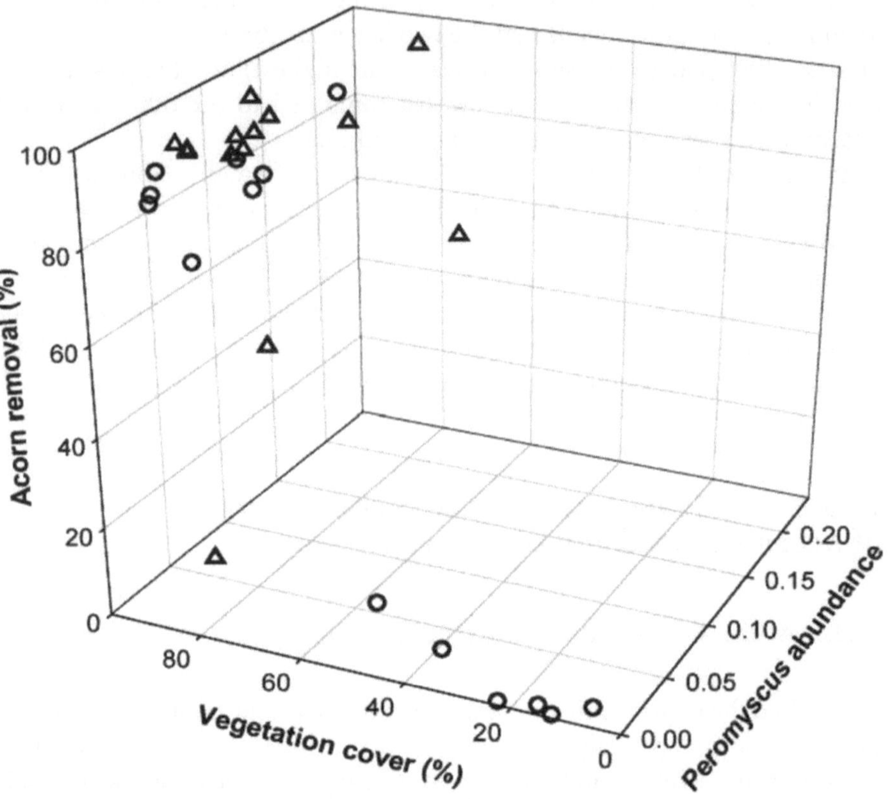

Fig. 13.4. Relationship between mean acorn removal, mean vegetation cover above piles of acorns, and mean abundance expressed as trapping success (captures/100 trap nights) of *Peromyscus* species captured immediately before and after seed removal experiments along sites with hard (*circles*) and soft (*triangles*) edges. Each data point represents the mean value of acorn removal (*Q. laurina*) from piles, generated by grouping data from both years of experiments, the three sites of each edge type, and the different distances from the edge where such piles were placed (for details, see López-Barrera 2003)

13.3.2 Acorn Dispersal

Many authors suggesting that low oak regeneration is due to high acorn predation assume that acorn removal is equal to acorn predation (Borchert et al. 1989; Callaway 1992; Plucinski and Hunter 2002). However, acorn consumers such as *Peromyscus* spp. and *Sciurus* spp. are known to store important amounts of acorns for later use in periods of food scarcity. Scatterhoarding (caches from one to seven acorns) has been documented in some studies for *Quercus liaotungensis* (Li and Zhang 2003), *Q. mongolica* (Miyaki and Kikuzawa 1988), *Q. laurina* and *Q. candicans* (López-Barrera 2003). As rodents rely on olfaction to relocate seeds, the probability of finding caches of single acorns is low (less odour vs. multiple seed caches), thus increasing the probability of escaping predation (Boucher 1981; Bonfil and Soberon 1999; Chap. 12).

Few studies have tagged acorns to precisely determine acorn fate within forests (Sork 1984; Miyaki and Kikuzawa 1988; Steele et al. 2001; Sone et al. 2002; Li and Zhang 2003; Iida 2004), and even fewer have tagged acorns to determine patterns of acorn dispersal into other, adjacent habitats such as grasslands or gaps (Jensen and Nielsen 1986; Kollmann and Schill 1996; Hubbard and McPherson 1999). In the highlands of Chiapas, acorns of *Q. laurina* and *Q. candicans* were metal tagged, placed along a hard and soft forest edge, and then relocated using a metal detector. Small mammals dispersed acorns up to 15 m into the grassland at the site with a soft edge, whereas the dispersal of seeds was mostly concentrated along the edge and within the forest at the site with a hard edge. Thus, acorn transportation into grasslands by rodents is influenced by the vegetation structure of adjacent forest edges. In general, acorns were stored (63 % of the caches were comprised of intact acorns) in many small-scattered caches (87.5 % were caches of one acorn, and the other caches had two acorns) located under dense shrub cover and leaf or root litter in the grassland. Such sites were more likely to foster germination and seedling emergence as they are less susceptible to rot, insect damage, and are protected from strong humidity and temperature fluctuations. This interaction between vegetation and litter cover and small mammal foraging behaviour seems key in understanding acorn predation or dispersal patterns, and thus predicting the success of oak establishment.

13.4 The Trade-Off Between Acorn Perishability and Acorn Germination

In temperate forests, the timing of germination in oaks is considered the main factor affecting the foraging behaviour of squirrels and other small mammals (Smallwood et al. 2002), with germinating acorns consumed or partially consumed and then stored, whereas dormant acorns tend to be stored at longer distances from the parent tree (Smallwood et al. 1998). As a consequence, seedlings from dormant acorns are dispersed at greater distances from parent sources, whereas seedlings originating from acorns with immediate germination show a clumped distribution close to parent trees (Smallwood et al. 1998; Steele and Smallwood 2002). Similar mechanisms operating in the Neotropical montane forests of Mexico could explain the species-specific spatial distribution patterns of seedlings and adults with respect to the seed source we have observed. However, more research is needed to determine whether acorn dispersal patterns are in fact related to the timing of acorn germination.

In central Mexico, Steele et al. (2001) followed the fates of metal-tagged acorns within the forest and found that *Sciurus aureogaster* excised the embryos of 66.7 % of cached acorns of *Q. crassifolia* (earlier germination), and only 1 % of those acorns of *Q. candicans* (dormant acorns). In the Chiapas Highlands, López-Barrera et al. (2005) recorded that during a mast-seeding year, acorns of *Q. segoviensis* (early germination) were consumed in situ whereas dormant acorns of *Q. laurina* were left intact in experimental dishes (apparently by squirrels). During the non-mast seeding year, acorn removal patterns were similar for both *Q. crassifolia* (early germination) and *Q. candicans* (late germination), probable due to changes in the abundance and composition of acorn consumers. However, this may also indicate that small mammals forage more selectively, depending on seed attributes and habitat during mast-seeding years, compared to years of low food abundance when they remove (and presumably eat) all seeds and take greater risks to forage in unsafe habitats (Jansen and Forget 2001; López-Barrera et al. 2005).

13.5 Forest Fragmentation and Perspectives for Conservation of Montane Oak Forest

The presence of small-scale landscape mosaics in the montane forest appears important in maintaining masting as a viable reproductive strategy for oaks, thereby promoting both oak tree recruitment and the conservation of small mammal populations, especially those species which are interior forest specialists. However, fragmentation patterns in the region are producing ever more isolated, small forest fragments (Ochoa-Gaona 2001). Mast seeding and predator satiation as a reproductive strategy appear particularly vulnerable to disruption by forest fragmentation (Curran et al. 1999). Lower densities of reproductive oak trees due to thinning, clearing and firewood harvest, together with the isolation of forest fragments, are likely to alter the spatial structure of oak populations, and interfere with reproduction and acorn survival and germination (Knapp et al. 2001). If remaining oak trees in isolated, small forest fragments in the study area produce fewer acorns or present asynchronous fruiting, the potential for oak regeneration will be very low and, given the importance of acorns to the entire forest ecosystem, lower acorn production may decrease the occurrence of several species of animals, which would have cascading effects on the regeneration of other forest tree species.

Long-term assessments of the effects of fragmentation and disturbance on small mammal populations in the highlands of Chiapas are largely lacking (Horvath et al. 2001). Studies of small mammals inhabiting forest fragments suggest that some threshold values of forest patch size and degree of isolation exist beyond which population sizes and community composition are influenced (Schweiger et al. 2004). Short-term consequences of forest isolation and increased edge area on small mammal populations are decreased animal emigration from the fragment, and higher animal concentration along the edges (Kozakiewicz 1993). If the size and use intensity of open patches in the Chiapas Highlands continue to increase, then small mammals may fail to cross open areas with homogeneous vegetation, due to the higher visibility of predators and the higher risk of being discovered (Mills 1995). Increased populations of small mammal seed consumers in isolated fragments may have a negative impact on seeds or seedling survival (Ostfeld et al. 1997; Manson et al. 2001). Predator satiation and oak reproduction may fail under these conditions (Santos and Telleria 1997; Guariguata et al. 2002; Gomez et al. 2003). The challenge in these systems is to identify those thresholds in fragment size, matrix composition, and isolation at which the occurrence of vertebrates and, therefore, the probability of oak expansion and forest regeneration diminish.

13.6 Conclusions

Rodents have only recently been investigated as acorn predators and/or dispersers in the Neotropics, despite their importance in montane forests. The data presented here indicate that acorn dispersal may occur at rates depending on the occurrence of mast-seeding years, which markedly affects small mammal populations and, in turn, determines acorn removal rates and survival. Also, small mammals may move and hide single acorns under leaf litter or grass roots, these being more suitable sites for acorn germination and seedling establishment (López-Barrera and Newton 2005). However, these movements are determined by edge type (Figs. 13.1 and 13.2).

This chapter provides baseline data highlighting the importance of thorough investigations in Neotropical montane forests of how vegetation cover, landscape features, fragmentation patterns, and variability in acorn production may interact to affect the foraging behaviour of acorn consumers and/or dispersers, and thus the probabilities of acorn dispersal and oak regeneration. Further long-term research is needed to elucidate the effects of population fluctuations of small mammals (due to oaks masting) on non-oak seed and seedling survival (Schnurr et al. 2002). Long-term assessment of key processes discussed in this chapter (Fig. 13.2), and occurring during the year preceding masting and during the following years should allow us to better understand how masting maintains forest tree and small mammal communities.

Acknowledgements We would like to thank A. Newton, C. Legg, M. González Espinosa, N. Ramírez-Marcial, D. Golicher, P. Quintana-Ascencio and M. Steele, who offered relevant information and valuable comments. We are extremely grateful to the many people who helped with the fieldwork, and to the owners of the study plots in Rancho Merced Bazom. CONACYT and The British Council provided a graduate scholarship to F. López-Barrera (ref. nos. 131197 and MEX2900177 respectively). Additional financial support was provided by the European Commission under the INCO-DC programme (framework 4) as part of the SUCRE project (ERBIC-18 CT 97-0146) and the BIOCORES project (PL ICA4-2000-10029).

References

Bonfil C, Soberon J (1999) *Quercus rugosa* seedling dynamics in relation to its re-introduction in a disturbed Mexican landscape. Appl Veg Sci 2:189–200

Borchert MI, Davis FW, Michaelsen J, Dee-Oyler L (1989) Interactions of factors affecting seedling recruitment of blue oak (*Quercus douglasii*) in California. Ecology 70:389–404

Boucher DH (1981) Seed predation by mammals and forest dominance by *Quercus oleoides*, a tropical lowland oak. Oecologia 49:409–414

Callaway RM (1992) Effect of shrubs on recruitment of *Quercus douglasii* and *Quercus lobata* in California. Ecology 73:2118–2128

Crawley MJ, Long CR (1995) Alternate bearing, predator satiation and seedling recruitment in *Quercus robur* L. J Ecol 83:683–696

Crow TR (1988) Reproductive mode and mechanisms for self-replacement of northern red oak (*Quercus rubra*): a review. For Sci 34:19–40

Curran LM, Caniago I, Paoli GD, Astianti D, Kusneti M, Leighton M, Nirarita CE, Haeruman H (1999) Impact of El Niño and logging on canopy tree recruitment in Borneo. Science 286:2184–2188

Figueroa-Rangel B, Olvera-Vargas M (2000) Regeneration patterns in relation to canopy species composition and site variables in mixed oak forests in the Sierra de Manantlán Biosphere Reserve, Mexico. Ecol Res 15:249–261

Gomez JM (2003) Spatial patterns in long-distance dispersal of *Quercus ilex* acorns by jays in a heterogeneous landscape. Ecography 26:573–584

Gomez JA, Garcia D, Zamora R (2003) Impact of vertebrate acorn- and seedling-predators on a Mediterranean *Quercus pyrenaica* forest. For Ecol Manage 180:125–134

González-Espinosa M, Quintana-Ascencio PF, Ramírez-Marcial N, Gaytan-Guzman P (1991) Secondary succession in disturbed *Pinus-Quercus* forest in the highlands of Chiapas, Mexico. J Veg Sci 2:351–360

Guariguata MR, Saenz GP (2002) Post-logging acorn production and oak regeneration in a tropical montane forest, Costa Rica. For Ecol Manage 167:285–293

Guariguata MR, Arias-Le Claire H, Jones G (2002) Tree seed fate in a logged and fragmented landscape, northeastern Costa Rica. Biotropica 34:405–415

Guevara S, Laborde J (1993) Monitoring seed dispersal at isolated standing trees in tropical pastures – consequences for local species availability. Vegetatio 108:319–338

Horvath A, March IJ, Wolf JHD (2001) Rodent diversity and land use in Montebello, Chiapas, Mexico. Stud Neotrop Fauna Environ 36:169–176

Hubbard JA, McPherson GR (1999) Do seed predation and dispersal limit downslope movement of a semi-desert grassland/oak woodland transition? J Veg Sci 10:739–744

Iida S (2004) Indirect negative influence of dwarf bamboo on survival of *Quercus* acorn by hoarding behaviour of wood mice. For Ecol Manage 202:257–263

Jansen PA, Forget PM (2001) Scatterhoarding rodents and tree regeneration. In: Bongers F (ed) Dynamics and plant-animal interactions in a neotropical rainforest. Kluwer, Dordrecht, pp 275–288

Janzen DH (1971) Seed predation by animals. Annu Rev Ecol Syst 2:465–492
Jensen TS (1982) Seed production and outbreaks of non-cyclic rodent populations in deciduous forests. Oecologia 54:184–192
Jensen TS, Nielsen OF (1986) Rodents as seed dispersers in a heath-oak wood succession. Oecologia 70:214–221
Knapp EE, Goedde MA, Rice KJ (2001) Pollen-limited reproduction in blue oak: implications for wind pollination in fragmented populations. Oecologia 128:48–55
Kollmann J, Buschor M (2002) Edge effects on seed predation by rodents in deciduous forests of northern Switzerland. Plant Ecol 164:249–261
Kollmann J, Schill HP (1996) Spatial patterns of dispersal, seed predation and germination during colonization of abandoned grassland by *Quercus petraea* and *Corylus avellana*. Vegetatio 125:193–205
Kozakiewicz M (1993) Habitat isolation and ecological barriers: the effect on small mammal populations and communities. Acta Therio 38:1–30
Li H-J, Zhang Z-B (2003) Effects of rodents on acorn dispersal and survival of the Liaodong oak (*Quercus liaotungensis* Koidz.). For Ecol Manage 176:387–396
López-Barrera F (2003) Edge effects in a forest mosaic: implications for the oak regeneration in the Highlands of Chiapas, Mexico. PhD Thesis, Institute of Atmospheric and Environmental Science, University of Edinburgh, Edinburgh
López-Barrera F, Newton AC (2005) Edge type effect on germination of oak tree species in the Highlands of Chiapas, Mexico. For Ecol Manage 217:67–79
López-Barrera F, Newton AC, Manson RH (2005) Edge effects in a tropical montane forest mosaic: experimental tests of post-dispersal acorn removal. Ecol Res 20:31–40
Lorimer CG, Chapman JW, Lambert WD (1994) Tall understorey vegetation as a factor in the poor development of oak seedlings beneath mature stands. J Ecol 82:227–237
Manson RH, Ostfeld RS, Canham CD (2001) Long-term effects of rodent herbivores on tree invasion dynamics along forest-field edges. Ecology 82:3320–3329
Margaletic J, Glavas M, Baumler W (2002) The development of mice and voles in a oak forest with a surplus of acorns. J Pest Sci 75:95–98
McShea WJ (2000) The influence of acorn crops on annual variation in rodent and bird populations. Ecology 81:228–238
Mills LS (1995) Edge effects and isolation: Red-Backed voles on forest remnants. Conserv Biol 9:395–402
Miyaki M, Kikuzawa K (1988) Dispersal of *Quercus mongolica* acorns in a broadleaved deciduous forest. 2. Scatterhoarding by mice. For Ecol Manage 25:9–16
Nixon KC (1993) The genus *Quercus* in Mexico. In: Ramamoorthy TP, Bye R, Lot A, Fa J (eds) Biological diversity of Mexico: origins and distribution. Oxford Univ Press, New York, pp 447–458
Ochoa-Gaona S (2001) Traditional land-use systems and patterns of forest fragmentation in the Highlands of Chiapas, Mexico. Environ Manage 27:571–586
Ostfeld RS, Manson RH, Canham CD (1997) Effects of rodents on survival of tree seeds and seedlings invading old fields. Ecology 78:1531–1542
Ostfeld RS, Manson RH, Canham CD (1999) Interactions between meadow voles and white-footed mice at forest–old field edges: competition and net effects on tree invasion of old fields. In: Barrett GW, Peles JD (eds) Landscape ecology of small mammals. Springer, Berlin Heidelberg New York, pp 227–247
Plucinski KE, Hunter ML (2002) Spatial and temporal patterns of seed predation on three tree species in an oak-pine forest. Ecography 24:309–317
Price MV, Jenkins SH (1986) Rodents as seed consumers and dispersers. In: Murray DR (ed) Seed dispersal. Academic Press, Sidney, pp 191–235
Quintana-Ascencio PF, Gonzalez-Espinosa M, Ramirez-Marcial N (1992) Acorn removal, seedling survivorship, and seedling growth of *Quercus crispipilis* in successional forests of the highlands of Chiapas, Mexico. Bull Torrey Bot Club 119:6–18

Santos T, Telleria JL (1997) Vertebrate predation on holm oak, *Quercus ilex*, acorns in a fragmented habitat: effects on seedling recruitment. For Ecol Manage 98:181–187
Schnurr JL, Ostfeld RS, Canham CD (2002) Direct and indirect effects of masting on rodent populations and tree seed survival. Oikos 96:402–410
Schweiger EW, Holt RD, Pierotti R, Diffendorfer J (2004) The relative importance of small-scale and landscape-level heterogeneity in structuring small mammal distributions. In: Barrett GW, Peles JD (eds) Landscape ecology of small mammals. Springer, Berlin Heidelberg New York, pp 175–211
Smallwood PD, Steele MA, Ribbens E, McShea WJ (1998) Detecting the effect of seed hoarders on the distribution of seedlings of tree species: Gray Squirrels (*Sciurus carolinensis*) and oaks (*Quercus*) as a model system. In: Steele MA, Merritt JF, Zegers DA (eds) Ecology and evolutionary biology of tree squirrels. Virginia Museum of Natural History, Virginia, pp 211–221
Smallwood PD, Steele MA, Faeth SH (2002) The ultimate basis of the caching preferences of rodents, and the oak-dispersal syndrome: Tannins, insects, and seed germination. Am Zool 41:840–851
Sone K, Hiroi S, Nagahama D, Ohkubo C, Nakano E, Murao S, Hata K (2002) Hoarding of acorns by granivorous mice and its role in the population processes of *Pasania edulis* (Makino). Ecol Res 17:553–564
Sork VL (1984) Examination of seed dispersal and survival in red oak, *Quercus rubra* (Fagaceae) using metal-tagged acorns. Ecology 65:1020–1022
Sork VL (1993) Evolutionary ecology of mast-seeding in temperate and tropical oaks (*Quercus* spp.). Vegetatio 107/108:133–147
Steele MA, Smallwood PD (2002) Acorn dispersal by birds and mammals. In: McShea WJ, Healy WM (eds) Oak forest ecosystems: ecology and management for wildlife. John Hopkins University Press, Baltimore, pp 182–195
Steele MA, Turner G, Smallwood PD, Wolff JO, Radillo J (2001) Cache management by small mammals: experimental evidence for the significance of acorn-embryo excision. J Mammal 82:35–42
Tovar-Sanchez E, Cano-Santana Z, Oyama K (2003) Canopy arthropod communities on Mexican oaks at sites with different disturbance regimes. Biol Conserv 115:79–87
Wolff JO (1996) Population fluctuations of mast-eating rodents are correlated with production of acorns. J Mammal 77:850–856
Yu X, Zhou H, Luo T (2003) Spatial and temporal variations in insect-infested acorn fall in a *Quercus liaotungensis* forest in North China. Ecol Res 18:155–164

14 Establishment, Survival and Growth of Tree Seedlings Under Successional Montane Oak Forests in Chiapas, Mexico

N. Ramírez-Marcial, A. Camacho-Cruz, M. González-Espinosa, and F. López-Barrera

14.1 Introduction

Tropical montane forests are considered globally under threatened status due to their high rates of deforestation (Webster 1995; Brown and Kappelle 2001; Bubb et al. 2004; Kappelle 2004). The current land-use pattern in many montane forests in southern Mexico has created a variety of successional habitats where relatively high plant diversity levels persist. However, it appears that thresholds exist in the intensity and frequency of human disturbance to these habitats. Such thresholds impede successful establishment of seedlings of original tree species (Ramírez-Marcial 2003). Forest recovery under these circumstances may be limited by the availability of seeds, high seed and seedling predation, and competition (Holl et al. 2000; see Chap. 13). A gradual elimination of reproductive adults of understorey broad-leaved tree species has been documented in many of these successional habitats (Ramírez-Marcial et al. 2001; Quintana-Ascencio et al. 2004). Absence of reproductive adults of understorey trees results in scarce or null establishment of their seedlings in large, open, disturbed forest areas (Camacho-Cruz et al. 2000), suggesting that in the short term their regeneration may be seriously affected. Therefore, forest restoration and rehabilitation practices may be regarded as valuable options (Ramírez-Marcial et al. 2005). If forest recovery is seriously hindered by a limited availability of seeds, restoration could be achieved by artificial introduction of juveniles that can later accelerate natural regeneration. Information about tree growth, survival, and regeneration along successional and light gradients is urgently needed.

Seedlings are considered a crucial life stage of tree regeneration, and their populations are highly dynamic, providing opportunities to gather meaningful data in relatively short time (Turner 2001). It is well known that seedling

Ecological Studies, Vol. 185
M. Kappelle (Ed.) Ecology and Conservation of Neotropical Montane Oak Forests
© Springer-Verlag Berlin Heidelberg 2006

mortality tends to decrease with age, but may be determined by the species response to environmental conditions and disturbance regimes. Variation in the intensity, frequency, and spatial distribution of disturbance represents one of the main forces determining patterns of regeneration along successional gradients (Guariguata and Ostertag 2001), including factors such as drought (Poorter and Hayashida-Oliver 2000; Pearson et al. 2003), wind and fire (Bellingham et al. 1996), and biotic interactions such as competition or depredation (Osunkoya et al. 1992; Pearson et al. 2003). Yet, in less disturbed conditions, seedling survival and growth may be determined by different pressures and other physical constraints (Ramírez-Marcial 2003).

In this chapter, we provide information on natural recruitment and establishment of native tree species in different seral stages of the montane pine-oak forests of Chiapas. We include data on survival and growth (relative growth rates with respect to height and basal stem diameter) of seedlings and saplings of 54 native tree species planted under field conditions (successional habitats) and greenhouse controlled experiments (shade houses), collected over several years. Seedlings and saplings are grouped based on taxonomical and morphological traits that they share under similar environmental conditions, resulting in four ecological groups: conifers, oaks, shade-intolerant, and shade-tolerant broad-leaved trees. Determining whether or not different species may have similar performance in contrasting environments (successional habitats) is important to build a model on possible mechanisms for their coexistence, with implications for conservation and restoration of the Neotropical montane forest (McDonald et al. 2003).

14.2 Montane Pine-Oak Forest in Chiapas

Montane pine-oak forest (MPOF) is included within the Tropical Montane Rain Forest (Breedlove 1981) or Tropical Montane Cloud Forest (Hamilton et al. 1995; Kappelle 2004). In Chiapas, it includes several plant associations such as the pine-oak forest, pine-oak-*Liquidambar* forest, and oak forest (see Chap. 16). Typical landscapes in the highlands of Chiapas include sloping lands between valleys and a karstic plateau; forested patches persist but are frequently poor in diversity of tree species (Ochoa-Gaona et al. 2004). Many areas of MPOF are secondary growth stands, surrounded by agricultural fields, grazing lands, and more recently, some dispersed human settlements (Ochoa-Gaona and González-Espinosa 2000). The high population growth in the highlands during the second half of the last century is considered to indicate that the region is being devastated by a rapid and irreversible process of deforestation.

In the MPOF in Chiapas, tree richness has been estimated in 170–190 species, including 35–40 canopy and understorey trees (González-Espinosa et al. 2005). *Quercus* and *Pinus* represent two coexisting and ecologically distinct genera of canopy trees, and changes in their dominance may promote significant changes in the understorey composition of shade-intolerant and shade-tolerant broad-leaved trees. Richness and density of naturally established understorey tree seedlings may decrease strongly when pines dominate the forest canopy, in contrast to the higher diversity observed in oak-dominated stands (González-Espinosa et al. 1991; Ramírez-Marcial et al. 2001; Galindo-Jaimes et al. 2002).

14.3 Ecological Niche and Performance of Seedlings

The ecological niche generally describes the integrated tolerances and requirements of an organism in the habitats in which it lives, grows, and reproduces (Townsend et al. 2003). Although the niche has been considered a concept, rather than a site as such, it is very useful when attempting to explain the performance of many tree species along ecological gradients (Grubb 1977; Prinzing et al. 2002; Pearson et al. 2003).

Performance of tree seedlings in tropical forests has been commonly evaluated using light gradients within gap-phase dynamics (Augspurger 1984; Turner 1990; Dalling et al. 1999). Species differences in performance along light gradients contribute to the maintenance of forest species diversity (Wright 2002). The spatial patterns of light availability within forest stands are likely to influence stand level regeneration patterns of woody species. In lowland tropical forests, natural disturbances such as treefall gaps are considered fine-scale disturbances necessary for the regeneration of canopy species (Denslow 1987; Popma and Bongers 1988). Increment in light levels at the forest floor after gap formation may increase the probability of establishment of some pioneer trees (Turner 1990; Davies 2001). However, gap partitioning among pioneer tree species arises directly from morphological and biochemical specialization to particular light gap environments, and in turn may result from a trade-off between seedling growth and survival (Kitajima 1994; Dalling et al. 1999; Kobe 1999; Koslowski and Pallardy 2002).

14.4 Survival and Growth of Tree Seedlings

14.4.1 Naturally Established Seedlings

Natural establishment of tree seedlings in some MPOF in Chiapas may be highly variable in species richness (10–40 species) and seedling density

(0.8–6.4 ind m^{-2}; González-Espinosa et al. 1991; Camacho-Cruz et al. 2000; Galindo-Jaimes et al. 2002). By contrast, in pine-oak-*Liquidambar* forests (POLF) under human disturbance, seedling species richness may decrease from 26 to four tree species, and density varies from 0.5 to 2.6 seedlings m^{-2}; in highly disturbed forests, only seedlings of pines and oaks have been recorded. After 3 years of evaluation of an old-growth *Quercus-Podocarpus* forest, survival of naturally established seedlings decreased to 40 %, whereas in POLF and pine forest this proportion was 64 % and 67 %, respectively (Ramírez-Marcial 2003).

14.4.2 Transplanted Seedlings

All study sites were located in the municipalities of San Cristóbal de Las Casas, Huistán, and Pueblo Nuevo Solistahuacán, covering an area of 3,220 km^2 in the central and northern highland regions of Chiapas. Climate is temperate sub-humid with summer rains. Site characteristics include a wide range of elevations (1,720–2,500 m) on relatively poor soils and steep slopes (Table 14.1). There is a different number of successional habitats at each study site, due to variations in area. We used the following four contrasting successional habitats: open areas (OA, mostly grasslands), shrublands (SHR), early-successional forests (ESF), and mid-successional forests (MSF; Table 14.1). In many of these habitats, a variable number of seedlings and saplings (25–300) of different species (4–25) were planted, with time reaching a total of 40 species, corresponding to 7,183 individuals (Table 14.1).

14.4.3 Greenhouse Experiment

We conducted a greenhouse experiment using 33–285 seedlings of 42 native tree species. Only 27 species used under greenhouse conditions had at least one replicated habitat under field conditions (Table 14.2). All plants were produced in a nursery (Ramírez-Marcial et al. 2003). Black houses were used to create four shade treatments, i.e., 0, 50, 75, and 90 % shade, representing mean levels of light recorded in previous studies in the successional habitats OA, SHR, ESF, and MSF, respectively (Quintana-Ascencio et al. 1992, 2004; Ramírez-Marcial et al. 1996). Seedlings of each species (total of 3,881 individuals) were randomly placed in black houses in order to have four replicates for each shade treatment (except in the non-shaded treatment, where only three replicates were used). Seedlings were homogeneously watered in order to avoid desiccation, and periodically rotated. Survival and relative growth rates were evaluated after one year.

Table 14.1. Description of the study sites. Successional habitats: *OA* open area, *SHR* shrubland, *ESF* early-successional forest, *MSF* mid-successional forest. Light availability under greenhouse (*GH*) conditions was simulated by shade treatment: *0* non-shaded, *1* 50 %, *2* 75 %, and *3* 90 %

Study site (area in m^2)	Mean elevation (m)	Mean annual rainfall (mm)	Mean temperature (°C)	Slope	Soil type	Successional habitats	Number of species	Study period (months)
S1 (600)	2,200	1,200	13	Flat	Luvisol-rendzines	ESF, MSF	9	48
S2 (5,000)	2,250	1,200	13	Flat–steep	Luvisol-rendzines	ESF	11	48
S3 (3,200)	2,300	1,200	13	Flat–steep	Luvisol-rendzines	MSF	5	27
S4 (1,600)	2,380	1,100	15	Flat	Cambisol	OA, SHR	4	12
S5 (7,200)	1,720	1,700	15	Steep	Luvisol, lithosols and rendzines	OA	16	34
S6 (1,200)	2,500	1,300	13	Steep	Cambisol	OA, ESF, MSF	7	96
S7 (3,200)	2,130	1,200	13	Steep	Lithosols	OA	25	34
S8 (400)	2,400	1,400	14	Flat–steep	Luvisol-rendzines	ESF, MSF	9	48
GH (800)	2,120	1,200	16	Flat	Forest soil mixture	0, 1, 2, 3	42	12

Table 14.2. Initial number of seedlings of 54 native tree species within four ecological groups used in experiments conducted in eight different successional habitats and in greenhouse controlled conditions (GH) in the northern and central highlands of Chiapas, Mexico (see Table 14.1 for details on study sites)

Species groups (with *n* species)	Study sites S1	S2	S3	S4	S5	S6	S7	S8	GH	Total
Conifers (*n*=6)	–	–	–	640	379	398	94	–	720	2,231
Oaks (*n*=11)	164	192	1,470	640	25	–	140	107	1,130	3,868
Shade-intolerant broad-leaved trees (*n*=18)	108	270	–	–	245	83	493	72	846	2,117
Shade-tolerant broad-leaved trees (*n*=19)	214	339	–	–	328	246	393	143	1,185	2,848
Total (*n*=54)	486	801	1,470	1,280	977	727	1,120	322	3,881	11,064

14.4.4 Species Grouping

An a priori classification of species (including 40 in successional habitats, and 42 under greenhouse conditions) resulted in the following four groups (Table 14.2): conifers (four *Pinus* spp. in addition to *Abies guatemalensis* and *Podocarpus matudai*), oaks (11 *Quercus* species), shade-intolerant broad-leaved trees (18 species), and shade-tolerant broad-leaved trees (19 species). The latter two groups are defined on basis of structural data reported in González-Espinosa et al. (1991, 2005; Chap. 16). This classification is useful to understand the effects of light variation associated with canopy cover along successional gradients in MPOF in Chiapas. Since our design is evidently non-balanced, having unequal numbers of species, different seedling ages, and different periods of evaluation, we present an exploratory analysis considering the overall response of the four groups of species, rather than individual species responses in each successional habitat or shade treatment. Mean seedling survival, and relative growth rates of height and basal stem diameter were obtained for all individual species, and these data then combined for each of the four ecological groups.

14.4.5 Natural vs. Greenhouse Survival

Under field conditions, the survival of seedlings of the four ecological groups showed high variation among successional habitats (Fig. 14.1). The conifer group had highest survival in the SHR (no plants of broad-leaved trees were available for this habitat), followed by the OA, and ESF habitats (50 and 67 %, respectively). Only 24 % of the conifer species planted in the MSF survived after 96 months (Quintana-Ascencio et al. 2004). As a group, oaks presented the highest survival (67–84 %) across all successional habitats. Both groups of broad-leaved trees, shade-intolerant and shade-tolerant species, showed higher survival under ESF and MSF (42–50 and 70–72 %, respectively). Less than 32 % of the shade-intolerant species, and less than 13 % of the shade-tolerant species survived in OA habitats (Fig. 14.1). By contrast, the mean survival of seedlings under greenhouse conditions showed different trends. In general, after a year of evaluation, the survival of the four groups of species was higher with increasing shade density (Fig. 14.1). In the four shade treatments (0, 50, 75, and 90 %), mean survival was 64, 93, 98, and 93 % (conifers), 82, 95, 96, and 95 % (oaks), 50, 68, 83, and 84 % (shade-intolerant broad-leaved trees), and 51, 86, 89, and 96 % (shade-tolerant broad-leaved trees), respectively.

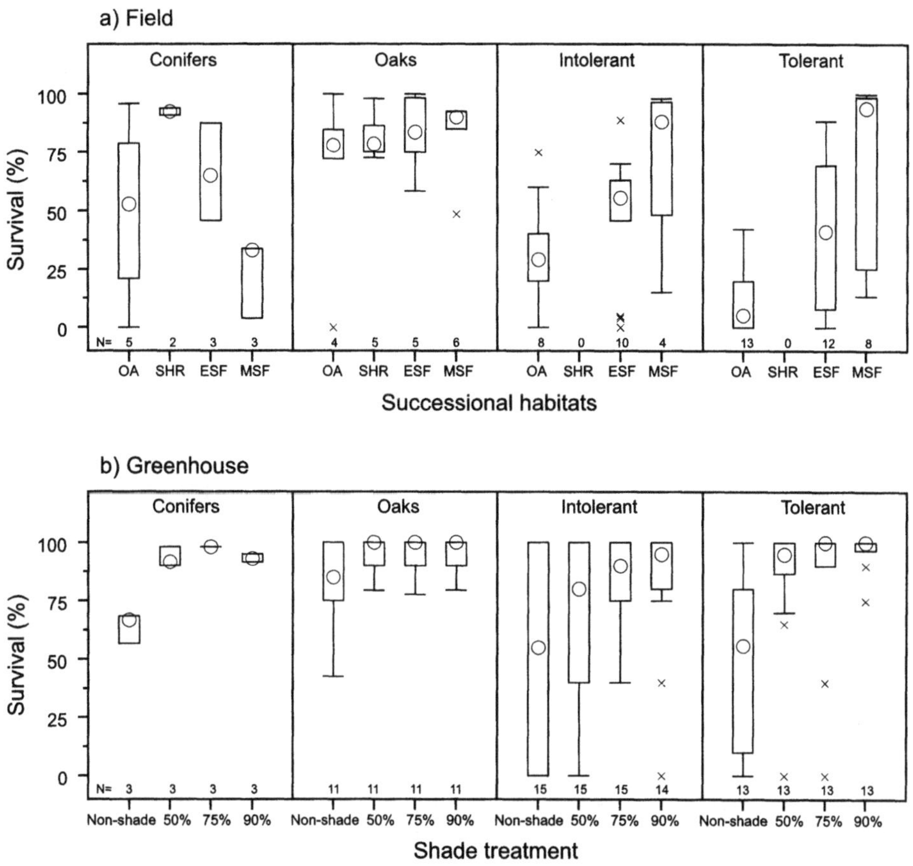

Fig. 14.1a, b. Box plot (median indicated by *open circle*+upper and lower quartiles; *crosses* are extreme values) for seedling survival in four ecological groups: conifers, oaks, and shade-intolerant and shade-tolerant broad-leaved trees under field (**a**), and greenhouse (**b**) conditions. Successional habitats are: *OA* open areas, *SHR* shrubland, *ESF* early-successional forest, *MSF* mid-successional forest. Shade treatments under greenhouse conditions roughly represent mean values of canopy cover of each corresponding successional habitat. *N* Number of species included within each group

14.4.6 Relative Growth Rates

Seedlings of the four ecological groups seem to grow similarly in the four successional habitats. Conifers showed the highest growth in OA and SHR habitats; their lowest value was recorded in MSF. Oaks and the shade-tolerant broad-leaved seedlings grew more in the ESF, whereas the shade-intolerant broad-leaved seedlings did not show any trend; still, the variation within this group was less pronounced in the MSF (Fig. 14.2). Under greenhouse conditions, seedlings of all ecological groups increased in height under intermedi-

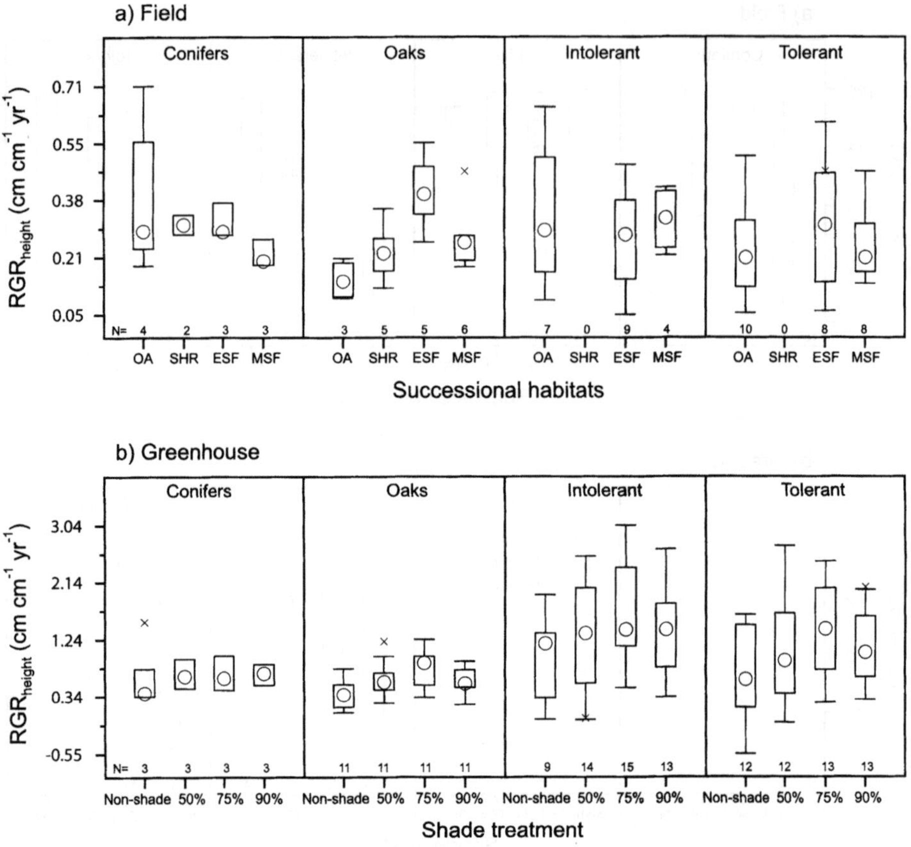

Fig. 14. 2a, b. Box plot for the relative growth rate of stem height (RGR_{height}, cm cm^{-1} year^{-1}) of tree seedlings of four ecological groups (conifers, oaks, shade-intolerant broad-leaved trees, and shade-tolerant broad-leaved trees) under field (**a**) and greenhouse (**b**) conditions. Successional habitats and shade treatments as in Fig. 14.1

ate levels of shade (50 and 75 % shade, Fig. 14.2). Oaks and conifers showed an increase in basal diameter in the OA and SHR habitats, the latter being where conifers had their greatest increase in basal diameter (Fig. 14.3). Under greenhouse conditions, the diameter growth rate of conifers and oaks was relatively low and similar in all shade treatments, whereas shade-intolerant and shade-tolerant broad-leaved trees increased their diameter under intermediate shade (50 %; Fig. 14.3). Among the four ecological groups evaluated in the greenhouse, conifers presented the lowest diameter increases – probably because the seedlings were smaller than those of the other groups. It is well known that seedlings of conifers grow less efficiently than those of angiosperms, at least during the initial stages of growth (Bond 1989).

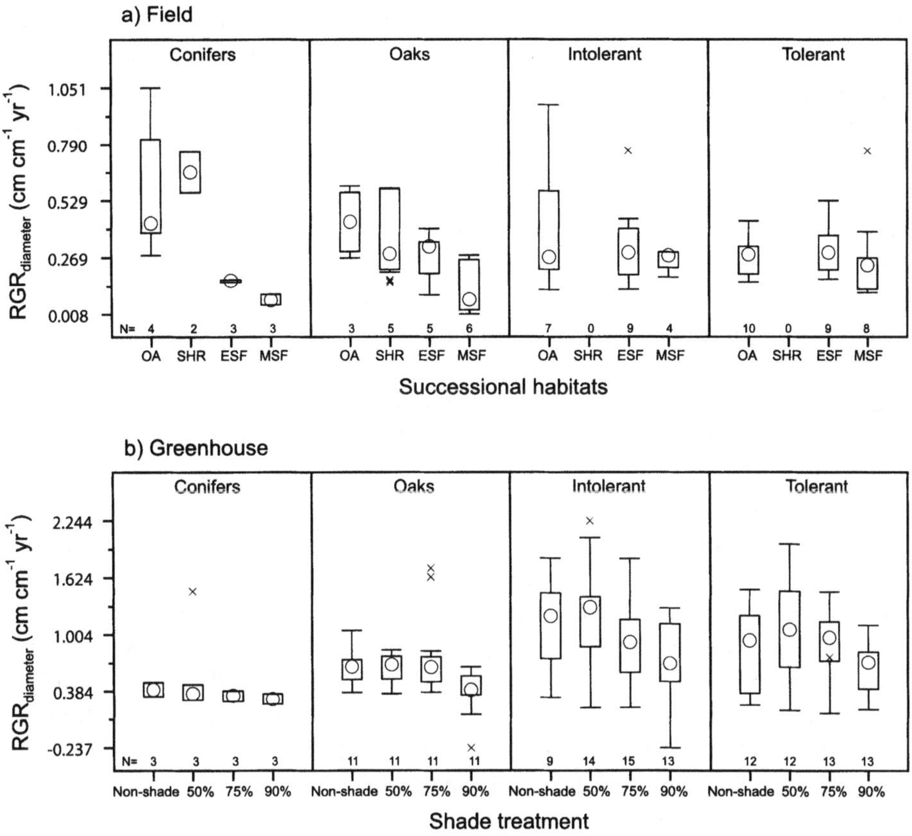

Fig. 14.3a, b. Box plot for the relative growth rate of basal stem diameter ($RGR_{diameter}$, cm cm^{-1} $year^{-1}$) of tree seedlings of four ecological groups (conifers, oaks, shade-intolerant broad-leaved trees, and shade-tolerant broad-leaved trees) under field (**a**) and greenhouse (**b**) conditions. Successional habitats and shade treatments as in Fig. 14.1

14.5 Conservation and Restoration Implications

Many of the ecological attributes of a community that are lost under the impacts of natural perturbations can eventually be recovered after disturbance (Brown and Lugo 1994). However, due to the high intensity and frequency of human disturbances in the MPOF of Chiapas, natural recovery can be slow. The cumulative effect of this type of disturbance tends to favor the regeneration of some pioneer species such as pines, oaks, *Alnus acuminata*, *Arbutus xalapensis*, *Buddleja cordata*, and *Prunus serotina*, which showed good performance in open areas or shrublands. However, shade-tolerant species such as *Clethra pachecoana*, *Cleyera theaeoides*, *Oreopanax xalapensis*, *Prunus rhamnoides*, *Psychotria galeottiana*, *Styrax magnus*, *Symplocos limoncillo*, and *Zanthoxylum melanostictum* had poor development in these open habitats. Diurnal patterns of solar radiation, temperature, and relative humidity display more extreme variations in open pine forests than in closed-canopy old-growth oak forests. Summer temperatures at forest-floor level can rise to 45 °C, and winter temperature decreases may cause frost. Such extreme temperature fluctuations, and additional water deficits occurring during the dry season represent a set of major constraints that reduce seedling performance (Ramírez-Marcial 2003; Ramírez-Marcial et al. 2005).

Restoration of native plant communities requires identifying those factors that favor or impede natural colonization by native tree species (Denslow 1996). In the study region, successful colonization of native trees in early seral stages is limited not only by physical conditions such as light intensity and related temperature (e.g., McCormick 1996), but also by arrival of the propagules. Therefore, a viable strategy for MPOF recovery in Chiapas seems to be through the reintroduction of seedlings in appropriate successional habitats. From a biological viewpoint, there could be two routes for forest restoration: (1) the suppression of disturbance in forested areas, which would promote natural regeneration, and (2) the establishment of enrichment planting in poor-species forested areas. Based on this and previous evidence, we consider feasible the reintroduction of pines and oaks, and some other shade-intolerant broad-leaved species in open and shrubland habitats, and of seedlings of shade-tolerant broad-leaved species in pinelands or other successional forests with intermediate light levels under the canopies.

14.6 Conclusions

Changes in floristic composition and forest structure in the study region have promoted the gradual elimination of reproductive individuals of many broad-leaved species. As a consequence, only low numbers of naturally recruited seedlings of these species are observed in the remaining secondary forests, which may lead to their local extinction.

Our classification of plant functional types (sensu Denslow 1996) based on a combination of taxonomical and ecological traits will help us to understand the interaction between physical factors (mostly light conditions) and successional gradients in MPOF in Chiapas. Grouping of species is useful to identify which species show similar responses to specific environmental conditions. Comprehensive management of montane forests in Chiapas requires recognition of yet wider and more detailed sets of the species' ecological attributes, in addition to social, economic, and cultural issues involved in local restoration programs.

Acknowledgements We thank Liliana Mascarúa-López, Magala Alcázar-Gómez, Alfonso Luna-Gómez, Pedro Girón-Hernández, and Miguel Martínez-Icó for their assistance in the field, greenhouse, and nursery. Duncan Golicher, Pedro F. Quintana-Ascencio, and Luis Galindo-Jaimes provided useful comments that improved the manuscript. This study was supported by the European Commission (BIOCORES Project, INCO Framework Programme 5, contract no. ICA4-CT-2001-10095), CONACYT (grant no. 020395 to NRM), COCYTECH (Project FOMIX-CHIS-2002-C01-4640), SEMARNAT (Project 2002-01-C01-048) and federal subsidies to ECOSUR.

References

Augspurger CK (1984) Light requirements of neotropical tree seedlings: a comparative study of growth and survival. J Ecol 72:777–795

Bellingham PJ, Tanner EVJ, Rich PM, Goodland TCR (1996) Changes in light below the canopy of a Jamaican montane rainforest after a hurricane. J Trop Ecol 12:699–722

Bond WJ (1989) The tortoise and the hare: ecology of angiosperm dominance and gymnosperm persistence. Biol J Linn Soc 36:227–249

Breedlove DE (1981) Flora of Chiapas, part I. Introduction to the flora of Chiapas. California Academy of Sciences, San Francisco

Brown AD, Kappelle M (2001) Introducción a los bosques nublados del Neotrópico: una síntesis regional. In: Kappelle M, Brown AD (eds) Bosques nublados del Neotrópico. Instituto Nacional de Biodiversidad (INBio), Santo Domingo de Heredia, Costa Rica, pp 25–40

Brown S, Lugo A (1994) Rehabilitation of tropical lands: a key to sustaining development. Rest Ecol 2:97–111

Bubb P, May I, Miles L, Sayer J (2004) Cloud forest agenda. UNEP-WCMC, Cambridge, UK. http://www.unep-wcmc.org/resources/publications/UNEP_WCMC_bio_series/20.htm

Camacho-Cruz A, González-Espinosa M, Wolf JHD, de Jong BHJ (2000) Germination and survival of tree species in disturbed forests of the highlands of Chiapas, Mexico. Can J Bot 78:1309–1318

Dalling JW, Lovelock CE, Hubbell SP (1999) Growth responses of seedlings of two neotropical pioneer species to simulated forest gaps environments. J Trop Ecol 15:827–839

Davies SJ (2001) Tree mortality and growth in 11 sympatric *Macaranga* species in Borneo. Ecology 82:920–932

Denslow JS (1987) Tropical treefall gaps and tree species diversity. Annu Rev Ecol Syst 18:431–451

Denslow JS (1996) Functional group diversity and responses to disturbance. In: Orians G, Dirzo R, Cushman JH (eds) Biodiversity and ecosystem processes in tropical forests. Springer, Berlin Heidelberg New York, pp 127–151

Galindo-Jaimes L, González-Espinosa M, Quintana-Ascencio PF, García-Barrios LE (2002) Tree composition and structure in disturbed stands with varying dominance by *Pinus* spp. in the highlands of Chiapas, Mexico. Plant Ecol 162:259–272

González-Espinosa M, Quintana-Ascencio PF, Ramírez-Marcial N, Gaytán-Guzmán P (1991) Secondary succession in disturbed *Pinus-Quercus* forests of the highlands of Chiapas, Mexico. J Veg Sci 2:351–360

González-Espinosa M, Ramírez-Marcial N, Méndez-Dewar G, Galindo-Jaimes L, Golicher D (2005) Riqueza de especies de árboles en Chiapas: variación espacial y dimensiones ambientales asociadas al nivel regional. In: González-Espinosa M, Ramírez-Marcial N, Ruiz-Montoya L (eds) Diversidad biológica en Chiapas. Plaza & Valdés, Mexico City, pp 81–125

Grubb PJ (1977) The maintenance of species-richness in plant communities: the importance of the regeneration niche. Biol Rev 52:107–145

Guariguata MR, Ostertag R (2001) Neotropical secondary forest succession: changes in structural and functional characteristics. For Ecol Manage 148:185–206

Hamilton LS, Juvik JO, Scatena FN (1995) Tropical montane cloud forest. Springer, Berlin Heidelberg New York

Holl KD, Loik ME, Lin EHV, Samuels IA (2000) Tropical montane forest restoration in Costa Rica: overcoming barriers to dispersal and establishment. Rest Ecol 8:339–349

Kappelle M (2004) Tropical montane forests. In: Burley J, Evans J, Youngquist JA (eds) Encyclopedia of forest sciences, vol 4. Elsevier, Oxford, UK, pp 1782–1793

Kitajima K (1994) Relative importance of photosynthetic traits and allocation patterns as correlates of seedling shade tolerance of 13 tropical trees. Oecologia 98:419–428

Kobe RK (1999) Light gradient partitioning among tropical tree species through differential seedling mortality and growth. Ecology 80:187–201

Koslowski TT, Pallardy SG (2002) Acclimation and adaptive responses of woody plants to environmental stresses. Bot Rev 68:270–334

McCormick JF (1996) A review of the population dynamics of selected tree species in the Luquillo experimental forest, Puerto Rico. In: Lugo A, Lowe C (eds) Tropical forests: management and ecology. Springer, Berlin Heidelberg New York, pp 224–257

McDonald MA, Hofny-Collins A, Healey JR, Goodland TCR (2003) Evaluation of trees indigenous to the montane forest of the Blue Mountains, Jamaica, for reforestation and agroforestry. For Ecol Manage 175:379–401

Ochoa-Gaona S, González-Espinosa M (2000) Land use and deforestation in the highlands of Chiapas, Mexico. Appl Geogr 20:17–42

Ochoa-Gaona S, González-Espinosa M, Meave JA, Sorani-Dalbon V (2004) Effect of forest fragmentation on the woody flora of the highlands of Chiapas, Mexico. Biod Conserv 13:867–884

Osunkoya OO, Ash JE, Hopkins MS, Graham AW (1992) Factors affecting survival of tree seedlings in north Queensland rainforests. Oecologia 91:569–578

Pearson TRH, Burslem DFRP, Goeriz RE (2003) Regeneration niche partitioning in neotropical pioneers: effects of gap size, seasonal drought and herbivory on growth and survival. Oecologia 137:456–465

Poorter L, Hayashida-Oliver Y (2000) Effects of seasonal drought on gap and understorey seedlings in a Bolivian moist forest. J Trop Ecol 16:481–498

Popma J, Bongers F (1988) The effect of canopy gaps on growth and morphology of seedlings of rain forest species. Oecologia 75:625–632

Prinzing A, Durka W, Klotz S, Brandl R (2002) Geographic variability of ecological niches of plant species: are competition and stress relevant? Ecography 25:721–729

Quintana-Ascencio PF, González-Espinosa M, Ramírez-Marcial N (1992) Acorn removal, seedling survivorship, and seedling growth of *Quercus crispipilis* in successional forests of the highlands of Chiapas, Mexico. Bull Torrey Bot Club 119:6–18

Quintana-Ascencio PF, Ramírez-Marcial N, González-Espinosa M, Martínez-Icó M (2004) Sapling survival and growth of conifer and broad-leaved trees in successional habitats in the highlands of Chiapas, Mexico. Appl Veg Sci 7:81–88

Ramírez-Marcial N (2003) Survival and growth of tree seedlings in anthropogenically disturbed Mexican montane rain forests. J Veg Sci 14:881–890

Ramírez-Marcial N, González-Espinosa M, García-Moya E (1996) Establecimiento de *Pinus* spp. y *Quercus* spp. en matorrales y pastizales de Los Altos de Chiapas. Agrociencia 30:249–257

Ramírez-Marcial N, González-Espinosa M, Williams-Linera G (2001) Anthropogenic disturbance and tree diversity in the montane rain forest in Chiapas, Mexico. For Ecol Manage 154:311–326

Ramírez-Marcial N, Camacho-Cruz A, González-Espinosa M (2003) Guía para la propagación de especies leñosas nativas de Los Altos y Montañas del Norte de Chiapas. El Colegio de la Frontera Sur (ECOSUR), San Cristóbal de Las Casas, Chiapas, Mexico

Ramírez-Marcial N, Camacho-Cruz A, González-Espinosa M (2005) Potencial florístico para la restauración de bosques en Los Altos y las montañas del norte de Chiapas. In: González-Espinosa M, Ramírez-Marcial N, Ruiz-Montoya L (eds) Diversidad biológica en Chiapas. Plaza & Valdés, Mexico City, Mexico, pp 325–369

Townsend CR, Begon M, Harper JL (2003) Essentials of ecology, 2nd edn. Blackwell, Oxford

Turner IM (1990) Tree seedling growth and survival in a Malaysian rain forest. Biotropica 22:146–154

Turner IM (2001) The ecology of trees in the tropical rain forest. Cambridge Univ Press, Cambridge, UK

Webster GL (1995) The panorama of neotropical cloud forest. In: Churchill SP, Balslev H, Forero E, Luteyn JL (eds) Biodiversity and conservation of neotropical montane forests. New York Botanical Garden Press, Bronx, NY, pp 53–77

Wright SJ (2002) Plant diversity in tropical forests: a review of mechanisms of species coexistence. Oecologia 130:1–14

15 Population Structures of Two Understory Plant Species Along an Altitudinal Gradient in Costa Rican Montane Oak Forests

T.V.M. Groot, M. Stift, J.G.B. Oostermeijer, A.M. Cleef, and M. Kappelle

15.1 Introduction

The altitudinal distribution of vascular plant species and communities has been intensively studied in neotropical montane forests (e.g., Cleef et al. 1984; Kappelle and Zamora 1995; Kappelle et al. 1995a; Lieberman et al. 1996; Chap. 4). However, still little is known about the population ecology of individual species and their responses to environmental factors along elevational gradients. Population structure studies of single species in relation to environmental factors have proven to be powerful in explaining ecological behavior in temperate regions (Wassen et al. 1990; Oostermeijer et al. 1994; Hegland et al. 2001), but such data are relatively scarce for tropical regions.

So far, the main emphasis in the tropics has been on population dynamics of lowland rainforest canopy trees (e.g., Arriaga et al. 1988; Martínez-Ramos et al. 1988; Álvarez-Buylla and Martínez-Ramos 1992; Olmsted and Alvarez-Buylla 1995), rather than on non-canopy species (Oyama 1990). Tropical studies directly relating population structure to environmental variables have not received much attention from scientists (Witkowski and Liston 1997; Hicks and Mauchamp 2000). To date, research papers on population dynamics of highland forest plant species are even rarer (but see Wesselingh et al. 1999, and Chaps. 14 and 18). In order to gain further insight into environmental factors (e.g., temperature, light availability) that may determine the population structure of vascular plant species on tropical mountains, we studied the relative proportions of different life stages of two tropical montane understory monocot species in response to environmental changes occurring along an altitudinal gradient in Costa Rican montane oak forests.

Ecological Studies, Vol. 185
M. Kappelle (Ed.) Ecology and Conservation of Neotropical Montane Oak Forests
© Springer-Verlag Berlin Heidelberg 2006

15.2 Study Area

This study was conducted in undisturbed mature, old-growth *Quercus*-dominated tropical montane cloud forest along the Pacific slopes (2,000–3,000 m altitude) of the Costa Rican Cordillera de Talamanca. Two mountains were investigated: (1) the Cerro de la Muerte (CM; peak at 3,491 m; coordinates 09°34'N and 83°46'W), which borders the southern limit of the 58,495-ha Tapantí-Macizo de la Muerte National Park and the northeastern sector of the 62,000-ha Los Santos Forest Reserve; and (2) the Cerro Chirripó (CC; peak at 3,819 m; coordinates 09°30'N and 83°30'W), the highest mountain in southern Central America, located in the center of the 50,920-ha Chirripó National Park.

Both mountains are within the same, large forest tract, as the Los Santos Forest Reserve and Chirripó National Park are connected by the undisturbed Tapantí-Macizo de la Muerte National Park. The importance of this large forest tract for biodiversity conservation – including that of the La Amistad International Park to the east – is globally recognized, as it has been designated as a Biosphere Reserve, a World Heritage Site, an Endemic Bird Area, and a Center of Plant Diversity (Kappelle 1996; Chaps. 10 and 30).

Mean annual temperatures range from 15 °C at 2,000 m and 9.5 °C at 3,000 m, and mean annual vertical precipitation oscillates around 2,500 and 3,000 mm, respectively (Kappelle et al. 1996). The area's montane oak forests receive a considerable amount of additional horizontal precipitation in the form of fog or mist, which significantly increase the total amount of intercepted water (Zadroga 1981; Chap. 21). Vascular and non-vascular epiphytes are abundant. Forest soils are generally wet, acid, humus-rich and clayey, and are developed in volcanic ashes (andosols) originating from volcanoes to the north (Kappelle et al. 1995a; Chap. 4).

15.3 Field Sampling

Two altitudinal transects were established: one along the SW slope of Cerro de la Muerte (CM), parallel to the San Gerardo de Dota–Cerro de la Muerte (Cerro Buenavista) hiking trail, and one along the SW slope of Cerro Chirripó (CC), parallel to the San Gerardo de Rivas–Cerro Chirripó hiking trail (also known as the Termómetro–Fila Cementerio de la Máquina–Base Crestones trail). The CM transect covered an altitudinal belt at 2,200–2,900 m, the CC transect at 2,000–2,900 m. Forest disturbance made mature forest sampling impossible at 2,000–2,200 m elevation on the CM trail.

Along transects, plots were established at 50-m altitudinal intervals. Two replicate plots were established at each altitudinal interval along each transect, totaling 30 plots along the CM transect and 40 along the CC transect. Plot size was 0.03 ha (15×20 m). Individual plots consisted of three quadrants, each covering 0.01 ha (5×20 m), and separated by buffer strips of similar shape and size for quadrant sampling without destructive trampling. Each quadrant was subdivided into four 5×5 m subplots.

In each plot, we measured altitude (using a Thomme 5000 altimeter), slope direction, canopy and understory densities, stem density (number of tree stems with diameter at breast height (DBH)≥25 cm), and percentage ground cover of living bryophytes, large litter (decaying trunks and branches), and uncovered forest floor (bare soil). Following Wolf's (1993) method, canopy coverage was estimated as the average percentage covered by black pixels (i.e., those that did not represent the sky) found on six scanned, computerized (Adobe Photoshop 5.5) vertical canopy photographs, which were taken in each plot using a 28-mm wide-angle lens on a 35-mm SLR camera at six fixed positions. Temperature (minimum and maximum values) was measured at 100-m intervals over 20-day periods.

15.4 Selected Study Species

Two indicator understory monocot plant species were selected for this study: the dwarf palm *Geonoma orbignyana* Naud. (previously identified as *Geonoma hoffmanniana* H. Wendl. ex Spruce, in Kappelle et al. 1995a; Kappelle 1996; Chap. 4), and the aroid herb *Anthurium concinnatum* Schott. These species can be used for feasible population structure analysis as they meet the following criteria: (1) high abundance along the entire altitudinal range, (2) easy identification of individuals during all life stages under harsh field conditions, (3) absence of clone reproduction, enabling identification of single individuals, and (4) easy age estimation based on growth of single-stemmed individuals, expressed in terms of stem length. For both species, representative specimens were collected, identified, and stored at the herbarium of INBio (INB) in Costa Rica.

Geonoma orbignyana (Arecaceae, Palmae), locally known as *súrtuba* or *súrtuba ratón*, is one of the most common and wide-spread palm species in Central American and Andean montane oak forests, and is the only palm that may reach elevations over 3,000 m (Henderson et al. 1995). This single-stemmed species may reach a maximum height of 4 m (Fig. 15.1). Adult individuals maintain up to 20 living leaves – though they generally have some 10–15 leaves – and can bear up to five inflorescences or infructescences at a time. Inflorescences are pollinated by a wide range of insects (Olesen and Balslev 1990; Listabarth 1993). Berries are dispersed mainly by birds (Zona and Henderson 1989).

Anthurium concinnatum (Araceae) is found in montane rainforests ranging from Costa Rica to Colombia, and probably also in Ecuador (Croat and

Fig. 15.1. Growth form of *Geonoma orbignyana*. Drawing by Francisco „Pancho" Quesada (INBio, Costa Rica)

Baker 1979). It grows at elevations of 2,000–3,000 m and is, within its genus, the species that occurs at the highest altitudes in Central America (Dortort 1980). *A. concinnatum* is a typical large-leaved, hemi-epiphytic understory species, rarely growing higher than 5 m above the forest floor. It produces a spadix 20–40 cm long, with up to 100 green oval fruits, each containing 1–3 seeds. Seeds are presumably dispersed by birds. It has multiple growth forms, ranging from terrestrial to epiphytic (Benzing 1990). Young plants commonly grow straight upward, but when they lack a suitable substrate to attach to, they eventually tend to fall down, after which they start to grow horizontally until they find a structure to climb on (Benzing 1990; Bown 2000). We identified four different types of growth forms (Fig. 15.2): (1) plants growing entirely upright and terrestrial (*standing terrestrial*); (2) plants growing (partly) horizontally and terrestrial (*lying terrestrial*); (3) plants growing entirely upright and epiphytic (*standing epiphytic*); (4) plants growing (partly) horizontally and epiphytic (*lying epiphytic*).

We estimated abundance values for both species, expressed in terms of their leaf ground cover proportions, applying conventional plant sociological

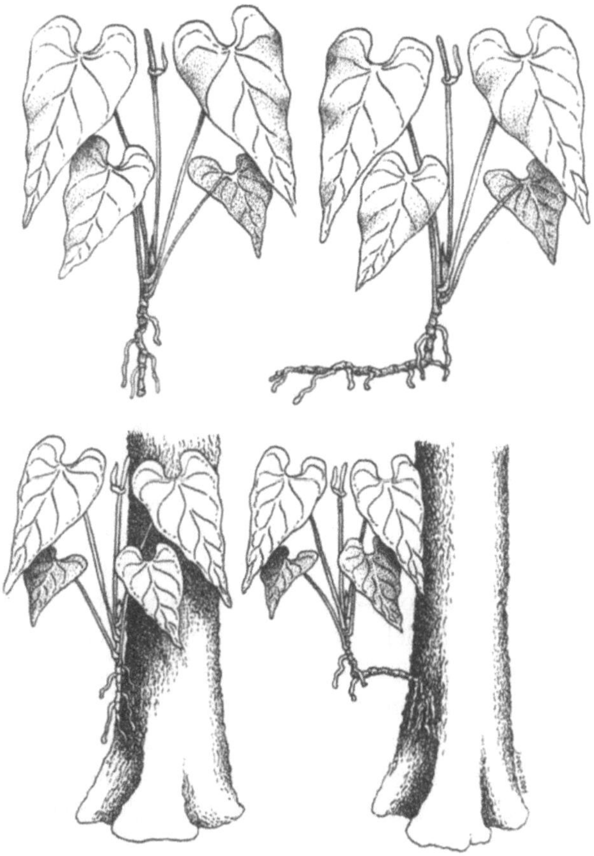

Fig. 15.2. Growth forms of *Anthurium concinnatum*: standing terrestrial (*upper left*), lying terrestrial (*upper right*), standing epiphytic (*lower left*), and lying epiphytic (*lower right*) individuals. Drawing by Francisco „Pancho“ Quesada (INBio, Costa Rica)

procedures (Kappelle 1996). For each individual of both species we assessed: total plant height, height up to the first leaf, and phenological state (fertile vs. sterile). For *G. orbignyana*, we also counted the total number of green leaves, and for *A. concinnatum* we additionally determined the type of growth form. Individuals of *G. orbignyana* less than 50 cm high and bearing less than five leaves were excluded from our analysis, as they could not be distinguished from other dwarf palm species present in this habitat.

In accordance with size and phenological state, each individual was assigned to one of the following criteria-based life stages, ranging from young to old:

- *G. orbignyana:* (1) sterile plant, total plant height<50 cm, total stem height <10 cm, and <6 leaves; (2) sterile plant, total plant height 50–100 cm, total stem height 10–40 cm, or 6–12 leaves; (3) sterile plant, total plant height 100–150 cm, total stem height 40–100 cm; (4) sterile plant, total plant height ≥150 cm, or total stem height ≥100 cm, or ≥12 leaves; and (5) fertile plant;

- *A. concinnatum:* (1) sterile plant, total plant height <30 cm, and length to first leaf <30 cm; (2) sterile plant, total plant height 30–80 cm, or length to first leaf 30–50 cm; (3) sterile plant, total plant height ≥80 cm, or length to first lea f≥50 cm; and (4) fertile plant.

15.5 Data Analysis

We calculated Spearman's rank correlation coefficients to assess statistical relations existing among the environmental variables. We used a sequential Bonferroni procedure to correct for multiple tests (Holm 1979). We performed stepwise, linear and non-linear (unimodal) multiple regressions using SPSS v. 9.0 software for MacIntosh (SPSS Inc. 1990), to test which environmental variables best explained the variation in population structure of both species. The environmental variables investigated in each plot were treated as independent variables, whereas the relative proportions of the life stages in the plots were treated as dependent variables. As the fitting of linear regression models proved superior in all cases, only these are presented here (Tables 1 and 2). The structure of *A. concinnatum* populations was also assessed independently for each of the four individual growth forms. Plots containing fewer than 40 individuals were excluded from the analysis, in order to eliminate sampling errors. Within-plot heterogeneity did not allow for statistical analysis of variations in absolute density for separate life stages.

15.6 Environmental Correlations

As we hypothesized and tested ($P<0.001$ after a Bonferroni correction for multiple tests), along both transects there was a strong positive correlation between minimum and maximum temperatures (CC: 0.886; CM: 0.679), and a clear negative correlation between both temperature measures and altitude (CC=–0.997, and CM=–0.704 for minimum temperature, and CC=–0.889 and CM=–0.991 for maximum temperature). Similarly, the percentage of uncovered forest floor and the percentage of large litter cover were highly inversely correlated in both transects (CC: –0.752; CM: –0.699). Along the CM transect, bryophyte ground cover was significantly positively correlated with altitude and, consequently, negatively with minimum and maximum temperatures (–0.849). Along the same transect, canopy cover was found to be positively correlated with the percentage uncovered forest floor (0.640). Hence, in plots with a dense canopy cover, the forest floor was relatively open.

Table 15.1. Multiple linear regression models for *G. orbignyana* and *A. concinnatum* along both transects. Only the significant models are presented (* $P<0.05$, ** $P<0.005$, *** $P<0.0005$)

Life stage	Cerro Chirripó transect: Variable	Beta	df	*F*	Cerro de la Muerte transect: Variable	Beta	df	*F*
G. orbignyana								
1	Slope dir.	0.450*	19	0.041	Canopy cov.	–1.000***	2	<0.001
2	–	–	–	–	Bryoph. cov.	1.129**	1	0.005
3	–	–	–	–	Min. temp.	0.993*	2	0.007
4	Tree s. den.	–0.427*	18	<0.001	–	–	–	–
4	Max. temp.	0.957*	18	<0.001	–	–	–	–
5	Altitude	0.446*	18	0.004	–	–	–	–
5	Canopy cov.	–0.392*	18	0.004	–	–	–	–
A. concinnatum								
1	–	–	–	–	Bryoph. cov.	–0.697**	25	<0.001
1	–	–	–	–	Max. temp.	–0.839***	25	<0.001
2	Uncov.	0.668**	21	0.001	–	–	–	–
3	Trees	0.372*	20	0.009	Canopy cov.	0.396*	24	0.001
3	Uncov.	–0.529*	20	0.009	Uncov.	–0.517**	24	0.001
3	–	–	–	–	Max. temp.	0.582**	24	0.001
4	–	–	–	–	Canopy cov.	0.407*	25	<0.001
4	–	–	–	-	Underst. den.	0.659***	25	<0.001

Table 15.2. Multiple linear regression models for four separate growth form types of *A. concinnatum* along both transects. Only the significant models are presented (* $P<0.05$, ** $P<0.005$, *** $P<0.0005$)

Life	Cerro Chirripó transect				Cerro de la Muerte transect			
stage	Variable	Beta	df	*F*	Variable	Beta	df	*F*
Standing epiphytic plants								
1	Uncov.	0.676**	17	0.002	Underst. den.	–0.459*	20	0.032
2	Canopy cov.	0.658**	15	<0.001	–	–	–	–
2	Slope dir.	–0.388*	15	<0.001	–	–	–	–
3	Altitude	–0.402**	15	<0.001	Underst. den.	0.424*	20	0.050
3	Canopy cov.	–0.450**	15	<0.001	–	–	–	–
3	Uncov.	–0.719***	15	<0.001	–		–	–
4	Uncov.	–0.491*	17	0.033	Altitude	0.463*	18	<0.001
4	–	–	–	–	Large litter	–0.385***	18	<0.001
4	–	–	–	–	Underst. den.	0.688***	18	<0.001
Lying epiphytic plants								
1	–	–	–	–	Bryoph. cov.	–0.781**	10	0.003
2	–	–	–	–	Underst. den.	–0.632*	10	0.028
3	Underst.den.	–0.656*	9	0.028	Canopy cov.	0.597*	9	0.014
3	–	–	–	–	Bryoph. cov.	0.952**	9	0.014
4	Uncov.	–0.794**	9	0.004	Altitude	0.316*	9	<0.001
4	–	–	–	–	Underst. den.	0.974***	9	<0.001
Standing terrestrial plants								
1	Canopy cov.	0.458*	17	0.048	Bryoph. cov.	–0.512*	18	0.003
1	–	–	–	–	Slope dir.	0.384*	18	0.003
1	–	–	–	–	Underst. den.	0.407*	18	0.003
2	Min. temp.	0.529*	17	0.020	–	–	–	–
3	–	–	–	–	Bryoph. cov.	0.541*	19	0.009
3	–	–	–	–	Slope dir.	–0.441	19	0.009

4	Uncov.	−0.686*	17	0.001	Canopy cov.	0.324	17	<0.001
4	–	–	–	–	Slope dir.	−0.339*	17	<0.001
4	–	–	–	–	Underst. den.	0.827***	17	<0.001
4	–	–	–	–	Min. temp.	−0.548***	17	<0.001
Lying terrestrial plants								
1	–	–	–	–	Canopy cov.	0.247*	7	<0.001
1	–	–	–	–	Uncov.	0.576***	7	<0.001
1	–	–	–	–	Slope dir.	0.677***	7	<0.001
2	Canopy cov.	0.773**	12	0.001	Underst. den.	−0.842**	9	0.001
3	–	–	–	–	Trees	0.298*	7	<0.001
3	–	–	–	–	Uncov.	−0.675***	7	<0.001
3	–	–	–	–	Slope dir.	−0.721***	7	<0.001
4	Canopy cov.	−0.842***	12	<0.002	Underst. den.	0.916***	9	<0.001

15.7 Abundance of Two Species

In all, 5,850 plant individuals of *G. orbignyana*, and 19,536 individuals of *A. concinnatum* were recorded along the two transects. *G. orbignyana* reached peak abundance (20–50 % leaf ground cover) at 2,300–2,500 m elevation along the CC transect, and at 2,350–2,600 m along the CM transect (Fig. 15.3a). Maximum abundances for *G. orbignyana* were much higher at the CC than at the CM site. *A. concinnatum* reached highest abundance values at 2,600–2,800 m along the CC transect, and at 2,350–2,850 m along the CM transect (Fig. 15.3b). Maximum abundances for *A. concinnatum* were lower along the CC transect than along the CM transect.

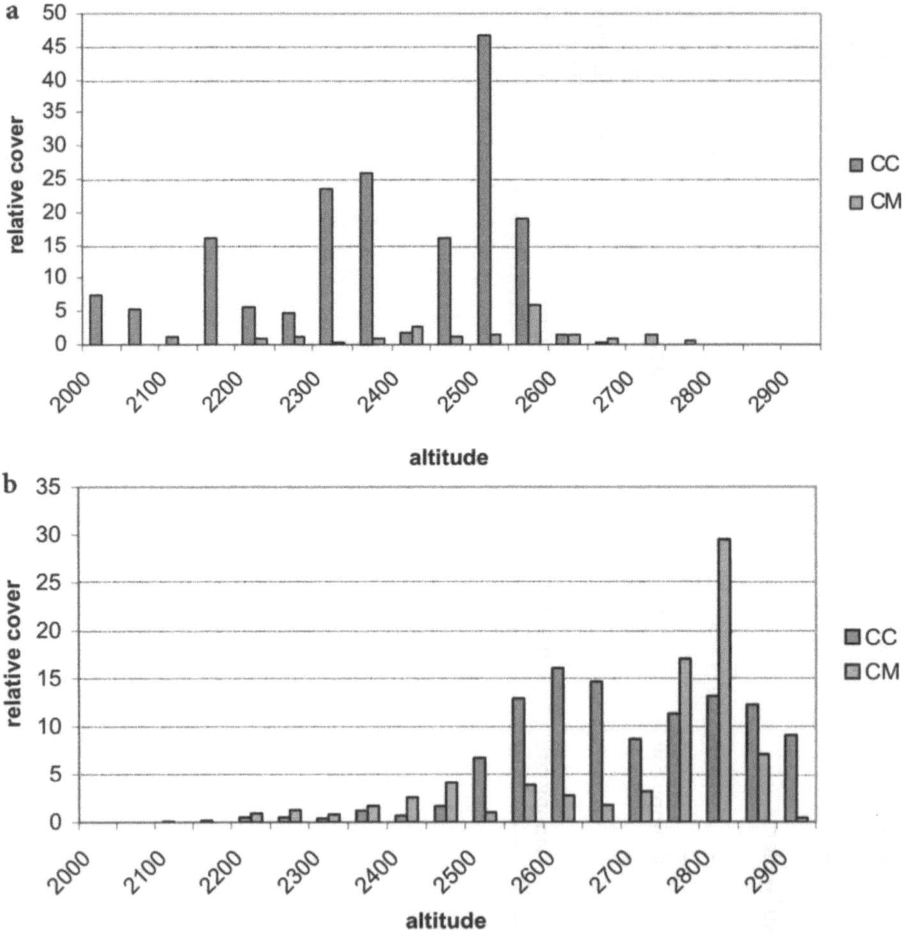

Fig. 15.3a, b. Relative ground leaf cover of **a** *Geonoma orbignyana*, and **b** *Anthurium concinnatum* along both transects. *CC* Cerro Chirripó transect, *CM* Cerro de la Muerte transect

a

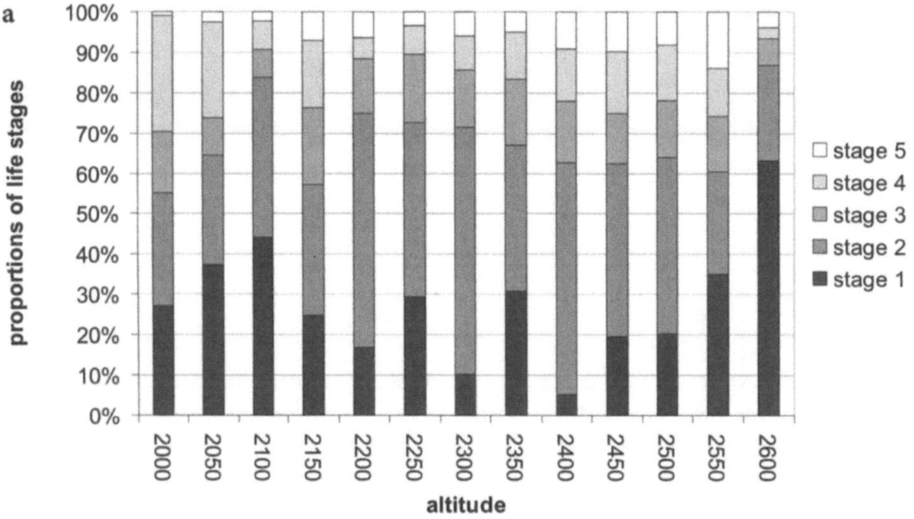

b

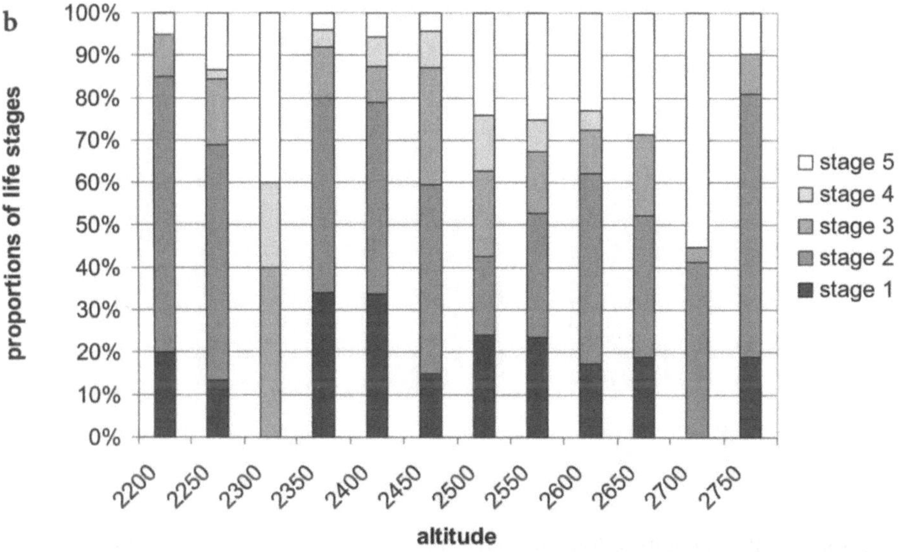

Fig. 15.4a, b. Proportions of five life stages of *Geonoma orbignyana* along the elevational gradient for the **a** CC transect (Cerro Chirripó), and **b** CM transect (Cerro de la Muerte)

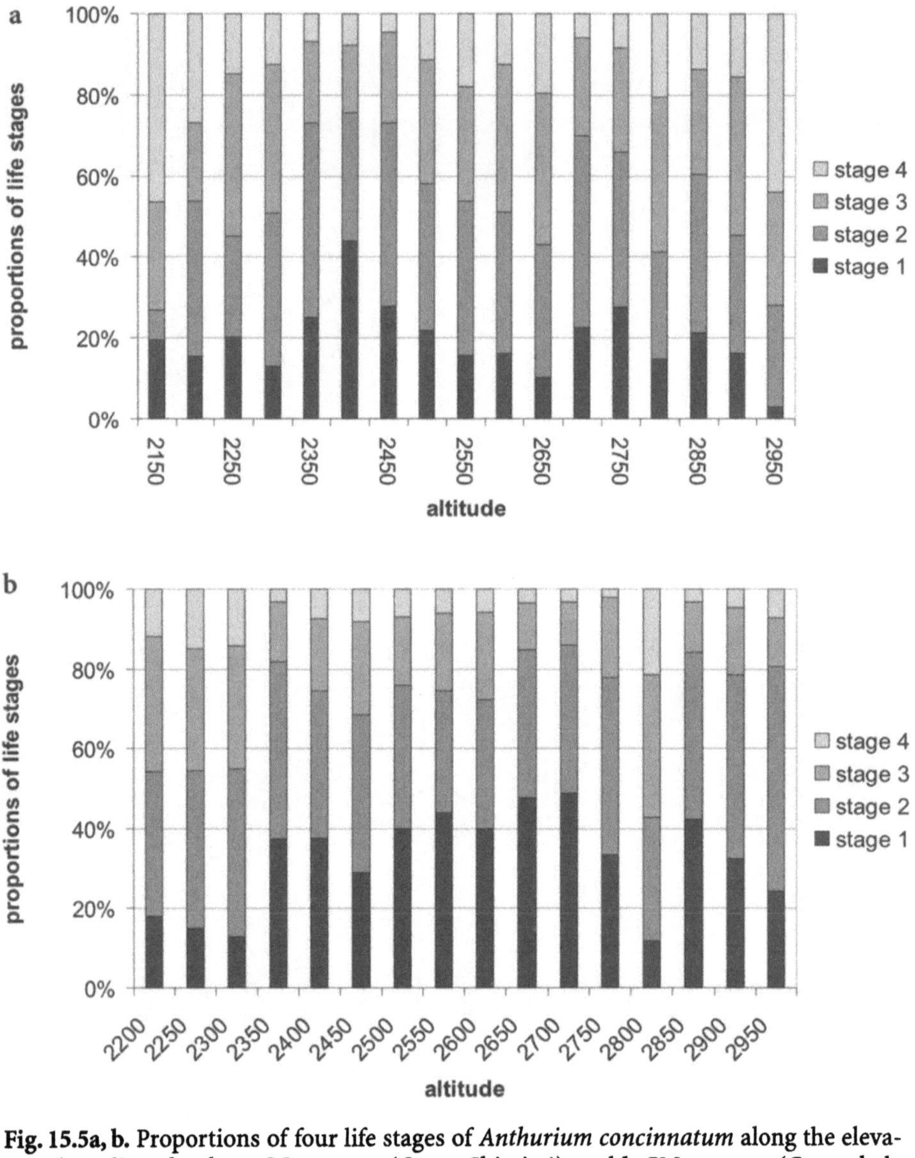

Fig. 15.5a, b. Proportions of four life stages of *Anthurium concinnatum* along the elevational gradient for the **a** CC transect (Cerro Chirripó), and **b** CM transect (Cerro de la Muerte). All plant individuals are included, regardless their specific growth form

The proportions of the different life stages of both *G. orbignyana* and *A. concinnatum* did not show any relationship with altitude (Figs. 15.4 and 15.5). Although the multiple regressions showed some significant correlations for *G. orbignyana*, none of these reflected a response in the structure of the population as a whole.

15.8 Life Stages and Growth Forms of *A. concinnatum*

Multiple regressions on life stages of *A. concinnatum*, irrespective of growth form, showed that along the CC transect, the percentage of uncovered forest floor was positively related to the proportion of life stage 2 (younger plants), whereas it was negatively related to life stage 3 (older plants). Along the CM transect, maximum temperature was negatively correlated to the proportion of *A. concinnatum*'s life stage 1. This phenomenon was reconfirmed by the positive correlation of maximum temperature to life stage 3. This result implies that more small individuals were observed in plots with a lower temperature (i.e., at higher altitude).

Multiple linear regression analyses of data collected along the CC transect for *A. concinnatum*'s standing epiphytic growth form showed that the percentage of uncovered forest floor was positively related to the proportion of life stage 1, whereas it was negatively related to the proportions of life stages 3 and 4. Likewise, along the CM transect the proportion of life stage 1 decreased with increasing undergrowth density, whereas the proportions of life stages 3 and 4 increased. The percentage uncovered forest floor and the density of the undergrowth are both measures of the structure of the understory vegetation. An open understory thus appears to promote the recruitment of standing epiphytes.

Along the CC transect, canopy cover was additionally positively correlated with the proportion of life stage 2, and negatively with the proportion of life stage 3. Here, multiple regressions on the lying epiphytic growth form showed no significant population responses. Along the CM transect, however, a denser understory was negatively correlated to the proportion of life stage 2, and positively correlated to the proportion of life stage 4. Along the same transect, the proportion of life stage 1 was negatively related to the percentage cover of bryophytes, the reverse being the case for the proportion of life stage 3.

For *A. concinnatum*'s standing terrestrial growth form, no significant population response was observed along the CC transect. Along the CM transect, by contrast, the percentage cover of bryophytes was negatively related to the proportion of life stage 1, and positively to the proportion of life stage 3. Along the latter transect, a more northern exposition was associated with a higher proportion of life stage 1, whereas at more southern expositions higher proportions of the old life stages 3 and 4 were observed.

Finally, understory density was negatively correlated with life stage 1, and positively with life stage 4. For *A. concinnatum*'s lying terrestrial growth form along the CC transect, canopy density was positively related to the proportion of life stage 2, and negatively to life stage 4. Along the CM transect, the proportion of life stage 1 increased both at higher percentages of uncovered forest floor and at more northern expositions, the reverse being the case for life stage 3. The density of the understory also seemed to play a role here: it was negatively correlated to the proportion of life stage 2, and positively to the proportion of life stage 4.

15.9 Conclusions

The variation in population structure of *G. orbignyana* was not explained by any of the environmental variables that we measured. We conclude that the population structure of *G. orbignyana* is controlled either by other environmental variables (e.g., thickness of humus layer, M. Kappelle, personal observations), or by factors that are not related to the physical environment (e.g., seed dispersal and seedling recruitment). A large number of palms that occur in the understory of tropical forests have been shown to respond strongly to the structure of the forest. Light availability has the effect that it largely enhances recruitment in those species (Oyama 1990). The genus *Geonoma* seems to be an exception, as it is reported to occur in undisturbed or late secondary forests only (Kappelle et al. 1995a, b). *Geonoma* species are often dominant in the understory (Chazdon 1986; Olesen and Balslev 1990; Listabarth 1993); they are shade-tolerant specialists that are highly adapted to the relative darkness of the understory (Chazdon 1986), and therefore may show at least some level of recruitment at any given light availability. Since our results show that a higher light availability does not seem to enhance recruitment, we may conclude that *G. orbignyana* is an understory specialist, too.

By contrast, *A. concinnatum* has been observed in forests of all successional stages (Kappelle et al. 1995b). Recruitment of this species is thus expected to take place under completely open as well as partially open canopies, e.g., within newly formed gaps after a tree fall. Its young life stages were associated with an open understory, with little competition for space and light, and old life stages with a relatively open canopy and a high bryophyte cover. It has been observed in tropical forests that bryophyte cover is positively correlated with air humidity (Pócs 1982). Assuming that the same is true for the forests we studied, we suggest that the higher proportion of large individuals in bryophyte-rich patches is explained by higher moisture availability. This implies that drought stress prevents plants from becoming large in patches with fewer bryophytes. Such drought stress was also reported for the epiphytic species of *Anthurium bredemeyeri* (Rada and Jaimez 1992). Since bryophyte ground cover was negatively correlated with maximum temperature along the CM transect, the drought sensitivity of adults might also explain the higher proportions of large individuals observed at lower maximum temperatures, as seen in the general data analysis for this transect.

The low proportion of large standing epiphytes under a denser canopy along the CC transect can be explained by a reduction in growth of adult plants under low light availability. This finding is at odds with our initial assumption of equal growth rates throughout transects. Nevertheless, observations on the distances between leaf scars on the stem suggest that in most plots size was indeed closely related to age (unpublished data).

Finally, competition for light and moisture availability seems important in determining the population structure of *A. concinnatum*. Higher levels of recruitment of this species appear to occur in a more open understory. Within the context of tropical montane oak forest dynamics, we may conclude that *A. concinnatum* exhibits a pioneer strategy sensu Alvarez-Buylla and Martínez-Ramos (1992), being able to colonize relatively open patches in the understory, and requiring a relatively open canopy for optimal growth.

Acknowledgements We thank IBED staff for support during the research design and data analysis phases. L. Köhler and I. Holz, University of Göttingen, and INBio staff in Costa Rica are gratefully acknowledged for assistance during fieldwork. We are very grateful to Francisco Quesada (INBio) for preparing the two drawings. The Martínez family is thanked for their generous hospitality. This project was funded by the Alberta Mennega Foundation, FONA at IBN-DLO (now WURC), Shell Netherlands BV, the Pro-Natura Foundation, and the University of Amsterdam.

References

Alvarez-Buylla ER, Martínez-Ramos M (1992) Demography and allometry of *Cecropia obtusifolia*, a neotropical pioneer tree – an evaluation of the climax pioneer paradigm for tropical rain forests. J Ecol 80:275–290

Arriaga L, Franco M, Sarukhán J (1988) Identification of natural groups of trees in uneven-aged forests using multivariate methods. J Ecol 76:1092–1100

Benzing DH (1990) Vascular epiphytes: general ecology and related biota. Cambridge Univ Press, Cambridge, UK

Bown D (2000) Aroids: plants of the Arum family, 2nd edn. Timber Press, Portland, OR

Chazdon RL (1986) Physiological and morphological basis of shade tolerance in rain forest understory palms. Principes 30:92–99

Cleef AM, van der Hammen T, Jaramillo R (1984) La vegetación de las selvas del transecto Buritaca. In: Van der Hammen T, Ruiz PM (eds) La Sierra Nevada de Santa Marta (Colombia), transecto Buritaca – La Cumbre. Cramer, Vaduz, Studies of Tropical Andean Ecosystems, vol 2, pp 276–406

Croat TB, Baker R (1979) The genus *Anthurium* (Araceae) in Costa Rica. Brenesia 16 Suppl 1:1–174

Dortort F (1980) In the forests of Costa Rica. Aroideana 3:39–48

Hegland SJ, van Leeuwen M, Oostermeijer JGB (2001) Population structure of the rare perennial *Salvia pratensis* in relation to vegetation and management of Dutch river valley grasslands. J Appl Ecol 38:39–48

Henderson A, Galeano G, Bernal R (1995) Field guide to the palms of the Americas. Princeton Univ Press, Princeton, NJ

Hicks DJ, Mauchamp A (2000) Population structure and growth patterns of *Opuntia echios* var. *gigantea* along an elevational gradient in the Galapagos Islands. Biotropica 32:235–243

Holm S (1979) A simple sequential rejective multiple test procedure. Scand J Stat 6:65–70
Kappelle M (1996) Los bosques de roble (*Quercus*) de la Cordillera de Talamanca, Costa Rica: biodiversidad, ecología, conservación y desarrollo. Instituto Nacional de Biodiversidad (INBio), Santo Domingo de Heredia, Costa Rica
Kappelle M, Zamora N (1995) Changes in woody species richness along an altitudinal gradient in Talamancan montane *Quercus* forests, Costa Rica. In: Churchill SP, Balslev H, Forero E, Luteyn JL (eds) Biodiversity and conservation of neotropical montane forests. New York Botanical Garden Press, Bronx, NY, pp 135–148
Kappelle M, van Uffelen JG, Cleef AM (1995a) Altitudinal zonation of montane *Quercus* forest along two transects in Chirripó National Park, Costa Rica. Vegetatio 119:119–153
Kappelle M, Kennis PAF, de Vries RAJ (1995b) Changes in diversity along a successional gradient in a Costa Rican upper montane *Quercus* forest. Biodiv Conserv 4:10–34
Kappelle M, Geuze T, Leal M, Cleef AM (1996) Successional age and forest structure in a Costa Rican upper montane *Quercus* forest. J Trop Ecol 12:681–698
Lieberman D, Lieberman M, Peralta R, Hartshorn GS (1996) Tropical forest structure and composition on a large-scale altitudinal gradient in Costa Rica. J Ecol 84:137–152
Listabarth C (1993) Pollination in *Geonoma macrostachys* and three congeners, *G. acaulis*, *G. gracilis*, and *G. interrupta.* Bot Acta 106:496–506
Martínez-Ramos M, Alvarez-Buylla E, Sarukhán J, Piñero D (1988) Treefall age determination and gap dynamics in a tropical forest. J Ecol 76:700–716
Olesen JM, Balslev H (1990) Flower biology and pollinators of the Amazonian monoecious palm *Geonoma macrostachys*: a case of Bakerian mimicry. Principes 34:181–190
Olmsted I, Alvarez-Buylla ER (1995) Sustainable harvesting of tropical trees: demography and matrix modes of two palm species in Mexico. Ecol Appl 5:484–500
Oostermeijer JGB, Van't Veer R, Den Nijs JCM (1994) Population structure of the rare, long-lived perennial *Gentiana pneumonanthe* in relation to vegetation management in The Netherlands. J Appl Ecol 31:428–438
Oyama K (1990) Variation in growth and reproduction in the neotropical dioecious palm *Chamaedorea tepejilote.* J Ecol 78:648–663
Pócs T (1982) Tropical forest bryophytes. In: Smith AJE (ed) Bryophyte ecology. Cambridge Univ Press, Cambridge, UK, pp 59–104
Rada F, Jaimez R (1992) Comparative ecophysiology and anatomy of terrestrial and epiphytic *Anthurium bredemeyeri* Schott in a tropical Andean cloud forest. J Exp Bot 43:723–727
SPSS Inc (1990) SPSS for the MacIntosh: operations guide. SPSS Inc, Chicago, IL
Wassen MJ, Barendrecht A, Palczynski A, de Smidt JT, de Mars H (1990) The relationship between fen vegetation gradients, ground water flow and flooding in an undrained valley mire at Biebrza, Poland. J Ecol 78:1106–1122
Wesselingh RA, Witteveldt M, Morissette J, Den Nijs HCM (1999) Reproductive ecology of understory species in a tropical montane forest in Costa Rica. Biotropica 31(4):637–645
Witkowski ETF, Liston RJ (1997) Population structure, habitat profile and regeneration of *Haworthia koelmanioum*, a vulnerable dwarf succulent, endemic to Mpumalanga, South Africa. S Afr J Bot 63:373–370
Wolf JHD (1993) Epiphyte communities of tropical montane rain forests in the northern Andes. I. Lower montane communities. Phytocoenologia 22:1–52
Zadroga F (1981) The hydrological importance of a montane cloud forest area of Costa Rica, In: Lal R, Russell EW (eds) Tropical agricultural hydrology. Wiley, New York, NY, pp 59–73
Zona S, Henderson A (1989) A review of animal-mediated seed dispersal of palms. Selbyana 11:6–21

Part V
Ecosystem Disturbance and Regeneration

16 Secondary Succession in Montane Pine-Oak Forests of Chiapas, Mexico

M. González-Espinosa, N. Ramírez-Marcial, and L. Galindo-Jaimes

16.1 Introduction

Montane forests of southern Mexico and Guatemala are highly diverse formations (Steyermark 1950; Miranda 1952; Breedlove 1981; Chaps. 8 and 9). In addition to occasional natural perturbations (landslides, windstorms, fire), these forests have been subjected for centuries to a wide range of human disturbances derived from slash-and-burn *milpa* agriculture - mixed shifting cultivation of maize, squash and beans. Traditional Maya land-use patterns drive secondary succession with impacts on forest composition, structure, and regeneration due to low-intensity, long-duration disturbance regimes, effected through practices such as sparse logging, extraction of saplings and lopping of hardwoods for fuelwood, and sporadic cattle grazing (Ramírez-Marcial et al. 2001; Barrón-Sevilla 2002). Similar disturbance patterns have been observed in other densely populated highland regions where traditional subsistence practices prevail (Hong et al. 1995; Vetaas 1997; Kappelle and Brown 2001; Holder 2004; Chap. 30).

The forest cover of the densely populated Central Highlands and Northern Mountains regions of Chiapas has been - and continues to be - severely reduced (Wagner 1963; Ochoa-Gaona and González-Espinosa 2000; Cayuela et al. 2005). Currently, social variables and environmental factors interact to determine fire incidence in several forest types in Chiapas (Román-Cuesta et al. 2003). Relatively large patches of forest cover are detected by satellite imagery (Palacio-Prieto et al. 2000), but degradation in forest structure and impoverished floristic composition are evidenced with field inventories (González-Espinosa et al. 1995; Ramírez-Marcial et al. 2001; Galindo-Jaimes et al. 2002). Vegetation studies in southern Mexico have focused on defining discrete physiognomical forest types and their relationships with the environment (Miranda 1952; Breedlove 1981). More recent work in these forests pin-

Ecological Studies, Vol. 185
M. Kappelle (Ed.) Ecology and Conservation of Neotropical Montane Oak Forests
© Springer-Verlag Berlin Heidelberg 2006

point to their extensive floristic overlap along environmental gradients (Ramírez-Marcial et al. 2001; Galindo-Jaimes et al. 2002).

We hypothesize that the floristic composition and forest structure of old-growth stands and their preceding successional stages may help to show both their connections and differences within a complex forest–agriculture landscape. We rely on primary and secondary information to propose major successional trends of montane pine-oak forests of Chiapas. Published information is still too incomplete to attain a detailed view of succession in these systems; yet, we have pooled available data to propose an account that is more inclusive than those reported to date. We at first offer an account of the major floristic, structural, and environmental attributes of a minimum set of clearly defined seral stages. When possible, we then relate species responses to environmental conditions to explain or propose their successional role. In addition to gaining understanding on natural relationships among landscape units, this analysis is important also because secondary succession is the major process underlying ecological restoration and other practices aiming at sustainable use of montane forests.

16.2 Sources of Information

Montane pine-oak forests (MPOF) in Chiapas are plant formations above 1,500 m elevation with dominant species of *Pinus* and *Quercus* in the canopy. However, species of these two genera appear also in forests at lower elevations. Under MPOF, we consider associations included by Breedlove (1981) as montane rain forest (MRF, higher-elevation associations only), evergreen cloud forest (ECF), pine-oak-*Liquidambar* forest (POLF), and pine-oak forest (POF). Among vegetation types identified in Chiapas by Miranda and Hernández-X (1963), we include *Selva Mediana o Baja Perennifolia* (ECF and MRF at high elevations), *Bosque Caducifolio* (POLF), *Bosque de Pino-Encino* (pine-oak forest, POF), *Encinares* (oak forest, OF), and *Pinares* (pine forest, PF).

We use data on chronosequences after agricultural abandonment, on forest structure and composition of human-disturbed forests, and on microclimates in cleared areas and different types of forest. We also use a database of 16,777 records of herbarium vouchers of tree taxa collected in Chiapas between 1864 and 1999, deposited at CAS, DS, MEXU, ENCB, CHAPA, XAL, and ECOSUR. This database includes 85–90 % of all vouchers available for Chiapas worldwide (González-Espinosa et al. 2004).

16.3 Pines and Oaks in the Forests of Chiapas

Breedlove (1981) does not mention *Pinus* within mid-elevation non-seasonal formations such as MRF, yet *Pinus* and *Quercus* are mentioned within the ECF. Among seasonal formations, pines and oaks are included only in POLF and POF. However, vouchers indicate that a number of *Quercus* (17–24) and *Pinus* (5–11) species have been collected from 500 m elevation upward, mostly above 800 m. This is not surprising, as *Quercus* includes at least 26 species in Chiapas; together with *Acacia* (26 spp.), *Ficus* (29 spp.), *Inga* (25 spp.), and *Lonchocarpus* (26 spp.), *Quercus* is one of the most diverse tree genera in Chiapas (Table 16.1).

In Chiapas, MPOF occurs under highly variable climatic conditions above 1,500 m elevation: mean annual temperature may be 14–25 °C, mean annual rainfall 900–3,700 mm, the dry season may last 0–6 months, and MPOF occupies a variety of landforms and soil types (González-Espinosa et al. 2005). Stands of MPOF usually include 1–3 species of pines, and 2–4 species of oaks; mean species richness of additional associated trees (±SE) is 7±0.7 at seasonal and variously disturbed PF, POF, and OF sites (N=36; Galindo-Jaimes et al. 2002), and 13±0.8 at OF, POLF, MRF, and ECF sites (N=81; Ramírez-Marcial et al. 2001). Most pines and oaks can be considered as early-successional species, as they require relatively large forest gaps or clearings for recruitment

Table 16.1. Species of *Quercus*, *Pinus*, other broad-leaved tree species, and estimated tree species richness in the floristic pool associated to montane pine-oak forests within 500-m-wide elevational belts in Chiapas. See text for details on the database of herbarium vouchers. A list of species and number of vouchers is available on request

	500–999 m	1,000–1,499 m	1,500–1,999 m	2,000–2,499 m	2,500–3,000 m
Quercus species	17	24	23	20	8
Pinus species	5	6	10	11	4
Quercus and *Pinus* species	22	30	33	31	12
Trees other than *Quercus* and *Pinus*					
Early-successional	103	163	158	88	33
Mid-successional	117	154	149	132	37
Late-successional	24	36	71	68	31
Total	244	353	378	288	101
Estimated tree species richness in the floristic pool associated to montane pine-oak forests in Chiapas	266	383	411	319	113

(Quintana-Ascencio et al. 2004; Chap. 14); some can regenerate at forest edges (López-Barrera 2003; Chap. 13) and under open-crowned adults, but no species of these two genera recruits mostly under closed canopy.

16.4 Post-agricultural Succession in Montane Habitats of Chiapas

No long-term monitoring of ecological succession has been conducted in montane forests of Chiapas. Available studies on succession use chronosequences, particularly after agricultural field abandonment. We draw upon data obtained with different methods at many sites (about 300 stands over a period of 10 years). The dataset comes mostly from the highlands of Chiapas, but major trends appear similar to those reported or suggested for neighboring montane regions (Steyermark 1950; Córdova and del Castillo 2001; Chaps. 8 and 9).

16.4.1 Old-Field Fallow (FF)

The old-field fallow (FF) stage is established after *milpa* abandonment, and lasts 3–4 years (Fig. 16.1). Composition and species richness depends on prior use, soil type, and regional climate (Montes-Avelar 2001). Species richness may be as low as 30 taxa, but can reach 85 taxa (mean±SE is 50±3.6 species; N=22 sites); a floristic pool of 290 taxa has been recorded at FF sites. Annual and perennial herbs are dominant but only a few species may be abundant (mostly grasses, composites, and *Hedyotis serpyllacea*, *Lepechinia schiedeana*, and *Oxalis corniculata*); many of these species may already be present in agricultural fields as weeds. Seedlings and juveniles of shrubs and trees (e.g., *Baccharis vaccinioides*, *Fuchsia* spp., *Rubus* spp., *Solanum* spp.), in addition to *Pinus* spp. and *Quercus* spp. may be present (Table 16.2). Bare ground may be ca. 40 %.

16.4.2 Grassland (GRA)

The grassland (GRA) stage follows the FF stage, provided that grazing by sheep and cattle is continuous (Fig. 16.1). Its duration may be only a few years under light grazing, but it may persist for decades under frequent and not too intensive use by domestic animals. Species richness is lower (42±1.7 species, N=24 sites) and less variable than for FF (30–57). A floristic pool of 195 species has been recorded at GRA sites (Montes-Avelar 2000), mostly perennial grasses and composites. Grazing and trampling damage seedlings of

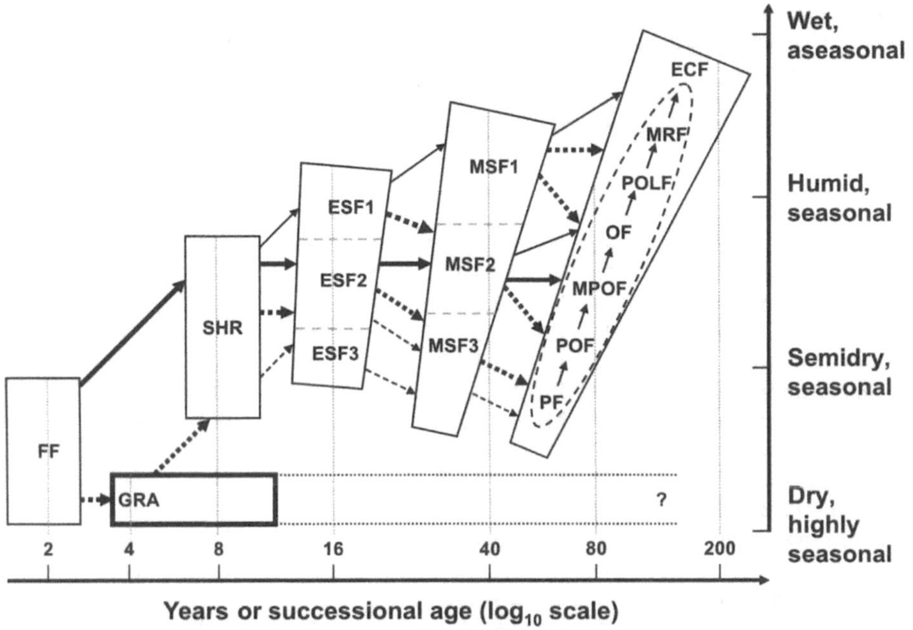

Fig. 16.1 Relationships among post-agriculture seral stages as a function of time ($\log_{10}$ scale), annual rainfall regime (900–3,700 mm), and seasonality (rainy season of 6–12 months). *FF* Old-field fallow, *GRA* grassland (*pointed horizontal lines* represent unknown duration), *SHR* shrubland, *ESF* early-successional forest (*1–3* indicate increasing dryness), *MSF* mid-successional forest (*1–3* as above), *PF* pine forest, *POF* pine-oak forest, *MPOF* montane pine-oak forest (*dashed ellipse in quadrangle at right*), *OF* oak forest, *POLF* pine-oak-*Liquidambar* forest, *MRF* montane rain forest, *ECF* evergreen cloud forest. *Thick solid arrows* represent the natural sequence leading to MPOF, *punctuated thick arrows* deviations due to low-intensity and long-duration human disturbance, and *thin arrows* non-anthropogenic alternative pathways at mesic (*solid*) or dry sites (*dashed*). Arrows connecting old-growth forest types (*quadrangle at right*) represent potential successional progression if human disturbance is removed or reduced. Size of *solid line quadrangles* is proportional to known (or estimated) richness of vascular species pools (epiphytes excluded). *Dashed lines inside the quadrangles* indicate overlap in floristic composition among component cells. Location of acronyms represents their relative position in time, but the length of the arrows connecting them cannot represent the true duration of the transitions

shrubs and trees, and may arrest successional change (Ramírez-Marcial et al. 1996). Exotic species such as *Bromus carinatus*, *Pennisetum clandestinum*, *Poa annua*, and *Prunella vulgaris* may be common. The soil is almost completely covered (>97 %).

An 8-year experiment showed that *Pinus* spp., *Abies guatemalensis*, and broad-leaved shade-tolerant species such as *Rapanea juergensenii*, *Rhamnus sharpii* and *Ternstroemia lineata* grew best in GRA, but survival was high only for pines, and it was quite low for the other species (Quintana-Ascencio et al.

Table 16.2. Successional status (*SS*) and seral stages of old-growth montane forest types where tree species are associated to *Pinus* or *Quercus* within 500-m-wide elevational belts in Chiapas

Species	*N*[a]	SS	RQI	Altitudinal belt (m) 500–999	1,000–1,499	1,500–1,999	2,000–2,499	2,500–3,000
Acacia angustissima	105	E	1,2,4,5,6,7	*	*	(+)	(+)	-
Acacia pennatula	58	E	3,4,5,6	*	*	(+)	*	-
Acer negundo	24	L	1,2,4,5,7	-	-	(+)	(+)	*
Ageratina nubigena	23	L	2,3,4,5,6,7	-	*	+	(+)	*
Alnus acuminata	52	E	1,2,3,4,7	*	*	+	(+)	*
Arbutus xalapensis	60	E	1,2,3,4,5,6	-	*	(+)	(+)	+
Brunellia mexicana	29	M	4,5,6,7	*	*	(+)	(+)	*
Buddleja nitida	50	M	1,2,4,6,7	-	-	*	(+)	*
Buddleja skutchii	67	E	2,4,5,6,7	-	-	*	(+)	*
Carpinus caroliniana	29	L	4,5,6,7	-	*	(+)	*	*
Clethra suaveolens	38	M	1,3,4,5,7	*	*	(+)	(+)	-
Cleyera theaeoides	88	M	2,3,4,5,6,7	-	*	(+)	(+)	+
Comarostaphylis discolor	39	E	2,4,5,6,7	-	-	*	(+)	*
Cornus disciflora	56	M	2,4,5,6,7	-	*	(+)	(+)	+
Cornus excelsa	43	E	1,2,3,4,5,7	-	-	(+)	(+)	+
Crataegus pubescens	61	E	1,2,3,4,5,7	-	*	(+)	(+)	*
Cupressus lusitanica	32	E	3,4,5,6,7	-	*	*	(+)	*
Dendropanax arboreus	88	M	4,5,6,7	*	*	(+)	*	-
Drimys granadensis	63	L	4,5,6,7	-	*	*	(+)	+
Fraxinus uhdei	24	M	2,4,5,6,7	-	*	(+)	*	-
Fuchsia paniculata	99	M	4,5,7	-	*	(+)	(+)	*
Garrya laurifolia	58	M	1,2,4,5,6,7	-	*	(+)	(+)	*
Hauya elegans	69	M	4,5,6,7	*	*	(+)	-	-
Hedyosmum mexicanum	38	L	4,5,6,7	*	*	(+)	*	-
Ilex vomitoria	44	M	4,5,7	-	*	+	(+)	*
Juniperus gamboana	27	E	1,2,4	-	*	(+)	*	-
Liquidambar styraciflua	42	M	2,4,5,6,7	-	+	(+)	(+)	-
Litsea glaucescens	66	M	1,2,3,4,5,6,7	*	*	(+)	(+)	*
Litsea neesiana	42	M	1,2,3,4,5,6,7	-	-	(+)	(+)	*
Magnolia sharpii	28	L	4,5,6,7	-	-	+	(+)	-
Miconia glaberrima	164	M	1,2,4,5,6,7	*	*	(+)	(+)	*
Myrica cerifera	153	E	1,2,3,4,5,6,7	*	*	(+)	(+)	*
Nyssa sylvatica	35	E	2,4,5	-	+	(+)	*	-
Oreopanax xalapensis	83	M	2,4,5,6,7	+	+	+	(+)	+
Ostrya virginiana	58	E	4,5,6,7	*	+	(+)	(+)	*
Persea americana	58	M	4,5,6,7	*	+	+	(+)	+
Podocarpus matudai	27	L	4,5,6,7	-	*	(+)	(+)	-
Prunus serotina	42	E	1,2,3,4,5,7	-	*	(+)	(+)	*
Psychotria chiapensis	51	E	2,6,7	*	*	(+)	-	-
Quetzalia (Microtropis) contracta	21	L	4,7	-	-	-	(+)	+
Rapanea juergensenii	58	M	1,2,3,4,5,6,7	-	*	(+)	(+)	*
Rapanea myricoides	69	M	1,2,4,5,6,7	*	*	(+)	*	-
Rhamnus capraeifolia	35	E	1,2,4,5,6,7	-	*	(+)	(+)	*

Table 16.2. (*Continued*)

Species	N^a	SS	RQI	Altitudinal belt (m) 500–999	1,000–1,499	1,500–1,999	2,000–2,499	2,500–3,000
Rhamnus sharpii	42	L	1,2,4,5,6,7	-	*	(+)	(+)	*
Styrax argenteus	46	L	2,3,4,5,7	*	*	(+)	(+)	+
Symplocos limoncillo	35	L	2,4,5,6,7	*	*	(+)	(+)	+
Synardisia venosa	22	L	4,6,7	-	-	(+)	*	-
Ternstroemia lineata	44	M	2,3,4,5,7	-	-	*	(+)	+
Ternstroemia oocarpa	83	M	2,4,5,6,7	-	*	(+)	*	-
Verbesina perymenioides	43	E	1,2,4,5,6,7	-	*	*	(+)	*
Viburnum elatum	28	E	1,2,4	-	*	(+)	(+)	
Viburnum jucundum	82	M	2,3,4,5,6,7	-	*	(+)	(+)	+
Weinmannia pinnata	24	L	1,2,4,5,6,7	-	*	(+)	(+)	*

[a] *N*, number of herbarium vouchers (only species with *N*>20 vouchers are included; a complete list is available on request); E, early-successional (regeneration occurs in clearings and forest gaps); M, mid-successional (regeneration in forest edges and under open canopies); L, late-successional (regeneration under closed canopies). RQI refers to habitats where the taxon has actually been recorded in quantitative inventories: 1, old-field, fallow grassland or shrubland; 2, early- or mid-successional forest; 3, pine forest; 4, pine-oak or oak forest; 5, pine-oak-*Liquidambar* forest; 6, montane rain forest; 7, evergreen cloud forest. *, species collected within the indicated altitudinal belt, but without explicit association to either a *Pinus* or *Quercus* specimen; +, genus mentioned in the voucher label as associated to either a *Pinus* or *Quercus* specimen (*N*=1,669 records); (+), quantitative abundance data available within the altitudinal belt

2004). Temperatures in GRA are more extreme than in nearby forests: values may drop below 0 °C in winter and rise up to >40 °C in spring, which is 10–15 °C higher than top values under canopies; daily and weekly temperature oscillations in GRA may be up to 100 % larger than in forests. Minimum relative humidity in GRA frequently drops to 20–40 %, whereas in pine stands it rarely reaches 20 % (Romero-Nájera 2000; Ramírez-Marcial 2003).

16.4.3 Shrubland (SHR)

Whenever grazing is light or absent, GRA is replaced by a shrubland (SHR) community including herbs and a sparse, tall woody layer. Richness varies in the range 51–82 (76±3.3 species, *N*=17 sites); a total pool of 321 species has been recorded at SHR sites. Many tree species may appear in addition to *Pinus* and *Quercus* (Table 16.2), the latter frequently growing from resprouting stumps (N. Ramírez-Marcial, personal communication). The soil is completely covered by prostrate herbs and litter.

Survival of transplanted juveniles of *Pinus ayacahuite*, *P. pseudostrobus*, and *Quercus rugosa*, as well as seedlings of *Q. crassifolia* was higher in grazed SHR sites than in grazed GRA, and similar to survival in GRA enclosures. Natural recruitment from seed was recorded for nine tree species, and 86 % of 139 natural and transplanted seedlings were located less than 1 m from the shrub *Baccharis vaccinioides*. This relationship suggests a mechanism of oak facilitation by *Baccharis*, which hinders browsing and grazing, and provides some shade (SHR cover is 27±2.3 %) that reduces soil dryness (20–26 % of available water at GRA sites vs. 65–76 % in SRH; Ramírez-Marcial et al. 1996; Chap. 14).

16.4.4 Early-Successional Forest (ESF)

A floristic pool of up to 328 species has been recorded at early-successional forest (ESF) sites (Fig. 16.1). The ESF stage may last 20–25 years, and has three strata: (1) a herbaceous layer with mostly perennial herbs, vines and lianas, ferns, and seedlings of shrubs and trees (47±2.3 species; *N*=25 sites); (2) a shrubby layer with large ferns, adult shrubs, and saplings of tree species (up to 50 species, mostly early- and mid-successional species; Table 16.2); (3) a dense and diverse tree layer (7,300–8,500 stems ha^{-1}; González-Espinosa et al. 1991) with understory and canopy trees up to 5–10 m high (27±1.2 species; *N*=25 sites). Young (diameter at breast height DBH<15 cm) *Quercus* and *Pinus* usually are the only components of a closed canopy. Cover at forest floor level may be 83±1.4 % (Quintana-Ascencio et al. 1992); light levels at the forest floor may be as low as 19 $\mu mol\ m^{-2}\ s^{-1}$ (Quintana-Ascencio et al. 2004).

Dead trees may be abundant at ESF sites (360 stems ha^{-1}). Shrub or tree species with a higher than expected contribution to the total number of dead stems in the stand, interpreted as being removed from ESF and replaced during succession, include *B. vaccinioides*, *Prunus serotina*, *Viburnum* spp., and species of *Pinus* (χ^2=253.3, *P*<0.001). A reverse trend has been found for *Q. laurina* and *Q. rugosa* (χ^2=13.2, *P*<0.001; González-Espinosa et al. 1991), and in other less abundant species (*Rhamnus sharpii*, *Cleyera theaeoides*, *Magnolia sharpii*, *Miconia glaberrima*, *Oreopanax xalapensis*, *Rapanea juergensenii*, and *Ternstroemia lineata*).

16.4.5 Mid-Successional Forest (MSF)

The mid-successional forest (MSF) community shares >50 % of its floristic composition with both ESF and old-growth stands, and has the final four-layered forest structure. Understory (15–20 species; 10–12 m high, rarely >15 m) and canopy (5–8 species; >20 m high, rarely >25 m) tree species can be distinguished. Most canopy trees are pines and oaks, but occasional *Arbutus xalapensis*, *Clethra* spp., *Chiranthodendron pentadactylon*, and *Persea americana* may be present. Results of a self-thinning process between ESF and MSF are evident (60–65 % stem density decrease). The forest floor may be more heterogeneous and receive more direct light than is the case for ESF (8–10 % more cover, and 58 μmol m^{-2} s^{-1}, respectively; Quintana-Ascencio et al. 2004), which may correspond to only 8–10 % of full sunlight in open areas. Soil moisture and temperatures in MSF are similar to those in old-growth stands (Camacho-Cruz et al. 2000).

16.4.6 Old-Growth Montane Pine-Oak Forest Associations

Several montane formations develop in Chiapas within a range of elevations (1,500–3,100 m), climates, seasonality patterns, and soil conditions (Fig. 16.1). Their structure may be different but they share a large core floristic set. Species richness in old-growth stands may be the highest along the successional series (cf. floristic pool of up to 200 taxa), with at least 35–40 understory and canopy tree species (González-Espinosa et al. 1995).

Basal area may be high in old-growth pinelands (PF; 70–85 m^2 ha^{-1}). Values for old-growth POF, MRF and ECF are smaller (60–78 m^2 ha^{-1}), and for POLF even more so (47–55 m^2 ha^{-1}). The lowest values have been recorded in secondary pinelands on steep slopes (24–30 m^2 ha^{-1}; Ramírez-Marcial et al. 2001; Galindo-Jaimes et al. 2002). Maximum light levels in OF and ECF stands may be as low as 10–30 μmol m^{-2} s^{-1} (Quintana-Ascencio et al. 2004). Of all successional stages, old-growth POF, OF, and ECF show lowest temperatures, highest relative humidity, and smallest yearly oscillations; freezing temperatures at forest floor level do not occur under broad-leaved canopies (Romero-Nájera 2000; Ramírez-Marcial 2003). Compared to PF, air temperatures may be 10–15 °C lower in POF or OF stands, and frost may be common under their sparser canopies. Litter accumulation in POF and ECF may be 10–20 % higher (in terms of weight or volume) than in MSF stands (Camacho-Cruz et al. 2000).

16.5 Relationships Among Seral Stages

We use floristic replacement as the major variable accounting for successional change, and can only suggest trends in other structural and functional attributes deserving further study. Plant species assemblages of old-growth montane forests in Chiapas overlap extensively. This is more evident when we consider seral stages, and the long-term intermingling of different forest cover patches and open areas within a complex human-influenced landscape.

We suggest a regional pool of widespread secondary species and few exclusive tree species in the old-growth forests. Table 16.2 shows that all species recorded in PF stands have also been documented in POF, OF, POLF, or ECF, and most species recorded in POF or OF have also been found in MRF or ECF. Trees in open areas or in MSF stands may also be found in gaps within old-growth stands. Given the floristic richness of Chiapas (1,400–1,500 tree taxa; González-Espinosa et al. 2004), it is worth noting that restoration efforts may currently be focusing only on a relatively small set of species (Ramírez-Marcial et al. 2005).

Structure and composition of MSF stands are altered by long-term sparse logging of pre-reproductive *Quercus* spp. (or lopping of resprouted stumps). Whereas oaks are the preferred trees for firewood and are not used or sold as lumber, pines are not highly valued as fuel - except for occasional harvesting of resin-rich pitch bark - and are not logged until they attain appropriate sizes for lumber (20–25 cm DBH). As a result of the different morphologies, tolerances, life histories, dispersal and colonizing abilities, and use patterns imposed on pines and oaks, MSF stands may not progress to old-growth POF or OF at mesic sites, but rather toward species-poor pinelands. A truncated distribution of diametrical classes (particularly for *Pinus* spp., rarely with DBH>70 cm, 60–70 years old) suggests that at many sites such pinelands are a successional stage maintained by human disturbance (González-Espinosa et al. 1995; Galindo-Jaimes et al. 2002).

Increasing pine abundance has been pinpointed within the context of species invasions and naturalization where pines are non-native (Richardson and Rejmánek 2004). Yet, increase of native pines beyond their possibly originally poor sites due to human disturbance has received less attention; indeed, it is interesting to note the absence in the wild of invasive trees in Chiapas. In addition to light availability and shade tolerance, the observed pine rise seems to depend on the availability of water and soil nutrients (Galindo-Jaimes et al. 2002). Soils in pinelands may be more compacted, less acid, and with lower cation exchange capacity (CEC) as well as lower nitrogen (N) and organic matter contents than soils in nearby old-growth POF and OF stands. In turn, this may result in lower site productivity, and contribute to accelerate rotation cycles in an increasingly disturbed forest–agriculture landscape (García-Barrios and González-Espinosa 2004).

16.6 Conclusions

Only chronosequence studies are available for the region, yet floristic replacement delineates major successional trends toward a closely related set of old-growth forest types within a range of altitudinal, climatic and edaphic conditions. Species richness, basal area, and accumulated litter increase with successional age, yet canopy cover peaks at both ESF and old-growth stages. Oscillations of air temperature and relative humidity decrease as stands become older. Traditional land use involving long-term, sparse, and differential use of pines and broad-leaved trees may drive succession toward species-poor pinelands at mesic sites where *Quercus*-dominated forests should prevail. Land use associated to a rapidly increasing human population strongly affects secondary succession processes; our understanding of these processes is of utmost importance for their sustainable use.

Acknowledgements We appreciate comments from A. Camacho, L. Cayuela, D. Golicher, L. García-Barrios, P.F. Quintana-Ascencio, and J.M. Rey-Benayas. This research was supported by CONACYT (2140-N9303), CONABIO (F-019), US-EPA (CR822200), NIH (ICBG-Maya project), the European Commission (INCO Programme, Framework 5, BIOCORES project, ICA4-CT-2001-10095), and ECOSUR.

References

Barrón-Sevilla JA (2002) Efecto del disturbio antropogénico sobre la estructura y riqueza arbórea en bosques de pino-encino de Los Altos de Chiapas, México. MSc Thesis, El Colegio de la Frontera Sur (ECOSUR), San Cristóbal de Las Casas, Chiapas, Mexico

Breedlove DE (1981) Flora of Chiapas, part 1. Introduction to the flora of Chiapas. California Academy of Sciences, San Francisco

Breedlove DE (1986) Listados florísticos de México. IV. Flora de Chiapas. Instituto de Biología, Universidad Nacional Autónoma de México, Mexico City

Camacho-Cruz A, González-Espinosa M, Wolf JHD, de Jong BHJ (2000) Germination and survival of tree species in disturbed forests of the highlands of Chiapas, Mexico. Can J Bot 78:1309–1318

Cayuela L, González M, Rey JM, Ramírez N, Martínez M (2005) Imágenes de satélite revelan cómo desaparece el bosque en Chiapas. Quercus 232:60–61

Córdova J, del Castillo RF (2001) Changes in epiphyte cover in three chronosequences in a tropical montane cloud forest in Mexico. Diss Bot 346:79–94

Galindo-Jaimes L, González-Espinosa M, Quintana-Ascencio P, García-Barrios L (2002) Tree composition and structure in disturbed stands with varying dominance by *Pinus* spp. in the highlands of Chiapas, México. Plant Ecol 162:259–272

García-Barrios L, González-Espinosa M (2004) Change in oak to pine dominance in secondary forests may reduce shifting agriculture yields: experimental evidence from Chiapas, Mexico. Agric Ecosys Environ 102:389–401

González-Espinosa M, Quintana-Ascencio PF, Ramírez-Marcial N, Gaytán-Guzmán P (1991) Secondary succession in disturbed *Pinus-Quercus* forests of the highlands of Chiapas, Mexico. J Veg Sci 2:351–360

González-Espinosa M, Ochoa-Gaona S, Ramírez-Marcial N, Quintana-Ascencio PF (1995) Current land-use trends and conservation of old-growth forest habitats in the highlands of Chiapas, Mexico. In: Wilson MH, Sader SA (eds) Conservation of Neotropical migratory birds in Mexico. Maine Agric For Exp Sta, Orono, Maine, Misc Publ 727, pp 190–198

González-Espinosa M, Rey-Benayas JM, Ramírez-Marcial N, Huston MA, Golicher D (2004) Tree diversity in the northern Neotropics: regional patterns in highly diverse Chiapas, Mexico. Ecography 27:741–756

González-Espinosa M, Ramírez-Marcial N, Méndez-Dewar G, Galindo-Jaimes L, Golicher D (2005) Riqueza de especies de árboles en Chiapas: variación espacial y dimensiones ambientales asociadas al nivel regional. In: González-Espinosa M, Ramírez-Marcial N, Ruiz-Montoya L (eds) Diversidad biológica de Chiapas. Plaza y Valdés, Mexico City, pp 81–125

Holder CD (2004) Changes in structure and cover of a common property pine forest in Guatemala, 1954-1996. Environ Conserv 31:22–29

Hong S-K, Nakagoshi N, Kamada M (1995) Human impacts on pine-dominated vegetation in rural landscapes in Korea and western Japan. Vegetatio 116:161–172

Kappelle M, Brown AD (eds) Bosques nublados del neotrópico. Instituto Nacional de Biodiversidad (INBio), Santo Domingo de Heredia, Costa Rica

Kappelle M, Geuze T, Leal ME, Cleef AM (1996) Successional age and forest structure in a Costa Rican upper montane *Quercus* forest. J Trop Ecol 12:681–698

López-Barrera F (2003) Edge effects in a forest mosaic: implications for oak regeneration in the highlands of Chiapas, México. PhD Thesis, University of Edinburgh, Edinburgh

Miranda F (1952) La vegetación de Chiapas, vol 1. Gob Estado, Tuxtla Gutiérrez, Chiapas

Miranda F, Hernández-X E (1963) Los tipos de vegetación de México y su clasificación. Bol Soc Bot Méx 28:133–176

Montes-Avelar CA (2001) Patrones de diversidad florística en el paisaje agrícola de Los Altos de Chiapas, México. BSc Thesis, Universidad Nacional Autónoma de México (UNAM), Mexico City

Ochoa-Gaona S, González-Espinosa M (2000) Land-use and deforestation in the highlands of Chiapas, Mexico. Appl Geogr 20:17–42

Palacio-Prieto JL, Bocco G, Velázquez A, Mas J-F, Takaki-Takaki F, Victoria A, Luna-González L, Gómez-Rodríguez G, López-García J, Palma-Muñoz M, Trejo-Vázquez I, Peralta-Higuera A, Prado-Molina J, Rodríguez-Aguilar A, Mayorga-Saucedo R, González-Medrano F (2000) La condición actual de los recursos forestales en México: resultados del Inventario Forestal Nacional 2000. Inv Geogr 43:83–203

Quintana-Ascencio PF, González-Espinosa M (1993) Afinidad fitogeográfica y papel sucesional de la flora leñosa de los bosques de pino-encino de Los Altos de Chiapas, México. Acta Bot Mex 21:43–57

Quintana-Ascencio PF, González-Espinosa M, Ramírez-Marcial N (1992) Acorn removal, seedling survivorship, and seedling growth of *Quercus crisipipilis* in successional forests of the highlands of Chiapas, Mexico. Bull Torrey Bot Club 119:6–18

Quintana-Ascencio PF, Ramírez-Marcial N, González-Espinosa M, Martínez-Icó (2004) Sapling survival and growth of coniferous and broad-leaved trees in successional highland habitats in Mexico. Appl Veg Sci 7:81–88

Ramírez-Marcial N (2003) Survival and growth of tree seedlings in anthropogenically disturbed Mexican montane rain forests. J Veg Sci 14:881–890

Ramírez-Marcial N, González-Espinosa M, García-Moya E (1996) Establecimiento de *Pinus* spp. y *Quercus* spp. en matorrales y pastizales de Los Altos de Chiapas, México. Agrociencia 30:249–257

Ramírez-Marcial N, González-Espinosa M, Williams-Linera G (2001) Anthropogenic disturbance and tree diversity in montane rain forests in Chiapas, Mexico. For Ecol Manage 154:311–326

Ramírez-Marcial N, Camacho-Cruz A, González-Espinosa M (2005) Potencial florístico para la restauración de bosques en Los Altos y Montañas del Norte de Chiapas. In:

González-Espinosa M, Ramírez-Marcial N, Ruiz-Montoya L (eds) Diversidad biológica de Chiapas. Plaza y Valdés, Mexico City, pp 325–369
Richardson DM, Rejmánek M (2004) Conifers as invasive aliens: a global survey and predictive framework. Diversity Distrib 10:321–331
Román-Cuesta, RM, Gracia M, Retana J (2003) Environmental and human factors influencing fire trends in ENSO and non-ENSO years in tropical Mexico. Ecol Appl 13:1177–1192
Romero-Nájera I (2000) Estructura y condiciones microambientales en bosques perturbados de Los Altos de Chiapas, México. BSc Thesis, Universidad Nacional Autónoma de México (UNAM), Mexico City
Steyermark JA (1950) Flora of Guatemala. Ecology 31:368–372
Vetaas OR (1997) The effect of canopy disturbance on species richness in a central Himalayan oak forest. Plant Ecol 132:29–38
Wagner PL (1963) Natural and artificial zonation in a vegetation cover: Chiapas, Mexico. Geogr Rev 52:253–274

17 Changes in Diversity and Structure Along a Successional Gradient in a Costa Rican Montane Oak Forest

M. KAPPELLE

17.1 Introduction

Tropical montane forests are among the most fragile of all ecosystems on Earth (Stadtmüller 1987; Hamilton et al. 1995; Kappelle and Brown 2001; Schneider et al. 2003). Following clearing, they recover extremely slowly, and it may take one to several centuries before their structure, composition, and function return to the original, pre-disturbance state (Ewel 1980; Hooftman 1998; Kappelle 2004). Southern temperate *Nothofagus* forest (González et al. 2002) and neotropical montane *Quercus* forests (Kappelle et al. 1994, 1995b, 1996; Ramírez-Marcial et al. 2001; Chaps. 14 and 16) appear to be no exception to this rule. To gain a deeper insight into the patterns and processes of forest recovery and resilience in Central American montane oak forests, we studied changes in structure and diversity in a series of successional vegetation patches along a time sequence, ranging from grazed and recently abandoned pastures to late-successional (35 years old) and old-growth (>200 years old) oak forest stands in the highlands of southern Costa Rica.

17.2 Study Area

The presented research on changes in vegetation structure and diversity along a successional gradient was conducted in the montane oak forest belt (2,000–3,000 m elevation) of the Cordillera de Talamanca in southern-central Costa Rica. Most sample plots were located along the Pacific slope of the largely deforested Los Santos Forest Reserve (Kappelle and Juárez 1995; Helmer 2000; Helmer et al. 2000), and on the Atlantic slope of the almost completely intact Tapantí–Macizo de la Muerte National Park. Plots are concen-

Ecological Studies, Vol. 185
M. Kappelle (Ed.) Ecology and Conservation of Neotropical Montane Oak Forests
© Springer-Verlag Berlin Heidelberg 2006

trated in the highland area between the villages of El Empalme to the northwest and Villa Mills to the southeast, both situated along the Panamerican Highway. The Tapantí–Macizo de la Muerte National Park is part of the 612,600-ha La Amistad Biosphere Reserve (a World Heritage Site), and the 62,000-ha Los Santos Forest Reserve serves as a buffer zone at the southwestern tip of the magnificent La Amistad Reserve, recognized for its amazing biodiversity by the UNESCO (Kappelle and Juárez 1994).

The Cordillera de Talamanca is made up of intrusive and Tertiary volcanic rocks alternated with marine sediments (Castillo 1984). Pleistocene glaciations have left their traces on peaks over 3,000 m (Horn 1990; Kappelle and Horn 2005). Soils are dark to pale brown, medium-textured, moderately fertile (Vasquez 1983), and very acid, with pH values of 4.5–6.5; andosols are common (van Uffelen 1991; Chap. 4). The climate is humid to super-humid, temperate to cold, and has a short dry season (January–April). The average annual temperature ranges from 10 °C at 3,000 m to about 14 °C at 2,000 m elevation (Herrera 1986). However, due to the diurnal climate reigning on tropical mountains, temperatures measured over 24 h may vary greatly, shifting from a maximum of 20–24 °C at noon to a minimum of 2–6 °C at night (Kappelle 1996; Chap. 4). Mean annual rainfall oscillates around 2,700 mm per year. Diurnal fog during most afternoons throughout the year turns the prevailing evergreen oak-dominated rainforest (1,800–3,100 m) into a true tropical montane cloud forest (TMCF; Kappelle 1992).

17.3 Plant Species Assemblages and Diversity

17.3.1 Classification of Successional Plant Communities

To date, at least 24 plant species communities have been identified in fragmented montane oak forest environments in Costa Rica's high Talamanca Range. Twelve communities correspond to mature old-growth forest (Kappelle et al. 1995a, 1989; Chap. 4), whereas another set of 12 are secondary plant communities growing at previously deforested sites (Kappelle et al. 1994, 1995b). The latter include six lower montane and six upper montane communities, with three grassland communities (grazed and recently abandoned, non-grazed pastures), one scrub association, and two 30–35 year old secondary forest types per altitudinal belt. Communities were classified on the basis of multivariate analysis of aerial crown cover data, estimated as proportions of plot area sensu Braun-Blanquet (1965) for terrestrial vascular plant species, and applying TWINSPAN classification software (Hill 1979a; Kent and Coker 1992). This multivariate analysis included over 120 plant sociological, randomly stratified sample plots (sizes: 0.005–0.1 ha), located at

2,000–3,400 m elevation in the western sector of the Cordillera de Talamanca, and previously identified on aerial photographs (scale 1:60,000, year 1992, stored at the Instituto Geográfico Nacional (IGN) archives).

The twelve successional plant communities were classified as a *Monochaetum neglectum–Rubus eriocarpus* vegetation complex (Kappelle et al. 1994). Grasslands were characterized and dominated by species such as *Ageratina subcordata*, *Bromus* sp., *Carex jamesonii*, *Cheilanthes notholaenoides*, *Geranium guatemalense*, *Gnaphalium americanum*, *Halenia rhyacophylla*, *Holcus lanatus*, *Lolium perenne*, *Muehlenbeckia tamnifolia*, *Oenothera epilobifolia*, *Orthrosanthus chimborasensis*, *Pennisetum clandestinum*, *Plantago australis*, *Rumex acetosella*, and *Thelypteris rudis*. Diagnostic species in shrubby scrublands – locally known as 'charrales' – were *Galium mexicanum*, *Polypodium macrolepis*, *Pteridium aquilinium* (bracken fern) and *Vaccinium consanguineum*. Secondary forest associations were characterized by *Abatia parviflora*, *Bocconia frutescens*, *Buddleja nitida*, *Chusquea tomentosa*, *Cornus disciflora*, *Freziera candicans*, *Fuchsia arborescens*, *Monnina crepinii*, *Oreopanax xalapensis*, *Quercus copeyensis* (now known as *Q. bumelioides* – K.C. Nixon, personal communication; Chap. 1), *Q. costaricensis*, *Q. seemannii*, *Verbesina oerstediana*, *Viburnum costaricanum*, *Weinmannia pinnata* and *Wercklea lutea* (van Velzen et al. 1993; Kappelle et al. 1994).

17.3.2 Ordination of Successional Plant Communities

We applied detrended correspondence analysis (DCA) to the plant sociological data for 12 0.1-ha plots along the successional sere, using the DECORANA software (Hill 1979b; Jongman et al. 1987). This multivariate analysis revealed the occurrence of five ecological species groups arranged along the time gradient (DCA axis 1): pioneer species, early-successional secondary species, late-successional secondary species, early-recovering old-growth species, and late-recovering old-growth species (Kappelle et al. 1995b). The second axis (DCA axis 2) correlated to a moisture gradient, with species such as *Jungia ferruginea*, *Hydrocotyle bowlesioides*, *Piper bredemeyeri*, *Senecio copeyensis* and *Solanum incomptum* inhabiting wetter sites.

17.3.3 Alpha Diversity

Along the successional gradient, a total of 176 vascular plant species in 122 genera and 75 families was identified (Kappelle et al. 1995b). Species were distributed over 52 trees, 19 shrubs, 52 herbs, 16 climbers, one bamboo, 34 ferns and two lycopods. The most speciose families were Asteraceae (20 species), followed by Polypodiaceae (10), Lomariopsidaceae (7), Rosaceae (7), Ericaceae (6), Solanaceae (6), Lauraceae (5), Myrsinaceae (5) and Piperaceae (5).

Surprisingly, terrestrial vascular species richness and density decreased with successional progress (Table 17.1). This may be due to the downslope migration of a number of herbaceous plant species that are common in the upslope subalpine and alpine paramo grassland environment (3,100–3,820 m), and invade deforested and early-successional montane habitats at lower elevation (2,300–3,100 m). Thus, alpha diversity measured using the Shannon-Wiener index (H'; see Magurran 1988) appeared to be far higher in secondary forests than in old-growth mature forest, for the 12 0.1-ha plots at 2,800–3,000 m. In fact, the Shannon-Wiener index for terrestrial vascular plant species dropped significantly (Tukey-Kramer's one-way ANOVA test, $p<0.05$) from 5.1 in early-successional forest to values below 4.5 in mature, old-growth forest (Kappelle et al. 1995b). Probably, in mature old-growth forest, larger plot sizes are needed to ensure that most of the terrestrial plant species are included in the analysis. Shannon-Wiener's diversity index was also calculated, for tree species only (42 species). The index values fluctuated in the range 2.9–3.9 but did not change significantly along the gradient (Student's *t* test, $p>0.02$; Kappelle et al. 1996).

17.3.4 Beta Diversity and the Minimum Time for Floristic Recovery

We were able to calculate beta diversity on the basis of a chronosequence of successional and old-growth montane oak forest plots (Fig. 17.1). Beta diversity analysis helped us to estimate the minimum time required for a previously forested but now cleared and abandoned site to recover to acceptable levels in terms of plant species composition and diversity. Acceptable levels are defined as those at which the flora of a recovering site has a similarity of 95% in comparison with the original flora of a non-cleared, pristine forest site. To calculate these levels, beta diversity was assessed as the degree of similarity between pairs of successional and mature, old-growth forest stands, using Sorensen's coefficient of community (CC) or similarity coefficient (Jongman et al. 1987). Subsequently, similarity values were extrapolated in time (period of recovery since abandonment) by fitting data to a linear regression equation. Beta diversity was found to decrease along the successional gradient, and the theoretical minimum floristic recovery time – including only terrestrial vascular plants – was estimated at approximately 65 years ($r^2=0.66$). Thus, about a third of the total variance of Y before regression remains unexplained in this case.

Table 17.1. Stand structure and diversity data for nine 0.1-ha plots in successional montane oak forest at ~2,950 m a.s.l. in Costa Rica. Values are based on plot data presented in Kappelle et al. (1995b, 1996). Similar data for mature old-growth oak forest are presented in Chap. 10

Variable[a]	1	2	3	4	5	6	7	8	9
Stand age (years following abandonment)	8	10	12	20	20	25	30	30	32
Canopy height (m)	8	8	9	14	11	11	16	17	18
Number of stems≥3 cm DBH per plot	193	167	219	325	313	229	302	216	327
Number of stems≥10 cm DBH per plot	5	12	13	71	39	23	84	59	85
Number of stems per diameter class									
Stems 3–5 cm DBH per plot	123	111	137	90	132	116	104	72	141
Stems 5–10 cm DBH per plot	65	44	69	164	142	90	114	85	101
Stems 10–20 cm DBH per plot	5	11	13	65	37	22	79	49	65
Stems 20–40 cm DBH per plot	0	1	0	6	2	1	5	10	20
Stems>40 cm DBH per plot	0	0	0	0	0	0	0	0	0
Stem density (stems≥3 cm DBH per ha)	1,930	1,670	2,190	3,250	3,130	2,290	3,020	2,160	3,270
Basal area for stems≥3 cm DBH (m^2 ha^{-1})	4.5	4.8	5.7	20.5	13.3	8.7	19.3	16.5	25.3
Species richness per plot (terr. vasc.)[b]	91	91	98	91	100	90	75	90	75
Species richness per plot (trees only)	18	21	15	21	30	18	20	19	24
Shannon-Wiener's index (terr. vasc.)	5.07	5.06	5.19	5.39	5.39	5.20	4.73	4.63	4.74
Shannon-Wiener's index (trees only)	3.32	3.23	2.96	3.56	3.90	3.47	2.92	3.29	2.90
Reciprocal Simpson's index (terr. vasc.)	17.2	15.5	11.2	24.4	22.4	17.7	11.7	7.9	6.3
Reciprocal Simpson's index (trees only)	6.90	5.41	5.68	9.62	9.26	8.39	3.88	5.39	3.92
Species density (terr. vasc.)	30.3	30.3	32.6	30.3	33.3	30.0	25.0	30.0	25.0
Evenness or equitability index (terr. vasc.)	0.78	0.78	0.78	0.83	0.81	0.80	0.76	0.71	0.76

[a]Shannon-Wiener's index, reciprocal Simpson's index, species density, and evenness index were measured following procedures presented in Magurran (1988)
[b]Terr. vasc., all terrestrial vascular plant species

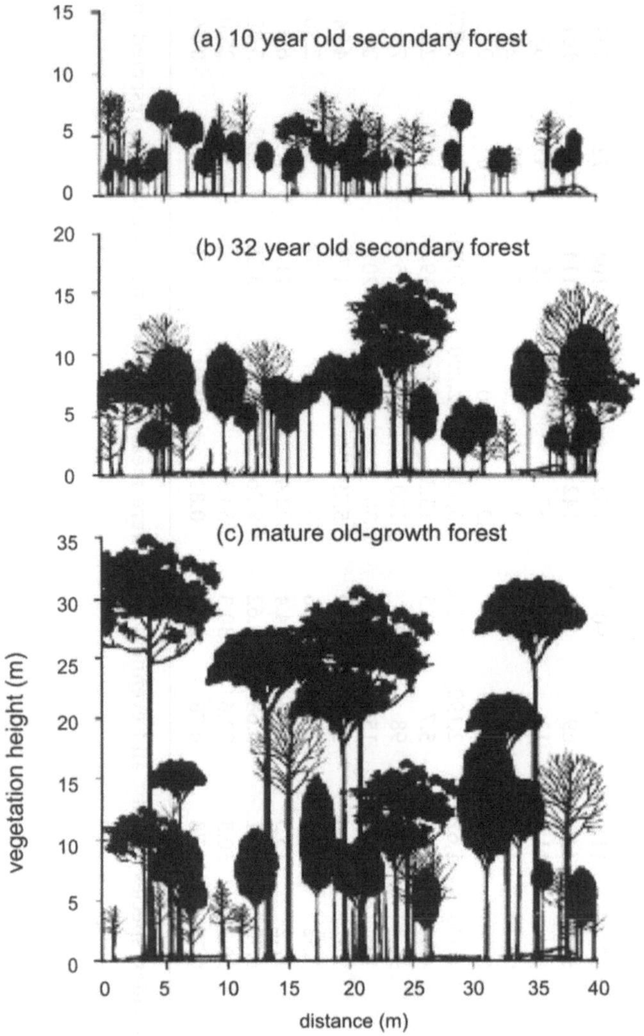

Fig. 17.1a–c. Schematic lateral profiles of three successional stages of tropical montane oak-bamboo forest at 2,700–2,900 m elevation in Costa Rica (Talamanca Mountains): **a** 10-year-old successional forest following clearing, grazing and abandonment; **b** 32-year-old successional forest following clearing, grazing and abandonment; and **c** >250 year old, mature old-growth oak-bamboo forest. Reproduced from Kappelle (2004), with permission from Elsevier

17.4 Stand Structure

17.4.1 Forest Layering

The maximum height of the closed forest canopy varied in the range ca. 5–8 to 14 m in early-successional forest, 11–18 m in late-successional forest, and 35–40 m in mature old-growth forest (Fig. 17.2; Kappelle et al. 1996; Chap. 10). Only mature old-growth forest stands showed stratification into two horizontal tree layers: a 20–40 m tall, uniform canopy layer dominated by oak, and a

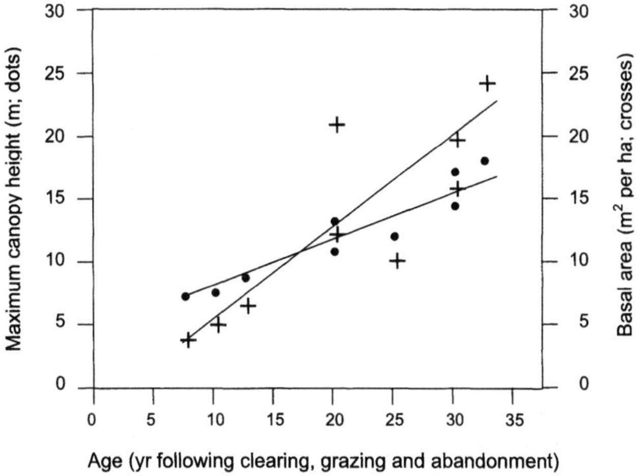

Fig. 17.2. Maximum canopy height (*dots*) and basal area (*crosses*) as functions of time of recovery following old-growth forest clearing, grazing and abandonment at nine 0.1-ha, 8–32 year old successional forest plots in San Gerardo de Dota, Costa Rica. Lines are drawn using coefficients of linear regressions for both functions

3–20 m tall, mixed subcanopy layer with species belonging to *Ardisia*, *Cleyera*, *Myrsine*, *Ocotea*, *Schefflera*, *Styrax*, *Vaccinium* and *Weinmannia*, together with younger and smaller oaks.

17.4.2 Stem Density and Basal Area

In all 2,854 stems ≥3.0 cm DBH (diameter at breast height) were counted, identified, and measured in 1.2 ha of successional and old-growth forest stands (Kappelle et al. 1996; Table 17.1). Numbers of stems decreased with increasing stem diameter for all successional forest phases. The proportion of tree stems≥10.0 cm DBH increased from 3% in late-successional forest to 40% in mature, old-growth forest. Stem density fluctuated between 1,670 stems per ha in 10-year-old early-successional forest and 3,270 stems per ha in 30–35 year old, late-successional forest. Stem density was significantly higher in late-successional forest than in mature, old-growth forest (Student's *t* test, $p<0.05$). Basal area increased linearly with plot age for successional forest stands, and ranged from 4.46 m^2 per ha in 8-year-old early-successional forest to 64.69 m^2 per ha in mature, old-growth forest (Kappelle et al. 1996). Basal area increased during succession (Fig. 17.2), and was significantly higher in mature, old-growth forest than in early- or late-successional forest (Student's *t* test, $p<0.001$).

17.4.3 Growth and the Minimum Time for Structural Recovery

Forest canopy height and DBH are significantly correlated for stems ≥3.0 cm DBH occurring along the assessed successional gradient (logarithmic regression, r^2=0.72–0.92, p<0.001). Similarly, for successional forest stands, forest canopy height, basal area and recovery time (plot age) revealed significant linear regressions (r^2=0.68–0.97, p<0.001). On the basis of the latter regression equation, the theoretical minimum time needed for forest recovery was estimated (1) by solving for X (age) where Y (canopy height) is the mean old-growth forest canopy height of 36 m, in which case the minimum time required for a successional forest to reach structural maturity was estimated at 79.5 years; and (2) by solving for X (age) where Y (basal area) is the mean old-growth forest basal area of 60.3 m^2 per ha, in which case the minimum time required for a successional forest to reach structural maturity was estimated at 89.0 years (Kappelle et al. 1996). In both cases, structural maturity does not account for the presence and abundance of epiphytes. To synthesize, an average recovery period of 84.3 years (r^2=0.68–0.86) can be calculated for a successional montane oak forest to become structurally similar to old-growth forest – excluding epiphytic richness and biomass, which may take many more decades to recover fully (Chaps. 7 and 21). Still, 15–35 % of the total variance of Y before these regressions remains unexplained.

17.5 Conclusions

An important trend that has been noted in this study is the significant decrease in vascular plant diversity as secondary succession advances. The outcomes of the diversity measures substantiate this trend, and confirm results from earlier research in the same study area that focused on tree species recovery (Kappelle 1993), ground cover recuperation (A. Schumacher, personal communication), and vascular plant recovery at the forest–pasture edge (Oosterhoorn and Kappelle 2000). These outcomes are concordant with the hypothesis that high species diversity in successional forests is commonly due to a high degree of vertical and horizontal micro-environmental heterogeneity (high niche differentiation) in young recovering forest (Bazzaz 1975).

Another explanation is provided by the theory that a large number of plant species naturally found in upslope, subalpine and alpine vegetation communities may disperse downslope into successional patches – a process locally known as 'paramization' (Kappelle and Horn 2005). Examples of these light-demanding species are the trees *Abatia parviflora*, *Buddleja nitida*, *Comarostaphylis arbutoides*, *Escallonia myrtillioides*, *Fuchsia arborescens*, *Garrya laurifolia* and *Verbesina oerstediana* – all very abundant in subalpine paramo communities at 3,200–3,400 m elevation (Kappelle et al. 1991). These species are all adapted to open, harsh environments, and apparently provide more favorable conditions at microsites for the establishment of shade-tolerant, late-successional and typical mature oak forest genera such as *Chusquea*, *Cleyera*, *Quercus*, *Vaccinium*, *Weinmannia*, and oak. As canopy closure occurs, these latter species may outcompete the light-demanding species of subalpine and alpine origin.

Finally, a third explanation for the high number of species in early phases, compared to old-growth forest, is provided by Denslow (1980). She suggests that species diversity in successional phases in tropical forests is initially high due to the high number of seedlings of possibly disturbance-adapted – in our case, subalpine and alpine – species in the early stages of recovery. Availability, dispersal, and germination of seeds of such species may indeed play a key role in speeding up forest recovery during initial stages at cleared sites (Chaps. 18, 23, 24 and 25).

Oak species commonly show high recovery capacity (e.g., Chaps. 13, 18 and 22). After clearing, burning, grazing, and subsequent abandonment, oaks with stems≥3 cm DBH appear in 8–10 year old, early-successional forest. As soon as acorn germination and establishment have been successfully completed, stem numbers of oak increase rapidly throughout the first successional stages. After 30–35 years stem density levels off, whereas stem size continues to increase and reaches highest values at the end of the successional pathway – when the tree falls, causing a gap and initiating a renewed phase of local forest-interior recovery (gap dynamics). When forest maturity has been reached, the distribution of oak stems (in numbers) based on DBH size class displays an inverted J-shaped curve – a model previously observed in other parts of the Talamancan montane oak forests (Jiménez et al. 1988).

Acknowledgements I wish to thank W.F. van Buuren, A.M. Cleef, T. Geuze, M.E. Juárez, O. Juárez, P. Kennis, M.E. Leal, L. Monge, M. Spreuwenberg, H.P. van Velzen, R.A.J. de Vries, W. Wijtzes, and the late A. Chaverri for scientific, technical and field support. I am much indebted to the many plant taxonomists who identified botanical specimens, in particular L.D. Gómez, B. Hammel, Q. Jiménez, L.J. Poveda, P.E. Sánchez, and N. Zamora. Funding was provided by The Netherlands Organization for Scientific Research (NWO-WOTRO grant W-84331), the University of Amsterdam, the University of Nijmegen, the Universidad Nacional at Heredia, the Hugo de Vries Foundation, the National Biodiversity Institute (INBio), and the WSO Foundation. Research permission was granted by Costa Rica's Ministry of Environment and Energy (MINAE).

References

Bazzaz FA (1975) Plant species diversity in old-field successional ecosystems in southern Illinois. Ecology 56:485–488

Braun-Blanquet J (1965) Plant sociology: the study of plant communities. Hafner, London

Castillo R (1984) Geología de Costa Rica: una sinopsis. Universidad de Costa Rica, San José, Costa Rica

Denslow JS (1980) Patterns of plant species diversity during succession under different disturbance regimes. Oecologia (Berlin) 46:18–21

Ewel J (1980) Tropical succession: manifold routes to maturity. Biotropica 12 Suppl 2:2–7
González ME, Veblen TT, Donoso C, Valeria L (2002) Tree regeneration responses in a lowland *Nothofagus*-dominated forest after bamboo dieback in South-Central Chile. Plant Ecol 161:59–73
Hamilton LS, Juvik JO, Scatena F (eds) (1995) Tropical montane cloud forests. Springer, Berlin Heidelberg New York
Helmer EH (2000) The landscape ecology of tropical secondary forest in montane Costa Rica. Ecosystems 3(1):98–114
Helmer EH, Brown S, Cohen WB (2000) Mapping montane tropical forest successional stage and land use with multi-date Landsat imagery. Int J Remote Sens 21(11):2163–2183
Herrera W (1986) Clima de Costa Rica. EUNED, San José, Costa Rica
Hill MO (1979a) TWINSPAN – a FORTRAN program for arranging multivariate data in an ordered two-way table by classification of individuals and attributes. Department of Ecology and Systematics, Cornell University, Ithaca, NY
Hill MO (1979b) DECORANA – a FORTRAN program for Detrended Correspondence Analysis and Reciprocal Averaging. Department of Ecology and Systematics, Cornell University, Ithaca, NY
Hooftman DAP (1998) Generic composition, structure and diversity of secondary forests at Amisconde, the Pacific slope of the Cordillera de Talamanca, Costa Rica. Rev Biol Trop 46(4):1069–1079
Horn SP (1990) Timing of deglaciation in the Cordillera de Talamanca, Costa Rica. Climate Res 1:81–83
Jiménez W, Chaverri A, Miranda R, Rojas I (1988) Aproximaciones silviculturales al manejo de un robledal (*Quercus* spp.) en San Gerardo de Dota, Costa Rica. Turrialba 38:208–214
Jongman RHG, Ter Braak CFJ, van Tongeren OFR (eds) (1987) Data analysis in community and landscape ecology. PUDOC Press, Wageningen, The Netherlands
Kappelle M (1992) Structural and floristic differences between wet Atlantic and moist Pacific montane *Myrsine–Quercus* forests in Costa Rica. In: Balslev H, Luteyn JL (eds) Páramo: an Andean ecosystem under human influence. Academic Press, London, pp 61–70
Kappelle M (1993) Recovery following clearing of an upper montane *Quercus* forest in Costa Rica. Rev Biol Trop 41(1):47–56
Kappelle M (1996) Los bosques de roble (*Quercus*) de la Cordillera de Talamanca, Costa Rica: biodiversidad, ecología, conservación y desarrollo. Instituto Nacional de Biodiversidad (INBio), Santo Domingo de Heredia, Costa Rica
Kappelle M (2004) Tropical montane forests. In: Burley J, Evans J, Youngquist JA (eds) Encyclopedia of Forest Sciences, vol 4. Elsevier, Oxford, UK, pp 1782–1793
Kappelle M, Brown AD (eds) (2001) Bosques nublados del Neotrópico. Instituto Nacional de Biodiversidad (INBio) and World Conservation Union (IUCN), Santo Domingo de Heredia, Costa Rica
Kappelle M, Horn SP (eds) (2005) Paramos de Costa Rica. Instituto Nacional de Biodiversidad (INBio), Santo Domingo de Heredia, Costa Rica
Kappelle M, Juárez ME (1994) The Los Santos Forest Reserve: a bufferzone vital for the Costa Rican La Amistad Biosphere Reserve. Environ Conserv 21(2):166–169
Kappelle M, Juárez ME (1995) Agroecological zonation along an altitudinal gradient in the montane belt of the Los Santos Forest Reserve in Costa Rica. Mount Res Dev 15(1):19–37
Kappelle M, Cleef AM, Chaverri A (1989) Phytosociology of montane *Chusquea-Quercus* forests, Cordillera de Talamanca, Costa Rica. Brenesia 32:73–105

Kappelle M, Zamora N, Flores T (1991) Flora leñosa de la zona alta (2000-3819 m) de la Cordillera de Talamanca, Costa Rica. Brenesia 34:121–144

Kappelle M, van Velzen HP, Wijtzes WH (1994) Plant communities of montane secondary vegetation in the Cordillera de Talamanca, Costa Rica. Phytocoenol 22(4):449–484

Kappelle M, van Uffelen JG, Cleef AM (1995a) Altitudinal zonation of montane *Quercus* forests along two transects in the Chirripó National Park, Costa Rica. Vegetatio 119:119–153

Kappelle M, Kennis PAF, de Vries RAJ (1995b) Changes in diversity along a successional gradient in a Costa Rican upper montane *Quercus* forest. Biodiv Conserv 4:10–34

Kappelle M, Geuze T, Leal ME, Cleef AM (1996) Successional age and forest structure in a Costa Rican upper montane *Quercus* forest. J Trop Ecol 12:681–698

Kent M, Coker P (1992) Vegetation description and analysis: a practical approach. Belhaven Press, London

Magurran AE (1988) Ecological diversity and its measurement. Croom Helm, London

Oosterhoorn M, Kappelle M (2000) Vegetation structure and composition along an interior-edge-exterior gradient in a Costa Rican montane cloud forest. For Ecol Manage 126:291–307

Ramírez-Marcial N, González-Espinosa M, Williams-Linera G (2001) Anthropogenic disturbance and tree diversity in montane rain forests in Chiapas, Mexico. For Ecol Manage 154(1/2):311–326

Schneider JV, Zipp D, Gaviria J, Zizka G (2003) Successional and mature stands in an upper Andean rain forest transect of Venezuela. J Trop Ecol 19:251–259

Stadtmüller T (1987) Los bosques nublados en el Trópico húmedo. Centro Agronómico Tropical de Investigación y Enseñanza (CATIE), Turrialba, Costa Rica

Van Uffelen JG (1991) A geological, geomorphological and soil transect study of the Chirripó massif and adjacent areas, Cordillera de Talamanca, Costa Rica. MSc Thesis, Wageningen Agricultural University, Wageningen, The Netherlands

Van Velzen HP, Wijtzes WH, Kappelle M (1993) Lista de especies de la vegetación secundaria del piso montano pacífico, Cordillera de Talamanca, Costa Rica. Brenesia 39-40:147–161

Vasquez A (1983) Soils. In: Janzen DH (ed) Costa Rican natural history. Univ Chicago Press, Chicago, IL, pp 63–65

18 Regeneration Dynamics in a Costa Rican Montane Oak Forest After Reduced-Impact Logging

M.R. Guariguata, G.P. Sáenz, and L. Pedroni

18.1 Introduction

With respect to their lowland counterparts, tropical montane forests are generally less well known in terms of their regeneration dynamics, post-logging, community-level responses, and the implications these may have for timber management and development of silvicultural treatments (e.g., Dawkins and Philip 1998). Neotropical montane oak forests are no exception. In order to partially fill this knowledge gap, an experimental harvesting scheme was implemented during the early 1990s in the extensive oak-bamboo forest stands in upland Costa Rica, whose canopies are typically dominated by the endemic *Quercus costaricensis* Liebmann and *Q. copeyensis* C.H. Müller (now known as Q. *bumelioides*; K.C. Nixon, personal communication; Chap. 1).

Alone, these two species may account for up to 90 % of both density and basal area for stems≥50 cm DBH (diameter at breast height), and thousands of juveniles may fit into a single hectare (Blaser and Camacho 1991; Guariguata and Sáenz 2002). Thus, oak-bamboo forests in this region are comprised of very large trees of *Quercus* (i.e., they harbor attractive commercial volumes), along with abundant natural regeneration. These two traits have the potential of facilitating timber management in the area, since a specific, ample resource can be targeted over large areas. In this chapter, we present the effects of a pilot reduced-impact logging operation on forest stand regeneration in the Cordillera de Talamanca.

18.2 Study Area and Logging Treatments

The natural vegetation of the study site is tropical montane rain forest (sensu Holdridge et al. 1971), which occurs in Costa Rica from 2,000 to 3,400 m ele-

Ecological Studies, Vol. 185
M. Kappelle (Ed.) Ecology and Conservation of Neotropical Montane Oak Forests
© Springer-Verlag Berlin Heidelberg 2006

vation (Kappelle 1996). The forest canopy is tall (35–40 m), and the understory is composed of several species of *Chusquea* bamboo, shrubs in the Ericaceae, Rubiaceae, Myrsinaceae, and Melastomataceae, and tree ferns (Cyatheaceae). Other common tree species in this forest are *Weinmannia pinnata* L., *Ocotea austinii* C.K. Allen, *Styrax argenteus* C. Presl., and *Drymis granadensis* L. Detailed floristic descriptions of the study area are given elsewhere in this volume.

Between 1991 and 1992, a strictly controlled logging operation was carried out in 23 ha of primary forest at the Villa Mills site (about 325 ha total; 9°34'N, 83°41'W), located at ~2,700 m elevation. The topography is moderate, and the current vegetation comprises old-growth, selectively logged, and patches of secondary forest. Annual average precipitation is ca. 2,600 mm (Camacho and Orozco 1998).

In the study area, 1-ha plots, separated by 20–25 m wide buffer strips, were laid out and subsequently, two harvesting intensity treatments were randomly assigned to four replicate plots per treatment (Fig. 18.1). The treatments consisted of extracting approximately 20 % („light intensity") and 30 % („moderate intensity") of the stand basal area for stems ≥10 cm DBH, on a tree-by-tree basis (Table 18.1). The harvesting operation involved careful design of logging roads and log-yards, and application of directional felling by locally trained sawyers, all in order to minimize both collateral tree damage and soil disturbance by machinery. Detailed information about planning, implementation,

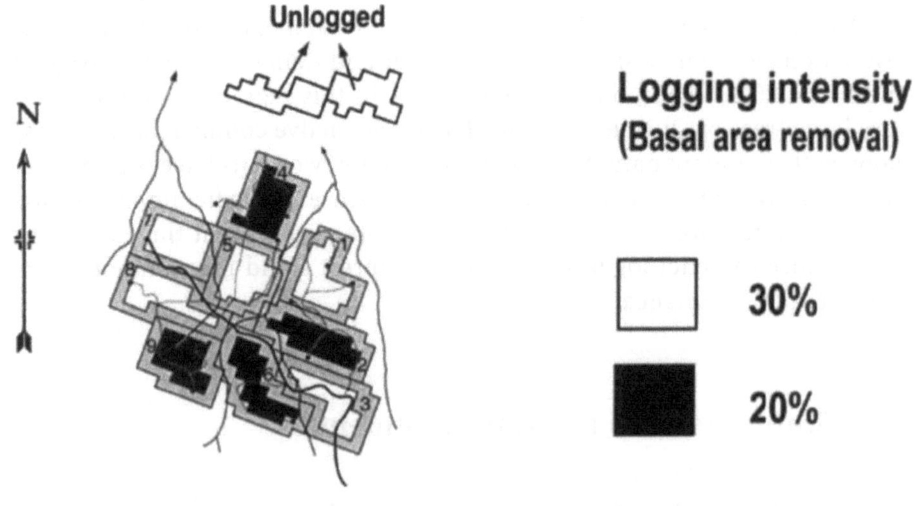

Fig. 18.1. Layout of the 1-ha plots under a given harvesting intensity (percent of basal area removed for stems ≥10 cm diameter at breast height), and the 1-ha unlogged plots at the Villa Mills experimental site in Costa Rica. Logged plots are separated by 25-m-wide buffer strips. Reprinted from Guariguata and Sáenz (2002), with permission from Elsevier

Table 18.1. Selected pre-logging structural attributes (stems ≥10 cm DBH) in four, 1-ha experimental plots under each of two different logging treatments (percent basal area removed for stems ≥10 cm DBH), and in two, 1-ha unlogged plots in the Talamanca Cordillera, Costa Rica. Basal area values in $m^2\ ha^{-1}$

	20 % basal area removal		30 % basal area removal		Unlogged	
	% Abundance	Basal area	% Abundance	Basal area	% Abundance	Basal area
Quercus costaricensis	24	17	22	17	16	19
Quercus copeyensis	13	11	9	9	35	24
Other commercial species (14)	34	10	34	10	49	8
Noncommercial species	29	2	35	3	–	–
Total	100	40	100	39	100	51

and the impacts of the experimental harvesting on the residual stand is found in Venegas and Louman (2001).

In both logging intensity treatments, harvesting consisted of commercial extraction, release of future crop trees from light competition, and elimination of over-mature and/or poorly formed individuals. Extraction levels for commercial size trees (≥30 cm DBH) were about 26 trees/ha (~51 $m^3\ ha^{-1}$) and 37 trees/ha (~62 $m^3\ ha^{-1}$) in the light and moderate intensity treatments, respectively. Note that these extracted volumes contrast to those from controlled logging operations in the neighboring lowlands, where they do not exceed 20 $m^3\ ha^{-1}$. No post-logging silvicultural treatments were applied.

18.3 Post-Logging Tree Juvenile Demography

We assessed growth in diameter and height, and individual survivorship of juveniles of the tree species *Quercus costaricensis*, *Q. copeyensis*, *Drymis granadensis*, *Ocotea austinii*, and *Weinmania pinnata* (all of local commercial value) in the treatment plots described above. Individual measurements were performed annually in the period 1993–1998 in two size classes: seedlings (stems ≥0.3 but <1.5 m tall) and saplings (stems ≥1.5 m tall but <10 cm DBH), tallied 6 months after logging in eight, 20×25 m plots randomly located within each of the 1-ha experimental plots. The results presented below, along with detailed methodological considerations, are derived from Sáenz and Guariguata (2001).

Overall, annual mortality rate for seedlings was higher under the light harvesting intensity (~5 %) than under the moderate harvesting intensity (~1 %;

$p<0.001$), probably related to differences in post-logging overhead light conditions. In particular, and under both harvesting intensities, seedlings of *Ocotea austinii* ranked first by having the highest mortality, compared to the other four study species. In contrast to seedlings, saplings showed no statistical differences in mortality across species, either among or within a given harvesting intensity ($p>0.1$). For all species combined, 5-year diameter and height median growth rates did not statistically differ between harvesting intensities for the seedlings ($p=0.5$). By contrast, both variables were significantly higher in the light and moderate harvesting intensity for saplings, suggesting their greater growth responsiveness to logging-mediated canopy opening.

When individual species were compared, however, both species of *Quercus* showed the largest, statistically significant differences in diameter growth rates among logging treatments (Table 18.2). Our results revealed that, at the applied harvesting intensities, juveniles of *Quercus* (particularly *Q. costaricensis*) performed better in terms of radial growth than the other study species. We suggest that any canopy manipulations aimed at maximizing juvenile recruitment may therefore target saplings, rather than seedlings. Low growth responsiveness in seedlings may be partially due to their susceptibility of being rapidly overtopped by fast-growing *Chusquea* bamboo, which thrives in canopy gaps in our area (Sáenz 1991; Widmer 1998).

18.4 Post-Logging Acorn Production and Early Seedling Establishment

Because of the high dominance of *Quercus* in the forest canopy at our site, we also sought to determine whether logging had any effect on stand-level acorn production with respect to unlogged tracts. For this, we took advantage of a mast seeding event in *Q. costaricensis* which occurred in the period 1999–2000. Previous observations at the site strongly suggest that *Q. costaricensis* shows synchronous, supra-annual flowering and fruiting patterns, spaced every 3–4 years (Camacho and Orozco 1998). Individuals of coexisting *Q. copeyensis* also seem to show masting but with a different pattern of among-year synchrony in flowering and fruiting (Camacho and Orozco 1998).

A total of 240, 1×1 m acorn traps were randomly distributed during the year 2000 in the eight plots under both logging intensities (cf. above), and in two, 1-ha unlogged plots (Table 18.1, Fig. 18.1). Trapped acorns of *Q. costaricensis* were collected monthly until no acorns were recorded in any trap at a given census. Acorns were examined by eye, and subdivided into the categories „sound“, „immature“ (based on external color), and „obviously damaged and/or insect infested“. To assess whether acorn production at the stand

Table 18.2. Annualized stem diameter growth rates (median and maximum in mm/year) of surviving individuals over a 5-year period in juveniles of five tree species under each of two different logging treatments (percent basal area removed for stems≥10 cm DBH) in the Cordillera Talamanca, Costa Rica (see text for plant size descriptions). Probability values denote comparisons among harvesting intensities. Reprinted from Saenz and Guariguata (2001), with permission from Elsevier

Harvest	Drymis granadensis				*Ocotea austinii*				*Quercus copeyensis*				*Q. costaricensis*				*Weinmannia pinnata*			
intensity	N	med.	max.	p	N	med.	max.	p	N	med.	max.	p	N	med.	max.	p	N	med.	max.	p
Seedlings				0.6				0.3				0.001				0.02				0.3
20 %	22	1.8	3.6		51	0.8	4.2		79	2.5	6		144	1.8	8		13	2	4	
30 %	20	1.6	3.8		99	0.8	3		84	3.2	8.4		191	2.4	7.6		9	2.4	2.7	
Saplings												0.001				0.001				0.03
20 %	46	2.4	6.2	0.4	35	1.4	5.6	0.09	53	2.6	9.8		186	2.7	7.8		56	1.8	6.2	
30 %	32	2.4	7		49	1	7.8		32	3.3	9.4		165	3.7	9.8		48	2.3	7	

level varied in terms of logging intensity, we censused all *Q. costaricensis* individuals ≥10 cm DBH present in all 10 study plots, and distinguished those bearing fruit from non-reproducing individuals, and recorded their crown illumination index (using a 1–5 scale sensu Dawkins and Field 1978). To further ascertain whether any difference in fruit production among treatment plots would be reflected in densities of oak natural regeneration, we followed acorn germination in logged vs. unlogged plots, and quantified the abundance of *Q. costaricensis* seedlings from the year 2000 acorn crop. Guariguata

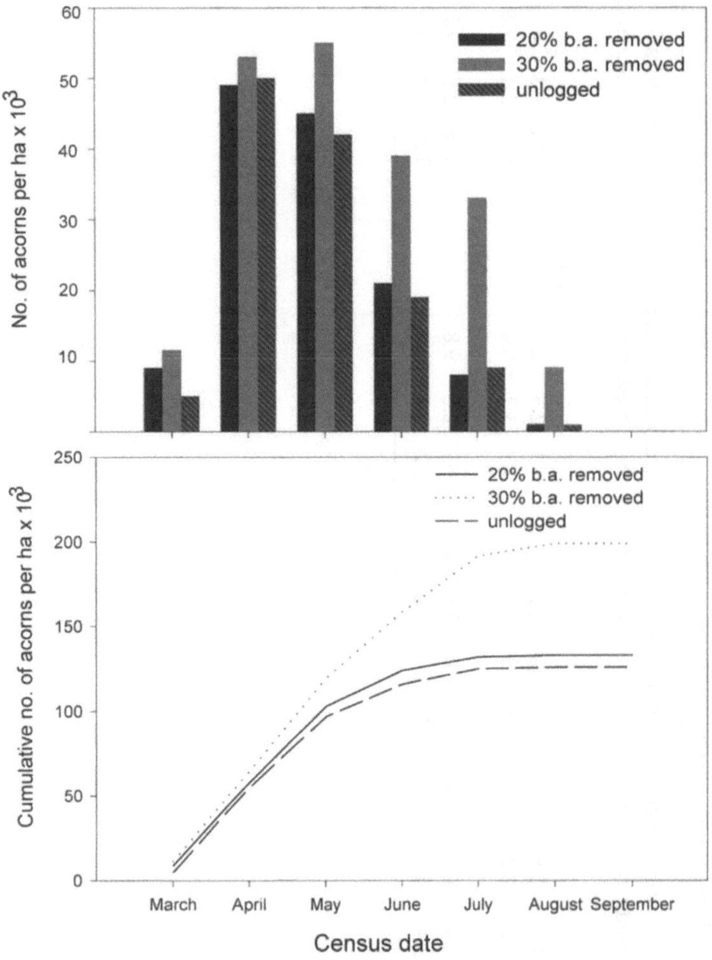

Fig. 18.2. Production of "sound", mature acorns of *Quercus costaricensis* as a function of census period (*top*) and on a cumulative basis (*bottom*) during the year 2000 under two logging treatments (percent basal area removed for stems≥10 cm diameter at breast height), and in a control, unlogged area. Modified from Guariguata and Sáenz (2002), with permission from Elsevier

and Sáenz (2002) provide further details on the methods and results of this particular research.

Experimental logging at our study site seems to have stimulated stand-level acorn production in *Q. costaricensis.* Between March and August 2000, an estimated total of 180,000 („sound") acorns/ha fell in those plots logged under moderate intensity, substantially more than in unlogged plots (100,000 acorns/ha), whereas acorn production in the lightly logged plots was about 120,000 acorns/ha (Fig. 18.2). Our findings suggest that changes in stand-level acorn production were related to logging-mediated canopy opening: (1) for a given size class, most trees had their crowns less well illuminated in unlogged plots, compared to logged plots (Table 18.3); and (2) within plot types, trees bearing acorns had their crowns significantly better illuminated than those that did not set fruit ($p<0.001$). In logged plots, more acorns per unit area may have resulted from the presence of small (i.e., 10–30 cm DBH) individuals that fruited only there (Guariguata and Sáenz 2002), and probably was also due to larger, per capita acorn crops (for which we had no data).

Table 18.3. Crown illumination index (sensu Dawkins and Field 1978) in *Quercus costaricensis* trees located in experimental plots under each of two different logging treatments (percent basal area removed for stems≥10 cm DBH) in the Talamanca Cordillera, Costa Rica. Index values close to 1 indicate crown fully exposed to sunlight, and those close to 5 indicate no direct illumination. The unlogged treatment corresponds to two, 1-ha plots (see Table 18.1). Values are means and ranges (in parentheses). Reprinted from Guariguata and Saenz (2002), with permission from Elsevier

Diameter class (cm)	Unlogged	Logged 20 %	Logged 30 %	P
10-19.9	3.6 (2-4)	2.9 (2-4)	2.8 (1-4)	<0.001
N	51	166	138	
20-29.9	3.5 (2-4)	2.6 (1-4)	2.5 (1-3)	<0.001
N	19	70	39	
30-39.9	3.1 (2-4)	2.4 (1-3)	2.2 (1-3)	<0.01
N	12	36	37	
40-49.9	2.9 (2-4)	2.4 (2-4)	2.1 (2-3)	<0.001
N	20	24	16	
50-59.9	3.2 (2-4)	2 (1-3)	2.1 (2-3)	<0.001
N	12	24	15	
60-69.9	2.4 (1-4)	2 (1-3)	1.9 (1-3)	0.28
N	14	21	21	
70-79.9	2.3 (1-4)	1.7 (1-3)	1.6 (1-3)	0.13
N	15	21	31	
80-89.9	2.7 (2-4)	1.8 (1-2)	1.4 (1-3)	<0.001
N	11	5	16	
90-129.9	3.4 (1-4)	1.7 (1-2)	1.2 (1-2)	<0.001
N	11	10	10	

As masting in oaks usually „satiates“ seed predators, leaving more acorns to escape consumption and thus to germinate (see review in Sork 1993), we also compared the likelihood of acorn removal (a surrogate of vertebrate seed predation) among plots logged under moderate intensity – the plot type where more acorns fell per unit area – and unlogged plots. As expected, „satiated“ vertebrate predators left proportionately more acorns to germinate in these logged plots (30 days after placement, about 30 % removal) than was the case in unlogged plots (about 62 % removal; $p<0.05$). Levels of naturally established *Q. costaricensis* seedlings mirrored the observed differences both in stand-level acorn production and vertebrate predation among these plot types. Indeed, we detected a statistically significant difference in seedling densities ($N=200$, 1-m^2 randomly placed quadrats per plot type) for the year 2000 acorn crop, between logged plots (mean: 3.4 seedlings/m^2, ~34,000/ha) and unlogged plots (mean: 1.1 seedlings/m^2, ~11,000/ha). Germination percentages (mean: 94 %) and percent seedling establishment (mean: 54 %) of surface-sown acorns were not statistically different between plot types.

18.5 Conclusions

The collective results described above suggest that reduced-impact logging applied in this oak-bamboo forest had enough effect to enhance juvenile growth and decrease mortality, particularly in the dominant *Quercus* species. Moreover, logging-mediated canopy opening further enhanced acorn production at the stand level – even one decade after harvesting was implemented – also decreased levels of vertebrate acorn predation, and ultimately increased the likelihood of juvenile recruitment. These pilot findings also suggest that reduced-impact logging may be the only silvicultural treatment needed in this forest type, which contrasts at least with many temperate oak species that, despite having high understory abundances, may require more intensive treatments than those presented here for stimulating recruitment into the sapling size class (e.g., Lorimer et al. 1994; Brose et al. 1999).

Although it appears that Costa Rican oak bamboo forests on gentle slopes have the biophysical attributes for sustainable timber management, a large part of the Talamanca Cordillera, where these forests are located, was declared as Forest Reserve by the Costa Rican Government in 1999. This means that, under the current law, any management plan for timber to be implemented here must include an environmental impact assessment, thus adding substantial costs to the forest owner. As most forest stands on relatively accessible terrain are small in size, investing in timber management becomes at best a financially unattractive land-use option. Since the late 1990s, rates of both submission by private owners and government approval of timber management plans have been exceedingly low in the area. There is therefore little room for further cross-validation of our results; but they may serve as initial guidance for developing silvicultural systems in other montane oak forests elsewhere in the neotropics.

Acknowledgements This work was funded by the Swiss Agency for Development and Cooperation (COSUDE) through the Centro Agronómico Tropical de Investigación y Enseñanza (CATIE), Costa Rica. We thank Lorena Orozco, Ligia Quirós, and Geoffrey Venegas for their assistance during manuscript preparation.

References

Blaser J, Camacho M (1991) Estructura, composición y aspectos silviculturales de un bosque de robles (*Quercus* spp.) del piso montano de Costa Rica. Centro Agronómico Tropical de Investigación y Enseñanza (CATIE), Turrialba, Costa Rica, Inf Téc 185

Brose P, van Lear D, Cooper R (1999) Using shelterwood harvests and prescribed fire to regenerate oak stands on productive upland sites. For Ecol Manage 113:125–141

Camacho M, Orozco L (1998) Patrones fenológicos de doce especies arbóreas del bosque montano de la Cordillera de Talamanca, Costa Rica. Rev Biol Trop 46:533–542

Dawkins HC, Field DRB (1978) A long-term surveillance system for British woodland vegetation. Department of Forestry, Oxford University, UK

Dawkins HC, Philip MS (1998) Tropical moist forest silviculture and management: a history of success and failure. Oxford Univ Press, Oxford, UK

Guariguata MR, Sáenz GP (2002) Post-logging acorn production and oak regeneration in a tropical montane forest, Costa Rica. For Ecol Manage 167:285–293

Holdridge LR, Grenke WC, Hatheway WH, Liang T, Tosi JA Jr (1971) Forest environments in tropical life zones. Pergamon Press, New York

Kappelle M (1996) Los bosques de roble (*Quercus*) de la Cordillera de Talamanca, Costa Rica: biodiversidad, ecología, conservación y desarrollo. Instituto Nacional de Biodiversidad (INBio), Santo Domingo de Heredia

Lorimer CG, Chapman JW, Lambert WD (1994) Tall understory vegetation as a factor in the poor development of oak seedlings beneath mature stands. J Ecol 82:227–237

Sáenz GP (1991) Densidad y dinámica de plántulas de *Quercus copeyensis* bajo dosel y en apertura, en el primer año después de la germinación en los robledales de Villa Mills, Costa Rica. MSc Thesis, CATIE, Turrialba, Costa Rica

Sáenz GP, Guariguata MR (2001) Demographic response of tree juveniles to reduced impact logging in a Costa Rican montane forest. For Ecol Manage 140:75–84

Sork VL (1993) Evolutionary ecology of mast-seeding in temperate and tropical oaks (*Quercus* spp.). Vegetatio 107/108:133–147

Venegas J, Louman B (2001) Aprovechamiento con tratamiento silvicultural de impacto reducido en un bosque montano de la Cordillera de Talamanca, Costa Rica. Centro Agronómico Tropical de Investigación y Enseñanza (CATIE), Turrialba, Costa Rica, Inf Tec 325

Widmer Y (1998) Pattern and performance of understory bamboos (*Chusquea* spp.) under different canopy closures in old-growth oak forests in Costa Rica. Biotropica 30:400–415

19 Growth and Physiological Responses of Oak, Pine and Shrub Seedlings to Edge Gradients in a Fragmented Mexican Montane Oak Forest

H. Asbjornsen, K.A. Vogt and P.M.S. Ashton

19.1 Introduction

Fragmented habitats are characterized by high amounts of edge environment (Forman and Godron 1986), which influence ecosystem recovery processes by directly altering microclimate conditions (Geiger 1957; Chen et al. 1993) and successional processes (Jose et al. 1996; Williams-Linera et al. 1997; Oosterhoorn and Kappelle 2000; Euskirchen et al. 2001). Edge effects on regeneration dynamics may be particularly severe in mountain ecosystems, as a result of extreme climatic conditions leading to slower rates of ecological processes (Leuschner 2000). Several studies have documented extremely slow recovery periods following severe disturbance in high-elevation ecosystems (Ewel 1980; Aide et al. 1996; Kappelle et al. 1996; Waide et al. 1998; Chaps. 16 and 17), which in some cases may result in arrested states of succession (Niering and Egler 1955; Putz and Canham 1992; Sarmiento 1997). In montane oak forests, recovery processes may be further delayed due to slow growth rates of montane oak species (Rzedowski 1978), long periods between mast years (Negi et al. 1996), and reliance on vegetative regeneration (Nixon 1993; Chap. 1).

The Mixteca Alta, located in the Sierra Madre del Sur mountain range in the southern Mexican state of Oaxaca, supports diverse oak forests that transition from seasonally dry deciduous woodlands at lower elevations to montane humid forests with a strong evergreen habit at the highest elevations (Rzedowski 1978). As with other neotropical mountain regions, the historically high human population densities in the Mixteca Alta have contributed to severe habitat fragmentation (Spores 1969; Kirkby 1972). However, in recent decades increasing out-migration by local people has resulted in the abandonment of agricultural and grazing lands (Stuart and Kearney 1981). This trend may be providing new opportunities for forest regeneration, if oak

Ecological Studies, Vol. 185
M. Kappelle (Ed.) Ecology and Conservation of Neotropical Montane Oak Forests
© Springer-Verlag Berlin Heidelberg 2006

seedlings can successfully establish and grow under current environmental conditions in these fragmented landscapes.

Although extensive research has examined the biogeographical distributions and phytosociological affinities of montane oak forests in Mexico (e.g., Gentry 1982; Rzedowski 1998), relatively few studies have assessed processes of secondary succession following land use disturbance and habitat fragmentation in this country (González-Espinosa et al. 1991; Chap. 16). The close relationship between plant physiological responses and long-term ecological patterns and processes (e.g., Lambers et al. 1998 and references therein) provides a powerful tool for elucidating mechanisms underlying ecological change, and predicting future trajectories in ecosystem development. However, few studies have explicitly examined the relationship between regeneration dynamics and physiological responses of individual plant species in neotropical oak forests (Kappelle and Leal 1996; Chap. 21).

This chapter examines whether diminishing land use pressures in the Mixteca Alta region may enable the reestablishment of oak forests, or alternatively, whether altered environmental conditions may inhibit regeneration of native oak species, thereby affecting successional trajectories. Specifically, we address the questions:

1. To what extent do edges in fragmented landscapes alter the regeneration environment for seedling establishment and growth?
2. How do seedling growth and physiological responses to edge environments vary between oaks and species from different successional seres?
3. How are microclimate and seedling responses influenced by the occurrence of a severe drought?

We conclude by discussing the potential long-term patterns for oak forest recovery in these landscapes, focusing on the role of edge environments and species' physiological and growth responses in determining successional processes.

19.2 Overview of the Research

This study was conducted in two adjacent communal forests having different degrees of fragmentation, belonging to the villages of San Pedro Cántaros Coxcaltepec and Santiago Huautlilla in the district of Nochixtlán, Oaxaca, at approximately 17° N and 2,500 m above sea level (m a.s.l.). Climate is seasonal, with a 6–8 month dry season (November–April) and a mean annual precipitation varying between 450–800 mm. Soils consist of rendzic lithosols. The dominant vegetation type is drought deciduous oak forest having an average canopy height of 10–15 m (Salas et al. 1994). In Cántaros ('high fragmentation'), extensive clearing of forests and conversion to agricultural and grazing lands, combined with harvesting of trees for fuelwood and charcoal production, had created a mosaic of forest patches embedded within a predominantly open landscape matrix of agricultural lands and secondary vegetation. By contrast, the forests in Huauclilla ('low fragmentation') were relatively intact, with only a few small agricultural clearings within a forest matrix.

The experimental design consisted of nine edge gradients occurring between forest patches and adjacent openings. Of these nine gradients, three were located in 'small' openings (<1 ha) and three in 'large' openings (>1 ha) within the high-fragmentation site, whereas the remaining three gradients were located in 'reference' openings (<1 ha) within the low-fragmentation site. Subplots were established in four microsites along each edge gradient on the south-facing edge, as follows: (1) approximate center of the clearing ('open'), (2) 5 m into the opening from the edge ('shrub'), (3) 5 m into the forest vegetation from the edge ('edge'), and (4) 30 m into the forest vegetation from the edge ('forest'). In June of 1996, 1-year-old seedlings of five native plant species were transplanted into each microsite: *Quercus acutifolia* and *Q. castanea* (mature canopy late-successional tree species), *Pinus oaxacana* (mid-successional tree species), and *Rhus virens* and *Dodonaea viscosa* (early-successional shrub species). The following parameters were measured: microclimate in each microsite (surface soil moisture, photosynthetically active radiation, and temperature), plant growth response (height, leaf number, mortality), and plant physiological response (phenology, pre-dawn moisture potential (ψ_{pd}), nutrient uptake and resorption). Above- and below-ground biomass was determined when seedlings were harvested in February of 1998. A detailed description of the site, methods, and statistical analyses are presented in Asbjornsen et al. (2004a, b).

19.3 Effects of Habitat Fragmentation on the Regeneration Environment

Changes in surface soil moisture in response to forest cover loss vary distinctly with climate: in drier environments, surface soil moisture generally increases following forest removal (Joffre and Rambal 1988; Belsky et al. 1989; Veenendaal et al. 1995; Breshears et al. 1997; Asbjornsen et al. 2004a), whereas in moist climates the reverse usually occurs (Mladenoff 1987; Ashton 1992; McDonald et al. 2002). These patterns may be attributed to differences in trade-offs between the ameliorative effects of shading by trees on reducing moisture loss (more important in drier environments) and increased water loss through uptake and transpiration by deeply rooted trees (Belsky et al. 1993; Asbjornsen et al. 2004a). In the Mixteca Alta study, soil moistures were consistently lower for all microsites in the highly fragmented landscape, compared to the reference landscape. Along the edge gradients within each landscape, soil moisture also increased sharply from the open environment into the forest understory, which corresponded to a decrease in temperature and photosynthetically active radiation (PAR) along this same gradient (Asbjornsen et al. 2004a). Further, the steepness of the microclimate edge gradient, as well as differences between the fragmented and reference landscapes, diminished during a severe drought. Combined, these results highlight the important function of forest cover in buffering environmental change across edge environments. This buffering capacity may be reduced with increasing habitat fragmentation, and may also be amplified by disturbance events such as drought.

19.4 Seedling Biomass and Mortality in Response to Edge Gradients

In seasonally dry environments, moisture is often the most important factor controlling plant growth, both through direct effects on water availability (Stephenson 1990) and indirect effects on nutrient availability (Evans and Ehleringer 1994). Studies conducted in tropical montane oak forests located in moister climates have documented contrasting patterns for oak regeneration. In Chiapas (1,100–1,200 mm annual rainfall; 2,300–2,400 m a.s.l.), *Q. crispipilis* seedlings planted into early-successional and mature forest stands had 100 % mortality, whereas 30–40 % of oak seedlings planted to grassland and shrubland habitat survived (Quintana-Ascencio et al. 1992). Of these surviving oak seedlings, leaf production and height growth were also greatest for seedlings growing in the more open sites. Similarly, in montane oak forests in the Central Himalayas (1,200 mm annual rainfall; 1,800–2,200 m a.s.l.), Thadani and Ashton (1995) reported fewer oak seedlings in the forest understory compared to open, more disturbed sites, despite high grazing and harvesting pressures in the open, suggesting poor oak regeneration capacity in shade microsites. However, Oosterhoorn and Kappelle (2000), working in tropical montane oak forests in the Cordillera de Talamanca (2,100–3,000 mm annual rainfall; 2,300–2,800 m a.s.l.) reported the presence of oak seedlings and saplings in both open pasture land and mature forest understory.

In our Mixeca Alta study, oak seedling mortality was significantly greater in the open (78.0–91.2 %) than in the understory (20.6–57.0 %) microsites (Asbjornsen et al. 2004b; Table 19.1). Further, average mortality and total biomass production varied distinctly among the five species studied: *D. viscosa* and *R. virens* had the lowest mortality rates (3–13 %) and greatest biomasses (5.2–8.6 g), whereas *P. oaxacana* and the two oak species (*Q. acutifolia* and *Q. castanea*) had the greatest mortality rates (cf. above) and lower biomasses (1.1–1.9 g). Interestingly, seedlings of all species that survived in the open microsites had significantly higher biomasses than was the case for seedlings

Table 19.1. Seedling mortality, and total biomass per species and microsite, for seedlings growing at different microsites along edge gradients in the Mixteca Alta, Oaxaca (mean±SE). Significant differences among species are denoted by lowercase letters and based on ANOVA applying the Tukey means comparison test. For complete species names, see details in the main text

Species	*Q. acutifolia*	*Q. castanea*	*P. oaxacana*	*D. viscosa*	*R. virens*
Mortality (%)					
Open	84.8±5.9	91.2±4.0	87.0±4.3	7.4±2.5	30.6±10.0
	a	a	a	c	b
Shrub	78.1±8.1	87.2±6.2	74.3±8.5	3.4±1.7	30.0±8.9
	a	a	a	c	b
Edge	23.8±6.4	57.0±6.8	26.1±5.0	4.3±2.6	12.0±5.6
	b	a	b	c	c
Forest	20.6±8.7	43.2±6.6	16.0±7.6	2.5±1.6	0.6±0.6
	b	a	b	c	c
Biomass (g)					
Open	1.7±1.0	1.3±0.2	8.1±5.6	8.6±1.8	8.1±1.1
	b	b	a	a	a
Shrub	1.0±0.2	0.7±0.0	1.9±0.2	7.3±0.9	5.2±0.4
	bc	ne	b	a	a
Edge	1.3±0.2	6.4±0.7	1.9±0.4	6.4±0.7	7.7±0.8
	b	c	b	a	a
Forest	1.1±0.1	0.6±0.1	1.4±0.2	6.1±0.5	6.1±0.4
	b	c	b	a	a

growing in the forest understory environments. These patterns likely reflect the differential access among species to soil moisture at greater depths in the soil profile. After 2 years of growth, oak and pine seedlings had relatively shallow roots (maximum lengths of 7.9–10.3 and 10.1–14.6 cm, respectively), compared to the deeper roots (12.3–14.6 cm) of the shrubs (Asbjornsen et al. 2004b). Soil moisture generally increases with depth in open environments as a result of lower water uptake and evapotranspiration rates by relatively shallow-rooted herbaceous plants (Breshears et al. 1997). Thus, seedlings surviving in the open had probably established longer root systems capable of accessing deeper water supplies, and were thereby able to benefit from the greater photosynthetic potential in the high-light environment.

These contrasting patterns may reflect diverging successional dynamics for montane oak forests occurring under differing climatic regimes. In dry seasonal montane environments where periodic droughts can pose a constraint to regeneration, forest regrowth may be strongly determined by patterns of surface moisture availability and associated vegetative cover. For example, studies on oaks in seasonally dry temperate savanna and woodland ecosystems confirm that a critical factor determining seedling survival during the initial 2–3 year establishment phase is the ability of oak seedlings to develop root systems capable of accessing deeper water sources (Pase 1969; Brown and Archer 1990; Weltzin and McPherson 1997). Thus, regeneration of open environments in the Mixteca Alta may be restricted to narrow windows of opportunity when consecutive years of adequate rainfall occur.

19.5 Seedling Physiological Responses

19.5.1 Leaf Phenology

Leaves are one of the most dynamic and sensitive organs of perennial plants, as they integrate responses across multiple scales (from cellular to organismal to environmental), and can respond rapidly to environmental changes on the scale of seconds (e.g., photosynthesis, respiration) to days (e.g., flushing, senescence; Field 1991). Because phenological responses vary strongly among species and functional groups, they provide an important indicator of intrinsic ecophysiological relationships (Eamus and Prior 2001). In our Mixteca Alta study, oak seedlings dropped their leaves during the drought of 1998, but did not flush new leaves until the beginning of the following rainy season. By contrast, fast leaf turnover in *D. viscosa* was closely linked to soil moisture availability: leafs were rapidly shed in response to drought, whereas leaf flushing occurred when conditions became more favorable after intermittent rainfall events. This high phenological plasticity may present an important drought avoidance strategy in *D. viscosa*, enabling the rapid reduction of transpiring leaf surface area during periods of moisture stress, while maximizing resource acquisition for growth during favorable periods. This agrees with general patterns that mature forest species usually have relatively high specific leaf area (SLA) and leaf longevity, in contrast with lower SLA and faster leaf turnover rates typical of early-successional, disturbance-maintained species (Reich et al. 1995).

19.5.2 Seedling Moisture Stress

Mature oak trees can generally sustain stomatal conductance under low ψ_{pd} and high moisture stress conditions (Parker et al. 1982), although poor stomatal control may increase susceptibility to extreme moisture stress (Abrams 1990). Oak seedlings in our Mixteca Alta study consistently maintained the lowest ψ_{pd} (−3.7 to −4.0 MPa), suggesting that moisture stress likely contributed to their high mortality (Table 19.2; Asbjornsen et al. 2004b). Pine seedlings maintained the highest ψ_{pd} of all species (−2.1 MPa), which agrees with pine species' conservative water use strategy (Royce and Barbour 2001)

Table 19.2. Pre-dawn moisture potentials, and total root length for seedlings growing in open and forest microsites along edge gradients in the Mixteca Alta, Oaxaca (mean±SE). For complete species names, see details in the main text

Species	*Q. acutifolia*	*Q. castanea*	*P. oaxacana*	*D. viscosa*	*R. virens*
ψ_{pd} (MPa)					
Open	−4.0±0.2	−3.8±0.5	−2.2±0.1	−2.4±0.1	−2.9±0.2
Forest	−3.9±0.2	−3.7±0.2	−2.1±0.1	−3.5±0.1	−3.4±0.2
Root length (cm)					
Open	9.4±0.9	8.3±2.0	11.5±1.4	14.6±0.4	12.4±0.7
Forest	10.3±0.6	7.9±0.3	10.1±0.3	13.6±1.1	12.4±0.7

and high stomatal sensitivity to vapor pressure deficits (Marshall and Waring 1984). This also corroborates earlier observations that pines are likely to have an advantage over oaks in edge and open environments, since they are adapted to high-light and low-moisture conditions, whereas oaks are more tolerant of understory conditions. Shrub seedlings also maintained relatively high average ψ_{pd} (−2.9 MPa), which, together with their rapid growth rates, suggest that they likely had access to deeper water supplies (Asbjornsen et al. 2004b). Further, *D. viscosa* seedlings also exhibited significant differences in ψ_{pd} between the open (−2.4 MPa) and forest (−3.5 MPa) understory microsites. This pattern probably reflects the more limited water availability at lower soil depths in the forest understory, due to higher water uptake and transpiration rates by mature trees. Thus, the ability to rapidly shed leaves (see above) may enable shrub seedlings to persist in the forest understory in the short term but perhaps not on longer time scales, since eventually reduced photosynthetic gain – combined with high cost of continual leaf production – would result in an unfavorable carbon balance (Lambers and Poorter 1992).

19.5.3 Foliar Nutrient Status and Resorption

The availability of nutrients and water to plants is intricately linked, especially in dry climates where nutrient availability may fluctuate in direct response to changing soil moisture (Chapin and Moilanen 1991; Evans and Ehleringer 1994). Plants growing in seasonally dry climates have evolved a wide range of strategies to adapt to nutrient stress. In our study, foliar N concentrations were highest in species with the lowest SLA (*D. viscosa*, 1.49±0.02 %), and lowest in species with the highest SLA (*Quercus* spp. and *R. virens*: 1.26–1.31 and 0.99 %, respectively; Table 19.3). However, when assessed on an area basis, these differences were no longer evident (Table 19.3). These patterns agree with observations that foliar N concentration tends to be negatively correlated with leaf life span, whereas foliar N content on a mass basis is unrelated

Table 19.3. Mean foliar N and P concentrations, content, and resorption by senescing leaves for seedlings of five different species transplanted to different microsites along edge gradients in the Mixteca Alta, Oaxaca (data pooled across all microsites). Significant differences among species denoted by lowercase letters; based on ANOVA with Tukey means comparison test. For complete species names, see details in the main text

Species	*Q. acutifolia*	*Q. castanea*	*P. oaxacana*	*D. viscosa*	*R. virens*
Nutrient concentration (mass basis %)					
N	1.31±0.04	1.26±0.08	1.32±0.02	1.49±0.02	0.99±0.02
	ab	b	ab	a	c
P	0.054±0.001	0.061±0.003	0.085±0.002	0.087±0.002	0.061±0.003
	b	b	a	a	b
Nutrient content (area basis mg/cm^2)					
N	18.8±1.8	17.8±3.3	9.2±2.0	20.4±4.0	19.6±6.0
	a	a	b	a	a
P	0.91±0.12	1.03±0.33	0.59±0.01	1.19±0.04	1.27±0.09
	ab	ab	b	a	a
Nutrient resorption (%)					
N	38	37	42	48	41
P	28	43	56	51	5

to leaf life span, and rather a function of differences in leaf morphology and life-history strategy (Reich et al. 1992). Studying upper montane oak forests in Costa Rica, Kappelle and Leal (1996) also reported that mature forest species had greater leaf SLA and lower N concentrations, compared to plant species common to early-successional forests.

Nutrient resorption from leaves prior to abscission provides a mechanism by which plants can increase the amount of carbohydrate production per unit of nutrient (Chapin 1980; Aerts 1996). Although few studies have measured nutrient resorption for co-existing species from different successional seres, evidence suggests that early-successional species have greater nutrient resorption levels than is the case for late-successional species (Reich et al. 1995). Our data support this observation, as the early-successional species *D. viscosa* resorbed higher percentages of both N and P (48 and 51 %, respectively) from leaves prior to abscission, compared to the late-successional oak species (37–38 and 28–43 %, respectively; Table 19.3). These patterns are congruent with the theory that the primary mechanism by which later-successional species conserve nutrients is through extended leaf life span, which enables greater carbon fixation per unit of nutrient over a longer period of time (Chabot and Hicks 1982). By contrast, species having faster growth rates and rapid leaf turnover rates may rely more on internal cycling of nutrients (i.e., leaf resorption) to maintain sufficient nutrient supplies for continued leaf production (Silla and Escudero 2003).

19.6 Edges: Facilitative Effects or Regeneration Barriers?

The presence of shrubs and trees in edge environments may facilitate seedling establishment through ameliorative effects on local abiotic and biotic conditions influencing seedling establishment (Callaway 1992; Owens et al. 1995; Raffaele and Veblen 1998; Asbjornsen et al. 2004a, b). Facilitative effects of vegetative cover may be especially important in more arid climates, since effects on reducing radiation loads and protecting against soil moisture loss may outweigh moisture losses resulting from greater transpiration by large, deep-rooted trees. This pattern is supported by studies documenting greater oak seedling mortality in more open environments in drier climates (Callaway 1992; Negi et al. 1996; Pugnaire and Haase 1996; Castro et al. 2002), the reverse having been reported for moist climates (Lorimer et al. 1994; Thadani and Ashton 1995). Our seedling growth and physiological data for species from contrasting successional seres in the Mixteca Alta suggest that oak establishment is not improved in shrub microsites located adjacent to the forest edge compared to open microsites, while establishment is favored within inner edge and forest understory environments having greater vegetative cover (Asbjornsen et al. 2004b). This suggests that expansion of existing oak forests into open abandoned lands may be impeded, potentially due to reduced vegetative cover and increased radiation and temperature loads within shrub vegetation common to outer-edge and open environments. However, oak establishment may significantly improve in more open environments if seedlings can form a deep root system capable of competing for limited water supplies, which may only occur during climatic periods characterized by unusually high rainfall.

Regeneration patterns in these seasonally dry montane oak forests in the Mixteca Alta appear to differ from those observed in more moist montane oak forests. In the Chiapas highlands, pine and oak species regenerate extensively in exposed sites where former oak forests have been cleared for agricultural production and later abandoned, whereas neither oaks nor pines appear to regenerate successfully in the understory of mature or successional forests (González-Espinosa et al. 1991; Chaps. 14 and 16). Similarly, banj oak (*Q. leucotrichophora*) seedlings in central Himalayan forests were most abundant in partial canopy openings and nearly absent in closed canopy forests (Thadani and Ashton 1995). In disturbed montane oak forests in Costa Rica (2,812 mm annual rainfall; 2,900–3,000 m a.s.l.), dominant oak species established successfully both within early-successional shrub communities and across forest–pasture borders (Kappelle et al. 1996; Oosterhoorn and Kappelle 2000).

A possible explanation for these apparently divergent patterns is that the drier climate at the Mixteca Alta site may restrict oak survival and growth during the initial seedling stage, and may be more similar to regeneration dynamics common to drier oak savanna and woodland ecosystems where oak regeneration is often poor (Gordon and Rice 1993). Thus, in highly fragmented landscapes in the Mixteca Alta, edges may function as effective barriers to the re-establishment of oak forests, thereby reinforcing the persistence of forest patches separated by distinct ecological and microclimate boundaries. The ability of oak seedlings to persist in the understory of existing forest remnants may enable the maintenance of these remnants through gap dynamic processes, as old trees die (or are removed by humans) and allow oak saplings to grow into the canopy. By contrast, establishment of successional vegetation in edge and adjacent open environments appears to be favored in this landscape, and may eventually result in a gradual transition toward more late-successional oak forest communities, but probably on longer time scales than observed in more moist montane oak forest regions. The exact mechanisms underlying facilitative and competitive interactions, and their influence on long-term oak forest regeneration in these seasonally dry oak forests require further study. In particular, shrub densities in the edge environments may have been too great to support facilitation of oak and pine regeneration, whereas more sparsely distributed individuals in open areas may increase facilitative effects. Further, it remains to be determined whether successful oak seedling establishment in open and shrub environments may occur under conditions of normal or above average rainfall over a period of several years. Finally, the influence of seed dispersal dynamics on oak seedling recruitment as related to distance from the forest edge needs to be further explored.

19.7 Conclusions

Recovery of dry deciduous oak forests in highly fragmented landscapes in the Mixteca Alta, where much of the original forest cover had been converted to agricultural crops or pasture production for long time periods prior to abandonment in recent decades, will likely occur extremely slowly because of the severe climatic constraints on oak regeneration. Conservation efforts should target the protection of areas that are recovering from grazing and other chronic human disturbances, in order to maximize the potential of oak seedlings to successfully establish during favorable periods with adequate rainfall. Recovery of oak forests on the landscape may be enhanced through more active management practices such as planting oak seedlings having well-developed root systems, and targeting open areas of sparse tree and shrub cover.

References

Abrams MD (1990) Adaptations and responses to drought in *Quercus* species of North America. Tree Physiol 7:227–238
Aerts R (1996) Nutrient resorption from senescing leaves of perennials: are there general patterns? J Ecol 84:597–608
Aide TM, Zimmerman JK, Rosario M, Marcano H (1996) Forest recovery in abandoned cattle pastures along an elevational gradient in northeastern Puerto Rico. Biotropica 28(4a):537–548
Asbjornsen H, Ashton PMS, Vogt DJ, Palacios S (2004a) Effects of habitat fragmentation on the buffering capacity of edge environments in a seasonally dry tropical oak forest ecosystem in Oaxaca, Mexico. Agric Ecosyst Environ 103:481–495
Asbjornsen H, Vogt KA, Ashton PMS (2004b) Synergistic responses of oak, pine and shrub seedlings to edge environments and drought in a fragmented tropical oak forest, Oaxaca, Mexico. For Ecol Manage 192:313–334
Ashton PMS (1992) Some measurements of the microclimate within a Sri Lankan tropical rainforest. Agric For Meteorol 59:217–235
Belsky AJ, Amundson RG, Duxbury JM, Riha SJ, Ali AR, Mwonga SM (1989) The effects of trees on their physical, chemical, and biological environments in a semi-arid savanna in Kenya. J Appl Ecol 26:1005–1024
Belsky AJ, Mwonga SM, Amundson RG, Duxbury JM, Ali AR (1993) Comparative effects of isolated trees on their undercanopy environments in high- and low-rainfall savannas. J Appl Ecol 30(1):143–155
Breshears DD, Rich PM, Barnes FJ, Campbell K (1997) Overstory-imposed heterogeneity in solar radiation and soil moisture in a semiarid woodland. Ecol Appl 7:1201–1215
Brown JR, Archer SR (1990) Water relations of a perennial grass and seedling vs. adult woody plants in a subtropical savanna, Texas. Oikos 57:366–374
Callaway RM (1992) Effect of shrubs on recruitment of *Quercus douglasii* and *Quercus lobata* in California. Ecology 73(6):2118–2128
Castro J, Zamora R, Hódar JA, Gómez JM (2002) Use of shrubs as nurse plants: a new technique for reforestation in Mediterranean mountains. Restor Ecol 10(2):297–305
Chabot BF, Hicks DJ (1982) The ecology of leaf life-spans. Annu Rev Ecol Syst 13:229–259
Chapin FS III (1980) The mineral nutrition of wild plants. Annu Rev Ecol Syst 11:233–260
Chapin FS III, Moilanen L (1991) Nutritional controls over nitrogen and phosphorus fractions and autumn retranslocation in evergreen and deciduous taiga trees. Ecology 64:376–391
Chen J, Franklin JF, Spies TA (1993) Contrasting microclimates among clearcut, edge, and interior of old-growth Douglas-fir forest. Agric For Meteorol 63:219–237
Eamus D, Prior L (2001) Ecophysiology of trees of seasonally dry tropics: comparisons among phenologies. Adv Ecol Res 12:113–197
Euskirchen ES, Chen J, Bi R (2001) Effects of edges on plant communities in a managed landscape in northern Wisconsin. For Ecol Manage 148:93–108
Evans RD, Ehleringer JR (1994) Water and nitrogen dynamics in an arid woodland. Oecologia 99:233–242
Ewel J (1980) Tropical succession: manifold routes to maturity. Biotropica 12(2):2–7
Field CB (1991) Ecological scaling of carbon gain to stress and resource availability. In: Mooney HA, Winner WE, Pell EJ (eds) Response of plants to multiple stresses. Academic Press, San Diego, pp 35–65
Forman RTT, Godron M (1986) Landscape ecology. Wiley, New York
Geiger R (1957) The climate near the ground. Harvard Univ Press, Cambridge, MS

Gentry AH (1982) Neotropical floristic diversity: phytogeographical connections between central and south America, Pleistocene climatic fluctuations or an accident of the Andean orogeny? Ann Mo Bot Gard 63:787–842

González-Espinosa M, Quintana-Ascencio PF, Ramirez-Marcial N, Gaytan-Guzman P (1991) Secondary succession in disturbed *Pinus-Quercus* forests in the highlands of Chiapas, Mexico. J Veg Sci 2:351–360

Gordon DR, Rice KJ (1993) Competitive effects of grassland annuals on soil water and blue oak (*Quercus douglasii*) seedlings. Oecologia 79:533–541

Joffre R, Rambal S (1988) Soil water improvement by trees in the rangelands of southern Spain. Acta Oecol 9(4):405–422

Jose S, Gillespie AR, George SJ, Kumar BM (1996) Vegetation responses along edge-to-interior gradients in a high altitude tropical forest in peninsular India. For Ecol Manage 87:51–62

Kappelle M, Leal ME (1996) Changes in leaf morphology and foliar nutrient status along a successional gradient in a Costa Rican upper montane *Quercus* forest. Biotropica 28(2):331–344

Kappelle M, Geuze T, Leal ME, Cleef AM (1996) Successional age and forest structure in a Costa Rican upper montane *Quercus* forest. J Trop Ecol 12:681–698

Kirkby M (1972) The physical environment of the Nochixtlan Valley, Oaxaca. Vanderbilt University, Nashville, Publications in Anthropology

Lambers H, Poorter H (1992) Inherent variation in growth rate between higher plants: a search for physiological causes and ecological consequences. Adv Ecol Res 23:187–261

Lambers H, Poorter H, van Vuuren MMI (1998) Inherent variation in plant growth: physiological mechanisms and ecological consequences. Backhuys, Leiden

Leuschner C (2000) Are high elevations in tropical mountains arid environments for plants? Ecology 81(5):1425–1436

Lorimer CG, Chapman JW, Lambert WD (1994) Tall understory vegetation as a factor in the poor development of oak seedlings beneath mature stands. J Ecol 82:227–237

Marshall JD, Waring RH (1984) Conifers and broadleaf species: stomatal sensitivity differs in western Oregon. Can J For Res 14:905–908

McDonald MA, Healy JR, Stevens PA (2002) The effects of secondary forest clearance and subsequent land-use on erosion losses and soil properties in the Blue mountains of Jamaica. Agric Ecosyst Environ 92(1):1–19

Mladenoff DJ (1987) The relationship of the soil seed bank and understory vegetation in old-growth northern hardwood hemlock treefall gaps. Can J Bot 68:2714–2721

Negi AS, Negi GCS, Singh SP (1996) Establishment and growth of *Quercus floribunda* seedlings after a mast year. J Veg Sci 7:559–564

Niering WA, Egler FE (1955) A shrub community of *Viburnum lentago*, stable for twenty-five years. Ecology 36:356–360

Nixon KC (1993) The genus *Quercus* in Mexico. In: Ramamoorthy TP, Bye R, Lot A, Fa J (eds) Biological diversity of Mexico: origins and distribution. Oxford Univ Press, New York, pp 447–458

Oosterhoorn M, Kappelle M (2000) Vegetation structure and composition along an interior-edge-exterior gradient in a Costa Rican montane cloud forest. For Ecol Manage 126:291–307

Owens MK, Wallace RB, Archer SR (1995) Landscape and microsite influences on shrub recruitment in a disturbed semi-arid *Quercus-Juniperus* woodland. Oikos 74:493–502

Parker WC, Pallardy SG, Hinckley TM, Teskey RO (1982) Seasonal changes in tissue water relations of three woody species of the *Quercus-Carya* forest type. Ecology 77(5):1259–1267

Pase CP (1969) Survival of *Quercus turbinella* and *Quercus emoryi* in an Arizona chaparral community. Southwest Nat 14:149–156

Pugnaire FI, Haase P (1996) Facilitation between higher plant species in a semiarid environment. Ecology 77(5):1420–1426

Putz FE, Canham CD (1992) Mechanisms of arrested succession in shrublands: root and shoot competition between shrubs and tree seedlings. For Ecol Manage 49:267–275

Quintana-Ascencio PF, Gonzalez-Espinosa M, Ramirez-Marcial N (1992) Acorn removal, seedling survivorship, and seedling growth of *Quercus crispipilis* in successional forests of the highlands of Chiapas, Mexico. Bull Torrey Bot Club 119(1):6–18

Raffaele E, Veblen TT (1998) Facilitation by nurse shrubs of resprouting behavior in a post-fire shrubland in northern Patagonia, Argentina. J Veg Sci 9:693–698

Reich PB, Walters MB, Ellsworth DS (1992) Leaf lifespan in relation to leaf, plant and stand characteristics among diverse ecosystems. Ecol Monogr 62:365–392

Reich PB, Ellsworth DS, Uhl C (1995) Leaf carbon and nutrient assimilation and conservation in species of differing successional status in an oligotrophic Amazonian forest. Funct Ecol 9:65–76

Royce EB, Barbour MG (2001) Mediterranean climate effects. I. Conifer water use across a Sierra Nevada ecotone. Am J Bot 88(5):911–918

Rzedowski J (1978) Vegetation of Mexico. Limusa, Mexico DF, Mexico

Rzedowski J (1998) Diversidad y orígenes de la flora fanerogámica de México. In: Ramamoorthy TP, Bye R, Lot A, Fa J (eds) Biological diversity of Mexico: origins and distribution. Oxford Univ Press, New York, pp 45–60

Salas SH, Ramírez G, Schibli L, de Avila A, Aguilar R (1994) Análisis de la vegetación y el uso de la tierra actual en el estado de Oaxaca. II. Regiones de valles centrales, Sierra norte, y Papaloapan. La Sociedad para Estudios de los Recursos Bióticos de Oaxaca, AC, Costa Rica, pp 1–19

Sarmiento FO (1997) Arrested succession in pastures hinders regeneration of Tropandean forests and shreds mountain landscapes. Environ Conserv 24(1):14–23

Silla F, Escudero A (2003) Uptake, demand and internal cycling of nitrogen in saplings of Mediterranean *Quercus* species. Oecologia 136:28–36

Small E (1972) Photosynthetic rates in relation to nitrogen recycling as an adaptation to nutrient deficiency in peat bog plants. Can J Bot 50:2227–2233

Spores R (1969) Settlement, farming technology, and environment in the Nochixtlan valley. Science 166:557–569

Stephenson NL (1990) Climatic control of vegetation distribution: the role of the water balance. Am Nat 135(5):649–670

Stuart J, Kearney M (1981) Causes and effects of agricultural labor migration from the Mixteca of Oaxaca to California. University of California, San Diego, Work Pap US-Mexican Stud no 28

Thadani R, Ashton PMS (1995) Regeneration of Banj oak (*Quercus leucotricophora* A. Camus) in the Central Himalaya. For Ecol Manage 78:217–224

Veenendaal EM, Swaine MD, Agyeman VK, Blay D, Abebrese IK, Mullins CE (1995) Differences in plant and soil water relations in and around a forest gap in west Africa during the dry season may influence seedling establishment and survival. J Ecol 83:83–90

Waide R, Zimmerman JK, Scatena FN (1998) Controls of primary productivity: lessons from the Luquillo Mountains in Puerto Rico. Ecology 79(1):31–37

Weltzin JF, McPherson GR (1997) Spatial and temporal soil moisture resource partitioning by trees and grasses in a temperate savanna, Arizona, USA. Oecologia 112:156–164

Williams-Linera G, Dominguez-Gastelu V, Garcia-Zurita ME (1997) Microenvironment and floristics of different edges in a fragmented tropical rainforest. Conserv Biol 12(5):1091–1102

20 Morphological Variations of Gall-Forming Insects on Different Species of Oaks (*Quercus*) in Mexico

K. Oyama, C. Scareli-Santos, M.L. Mondragón-Sánchez, E. Tovar-Sánchez, and P. Cuevas-Reyes

20.1 Introduction

Oaks (*Quercus*, Fagaceae) are common in temperate, tropical and semiarid regions from sea level to 3,100 m in altitude. Oaks can form dense stands in the main mountain chains of Mexico, such as the Sierra Madre Oriental, Sierra Madre Occidental, Eje Neovolcánico Transversal, and Sierra Madre del Sur (Rzedowski 1994), covering approximately 6.4 % of the Mexican territory (Inventario Nacional Forestal 2000). Mexico is considered one of the centers of species diversification of the genus *Quercus*, with 161 species including 109 endemics (Valencia-Ávalos 2004; Chap. 1), which correspond to ca. 30 % of the total number of species of the genus reported globally. The genus *Quercus* has an unusually high frequency of interspecific hybridization (e.g., Petit et al. 2003), suggesting the importance of this process in oak speciation (Manos et al. 1999; Chap. 1).

It is also well known that oak species interact with gall-forming insects, particularly with gallwasps (Hymenoptera, Cynipidae, Cynipini; e.g., Rokas et al. 2003). This group of insects has been intensively studied in Europe but not in the Neartic and Neotropical realms, despite the fact that the major lineages of gallwasps are thought to have diverged in Mexico and Guatemala (Kinsey 1936). Particularly in Mexico, only few studies have documented the ecological relationship between gallwasps and oak species (Tovar-Sánchez and Oyama 2004).

In this chapter, we report about the morphological variation in galls induced by insects in some species of *Quercus* in Mexico. Particularly, we describe the variety of galls, their distribution within host plants, and differences between internal and external gall morphology. In addition, we document the interaction of gall-forming insects in a hybrid complex of two oak species.

Ecological Studies, Vol. 185
M. Kappelle (Ed.) Ecology and Conservation of Neotropical Montane Oak Forests
© Springer-Verlag Berlin Heidelberg 2006

20.2 Gall Induction and Development of Galls

Gall-plant interaction is very important, and more than 15,000 gall-inducing insect species have been described, grouped in seven orders (Rohfritsch and Shorthouse 1982).

Gall formation on vegetative tissues is the result of chemical and ecological interactions between plants and various organisms. Gall-inducing insects alter plant growth, and produce galls and other malformations in undifferentiated meristems of the host plant in response to specific insect chemical stimuli (McCalla et al. 1962; Byers et al. 1976; Ananthakrishnan 1984; Meyer 1987). This induction also modifies the balance of energy allocation to host nutritional quality and plant secondary metabolites for protection against natural enemies (Cornell 1983; Waring and Price 1990; Fernandes and Price 1992; Hartley and Lawton 1992). However, the potential advantage of galls for protection against parasitoids and predators has been considered unclear and controversial. Some studies indicate that galls and others structural tissues do not confer mechanical protection against parasitoid wasps (Price et al. 1987; Hawkins 1988), whereas in other cases the induction of galls negatively affects the incidence of natural enemies (Schultz 1992; Foss and Rieske 2004). The adaptive significance of gall induction is not clear because most galls have nutritive tissues that provide high-quality nutrition but do not represent necessarily enemy-free sites for gall-forming insects (Stone and Schönrogge 2003).

Most of the cynipids that induce galls in oaks are restricted to one or few host species (Burk 1979; Abrahamson et al. 1998, 2003), suggesting a great specificity. Each species induces a highly characteristic and complex gall, leading to strong variations in shape, size and structure (Stone and Cook 1998). Furthermore, oakwasps present two cyclic generations, sexual and parthenogenetic, which also differ in both insect and gall morphologies (Rey 1992). Most recent studies indicate that gall morphology is controlled by the genotype of gall-forming insects (Nyman et al. 1998; Stone and Schönrogge 2003). Thus, a great diversity of gall forms is expected in groups as diverse as Mexican oaks.

20.3 Gall Morphology in Mexican Oaks

20.3.1 Introduction to Gall Morphology in Mexican Oaks

In this study, we report on the gall morphology of 22 oak species collected along a latitudinal gradient covering 11 states of Mexico. We observed 142 gall morphs on 22 species of *Quercus* along a latitudinal gradient. One oak species hosted more than one gallwasp species; *Q. laurina*, *Q. rugosa*, *Q. crassipes* and *Q. resinosa* had a greater number of galls than was the case for the other oak species. Sixty-four (45 %) gall morphs were observed on leaves, 52 (36.6 %) on stems, and 26 (17.6 %) on petioles; 51 species (41.5 %) of galls were present on leaf adaxial surfaces, another 51 (41.5 %) on abaxial surfaces, and 26 (17 %) on both these surfaces.

Degrer-Jauffret and Shorthouse (1992) reported that the majority of cynipid galls associated with *Quercus* occurs on leaves and stems. One possible explanation is that this pattern may be associated with differences in secondary metabolites and nutrients between different plant organs (Coley and Barone 1996), promoting the occurrence of potential colonization sites on leaves and stems.

We observed two different patterns of gall distribution within oaks. Most of the galls (91 species, 64.1 %) had an isolated or individual distribution on each plant organ, some were aggregated (38 species, 26.7 %), and only 13 (9.2 %) species had a mixed combination of isolated and aggregated distributions. The pattern of aggregation in insects is related to differences in pressures exerted by natural enemies on isolated and aggregated galls, the degree of parasitoid specialization on gall-forming insects, and the female parasitoid's foraging behavior for choosing galls (Price et al. 1987; Hawkins et al. 1997; Stone and Schönrogge 2003).

In several oak species, we found different gall morphs such as globoids, spherics, conics and ovoids (Fig. 20.1). The most predominant forms were globoid and spheric, followed by conic, elliptic and lenticular, whereas discoid, fusiform, ovoid, rosette and triangular gall morphs occurred only rarely. Similarly, we found a wide variety of gall colors, green and brown being more frequent. Colors varied with gall maturation. Gall textures were woody, spongy or fleshly. Most galls (120 species, 85.2 %) were directly attached to the plant surface, only a few being pedunculate.

20.3.2 External Gall Morphology

External gall morphology of Mexican oaks ranges in complexity from slight depressions to galls with glabrous and pilose surfaces. Most oak galls analyzed did not have trichomes (107 species, 74.7 %). Only 35 (23 %) species had hairy texture. We compared the structure of trichomes and stomata in non-infested leaves vs. leaves with galls of some oak species, using scanning electron microscopy. In *Q. resinosa*, abaxial and adaxial leaf surfaces had stellate trichomes, whereas in the globoid galls we observed similar stellate trichomes, but with fewer arms (Fig. 20.2). By contrast, in *Q. arizonica* we observed a marked differentiation of trichomes between ungalled leaves and galls; leaves without galls had glandular trichomes whereas galls on *Q. arizonica* leaves had stellate trichomes (see Fig. 20.2).

Fig. 20.1a–f. Different gall morphologies in some Mexican *Quercus* species: **a** globoid and spheric galls (*Q. affinis* and *Q. fulva*, respectively); **b** conic galls (*Q. arizonica*); **c** ovoid galls (*Q. crassipes* and *Q. eduardii*); **d** amorphic galls (*Q. laurina*); **e** pubescent galls (*Q. rugosa* and *Q. arizonica*); and **f** discoid galls (*Q. crassipes* and *Q. laurina*)

We did not observe stomata in galls, except for *Q. rugosa*, which presented differences in stoma structure between leaves and galls (see Fig. 20.2). It is not clear what the function and structure of stomata in gall tissues are, but Dreger-Jauffret and Shorthouse (1992) proposed that stomatic structure is not functional in galls, due to the prevailing hypertrophy during gall formation.

Plant cuticles are covered by waxes that form an essential structural element of their surface, and are of functional and ecological importance to the interaction between plants and their environment (Barthlott et al. 1998). Epi-

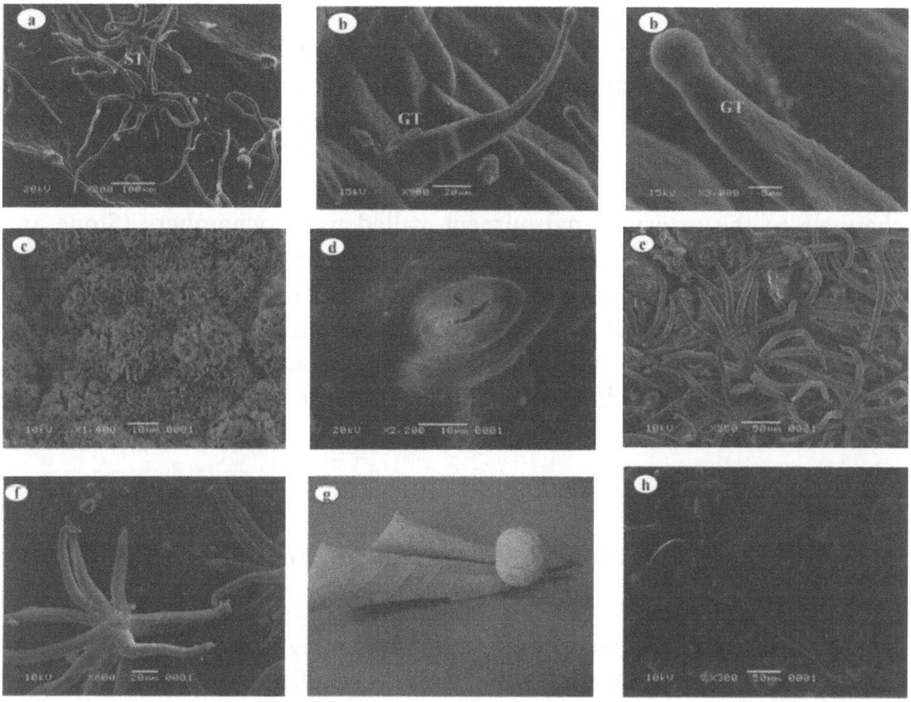

Fig. 20.2a–h. Scanning electron microscopy of ungalled leaves and galls: **a** glandular trichomes in ungalled leaves; **b** stellate trichomes in galls (*Q. arizonica*); **c** stoma structure in ungalled leaves, and **d** stoma structure in galls (*Q. rugosa*); **e** abaxial leaf surface and **f** adaxial leaf surface; **g** external gall morphology and **h** gall trichomes (*Q. resinosa*)

cuticular wax on gall surfaces determines the permeability of epidermic cells, affecting the performance and survival of gall larvae (e.g., Jeffree 1986). We observed that most galls had two wax types: crusts and films. Galls did not present waxes with prominent structures, which are typical for ungalled leaves. However, a great variation of wax types may occur on a single host plant. For example, *Q. arizonica* had platelet wax on leaf abaxial surfaces, and crust-type wax on the adaxial side, but galls had only thin wax films.

20.3.3 Internal Gall Morphology

Simple and complex galls are the result of biochemical and developmental processes that provide refuge and food to gall-forming insects (Bronner 1992). Internally, most galls had different kinds and distributions of tissues: a nutritive tissue lining the larval chamber; a lignified sclerenchymatous tissue with engrossed cellular walls; a parenchymatous cortex with vascular tissues connected to the host vascular system; and an epidermal tissue with rectangular cells. In general, we observed this structure in the galls collected. These different tissues change with larval age and life history of gall-forming insects. However, according to Weis et al. (1988), only parenchymatous and sclerenchymatous tissues change with gall age; the nutritive tissue constantly provides nutritive compounds such as nitrogen and lipids to gall larvae.

Some galls also have one or more larval chambers with communication channels to the external environment, called pseudochambers (Stone and Cook 1998). In Mexican oaks, we recorded only 16 species (11.3 %) of galls with pseudochambers. In addition, 101 galls (71.1 %) had only one larval chamber, and the rest (41 species, 28.9 %) had multilocular galls. In all cases, each gall chamber had only one larva. Exceptionally, in *Q. arizonica* we observed trichomes inside a gall chamber similar to those found at the external leaf surface.

20.4 The Role of Oak Hybridization in Gall-Forming Insects

Interspecific hybridization occurs frequently among oaks (Chap. 1). It is well documented that hybridization contributes to genetic diversity and plant speciation (Riesberg and Ellstrand 1993), but less is known about the effects of plant hybridization on the community structure and infestation of gall-forming insects. Diversity patterns and distribution of oak species affect the incidence and distribution of oak gall-forming insects (Stone et al. 2002). Mexico has the greatest richness of oak species (Valencia-Ávalos 2004; Chap. 1), with some cases of interspecific hybridization already documented (e.g., Bacon and Spellenberg 1996; González-Rodríguez et al. 2004; Tovar-Sánchez and Oyama 2004). Although we do not know how many gallwasp species occur in Mexican oaks, the greatest richness of oak gallwasps is found in the Neartic realm, with nearly 700 species in 29 genera (Weld 1957, 1960a, b).

Cynipid wasps in oak hybrid zones are highly sensitive to levels of introgression between host plant species (Boecklen and Spellenberg 1990). Distinct herbivore species respond in different ways to hybrids and their putative parent species (Morrow et al. 1994; Whitham et al. 1999). Host hybrid plants may support low, intermediate or high densities of herbivores in relation to their host parents (Fritz et al. 1994). Age and geographic range of hybrid zones, environmental gradients, microsite conditions, and genetic status of hybrids have been suggested as possible causes for these differences (Boecklen and Spellenberg 1990; Strauss 1994).

Floate et al. (1993) proposed that intermediate hybrid plants favor host shifts of endophagous insects from one host species to another. We tested this hypothesis for a hybrid complex of Mexican oaks including *Quercus crassifolia* and *Q. crassipes*. *Q. crassifolia* is distributed throughout the Sierra Madre Occidental, and *Q. crassipes* along the Sierra Madre Oriental. Both oak species

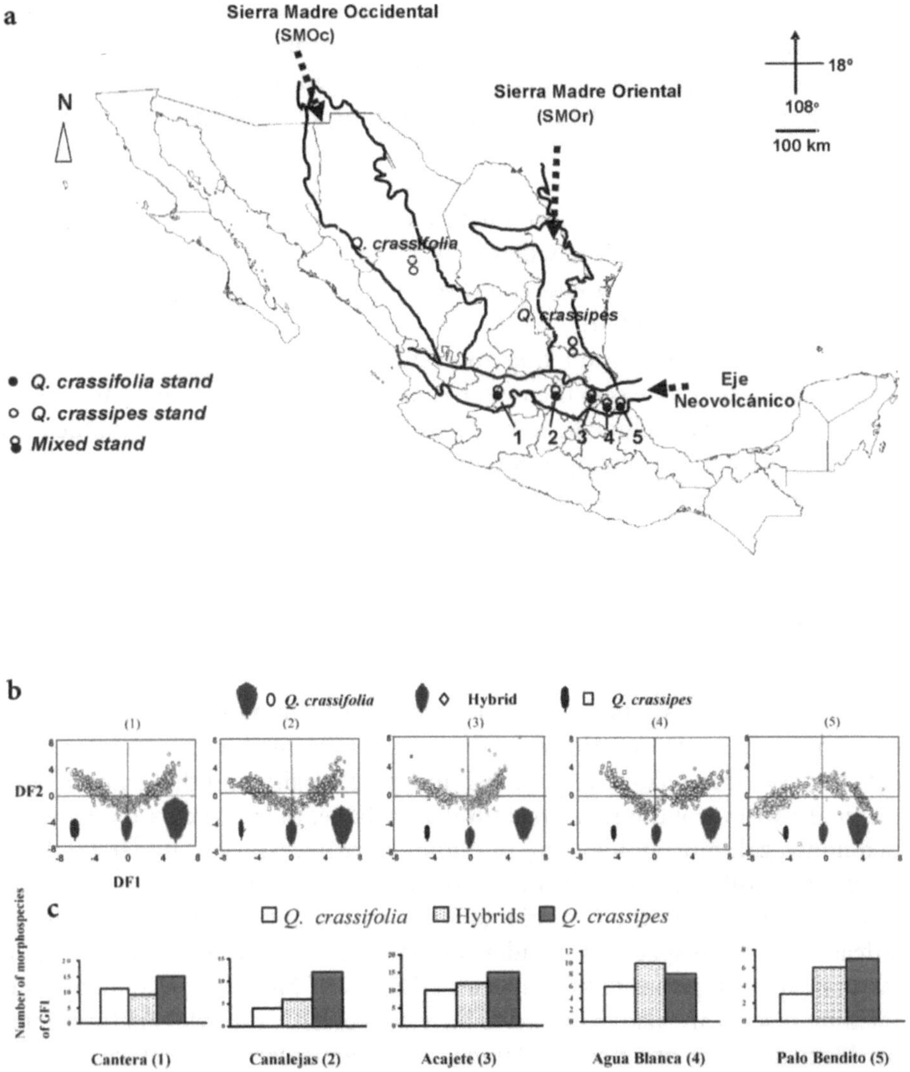

Fig. 20.3a–c. Hybrid complex of Mexican oaks. **a** Map of the locations of the five sampled populations of the *Q. crassifolia–Q. crassipes* complex; mixed stands are represented by numbers. **b** Discriminant function (DF) analysis of leaf morphology variation (17 measured characters). **c** Number of morphospecies of gall-forming insects (*GFI*) in five hybrid zones

overlap in distribution along the Neovolcanic Mountain Axis that crosses central Mexico in an east–west direction. Here, they form hybrids with intermediate characteristics (Fig. 20.3a).

The hybridization process in this oak complex was assessed using morphological characters (Fig. 20.3b) and DNA molecular markers (Tovar-

Sánchez and Oyama 2004). In five hybrid zones, we studied gall-forming wasps during 2 years, and we recorded 32 gall morphospecies (Fig. 20.3 c). We found that *Q. crassifolia* had only two specific gall morphs, and *Q. crassipes* had five specific morphs, while all gall morphospecies had already colonized the hybrid plants. Most interesting was that these hybrids had only one unique gall morph. This finding led us to assess the taxonomic distinctiveness of the two oak species and their putative hybrids. Based on the results of this assessment, we propose a theory in which hybrids act as a host bridge for gallwasps, and consequently create a new adaptive zone for gallwasp speciation (Tovar-Sánchez and Oyama 2005).

Currently, we are conducting molecular analyses using mitochondrial DNA of gallwasps to assess the patterns of molecular evolution. In this novel approach, we have included gallwasps collected from a wide geographic area in order to establish phylogeographic patterns and assess the genealogical relationships of gallwasps within this oak complex.

20.5 Conclusions

The knowledge of the interaction between gall-forming insects and oaks in Mexico is still very poor. However, the results presented in this chapter indicate that Mexican oaks have great gall morph diversity, and that external and internal gall morphology is structurally variable. More studies on gallwasp taxonomy, gall morphology and physiology are necessary to understand the adaptive significance of gall induction in oaks of temperate and tropical forests. Hybrids have the potential to act as new adaptive zones for the diversification of wasp species. Considering that hybridization is one of the main factors explaining the striking species richness of Mexican oaks, many gall-forming insects are also expected to have coevolved. Co-phylogenetic studies on Mexican oaks are now underway, and particularly aim at comparing endemic species with a restricted distribution to species with wider latitudinal and altitudinal distributions. This analysis is expected to contribute to our understanding of the co-evolutionary dynamics of this interaction at different spatial scales. Gall-forming insects associated to oaks may represent even greater species richness than found among oak species themselves. Therefore, this should be considered an important biodiversity component, which has its own specific conservation requirements within the diverse Mexican montane oak forests.

Acknowledgements We thank M.L. Herrera Arroyo, M.A. Pérez Pérez, J.C. Herrera-Flores, J.M. Cruz, and A. González-Rodríguez for field assistance, S. Zamudio and S. Valencia-Ávalos for plant species identification, and N. Pérez Nasser for technical support. This project was supported by grants from CONACYT (38550-V), CONACYT-SEMARNAT (2004-C01-97) and DGAPA-UNAM (IN229803) to K. Oyama.

References

Abrahamson WG, Melika G, Scrafford R, Csóka G (1998) Gall inducing insects provide insights plant systematic relationships. Am J Bot 85:1159–1165

Abrahamson WG, Hunter MD, Melika G, Price PW (2003) Cynipid gall-wasp communities correlate with oak chemistry. J Chem Ecol 29:209–223

Ananthakrishnan TN (1984) Adaptative strategies in cecidogenous insects. In: Ananthakrishnan TN (ed) The biology of gall insects. Oxford/IBH, New Delhi, pp 1–9

Bacon JR, Spellenberg R (1996) Hybridization in two distantly related Mexican black oaks *Quercus conzattii* and *Quercus eduardii* (Fagaceae: *Quercus*: section Lobatae). Sida 17:17–41

Barthlott W, Neinhuis C, Cutler D, Ditsch F, Meusel I, Theisen I, Wilhelmi H (1998) Classification and terminology of plant epicuticular waxes. Bot J Linn Soc 126:237–260

Boecklen WJ, Spellenberg R (1990) Structure of herbivore communities in two oaks (*Quercus* spp.) hybrid zones. Oecologia 85:92–100

Bronner R (1992) The role of nutritive cells in the nutrition of cynipids and cecidomyiids. In: Shorthouse JD, Rohfritsch O (eds) Biology of insect induced galls. Oxford Univ Press, Oxford, pp 118–140

Burks BD (1979) Superfamily Cynipoidea. In: Krombein KV, Hurd PD, Smith DR Jr, Burks BD (eds) Catalogue of Hymenoptera in America North of Mexico, vol I. Symphyta and Apocrita. Smithsonian Institution Press, Washington, DC, pp 1045–1107

Byers JA, Brewer JW, Denna DW (1976) Plant growth hormones in pinyon insect galls. Marcellia 39:125–134

Coley PD, Barone JA (1996) Herbivory and plant defenses in tropical forests. Annu Rev Ecol Syst 27:305–335

Cornell HV (1983) The secondary chemistry and complex morphology of galls formed by the Cynipidae (Hymenoptera): why and how? Am Midl Nat 2:225–235

Dreger-Jauffrete DF, Shorthouse JD (1992) Diversity of gall-inducing insects and their galls. In: Shorthouse JD, Rohfritsch O (eds) Biology of insect induced galls. Oxford Univ Press, Oxford, pp 8–33

Fernandes GW, Price PW (1992) The adaptive significance of insect gall distributions: survivorship of species in xeric and mesic habitats. Oecologia 90:14–20

Floate KD, Kearsley MJC, Whitham TG (1993) Elevated herbivory in plant hybrid zones: *Chrysomela confluens, Populus* and phonological sinks. Ecology 74:2056–2065

Foss LK, Rieske L (2004) Stem galls affects oak foliage with potential consequences for herbivory. Ecol Entomol 29:273–280

Fritz RS, Nichols-Orians CM, Brunsfeld SJ (1994) Interspecific hybridization of plants and resistance to herbivores: hypothesis, genetics, and variable responses in a diverse herbivore community. Oecologia 97:106–117

González-Rodríguez A, Arias DM, Valencia S, Oyama K (2004) Morphological and RAPD analysis of hybridization between *Quercus affinis* and *Q. laurina* (Fagaceae), two Mexican red oaks. Am J Bot 91:401–409

Hartley SE, Lawton JH (1992) Host-plant manipulation by gall-insects: a test of the nutrition hypothesis. J Anim Ecol 61: 113–119

Hawkins BA (1988) Do galls protect endophytic herbivores from parasitoids? A comparison of galling and non-galling Diptera. Ecol Entomol 13:473–477

Hawkins BA, Cornell HV, Hochberg ME (1997) Predators, parasitoids, and pathogens as mortality agents in phytophagous insect populations. Ecology 48:2145–2152

Inventario Nacional Forestal (2000) La condición actual de los recursos forestales en México: resultados del inventario forestal nacional 2000. UNAM Bol Inst Geogr 43:183–203

Jeffree CE (1986) The cuticle, epicuticular waxes and trichomes of plants, with reference to their structure, functions and evolution. In: Juniper BE, Southwood TRE (eds) Insects and the plant surface. Arnold, London, pp 23–64

Kinsey AC (1936) The origin of higher categories in Cynips. Indiana Univ Publ, Indiana

Manos PS, Doyle JJ, Nixon KC (1999) Phylogeny, biogeography, and process of molecular differentiation in *Quercus* subgenus *Quercus* (Fagaceae). Mol Phylog Evol 12:333–349

McCalla DR, Genthe MK, Hovanitz W (1962) Chemical nature of an insect gall growth-factor. Plant Physiol 37:98–103

Meyer J (1987) Plant galls and gall inducers. Borntraeger, Berlin

Morrow PA, Whitham TG, Potts BM, Ladiges P, Ashton DH, Williams JB (1994) Gall-forming insects concentrate on hybrid phenotypes of *Eucalyptus*. In: Price PW, Mattson WJ, Baranchikov YN (eds) The ecology and evolution of gall-forming insects. USDA Forest Service, North Central Forest Experimental Station, Rep GTR NC-174, pp 121–134

Nyman T, Roininen H, Vuorinen JA (1998) Evolution of different gall types in willow-feeding sawflies (Hymenoptera: Tenthredinidae). Evolution 52:465–474

Petit RJ, Bodénès C, Ducousso A, Roussel G, Kremer A (2003) Hybridization as a mechanism of invasion in oaks. New Phytol 161:151–164

Price PW, Fernandes GW, Waring GL (1987) Are galling insects better protected against parasitoids than exposed feeder? A test using tenthredinid sawflies. Ecol Entomol 13:195–205

Rey LA (1992) Developmental morphology of two types of hymenoptera galls. In: Shorthouse JD, Rohfritsch O (eds) Biology of insect induced galls. Oxford Univ Press, Oxford, pp 86–101

Riesberg L, Ellstrand NC (1993) What can molecular and morphological markers tell us about plant hybridization? Curr Rev Plant Sci 12:213–241

Rohfritsch O, Shorthouse JD (1982) Insect galls. In: Kahls G, Schell JS (eds) Molecular biology of plant tumors. Academic Press, New York, pp 131–152

Rokas A, Melika G, Abe Y, Nieves-Aldrey JL, Cook JM, Stone GN (2003) Lifecycle closure, lineage sorting, and hybridization revealed in a phylogenetic analysis of European oak gallwasps (Hymenoptera: Cynipidae: Cynipini) using mitochondrial sequence data. Mol Phylog Evol 26:36–45

Rzedowski J (1994) Vegetación de México. Editorial Limusa, México DF, Mexico

Schultz BB (1992) Insect herbivores as potential causes of mortality and adaptation in gall forming insects. Oecologia 90:297–299

Stone GN, Cook JM (1998) The structure of cynipid oak galls: patterns in the evolution of an extended phenotype. Proc R Soc Lond B 265:979–988

Stone GS, Schönrogge K (2003) The adaptive significance of insect gall morphology. Tree 18:512–522

Stone GN, Schönrogge K, Atkinson RJ, Pujade-Villar J (2002) The population biology of gall wasps (Hymenoptera: Cynipidae). Annu Rev Entomol 47:633–688

Strauss SY (1994) Levels of herbivory and parasitism in host hybrid zones. Tree 9:209–214

Tovar-Sánchez E, Oyama K (2004) Natural hybridization and hybrid zones between *Quercus crassifolia* and *Quercus crassipes* (Fagaceae) in Mexico: morphological and molecular evidence. Am J Bot 91:1352–1363

Tovar-Sánchez E, Oyama K (2005) Effect of hybridization of the *Quercus crassifolia* × *Quercus crassipes* complex on the community structure of endophagous insects. Oecologia (in press)

Valencia-Ávalos S (2004) Diversidad del género *Quercus* (Fagaceae) en México. Bol Soc Bot Méx 75:33–53

Waring GL, Price PW (1990) Plant water stress and gall formation (Cecidomyiidae: *Asphondylia* spp.) on creosote bush. Ecol Entomol 15:87–95

Weis AE, Walton R, Greco CL (1988) Reactive plant tissue sites and the population biology of gall makers. Annu Rev Entomol 33:467–486

Weld LH (1957) Cynipid galls of the Pacific Slope. Privately printed. Ann Arbor, Michigan

Weld LH (1960a) Cynipid galls of the Eastern United States. Privately printed. Ann Arbor, Michigan

Weld LH (1960b) Cynipid galls of the southwest. Privately printed. Ann Arbor, Michigan

Whitham TG, Martinsen GD, Floate KD, Dungey HS, Potts BM, Keim P (1999) Plant hybrid zones affect biodiversity: tools for a genetic-based understanding of community structure. Ecology 80:416–428

21 Above-Ground Water and Nutrient Fluxes in Three Successional Stages of Costa Rican Montane Oak Forest with Contrasting Epiphyte Abundance

L. Köhler, D. Hölscher, and C. Leuschner

21.1 Introduction

In the Cordillera de Talamanca (Costa Rica), a substantial part of the primary (old-growth) montane oak forest has been cut for timber and charcoal production or for conversion to crop fields during the last few decades (Kappelle and Juárez 1995; Chap. 30). In large areas, land is abandoned after several years and fields are subject to secondary succession (Kappelle et al. 1995, 1996; Chap. 17). Today, secondary forests are increasing in area throughout the tropics, and they accounted already in the 1980s for about 40 % of the total forest area (Brown and Lugo 1990). Conversion of old-growth forests leads to important changes in water and nutrient cycles at the ecosystem level (Bruijnzeel 1990). Whether these effects are reversed during forest recovery and, if so, how long it would take to recover from disturbance is largely unknown. In the present study, we investigated an old-growth and two nearby secondary forest stands of different age in the upper montane oak forest belt of the Cordillera de Talamanca, with the purpose to (1) analyse differences in nutrient fluxes associated with throughfall, stemflow (runoff along the outer bark) and litterfall, and (2) relate results to stand structure.

21.2 Study Sites

This study was conducted in the upper part of the Rio Savegre watershed on the Pacific slope of the Cordillera de Talamanca in Costa Rica (9°35'40"N, 83°44'30"W). Sites were situated in the upper montane forest belt of the Los Santos Forest Reserve at an elevation of ca. 2,900 m a.s.l. (above sea level). Average annual temperature at a weather station at 3,000 m a.s.l., 15 km east of the study sites near Villa Mills, is 10.9 °C. Here, average annual rainfall is

Ecological Studies, Vol. 185
M. Kappelle (Ed.) Ecology and Conservation of Neotropical Montane Oak Forests
© Springer-Verlag Berlin Heidelberg 2006

2,812 mm (IMN 1988). Studied stands included a 10–15 year old early-successional forest (ESF), a 40-year-old mid-successional forest (MSF), and an old-growth forest (OGF). The OGF had previously been described as 'primary forest' or 'mature forest' (Kappelle et al. 1995, 1996; Chap. 17).

The OGF and the MSF sites had already been studied in detail with respect to vascular plant diversity and forest structure (Kappelle et al. 1995, 1996; Chap. 17). They were largely dominated by *Quercus copeyensis* (now known as *Q. bumelioides* – K.C. Nixon, personal communication). Beside *Q. copeyensis*, some early-successional tree species were present in the ESF; a second oak species, *Q. costaricensis*, co-dominated the OGF. The height of the upper tree layer was about 30–35 m in OGF, 5–9 m in ESF, and 11–15 m in MSF. The leaf area index (LAI) in the three stands was analysed applying two different methods: (1) we conducted measurements with an optical canopy analyser (LAI 2000, Licor) with 30 replicate measurements per stand; and (2) we estimated leaf area from leaf life span as determined in individually marked leaves, annual leaf litter mass and specific leaf area (Roberts et al. 1996). In each stand, the epiphyte biomass and canopy humus were sampled in the canopy of six oaks (*Q. copeyensis*) which reached the upper canopy level. Samples were taken in a vertically stratified manner.

21.3 LAI and Epiphyte Biomass

There was no significant difference in the average leaf area index measured using the LAI 2000 system between ESF (3.8 $m^2 m^{-2}$) and OGF (3.8 $m^2 m^{-2}$). A significantly greater leaf area was found in MSF (4.7 $m^2 m^{-2}$). The second method revealed a similar pattern but considerably higher values (ESF: 8.8, MSF: 13.8, OGF: 7.7 $m^2 m^{-2}$). For other tropical montane forests, LAI values of 2.0–5.1 are reported (Steinhardt 1979; Tanner 1980; Weaver et al. 1986), whereas higher values are generally found in neotropical lowland forests (4.4–6.7; Roberts et al. 1996; Tobón-Marin 1999). High LAI in the young ESF stand suggests that leaf area recovers rapidly within the first years of tropical forest succession. This is in agreement with other studies in tropical forests (as reviewed in Hölscher et al. 2004a).

In all stands, epiphytic vegetation was dominated by non-vascular plants (bryophytes and lichens; Holz et al. 2002, and Chap. 7) which accounted for 99 % (ESF), 89 % (MSF) and 70 % (OGF) of the total epiphyte biomass (including canopy humus). At the stand level, the total epiphyte biomass increased with stand age, from 160 kg dry weight ha^{-1} in ESF to 520 kg ha^{-1} in MSF and 3,400 kg ha^{-1} in OGF. These estimates are at the lower range of epiphyte biomass data reported from tropical montane forests (370 kg ha^{-1}, Tanner 1980, 1985, to 44,000 kg ha^{-1}, Hofstede et al. 1993). We contend that the rather low biomass of epiphytes in OGF is mainly a consequence of the low frequency of clouds and fog, and the occurrence of a pronounced dry season in this region of the Pacific slope of the Cordillera de Talamanca. The low values measured in both secondary forests are likely to reflect young stand age because, even in cloud forest areas with more favourable conditions for epiphyte growth, recovery of epiphyte biomass is a slow process (Nadkarni et al. 2004).

21.4 Water and Nutrient Fluxes

Incident rainfall (gross precipitation) was collected during a 1-year period at two clearings using five bulk samplers per site. At the OGF site, we measured 2,830 mm year^{-1}, and in the clearing close to the secondary forest stands 2,900 mm year^{-1} (Table 21.1). Throughfall was determined with 30 fixed gauges per site; it did not show significant differences between stands, with annual totals averaging 69 % (ESF), 75 % (MSF) and 73 % (OGF) of incident rainfall. Stemflow was measured using ten collectors per site. In the MSF and OGF stands, stemflow samplers were installed on five trees of the upper tree layer, and on five trees of a lower stratum. Due to the less complex structure of the ESF stand, all stemflow collectors in this plot were installed on trees extending into the upper tree layer. At stand level, stemflow showed considerable differences among the stands, with values of 16 % (ESF), 17 % (MSF) and 2 % (OGF) of incident rainfall. Canopy interception reached 25 % in OGF, but only 15 % in ESF and 9 % in MSF. The interception measured in the OGF stand is relatively high in comparison to other neotropical upper montane forests (Bruijnzeel and Proctor 1995; Bruijnzeel 2001). This may reflect the low frequency of fog or low clouds, and a correspondingly high radiation load in this stand.

The nutrient concentration in incident rainfall showed only small differences among the three stands. Acidity (pH) of throughfall water (ESF: 6.5, MSF: 6.1, OGF: 6.0) was higher than in incident rainfall (ESF and MSF: 5.6,

Table 21.1. Gross precipitation (*Pg*), throughfall (*Tf*), stemflow (*Sf*) and interception (E_i) in the studied early-successional forest (*ESF*), mid-successional forest (*MSF*) and old-growth forest (*OGF*) over a 12-month period (June 1999–May 2000). *N* Number of collectors

	Pg (mm)			Tf (% Pg)			Sf (% Pg)			E_i (% Pg)
	Mean	SD	*n*	Mean	SD	*n*	Mean	SD	*n*	Mean
ESF	2,900	38	5	69	33	30	16	9	5–10	15
MSF	2,900	38	5	75	23	30	17	*[a]	10	9
OGF	2,830	68	5	73	19	30	2	*	10	25

[a]*, no standard deviation (SD) could be calculated due to sampling design

OGF: 5.3). Nutrient concentrations in throughfall water exceeded those measured in incident rainfall. In upper canopy trees, the average pH of stemflow water was 5.7 in ESF, 4.5 in MSF and 4.2 in OGF. Within the taller MSF and OGF stands, upper canopy trees usually had significantly higher nutrient concentrations in stemflow than was the case for lower canopy trees.

Element transportation within water fluxes in the canopy were calculated by multiplying measured water fluxes by the mean element concentration of the corresponding period. Element fluxes associated with incident rainfall in the Cordillera de Talamanca (Table 21.2) are within the range of literature values reported for tropical montane forests, as summarized by Hafkenscheid (2000). Only the input of Mg was found to be lower at sites analysed in the present study, compared to other tropical montane areas. Differences in below-canopy nutrient fluxes (stemflow plus throughfall, i.e. net precipitation) between secondary and old-growth forest stands in the Cordillera de Talamanca were only small. We therefore conclude that the size of canopy leaf area or specific structural components of the canopy, such as vascular or non-vascular epiphytes, have only a limited influence on nutrient fluxes through the canopy in these stands.

All nutrients (except NO_3-N) showed increasing flux rates from incident rainfall to net precipitation during tree crown passage. In the case of K, fluxes

Table 21.2. Nutrient fluxes in incident rainfall (*IR*), throughfall (*TF*) and stemflow (*SF*) in the studied early-successional forest (*ESF*), mid-successional forest (*MSF*) and old-growth forest (*OGF*). *Open area OGF* Measurements on clearing adjacent to OGF plot, *open area ESF* and *MSF* measurements on clearing close to ESF and MSF plots, *sum* net precipitation (throughfall+stemflow). After Hölscher et al. (2003), with permission from Cambridge University Press

	Flux	H_2O^a	K^b	Ca^b	Mg^b	Na^b	NH_4-N^b	NO_3-N^b
Open area								
ESF and MSF	IR	2,900	5.9	5.2	0.7	2.3	2.2	2.0
OGF	IR	2,830	5.8	4.0	0.7	3.0	1.4	1.7
Forest stands								
ESF	TF	2,003	54.7	12.8	4.1	2.3	2.4	0.8
	SF	466	11.3	1.6	0.2	0.9	0.3	0.1
	Sum	2,469	66.1	14.4	4.4	3.2	2.7	1.0
MSF	TF	2,171	41.1	12.4	3.4	4.1	5.7	1.0
	SF	480	14.5	6.0	1.7	0.7	0.4	0.1
	Sum	2,651	55.5	18.5	5.1	4.8	6.2	1.1
OGF	TF	2,058	61.8	14.8	5.4	3.2	3.4	0.6
	SF	63	3.5	0.8	0.2	0.2	<0.1	<0.1
	Sum	2,121	65.3	15.6	5.6	3.4	3.4	0.6

[a] Data given in mm $year^{-1}$

[b] Data given in kg ha^{-1} $year^{-1}$

increased by factors 11 (ESF), 9 (MSF) and 11 (OGF). The flux increase was less pronounced in the case of Na (1, 2 and 1), Ca (3, 4 and 4), Mg (6, 7 and 8) and NH_4-N (1, 3 and 2). Only NO_3-N (0.5, 0.6 and 0.4) showed a decreasing nutrient flux during crown passage, indicating a nitrate sink in the canopy. The enrichment factors for K, Na, Ca, Mg and NH_4-N are within the range observed in other tropical forest stands. Large below-canopy K fluxes in the ESF stand could result from high K concentrations in leaves of early-successional tree species (Kappelle and Leal 1996). By contrast, K fluxes were relatively small in the MSF stand, despite a large leaf area.

The relative proportion of stemflow to the nutrient fluxes with net precipitation showed large differences among the three forest stands. Stemflow accounted for 17 % of the K flux with net precipitation in ESF and for 26 % in MSF, but only for 5 % in OGF. Thus, stemflow was more important for nutrient fluxes in secondary forest stands than in OGF. When flux is expressed per single upper canopy tree stem, nutrient fluxes with stemflow were higher in OGF than in secondary forest stands. However, the much higher stem density in the secondary forests (diameter at breast height >3 cm: 5,400, 1,900 and 231 stems ha^{-1} in the ESF, MSF and OGF stands respectively) more than compensated for the lower specific stemflow rates.

We suggest that high water and nutrient fluxes with stemflow in the two secondary forest stands were due to a small canopy diameter and steep branching angles of young *Q. copeyensis* trees in these stands, which effectively channel rainwater to their lower trunks. By contrast, the large canopies and more horizontal branching patterns of tall adult Quercus trees in the OGF stand resulted in a smaller stemflow. A small contribution of stemflow to the water and nutrient fluxes, with values ranging between 1 % (Tobón-Marin 1999) and 8 % (Jordan 1978) of net precipitation, was also reported from other neotropical old-growth forests.

By contrast, high water and nutrient fluxes with stemflow (23 and 41 % of incident rainfall) were found by Hölscher et al. (1998) in two secondary forest stands in Eastern Amazonia; they were explained by high funnelling ratios of the standing vegetation. In Central Amazonia, Schroth et al. (2001) observed that stemflow was especially important for nutrient fluxes in vegetation with high stem densities, such as fallows. Similarly, in old-growth montane forests of Jamaica with high stem densities, Hafkenscheid (2000) measured large stemflow fractions of rainfall (13 and 18 %). Thus, high stemflow percentages do not seem to be characteristic for young secondary forests alone; rather, they are related to stands with high stem densities at the upper canopy level.

21.5 Litterfall and Associated Nutrient Fluxes

Litterfall was collected monthly over a period of 1 year with ten, systematically distributed litter traps (250-cm² aperture) in each stand. Litter was separated into four fractions, i.e. leaves, vascular epiphytes, non-vascular epiphytes (bryophytes and lichens) and trash (including twigs). The total annual litter production showed great differences between the three stands; it was lowest in ESF (9,333 kg ha^{-1} $year^{-1}$) and highest in MSF (17,197 kg ha^{-1} $year^{-1}$), with an intermediate value in OGF (12,870 kg ha^{-1} $year^{-1}$). Leaves dominated litter fractions (Fig. 21.1), which contributed to 84 % (ESF), 73 % (MSF) and 56 % (OGF) of total litter. In all stands, a distinct seasonal pattern of leaf litterfall was found, with low values in the rainy season and up to 12 times higher litterfall rates in the dry season. Leaf litter fractions were strongly dominated by *Quercus* leaves (mainly *Q. copeyensis*) in all three stands.

In the other three litter fractions, annual production was lowest in the ESF and highest in the OGF stand (Fig. 21.1). The contribution of epiphytes

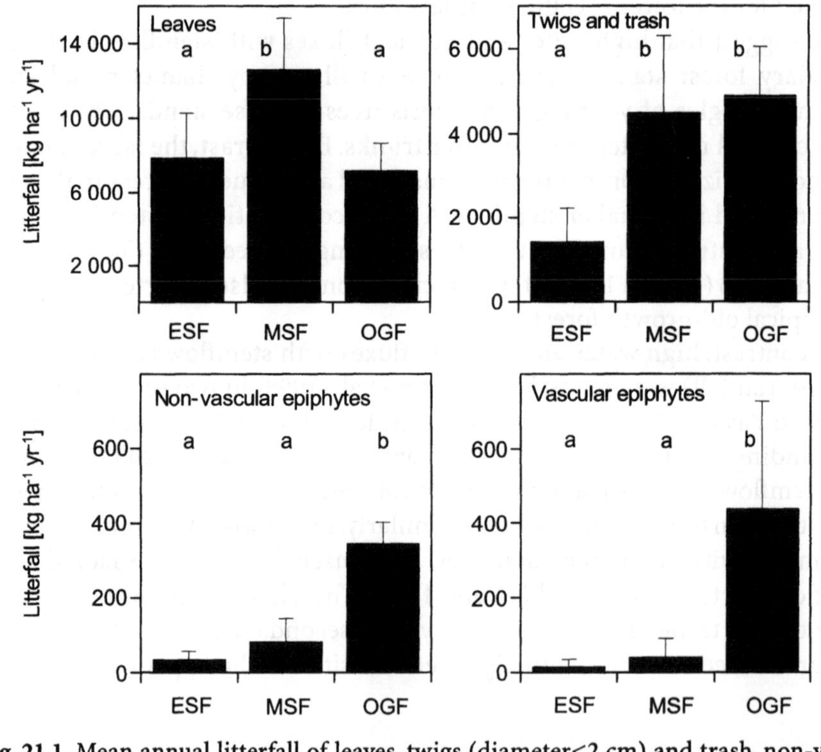

Fig. 21.1. Mean annual litterfall of leaves, twigs (diameter<2 cm) and trash, non-vascular epiphytes and vascular epiphytes in the studied early-successional forest (*ESF*), mid-successional forest (*MSF*) and old-growth forest (*OGF*). Significantly different means among the forest types are indicated by different letters. Vascular epiphytes: Wilcoxon U-test, $P<0.05$; other litter fractions: Scheffé-test, $P<0.05$

to total litterfall was low with 0.5 % in the ESF and 0.7 % in the MSF stands, but accounted for 6.1 % in OGF. Litterfall of epiphytes, twigs and trash showed a high temporal fluctuation with no marked seasonal pattern. A seasonal maximum in December 1999 was most likely caused by occasional high wind speeds during that month. Litterfall measurements in old-growth cloud forests or high montane forests (>2,500 m) in the tropics showed that the litterfall quantities in these forests are relatively low, compared to those found in lowland forests (Bruijnzeel and Proctor 1995). Our litterfall total for the OGF stand is the highest value so far reported for tropical old-growth montane forests. It is comparable to the litterfall measured by Lambert et al. (1980) in a lowland rainforest in Belize. Litterfall amounts in ESF are within the range of values reported for montane secondary forests (as reviewed by McDonald and Healey 2000). The high litterfall measured in the MSF stand is still smaller than the very high value of 27,000 kg ha^{-1} $year^{-1}$ found by Fournier and Comacho de Castro (1973) in a premontane secondary forest stand in Costa Rica.

We consider that the high litterfall values recorded in the Cordillera de Talamanca may be a consequence of the dominance of oak species in these forests. In our stands, *Quercus* accounted for over 90 % of the total leaf litter on many sampling occasions. In a Mexican montane forest, Williams-Linera and Tolome (1996) measured a higher litterfall in tree species from holarctic genera such as *Quercus* than in species belong to genera of tropical origin. Thus, the large dominance of oaks in the upper belt of the Cordillera de Talamanca possibly explains the high litter mass, which plays a major role in nutrient fluxes in these forests.

Fresh leaves from the sun canopy of *Quercus copeyensis* did not reveal significant differences in N, K and Mg concentrations among the three stands. By contrast, Ca concentration was significantly lower in OGF than in both secondary forest stands, and Mn showed significantly higher concentrations in MSF than in ESF and OGF. The nutrient concentrations in leaf litter from the Cordillera de Talamanca are largely within the range of values reported for other old-growth montane forests in the tropics (Heaney and Proctor 1989). Litter of the parasitic shrub *Phoradendron* revealed remarkably high leaf concentrations of K, this being more than 8 times higher than those of tree leaf litter.

Nutrient-return with leaf litterfall (trees and *Phoradendron* shrubs) was calculated by multiplying litter mass by the average nutrient concentration in the corresponding litter fraction. Element fluxes with litterfall of non-vascular epiphytes were calculated on the basis of average element concentrations of living bryophytes. For most elements, annual nutrient fluxes with leaf litterfall were highest in the MSF stand (Table 21.3). Nutrient-returns calculated for the three forest stands are very high in comparison with other tropical montane forests. This is mainly due to the high amounts of leaf litter in these stands. The particularly high Ca concentrations in the leaf litter of the ESF

Table 21.3. Nutrient fluxes with litterfall of tree leaves, litter of the parasitic shrub *Phoradendron* and litter of non-vascular epiphytes (bryophytes and lichens) in the studied early-successional forest (*ESF*), mid-successional forest (*MSF*) and old-growth forest (*OGF*). Data are given in kg ha^{-1} $year^{-1}$

	K	Ca	Mg	Mn	N
Tree leaf litter					
ESF	32.6	144.3	17.0	8.0	71.6
MSF	36.2	125.6	20.7	26.1	119.1
OGF	26.6	64.6	14.0	9.8	76.3
Litter of *Phoradendron*					
ESF	0.3	0.1	0.0	0.0	0.1
MSF	0.1	0.0	0.0	0.0	0.0
OGF	18.3	4.6	1.2	1.2	8.7
Litter of non-vascular epiphytes					
ESF	0.1	0.1	0.1	0.0	0.3
MSF	0.3	0.3	0.1	0.1	0.8
OGF	1.4	1.2	0.4	0.2	3.3

resulted in Ca transfers to the soil in this stand which exceeded those of the MSF and OGF considerably.

Nutrient fluxes with litterfall of non-vascular epiphytes (mainly bryophytes) were low for all elements. In the OGF stand, this flux represented only 5 % (K), 2 % (Ca), 3 % (Mg), 2 % (Mn) or 4 % (N) of the nutrient flux with leaf litterfall. In the secondary forests, none of the nutrient species accounted for more than 1 % of the corresponding nutrient flux with leaf litter. Compared with the nutrient input into the forest floor via net precipitation, the nutrient flux with leaf litterfall was 16–20 times larger for N, 4–10 for Ca, and 2–4 for Mg. Only the K flux, which is often underestimated due to leaching losses from litter traps (Proctor 1983; Hafkenscheid 2000), showed about 2 times larger fluxes via net precipitation than by leaf litter for all three stands.

21.6 The Influence of Epiphytes on Water and Nutrient Fluxes

The abundant epiphyte vegetation of tropical montane rainforests, often dominated by non-vascular epiphytes (bryophytes and lichens) in terms of biomass, is expected to influence the magnitude of canopy water and nutrient fluxes (throughfall, stemflow and rainfall interception; Pócs 1980; Veneklaas and Van Ek 1990; Coxson and Nadkarni 1995). We tested these predictions in a modelling study (Hölscher et al. 2004b). Remarkably, the relative importance of the mossy epiphyte component for total rainfall interception was less than 10 % and, thus, relatively small in the OGF stand, despite its considerable maximum water storage capacity. This is mainly explained by the fact that, under high rainfall conditions, only a small fraction of the total water storage capacity of the epiphytes is actually available at the beginning of the next rainfall event. Due to the much lower epiphyte biomass in secondary forests, the influence of epiphytes on the hydrological fluxes should be even lower in these stands.

Although epiphyte biomass differed more than 20-fold among the three stands, we did not observe a clear impact of epiphytes on nutrient fluxes in the canopy. The transfer of K with throughfall and stemflow from canopy to soil was not elevated in the epiphyte-rich OGF stand, which would be expected as a consequence of K leaching from epiphyte tissues in the canopy. The ab- or adsorption of nitrate during canopy passage, as indicated by the difference in NO_3-N fluxes with incident and net precipitation, was not higher in epiphyte-rich OGF than in the two secondary forests. Thus, factors other than epiphyte biomass are more important in controlling nitrate sink strength in the canopy. We conclude that the epiphytes have only a minor effect on the nutrient fluxes with throughfall and stemflow in these stands.

So far, only few studies have quantified litterfall of epiphytes. Highest values were measured in a montane forest in Costa Rica by Nadkarni and Matelson (1992; 500 kg ha^{-2} $year^{-1}$). Epiphyte litterfall measured in our OGF stand (785 kg ha^{-2} $year^{-1}$) was higher than that value, despite a relatively small epiphyte biomass in comparison to that reported from other tropical old-growth montane forests. However, biomass and annual litterfall of epiphytes in different stands are not necessarily related to each other. Litter mass is also dependent on species composition and species longevity, the structure of the host trees, the aerodynamic roughness of the stand, and wind speed. In our study, biomass and litterfall of epiphytes both increased with increasing stand age from ESF to OGF. The measured nutrient fluxes with epiphyte litterfall in our stands yielded rates comparable to those found in old-growth montane forests of Colombia (Veneklaas 1991) and Costa Rica (Monteverde: bryophytes only; Nadkarni and Matelson 1992), although different measuring approaches were used. Data on litterfall of epiphytes in secondary forests and associated nutrient fluxes are not yet available. We did not measure the nutrient fluxes with litterfall of vascular epiphytes because these organisms contributed only with little biomass to the epiphyte flora of the stands.

21.7 Conclusions

Compared to old-growth forests, early-successional tropical montane forests can have equally high or even higher leaf areas, but their epiphyte biomass is generally much smaller. Rainfall interception was found to be highest in the old-growth forest stand, though epiphytes contributed to less than 10 % of total canopy interception. Stemflow was elevated in those stands which had lowest epiphyte biomass and highest stem densities. Consequently, nutrient fluxes with stemflow were most important in the mid-successional forest stand where stemflow accounted for as much as 17 % of incident rainfall. Nutrient fluxes with litterfall were also highest in this stand for most nutrient species. Evidently, epiphytes had only little influence on within-stand nutrient fluxes with throughfall, stemflow and litterfall. We conclude that the cutting of old-growth forest, and its subsequent replacement by secondary forest result in substantial changes in water and nutrient fluxes at the stand level – changes which are not reversed even after 40 years of progressive forest succession. Such changes are mainly due to contrasts in stem density and canopy structure characteristics and, to a lesser extent, are determined by levels of epiphytic biomass which may need many more years to fully recover.

References

Brown S, Lugo AE (1990) Tropical secondary forests. J Trop Ecol 6:1–32

Bruijnzeel LA (1990) Hydrology of moist tropical forests and effects of conversion: a state of knowledge review. IHP-UNESCO Humid Tropical Programme, Paris

Bruijnzeel LA (2001) Hydrology of tropical montane cloud forests: a reassessment. Land Use Water Resource Res 1:1.1–1.18

Bruijnzeel LA, Proctor J (1995) Hydrology and biogeochemistry of tropical montane cloud forests: what do we really know? In: Hamilton LS, Juvik JO, Scatena FN (eds) Tropical montane cloud forests. Springer, Berlin Heidelberg New York, pp 38–78

Coxson DS, Nadkarni NM (1995) Ecological roles of epiphytes in nutrient cycles of forest ecosystems. In: Lowman MD, Nadkarni NM (eds) Forest canopies. Academic Press, San Diego, pp 495–543

Fournier LA, Comacho de Castro L (1973) Producción y descomposición del mantillo en un bosque secundario húmedo de premontano. Rev Biol Trop 21:59–67

Hafkenscheid RLLJ (2000) Hydrology and biogeochemistry of montane rain forests of contrasting stature in the Blue Mountains of Jamaica. PhD Thesis, Free University of Amsterdam

Heaney A, Proctor J (1989) Chemical elements in litter in forests on Volcán Barva, Costa Rica. In: Proctor J (ed) Mineral nutrients in tropical forest and savanna ecosystems. Blackwell, Oxford, pp 255–271

Hofstede RGM, Wolf JHD, Benzing DH (1993) Epiphytic biomass and nutrient status of a Colombian upper montane rain forest. Selbyana 14:37–45

Hölscher D, Sá TD, Möller RF, Denich M, Fölster H (1998) Rainfall partitioning and related hydrochemical fluxes in a diverse and in a mono-specific (*Phenakospermum guyanense*) secondary vegetation stand in eastern Amazonia. Oecologia 114:251–257

Hölscher D, Köhler L, Leuschner C, Kappelle M (2003) Nutrient fluxes in stemflow and throughfall in three successional stages of an upper montane rain forest in Costa Rica. J Trop Ecol 19:557–565

Hölscher D, Mackensen J, Roberts JM (2004a) Forest recovery in the humid tropics: changes in vegetation structure, nutrient pools and the hydrological cycle. In: Bon-

nell M, Bruijnzeel LA (eds) Forests-water-people in the humid Tropics. Cambridge Univ Press, Cambridge
Hölscher D, Köhler L, Van Dijk AIJM, Bruijnzeel LA (2004b) The importance of epiphytes to total rainfall interception by a tropical montane rain forest in Costa Rica. J Hydrol 292:308–322
Holz I, Gradstein SR, Heinrichs J, Kappelle M (2002) Bryophyte diversity, microhabitat differentiation and distribution of life forms in Costa Rican upper montane *Quercus* forest. Bryologist 105:334–348
IMN (1988) Catastro de las series de precipitaciones medidas en Costa Rica. Instituto Meteórologico Nacional, Ministerio de Recursos Naturales, Energía y Minas (MIRENEM), San José
Jordan CF (1978) Stemflow and nutrient transfer in a tropical rain forest. Oikos 31:257–263
Kappelle M, Juárez ME (1995) Agroecological zonation along an altitudinal gradient in the montane belt of the Los Santos Forest Reserve in Costa Rica. Mount Res Dev 15(1):19–37
Kappelle M, Leal ME (1996) Changes in leaf morphology and foliar nutrient status along a successional gradient in a Costa Rican upper montane *Quercus* forest. Biotropica 28:331–344
Kappelle M, Kennis PAF, De Vries RAJ (1995) Changes in diversity along a successional gradient in Costa Rican upper montane *Quercus* forest. Biodiv Conserv 4:10–34
Kappelle M, Geuze T, Leal ME, Cleef M (1996) Successional age and forest structure in a Costa Rican upper montane *Quercus* forest. J Trop Ecol 12:681–698
Lambert JDH, Arnason JT, Gale JL (1980) Leaflitter and changing nutrient levels in a seasonally dry tropical hardwood old forest, Belize. Plant Soil 55:429–443
McDonald MA, Healey JR (2000) Nutrient cycling in secondary forest in the Blue Mountains of Jamaica. For Ecol Manage 139:257–278
Nadkarni NM, Matelson TJ (1992) Biomass and nutrient dynamics of epiphytic litterfall in a Neotropical montane forest, Costa Rica. Biotropica 24:24–30
Nadkarni NM, Schaefer D, Matelson TJ, Solano R (2004) Biomass and nutrient pools of canopy and terrestrial components in a primary and a secondary montane cloud forest, Costa Rica. For Ecol Manage 198:223–236
Pócs T (1980) The epiphytic biomass and its effect on the water balance of two rain forest types in the Uluguru Mountains (Tanzania, east Africa). Acta Bot Acad Sci Hung 26:143–167
Proctor J (1983) Tropical forest litterfall. I. Problems of data comparison. In: Sutton SL, Whitmore TC, Chadwick AC (eds) Tropical rain forest: ecology and management. Blackwell, Oxford, pp 267–273
Roberts JM, Cabral OMR, da Costa JP, McWilliam ALC, de A Sá TD (1996) An overview of the leaf area index and physiological measurements during ABRACOS. In: Gash JHC, Nobre CA, Roberts JM, Victoria R (eds) Amazonian deforestation and climate. Wiley, Chichester, NY, pp 287–305
Schroth G, Elias MEA, Uguen K, Seixas R, Zech W (2001) Nutrient fluxes in rainfall, throughfall and stemflow in tree-based land use systems and spontaneous tree vegetation of central Amazonia. Agric Ecosyst Environ 87:37–49
Steinhardt U (1979) Untersuchungen über den Wasser- und Nährstoffhaushalt eines andinen Wolkenwaldes in Venezuela. PhD Thesis, University of Göttingen, Göttingen
Tanner EVJ (1980) Studies on the biomass and productivity in a series of montane rain forests in Jamaica. J Trop Ecol 68:573–588
Tanner EVJ (1985) Jamaican montane forests: nutrient capital and cost of growth. J Ecol 73:553–568

Tobón-Marin C (1999) Monitoring and modelling hydrological fluxes in support of nutrient cycling studies in Amazonian rain forest ecosystems. Tropenbos Foundation, Wageningen, Tropenbos Ser 17

Veneklaas EJ (1991) Litterfall and nutrient fluxes in two montane tropical rain forests, Colombia. J Trop Ecol 7:319–336

Veneklaas E, Van Ek R (1990) Rainfall interception in two tropical montane rain forests, Colombia. Hydrol Proc 4:311–326

Weaver PL, Medina E, Pool D, Dugger K, Gonzales-Liboy J, Cuevas E (1986) Ecological observations in the dwarf cloud forest of the Luquillo Mountains in Puerto Rico. Biotropica 18:79–85

Williams-Linera G, Tolome J (1996) Litterfall, temperate and tropical dominant trees, and climate in a Mexican lower montane forest. Biotropica 28:649–656

22 Changes in Fine Root System Size and Structure During Secondary Succession in a Costa Rican Montane Oak Forest

D. HERTEL, D. HÖLSCHER, L. KÖHLER and C. LEUSCHNER

22.1 Introduction

Destruction of pristine tropical forests and subsequent development of secondary forests is a widespread process in the humid tropics (Silver et al. 2000b; Achard et al. 2002). Conversion of primary old-growth rain forests to secondary forest stands not only leads to important changes in stand structure and species composition (Purata 1986; Kappelle et al. 1995, 1996), but it is also expected to have a significant effect on ecosystem functioning, such as for water, carbon, and nutrient cycling (Raich 1983; Vitousek et al. 1989; Bruijnzeel 1990; Ewel et al. 1991).

Tropical montane forests are considered to be particularly vulnerable to disturbance (Ewel 1980; Monasterio et al. 1987). They fulfil important functions in the regional water budget (Bruijnzeel and Proctor 1995), and are known to store considerably higher amounts of carbon in the soil than do lowland forests (Lugo and Brown 1993; Bird et al. 1994). Accordingly, destruction of the primary montane forests often results in significant changes in soil characteristics (Eden et al. 1991; Guariguata and Ostertag 2001). Whereas the conspicuous changes in above-ground structure of montane forests during secondary succession have been well described (Kappelle et al. 1996; Guariguata and Ostertag 2001), little is so far known about the fine root system in secondary and mature forests, and how this is related to successional changes in soil morphology and chemistry.

In a case study of the upper montane belt of the Cordillera de Talamanca, Costa Rica, we studied soil properties, above-ground structure, and rooting pattern of three forest stands differing in age and successional status. We analysed fine root morphology, distribution, and abundance in young and medium-aged secondary forests, and a nearby pristine old-growth *Quercus* forest to describe structural changes in the fine root system during secondary succession. The objective of this contribution is to compare the results of our

Ecological Studies, Vol. 185
M. Kappelle (Ed.) Ecology and Conservation of Neotropical Montane Oak Forests
© Springer-Verlag Berlin Heidelberg 2006

case study in Costa Rican montane oak forests to compilations of available data on the fine root system of other secondary and old-growth forests in the humid tropics.

22.2 Study Sites

This study was conducted in the Los Santos Forest Reserve located on the Pacific slope of the Cordillera de Talamanca in Costa Rica (9°35'40"N, 83°44'30"W). The sites were selected in the upper part of the Rio Savegre watershed at an elevation of 2,900 m a.s.l. in the vicinity of the Jaboncillo de Dota village. Three study plots were selected: a ca. 12-year-old early-successional forest (ESF), a 40-year-old mid-successional forest (MSF), and an old-growth forest (OGF).

Sites were chosen based on existing vegetation analyses in the region (Kappelle et al. 1995, 1996; Chap. 17). The OGF and MSF sites showed a high dominance of the late-successional species *Quercus copeyensis* (now known as *Q. bumelioides* – K.C. Nixon, personal communication, and Chap. 1), whereas the ESF site contained more than 15 tree and shrub species on a 40×40 m plot (e.g. *Quercus copeyensis*, *Q. costaricensis*, and species from the genera *Oreopanax*, *Viburnum*, *Myrsine*, *Weinmannia*, *Cleyera*, *Styrax*, *Cornus*). According to information obtained from local inhabitants, the OGF site has never been affected by major human impact. Presumably, this was the type of mature forest at the ESF and MSF sites prior to forest destruction, too. Since the most abundant species of the OGF stand, *Quercus copeyensis*, had already reached dominance in the 40-year-old MSF plot, one may speculate that a forest similar to the OGF stand represents the final stage of secondary succession at the ESF and MSF sites. Thus, ESF and MSF are assumed to represent two stages of a secondary succession after clear-cut, OGF being the putative terminal stage of this sequence.

The stand height of the successional stages ranged between 5–9 m (ESF), 11–15 m (MSF), and 30–35 m (OGF; Köhler 2002). Whereas stem density (trees≥3 cm DBH, diameter at breast height) decreased with stand age (5,730, 4,800 and 3,460 n ha^{-1} respectively), basal area increased markedly in the sequence ESF–MSF–OGF (19.3, 47.6 and 110.5 m^2 ha^{-1}). Despite these large differences in above-ground structural parameters, differences in leaf area were very small: optical LAI measurements (LAI-2000, Licor) showed essentially identical leaf areas in the ESF and OGF stands (3.8 m^2 m^{-2}), and a somewhat higher LAI value in the MSF stand (4.7 m^2 m^{-2}; Köhler 2002).

Average annual temperature at a weather station 15 km east of the study sites (Villa Mills station at 3,000 m a.s.l.) is 10.9 °C. Average annual rainfall is 2,812 mm (IMN 1988). The soils in the study region are derived from Tertiary intrusive rocks and volcanic ashes (Weyl 1980), and have principally been classified as humic andosols (FAO system) or hapludands (USDA system; van Uffelen 1991; Kappelle et al. 1995; Chap. 4).

A detailed description of the methods used to analyse the soil morphology and chemistry, and the fine root system is given in Hertel et al. (2003).

22.3 Soil Morphology and Chemistry

All three forest stands showed large differences in soil morphological and chemical properties. The organic material on the forest floor showed a large increase with increasing age of the forest stands (Table 22.1). The thickness of the organic layer increased by a factor of two between the two secondary forest stands, and by a factor of 4.5 between the ESF and the OGF stand, leading to a two-times higher and >10 times higher carbon storage in the organic layer of the old-growth forest, compared to the MSF and the ESF forest respectively. Similarly to the C pool, the pools of Ca, Mg and K, and the C/N ratio increased between the ESF and OGF sites. On the other hand, the mineral topsoil in the ESF forest showed significantly higher cation exchange capacity,

Table 22.1. Soil chemical parameters of the organic layer and the mineral topsoil (0–10 cm) of three forest stands (means±1 SE, n=10). Different letters indicate significant differences among the three stands (P<0.05, Wilcoxon U-test after Mann-Whitney). *ESF* Early-successional forest, *MSF* mid-successional forest, *OGF* old-growth forest, *CEC* cation exchange capacity

	ESF	MSF	OGF
Organic layer (above 0 cm)			
Thickness (cm)	4.2±0.4 a	9.9±0.9 b	18.8±1.5 c
Carbon pool (mol m^{-2})	80.2±2.6 a	252.5±5.6 b	533.2±9.1 c
C/N ratio (mol mol^{-1})	23.4±0.9 a	25.2±0.6 ab	27.6±0.7 b
N concentration (mol m^{-3})	84.4±3.2	101.2±1.8	103.2±2.6
P concentration (mol m^{-3})	2.73±0.08	2.71±0.15	2.58±0.15
Ca, Mg, K pool (mmol m^{-2})	1,464±84 a	1,991±247 a	2,067±245 a
pH (H_2O/KCl)	6.0/5.6 a	4.4/3.9 b	4.0/3.4 b
Mineral topsoil layer (below 0 cm: 0–10 cm)			
Carbon pool (mol kg^{-1})	8.6±0.7 a	11.6±1.3 ab	12.5±0.4 b
C/N ratio (mol mol^{-1})	16.1±0.6 a	22.1±1.0 b	25.0±0.7 c
N concentration (mol m^{-3})	148±13.7	141.0±16.4	144.2±3.4
P concentration (mol m^{-3})	16.9±0.6	12.5±0.9	11.7±0.7
CEC_e (μmol_c g^{-1})	250.0±25.3 a	173.2±9.8 b	158.1±14.5 b
Base saturation (%)	95.6±0.5 a	12.4±2.6 b	7.9±1.7 b
Percentage of Al in CEC_e (%)	0.0±0.0 a	75.1±3.0 b	76.0±5.5 b
pH (H_2O/KCl)	6.0/4.9 a	4.3/3.5 b	4.2/3.4 b

base saturation, and P concentration compared to the mid-successional forest and the old-growth forest. The mineral topsoil of the latter forest types had markedly higher loads of protons and aluminium, and showed a larger C/N ratio compared to the ESF stand.

The large storage of carbon in the organic layer of the old-growth forest compares well with other investigations on soil chemistry in the tropics: studies of Zinke et al. (1986), Dixon et al. (1994), Bird et al. (2001) and others show that tropical montane forests differ strongly from most lowland forests with regard to the carbon pools of the organic layer: they are significantly larger in tropical montane forests. This is largely due to lower decomposition rates of organic material at montane elevations, at which low temperatures, high rainfall amounts, and nutrient-poor soils often reduce soil microbial activity (Vitousek and Sanford 1986; Berendse et al. 1989; Brown and Lugo 1990). Consequently, destruction of old-growth forests leads to a substantial loss of carbon and nutrients from the soil. In our study, the clearing of old-growth forest reduced the carbon pool of the organic layer by a factor of ca. 7, from 530 to less than 80 mol C m^{-2}.

Forest conversion also affected chemical properties of the mineral soil. The cation exchange capacity, base saturation and pH value increased significantly, presumably due to burning of the woody debris after tree felling (as done traditionally in this region). A similar change in soil chemistry has been found in other tropical regions (Guariguata and Ostertag 2001). However, mineral soil chemical properties did not differ significantly between mid-successional (ca. 40 years old) and old-growth (>200 years old) forest in the Costa Rican Talamanca Mountains. According to our data, nearly half of the organic matter pool available on top of the OGF forest soil must have been accumulated over 40 years, i.e. about 6 mol C m^{-2} $year^{-1}$. This may be due to the fact, that after a few years, young tropical secondary forests establish high leaf area values, thus transferring high amounts of organic material to the soil via litter production (Brown and Lugo 1990; Richter et al. 1999).

22.4 Fine Root System Structure and Morphology

Analysis of soil samples from the organic layer and the upper mineral soil (0–10 cm) showed that the biomass of fine roots (roots <2 mm in diameter) increased significantly with increasing stand age of the three forest types, showing a seven-times higher biomass in the old-growth forest compared to the early-successional forest (Fig. 22.1). This was not only a result of the large increase in organic layer thickness, but also of an significant increase in fine root density (fine root biomass per soil volume) in the organic layer towards the older forest stands (Table 22.2). The main portion of the fine root biomass in the ESF stand was found in the mineral topsoil. In the MSF

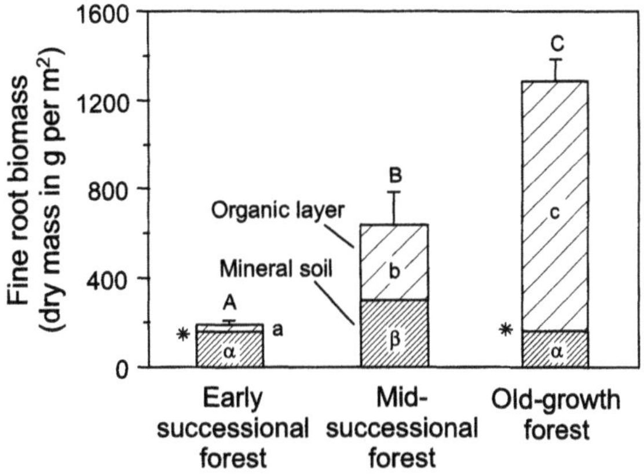

Fig. 22.1. Fine root biomass in the organic layer and the mineral topsoil (0–10 cm) of the three forests stands (means and standard errors, n=10). Different letters indicate significant differences among profile totals (*uppercase*), corresponding horizons (*lowercase Latin and Greek*), or between the two horizons of a given stand (*stars*); P<0.05, Wilcoxon U-test after Mann-Whitney

stand, fine root biomass was evenly distributed throughout both horizons, whereas in the old-growth forest the fine root biomass was located mainly in the organic layer.

The three forests differed not only in fine root biomass and density, but also in the surface area of the fine root system. A conspicuous increase in the fine root area index (RAI) was found: RAI values of ca. 9, 16 and 31 (m^2 root area m^{-2} ground area) in the ESF, MSF and OGF plots were observed. Similarly to the fine root biomass, this increase in fine root surface area was due to strongly increasing RAI values in the organic layer, whereas the root surface area in the mineral topsoil decreased in the same sequence (Table 22.2). This resulted only from the increase in fine root biomass, since the specific fine root surface area in the ESF stand was about twice as high as that in the MSF and OGF stands.

Other morphological parameters were also found to differ among the three stands: trees of the ESF stand tended to have fine roots with higher numbers of root tips per unit biomass than those of the MSF and OGF stands. This might be a result of differences in species composition and richness among the three forest stands: we distinguished eight different tree species based on root morphology in the ESF, whereas only two woody species were found in the MSF and OGF stands, roots of the *Quercus* species being the most abundant by far. This shift in tree species composition was also reflected by the percentage of rootlets infected by ectomycorrhizal fungi. In the ESF stand, only 24–34 % of the analysed rootlets showed clearly identifiable ectomycorrhizae, but 71–99 % of rootlets in the MSF and OGF stands were infected by ectomycorrhizal fungi (Table 22.2).

Although the specific root tip abundance decreased from the ESF to the OGF stand, the number of root tips per stand area was considerably higher in the old-growth forest (>4×10^6 m^{-2}) than in the MSF (>2×10^6 m^{-2}) and in the ESF stand (>0.6×10^6 m^{-2}; Fig. 22.2). Similarly to the root area index, this pattern was mainly a consequence of the increasing fine root biomass in the

Table 22.2. Fine root density, fine root area index (RAI), specific root surface area, specific root tip abundance, and relative abundance of rootlets infected by ectomycorrhizal fungi in the organic layer and the mineral topsoil (0–10 cm) of the three stands. Given are means±1 SE. *Numbers in brackets* Number of replications (*n*), *Latin or Greek letters* significant differences in organic layer or mineral topsoil among the three stands ($P<0.05$, Wilcoxon U-test after Mann-Whitney), *ESF* early-successional forest, *MSF* mid-successional forest, *OGF* old-growth forest

	ESF	MSF	OGF
Organic layer (above 0 cm)			
Fine root density (g dry mass m^{-3})	988±234 a (10)	3,927±1,237 b (10)	6,634±527 c (10)
Fine root area index (RAI, m^2 m^{-2})	1.8±0.4 a (10)	10.7±3.4 b (10)	27.8±2.2 c (10)
Specific root surface area (cm^2 g^{-1} dry mass)	591±31 a (57)	318±19 b (58)	247±12 c (63)
Specific root tip abundance (*n* mg^{-1} dry mass)	6.9±1.5 a (10)	4.6±0.7 a (10)	3.7±0.4 a (10)
Percentage of infected fine rootlets[a]	24.2±9.5 a (10)	80.9±9.8 (10)	98.6±0.6 b (9)
Mineral topsoil layer (below 0 cm: 0–10 cm)			
Fine root density (g dry mass m^{-3})	1,571±141 α (10)	2,988±534 β (10)	1,611±250 α (10)
Fine root area index (m^2 m^{-2})	7.2±0.7 α (10)	5.4±1.0 α β (10)	3.3±0.5 β (10)
Specific root surface area (cm^2 g^{-1} dry mass)	461±32 β (58)	180±11 β (63)	202±12 β (76)
Specific root tip abundance (*n* mg^{-1} dry mass)	3.4±0.5 β (10)	1.7±0.4 β (10)	1.4±0.7 , (10)
Percentage of infected fine rootlets[a]	33.8±12.1 α (10)	82.6±9.7 β (10)	71.2±7.4 β (9)

[a] Percentage (%) of fine rootlets infected by ectomycorrhizal fungi

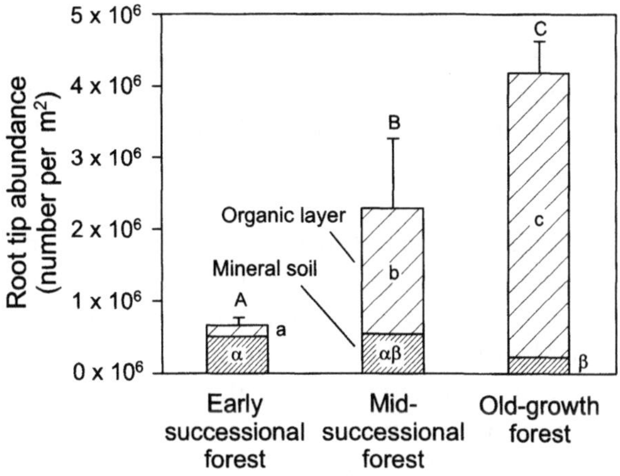

Fig. 22.2. Root tip abundance in the organic layer and the mineral topsoil (0–10 cm) of the three forests stands (means and standard errors, n=10). Different letters indicate significant differences among profile totals (*uppercase*), or corresponding horizons (*lowercase Latin and Greek*); $P<0.05$, Wilcoxon U-test after Mann-Whitney

organic layer with increasing age of the stands, whereas the root tip abundance decreased in the mineral topsoil, as did the fine root biomass in that horizon.

The three forest stands investigated differed markedly in their fine root biomass, fine root area and root tip abundance, the mature old-growth forest showing a ca. six-fold and two-fold increase in fine root system compared to that of the ESF and MSF respectively. This is notable, since differences in leaf area were much less. Given that in the same environment plant productivity is roughly proportional to the radiation absorbed by the leaf area in the canopy (Russell et al. 1989), and that nutrient demand is a function of production, annual nutrient uptake of the three successional stages should not be that different. If so, what might be the reasons for the large differences in the size of the fine root systems?

One explanation could be that the specific nutrient uptake rates of fine roots of the three successional stages may be different. Indeed, the trees of the ESF stand were found to possess roots of higher specific surface area and specific root tip abundance, compared to the MSF and OGF stands. Additionally, fine roots of the dominant tree species at the ESF site were infected mainly by arbuscular fungi whereas the *Quercus* trees of the MSF and OGF stands were infected by ectomycorrhizal fungi. One could speculate that these differences in specific fine root morphology as well as the mycorrhizal association of the roots may indicate a different nutrient absorption capacity. Still, this cannot explain the large contrast in fine root biomass, density, and surface area between the MSF and the OGF stand, where differences in tree species composition and type of mycorrhiza were negligible.

Alternatively, it could be assumed that the OGF and MSF stands might have greater nutrient demands compared to the early-successional forest, although the leaf area of the stands were similar. Our data show that fine root biomass and density increased strongly in the sequence ESF–MSF–OGF. In the same sequence, the organic layer became increasingly important as a preferred rooting medium. This is consistent with the finding that, in contrast to the MSF and OGF sites where the main pool of nutrients accumulated in the organic material of the forest floor, the mineral soil of the ESF site was richer in essential nutrients and had lower aluminium and proton concentrations.

Similar trends showing a strong preference of tree fine roots for the organic layer at nutrient-poor sites have been detected in many other tropical regions (e.g. Roy and Singh 1995; Cavelier et al. 1996; Priess et al. 1999; Chap. 19). Most likely, the roots are attracted by high supply rates of nitrogen and other nutrient elements in that horizon. A high fine root density in the organic layer may also cause a positive feedback loop by enhancing nutrient availability via a supply of substantial amounts of organic material derived from fine root decay (Ostertag and Hobbie 1999; Silver et al. 2000a). We conclude that the large fine root system in the organic layer of MSF and, above all, OGF stands plays a key role for the nutrition of these *Quercus* forests on nutrient-poor mineral soils in the Talamanca Mountains.

22.5 Does Tropical Rain Forest Fine Root Mass Generally Increase During Secondary Succession?

It has been stated that the fine root systems of early-successional forests can be as large as those of mature forests (Raich 1980; Guariguata and Ostertag 2001; Hölscher et al. 2005). On the other hand, there are reports that fine root mass increases with increasing age of the secondary forest (e.g. Berish 1982). Inspection of available data in the literature on fine root systems of tropical secondary forests showed that, beside our own work, there are only two other studies in which fine root mass was precisely separated into live and dead fractions. Moutinho et al. (2003) found a high fine root biomass value of 570 g m^{-2} in a 17-year-old secondary forest in Brazil. Berish and Ewel (1988) give a low fine root biomass value of 180 g m^{-2} for a 6-year-old lower montane secondary forest in Costa Rica. Both findings seem to support our results of an increase in fine root biomass with age, but more data are needed to fully assess this relationship.

Because of the paucity of fine root biomass data, we analysed the influence of stand age on total (live and dead) fine root mass in a set of seven studies with 17 fine root mass values of tropical secondary rain forests between one and 40 years of age. Fine root mass totals of these forests varied in a broad range of ca. 100–2,000 g m^{-2}. Despite this large variance, we detected a significant linear increase in fine root mass with increasing age of the secondary forests (Fig. 22.3A). These data suggest that late stages of succession tend to have larger fine root systems, since a third of the variation of the data was

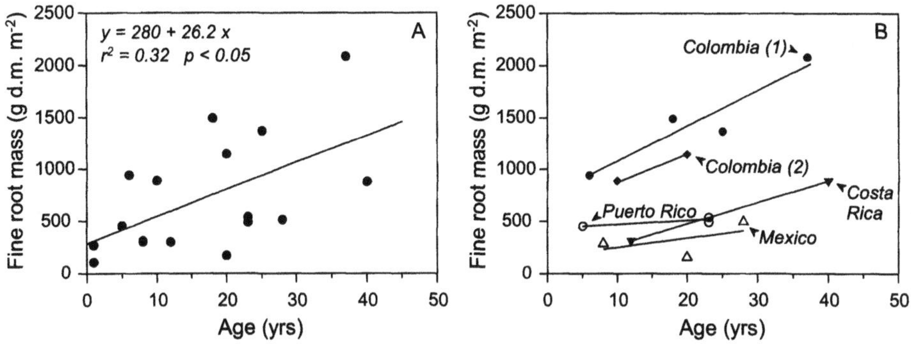

Fig. 22.3A, B. Relationship between stand totals of fine root mass (living and dead roots) and age (1–50 years) of secondary forests in the humid tropics. **A** Linear regression showing a significant increase in fine root mass with increasing secondary forest age (n=17). **B** Distinction between five location-specific data series of low and high pH values, easily distinguished by connection line slopes; *closed symbols* sites with soil pH of 3.8–4.4, *open symbols* sites with soil pH of 4.7–6.3. All data represent mass values of fine roots<2 mm in diameter. Note that the corresponding sampling depths differ among these studies. Data from Raich (1980, **A**), Berish (1982, **A**), Raich (1983, **A**), Lugo (1992, **A**, **B** *Puerto Rico*), Cavelier et al. (1996, **A**, **B** *Colombia 2*), Pavlis and Jenik (2000, **A**, **B** *Colombia 1*), Hertel et al. (2003, **A**, **B** *Costa Rica*), Jaramillo et al. (2003, **A**, **B** *Mexico*)

explained by the age of the stands. However, 68 % of the variation must be attributed to other factors. In particular, the inclusion of the dead root fraction may have obscured the relationship between live fine root mass and stand age. Since soil pH often decreases during secondary succession at clear-cut or burnt sites, the increase in total fine root mass may be due to an increase in necromass, and not in biomass, with stand development. Significant increases in fine root necromass with decreasing pH were indeed reported in tropical forests (e.g. Silver et al. 2000b).

Moreover, the increase in total fine root mass with stand age was much more pronounced at sites with lower soil pH (at least in the late-successional stages; Fig. 22.3B, filled symbols: pH 3.8–4.4), compared to sites with higher soil pH (Fig. 22.3B, open symbols: pH 4.7–6.3). On the other hand, tropical forests on acidic nutrient-poor soils often have larger masses of live fine roots than do forests on less acidic soils (Gower 1987; Cavelier 1992; Maycock and Congdon 2000). Thus, there is evidence that fine root biomass, and not only necromass, increases during secondary succession of tropical forests, at least at sites with low pH or showing a significant lowering of soil pH over time.

22.6 Are Large Fine Root Systems Characteristic for High-Elevation Tropical Rain Forests?

Our investigation revealed a very large fine root biomass (>1,300 g m^{-2}) in the old-growth forest (OGF) in the upper montane region of Costa Rica, in contrast to other mature forests in the humid tropics where typically less than 1,000 g m^{-2} was found (see Fig. 22.4). Are high amounts of fine root biomass a unique attribute of tropical forests at high elevation?

A precise analysis of the existing literature for data on mass of living roots <2 mm in diameter gave 55 different data points from mature tropical rain forest stands collected in 15 studies. The compilation of these data revealed a very broad range of fine root standing stocks of 8–1,440 g m^{-2} of biomass in tropical rain forests. Regression analysis showed a significant increase in fine root biomass with increasing elevation (Fig. 22.4). However, this non-linear increase was found to be less pronounced in the forest stands at lower altitudes, compared to higher-elevation forests. Thus, we subdivided the data into three elevational groups, to distinguish typical lowland forests (<400 m a.s.l.) from better drained hill and lower montane forest stands (400–2,000 m a.s.l.), and to combine all high-elevation forests (>2,000 m a.s.l.) with a less favourable temperature regime in a third class.The comparison of these three groups showed that mean fine root biomass was slightly higher in the lower

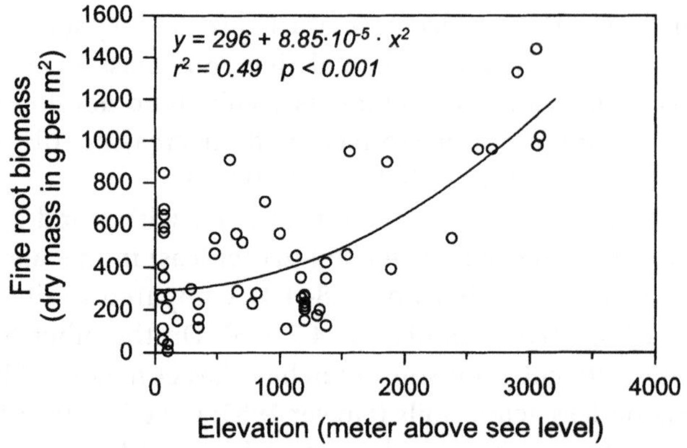

Fig. 22.4. Relationship between fine root biomass and elevation of old-growth forests in the humid tropics (n=55). All data represent mass of live fine roots<2 mm in diameter. Note that the corresponding sampling depths differ among these studies. Data from Gower (1987), Silver and Vogt (1993), Silver et al. (1996), Sundarapandian and Swamy (1996), Denslow et al. (1998), Herbert and Fownes (1999), Priess et al. (1999), Maycock and Congdon (2000), McGroddy and Silver (2000), Silver et al. (2000a), Ostertag (2001), Schuur and Matson (2001), Yavitt and Wright (2001), Kitayama and Aiba (2002), Hertel et al. (2003), Röderstein et al. (2005), Moser et al. (personal communication)

montane forests (399 g m^{-2}) than in the lowland forests (316 g m^{-2}). However, this difference was not significant. By contrast, fine root biomass average of the upper montane forests (1,032 g m^{-2}) was significantly larger than that of forest at lower elevation ($P<0.05$). Accordingly, the fine root biomass of the OGF stand from the Cordillera de Talamanca lies within the range found in other tropical upper montane forests.

What factors may be responsible for this apparent increase in fine root biomass with elevation? Tropical montane forests are characterized by low temperatures and high precipitation (Grubb 1977; Bruijnzeel and Veneklaas 1998). As a result, the soils are often water-saturated, oxygen-depleted and highly acidic (Vitousek and Sanford 1986; Tanner et al. 1998; Silver et al. 1999). Under such conditions, decomposition and mineralization processes are reduced, leading to lower rates of nutrient cycling and decreased nutrient availability (Meentemeyer 1977; Vitousek and Sanford 1986).

Studies by Marrs et al. (1988) and Kitayama and Aiba (2002) showed that N mineralization decreases with elevation in tropical forest soils. Although high soil acidity and nutrient limitation (e.g. by phosphorus) exist at tropical lowland sites, too (Vitousek and Sanford 1986; Cuevas and Medina 1988), there is evidence that low nutrient availability might be responsible for the development of large fine root systems in tropical montane forests. Nitrogen and phosphorus fertilization along an elevational gradient at Mt. Kinabalu, Malaysia, showed the largest stimulation of fine root growth at higher altitudes (Nomura and Kikuzawa 2003). Along a fertility gradient of tropical forests in Australia, fine root biomass was higher at poorer sites, which occurred at higher elevation (Maycock and Congdon 2000). Thus, lowered mineralization rates at higher elevation might well be one reason for a large fine root system in tropical upper montane forests.

22.7 Conclusions

This case study on successional forest stages in the Cordillera de Talamanca showed that conversion of the old-growth forest led to significant changes not only in above-ground stand structure, but also in the carbon and nutrient pools of the soil. Consequently, below-ground structure of the two secondary and the old-growth forest differed markedly: we found an increase in biomass, surface area, and root tip abundance of the fine root system with increasing age of the forests. Whether this is a consequence of an increasing nutrient demand of the mid- and late-successional forests remains unclear. A progressive decoupling of the nutrient cycle between trees and mineral soil, and the simultaneous build-up of large nutrient pools in the organic layer with secondary succession most likely is one cause of the enlargement of the fine root system in this organic horizon.

Review of available literature data showed that the total fine root mass (live and dead total) of secondary forests in the humid tropics seems to increase with stand age. This trend was found to be more pronounced at sites with more acidic soils. Increases in both fine root necromass and in fine root biomass with secondary succession are thought to be responsible for high root masses in late-successional forests.

Comparison with data from other tropical forests confirmed that a large fine root biomass, as recorded in the old-growth forest in this study, is a typical attribute of tropical high-elevation forests. Although high fine root biomasses are also found in certain tropical lowland forests, forests above 2,000 m a.s.l. had significantly higher biomasses than those at lower altitudes. It is hypothesised that reduced nutrient availability is a key factor for this elevational increase in fine root biomass.

References

Achard F, Eva HD, Stibig HJ, Mayaux P, Gallego J, Richards T, Malingreau J-P (2002) Determination of deforestation rates of the world's humid tropical forests. Science 297:999–1002

Berendse F, Bobbink R, Rouwenhorst G (1989) A comparative study on nutrient mineralization. Oecologia 78:338–348

Berish CW (1982) Root biomass and surface area in three successional tropical forests. Can J For Res 12:699–704

Berish CW, Ewel JJ (1988) Root development in simple and complex tropical successional ecosystems. Plant Soil 106:73–84

Bird MI, Haberle SG, Chivas AR (1994) Effect of altitude on the carbon-isotope composition of forest and grassland soils in Papua New Guinea. Global Biogeochem Cycles 8:13–22

Bird MI, Santruckova H, Lloyd J, Veenendaal E (2001) Global soil organic carbon pool. In: Schulze ED, Heimann M, Harrison SP et al. (eds) Global biogeochemical cycles in the climate system. Academic Press, New York, pp 185–199

Brown S, Lugo AE (1990) Tropical secondary forests. J Trop Ecol 6:1–32

Bruijnzeel LA (1990) Hydrology of moist tropical forests and effects of conversion: a state of knowledge review. IHP-UNESCO Humid Tropical Programme, Paris, France

Bruijnzeel LA, Proctor J (1995) Hydrology and biogeochemistry of tropical montane cloud forest: what do we really know? In: Hamilton LS, Juvik JO, Scatena FN (eds) Tropical montane cloud forests. Springer, Berlin Heidelberg New York, Ecological Studies 110, pp 38–78

Bruijnzeel LA, Veneklaas EJ (1998) Climatic conditions and tropical montane forest productivity: the fog has not lifted yet. Ecology 79:3–9

Cavelier J (1992) Fine-root biomass and soil properties in a semi-deciduous and a lower montane rain forest in Panama. Plant Soil 142:187–201

Cavelier J, Estevez J, Arjona B (1996) Fine root biomass in three successional stages of an Andean cloud forest in Colombia. Biotropica 28:728–736

Cuevas E, Medina E (1988) Nutrient dynamics within Amazonian forests. II. Fine root growth, nutrient availability and leaf litter decomposition. Oecologia 76:222–235

Denslow JS, Ellison AM, Sanford RE (1998) Treefall gap size effects on above- and below-ground processes in a tropical wet forest. J Ecol 86:597–609

Dixon RK, Brown S, Houghton RA, Solomon AM, Trexler MC, Wisniewski J (1994) Carbon pools and flux of global forest ecosystems. Science 263:185–190

Eden MJ, Furley PA, McGregor DFM, Milliken W, Ratter JA (1991) Effect of forest clearance and burning on soil properties in northern Roraima, Brazil. For Ecol Manage 38:283–290

Ewel JJ (1980) Tropical succession: manifold routes to maturity. Biotropica 12:2–7

Ewel JJ, Mazzarino MJ, Berish CW (1991) Tropical soil fertility changes under monocultures and successional communities of different structure. Ecol Appl 1:289–302

Gower ST (1987) Relations between mineral nutrient availability and fine root biomass in two Costa Rican tropical forests: a hypothesis. Biotropica 19:171–175

Grubb PJ (1977) Control of forest growth and distribution on wet tropical mountains: with special reference to mineral nutrition. Annu Rev Ecol Syst 8:83–107

Guariguata MR, Ostertag R (2001) Neotropical secondary forest succession: changes in structural and functional characteristics. For Ecol Manage 148:185–206

Herbert DA, Fownes JH (1999) Forest productivity and efficiency of resource use across a chronosequence of tropical montane soils. Ecosystems 2:242–254

Hertel D, Leuschner C, Hölscher D (2003) Size and structure of fine root systems in old-growth and secondary tropical montane forests (Costa Rica). Biotropica 35:143–153

Hölscher D, Mackensen J, Roberts JM (2005) Forest recovery in the humid tropics: changes in vegetation structure, nutrient pools and the hydrological cycle. In: Bonell M, Bruijnzeel LA (eds) Forests, water and people in the humid tropics. Cambridge Univ Press, Cambridge, UK, pp 598–622

IMN (1988) Catastro de las series de precipitationes medidas en Costa Rica. Instituto Meteorológico Nacional, Ministerio de Recursos Naturales, Energía y Minas (MIRENEM), San José, Costa Rica

Jaramillo VJ, Ahedo-Hernández R, Kauffman JB (2003) Root biomass and carbon in a tropical evergreen forest of Mexico: changes with secondary succession and forest conversion to pasture. J Trop Ecol 19:457–464

Kappelle M, Kennis PAF, de Vries RAJ (1995) Changes in diversity along a successional gradient in a Costa Rican upper montane *Quercus* forest. Biodiv Conserv 4:10–34

Kappelle M, Geuze T, Leal ME, Cleef AM (1996) Successional age and forest structure in a Costa Rican upper montane *Quercus* forest. J Trop Ecol 12:681–698

Kitayama K, Aiba SI (2002) Ecosystem structure and productivity of tropical rain forests along altitudinal gradients with contrasting soil phosphorus pools on Mount Kinabalu, Borneo. J Ecol 90:37–51

Köhler L (2002) Die Bedeutung der Epiphyten im ökosystemaren Wasser- und Nährstoffumsatz verschiedener Altersstadien eines Bergregenwaldes in Costa Rica. University of Göttingen, Göttingen, Germany, Ber Forschungszentr Waldökosyst A, 181

Lugo AE (1992) Comparison of tropical tree plantations with secondary forests of similar age. Ecol Monogr 62:1–41

Lugo AE, Brown S (1993) Management of tropical soils as sinks or sources of atmospheric carbon. Plant Soil 149:27–41

Marrs R, Proctor J, Heaney A, Mountford M (1988) Changes in soil nitrogen mineralization and nitrification along an altitudinal transect in tropical rain forest in Costa Rica. J Ecol 76:466–482

Maycock CR, Congdon RA (2000) Fine root biomass and soil N and P in north Queensland rain forests. Biotropica 32:185–190

McGroddy M, Silver WL (2000) Variations in belowground carbon storage and soil CO_2 flux rates along a wet tropical climate gradient. Biotropica 32:614–624

Meentemeyer V (1977) Climatic regulation of decomposition rates of organic matter in terrestrial ecosystems. In: Adrians DC, Brisbin IL (eds) Environmental chemistry and cycling processes. US Dept Energy Symp Ser CONF-760429, Washington, DC, pp 779–789

Monasterio M, Sarmiento G, Solbrig OT (1987) Comparative studies on tropical mountain ecosystems: planning for research. International Union of Biological Sciences (IUBC), Paris, Biol Int 12

Moutinho P, Nepstad DC, Davidson EA (2003) Influence of leaf-cutting ant nests on secondary forest growth and soil properties in Amazonia. Ecology 84:1265–1276

Nomura N, Kikuzawa K (2003) Productive phenology of tropical montane forests: fertilization experiments along a moisture gradient. Ecol Res 18:573–586

Ostertag R (2001) Effects of nitrogen and phosphorus availability on fine-root dynamics in Hawaiian montane forests. Ecology 82:485–499

Ostertag R, Hobbie SE (1999) Early stages of root and leaf decomposition in Hawaiian forests: effects of nutrient availability. Oecologia 121:564–573

Pavlis J, Jenik J (2000) Roots of pioneer trees in the Amazonian rain forest. Trees 14:442–455

Priess T, Then C, Fölster H (1999) Litter and fine-root production in three types of tropical premontane rain forest in Venezuela. Plant Ecol 143:171–187

Purata SE (1986) Floristic and structural changes during old-field succession in the Mexican tropics in relation to site history and species availability. J Trop Ecol 2:257–276

Raich JW (1980) Fine roots re-grow rapidly after forest felling. Biotropica 12:231–232

Raich JW (1983) Effects of forest conversion on the carbon budget of a tropical soil. Biotropica 15:177–184

Richter DD, Markewitz D, Trumbore SE, Wells CG (1999) Rapid accumulation and turnover of soil carbon in a re-establishing forest. Nature 400:56–58

Röderstein M, Hertel D, Leuschner C (2005) Above- and below-ground litter production in three tropical mountain forests (Ecuador). J Trop Ecol 21:483–492

Roy S, Singh JS (1995) Seasonal and spatial dynamics of plant-available N and P pools and N-mineralization in relation to fine roots in a dry tropical forest habitat. Soil Biol Biochem 27:33–40

Russell G, Jarvis PG, Monteith JL (1989) Absorption of radiation by canopies and stand growth. In: Russell G, Marshall B, Jarvis PG (eds) Plant canopies: their growth, form and function. Cambridge Univ Press, Cambridge, UK, pp 21–40

Schuur EAG, Matson PA (2001) Net primary productivity and nutrient cycling across a mesic to wet precipitation gradient in Hawaiian montane forest. Oecologia 128:431–442

Silver WL, Vogt KA (1993) Fine root dynamics following single and multiple disturbance in a subtropical wet forest ecosystem. J Ecol 81:729–738

Silver WL, Scatena FN, Johnson AH, Siccama TG, Watt F (1996) At what temporal scales does disturbance affect belowground nutrient pools? Biotropica 28:441–457

Silver WL, Keller M, Lugo AE (1999) Soil oxygen availability and biogeochemical cycling along elevation and topographic gradients in Puerto Rico. Biogeochemistry 44:301–328

Silver WL, Neff J, McGroddy M, Veldkamp E, Keller M, Cosme R (2000a) Effects of soil texture on belowground carbon and nutrient storage in a lowland Amazonian forest ecosystem. Ecosystems 3:193–209

Silver WL, Ostertag R, Lugo AE (2000b) The potential for carbon sequestration through reforestation of abandoned tropical agricultural and pasture lands. Restor Ecol 8:394–407

Sundarapandian SM, Swamy PS (1996) Fine root biomass distribution and productivity patterns under open and closed canopies of tropical forest ecosystems at Kodayar in Western Ghats, South India. For Ecol Manage 86:181–192

Tanner EVJ, Vitousek PM, Cuevas E (1998) Experimental investigation of nutrient limitation of forest growth on wet tropical mountains. Ecology 79:10–22

Van Uffelen JG (1991) A geological, geomorphological and soil transect study of the Chirripó Massif and adjacent areas, Cordillera de Talamanca, Costa Rica. MSc Thesis, Wageningen Agricultural University, Wageningen

Vitousek PM, Sanford RL (1986) Nutrient cycling in moist tropical forest. Annu Rev Ecol Syst 17:137–167

Vitousek PM, Matson PA, van Cleve K (1989) Nitrogen availability and nitrification during succession: primary, secondary and old-field series. Plant Soil 115:229–239

Weyl R (1980) Geology of Central America. Borntraeger, Stuttgart

Yavitt JB, Wright SJ (2001) Drought and irrigation effects on fine root dynamics in a tropical moist forest, Panama. Biotropica 33:421–434

Zinke PJ, Strangenberger AG, Post WM, Emanuel WR, Olson JS (1986) Worldwide organic soil carbon and nitrogen data. Rep NDP-018, Carbon Dioxide Information Centre, Oak Ridge National Laboratory, Oak Ridge, TN

23 Soil Seed Bank Changes Along a Forest Interior–Edge–Pasture Gradient in a Costa Rican Montane Oak Forest

M. ten Hoopen and M. Kappelle

23.1 Introduction

Past tropical deforestation has given rise to landscape mosaics of old-growth fragments, successional forest patches, crop fields and pasturelands (Laurance et al. 1997; Chap. 16). Differently shaped and sized patches are separated from each other by a variety of edges. At those edges, microclimatic variables such as light intensity and duration, relative humidity, air temperature, and soil factors differ significantly over short distances (Lovejoy et al. 1983; Williams-Linera 1990; José et al. 1996; Laurance et al. 1997; Chap. 19). As a result, plant community structure and composition change along gradients from the tropical forest interior, across the edge, into non-forest vegetation (Lopez de Casenave et al. 1995; José et al. 1996; Oosterhoorn and Kappelle 2000; Chaps. 13 and 19). The importance of soil seed banks as sources for tree recruits at such edges has long been recognized (e.g., Young et al. 1987; Garwood 1989; Parker et al. 1989; Teketay and Granström 1995; Quintana-Ascencio et al. 1996; Dalling and Denslow 1998; Cubiña and Aide 2001; Chap. 19).

However, since most studies have been conducted in tropical lowlands, it still remains unclear to what extent these trends fully apply to patchy landscapes in tropical montane forest regions. This is of special interest, recognizing that harsh environmental conditions on tropical mountains reduce forest recovery rates significantly (Ewel 1980). For instance, Kappelle et al. (1995, 1996, and Chap. 17) estimate that Costa Rican montane oak forests recovering from slash-and-burn and subsequent grazing will need at least about a century to return to their old-growth state, in terms of non-epiphytic structure and species composition – assuming forest seed sources are nearby. For this reason, we studied the size and composition of soil seed banks along a montane oak forest interior–edge–pasture gradient in Costa Rica. The insights

Ecological Studies, Vol. 185
M. Kappelle (Ed.) Ecology and Conservation of Neotropical Montane Oak Forests
© Springer-Verlag Berlin Heidelberg 2006

gained will be particularly useful to conservation efforts directed at the ecological restoration of fragile montane oak forests in the Neotropics (Holl and Kappelle 1999).

23.2 Study Area

The study was carried out in the montane cloud forest zone (2,300–2,800 m a.s.l.) of the Los Santos Forest Reserve (Dota County) and Tapantí-Río Macho National Park (El Guarco County) in the western part of Costa Rica's Cordillera de Talamanca. Average annual rainfall is 2,100–3,000 mm, and average annual temperature varies in the range 12–14.5 °C, depending on altitude. The dry season lasts from early January to late April (Herrera 1986). Soils are derived from volcanic ash, are acid (pH 3.5–5.5), and moderately fertile. Natural vegetation is 30–50 m tall, evergreen broad-leaved oak forest (Kappelle 1996; Chap. 10). Since the early 1950s, conversion of the region's oak forests into grasslands and croplands has led to a complex landscape mosaic (Helmer 2000). After abandonment of low-production pasturelands in the late 1970s and early 1980s, stands of successional forests have developed along edges of old-growth forests and productive pastures (Kappelle and Juárez 1995; Chap. 30).

23.3 Methods

23.3.1 Site Selection and Transect Establishment

From a set of black-and-white aerial photographs (1992; scale 1:60,000), we selected four 10-ha old-growth montane oak forest fragments bordering patches of grass-dominated, 30±5 year old pastures: Alto Roble de Copey de Dota (AR), La Damita del Guarco (LD), Providencia de Dota (PR), and San Gerardo de Dota (SG). Patches were located at 09°34'–09°42'N, 83°50'–83°55'W, and 2,380–2,710 m altitude. At each site, a 150-m-long transect was laid out, running from the forest interior (–50 m) through the abrupt edge (0 m) into the exterior (i.e., pasture; +100 m), perpendicular to the forest edge. As recommended by Fraver (1994) and José et al. (1996), the edge was delimited by the bases of bordering mature forest tree stems. Changes in standing vegetation structure and terrestrial vascular plant species composition along transects are published elsewhere (Oosterhoorn and Kappelle 2000).

23.3.2 Soil Seed Bank Sampling and Seed Germination

Topsoil samples were collected along four transects at a depth of 0–5 cm over an area of 35×35 cm at distance intervals of 5, 15, 30 and 50 m from the forest–pasture edge into the old-growth forest, and at 5, 15, 30, 50, 75 and 100 m from the edge into the pasture. An additional sample was taken at the edge itself, i.e., at a distance of 0 m, reaching 11 samples per transect and 44 samples in total. Samples were processed within 2–3 days of collection, and placed in an on-site (2,630 m altitude) greenhouse with a transparent plastic roof allowing for 60–80 % total sunlight. Following Thompson et al. (1996), each sample was thoroughly mixed, evenly divided, and spread in a 2.5-cm layer over two round 30-cm-diameter seed trays containing a layer of sterilized soil covered with a 1-cm layer of seed-free gray sand. Eight trays were filled only with sterilized soil and were used as controls to detect any seed contamination. Trays were hand-watered daily. Seedling mortality was negligible (<1 %).

23.3.3 Seedling Emergence Monitoring

Seedling emergence was monitored weekly over a 6-month period (April–September 1996). Although the soil was not checked for seeds that remained without germination in the period following the monitoring phase, seedling emergence after 6 months was extremely low. Germinated seedlings were counted, and carefully removed immediately after identification or after a morphologic species code was confidently assigned. Morphospecies were subsequently identified in the herbarium. Seedlings were removed to reduce shading and competition. Collected specimens were identified by specialists and deposited at herbaria (CR, INB, USJ). Seedling species were classified with respect to their life form in trees, shrubs, herbs (including grasses, sedges and forbs) and climbers. Species were categorized according to their dispersal mode (anemochorous, zoochorous, autochorous, hydrochorous, or unknown), on the basis of data presented by Wijtzes (1990), expert knowledge, and own observations. Ferns were excluded from all data analyses.

23.3.4 Quantitative Data Analysis

Counts data showed a Poisson distribution and were log transformed [log $(n+1)$] to normalize variance. Data were analyzed with a one-way or two-way ANOVA to test for differences between distance from edge and site location. Site–habitat interactions were not analyzed, as only one transect was run per site. Levels of seed bank diversity of both forest and pasture soils were estimated using Shannon-Wiener's diversity index:

$H' = -\Sigma\, (p_i)^*(\log_2 p_i)$

in which p_i is the proportion of emerged seedlings found in the *i*th species in a sample (Magurran 1988). Similarity between pairs of soil seed samples was assessed using Sørensen's quantitative similarity index for presence–absence data (Magurran 1988). Applying the same similarity index, soil seed bank composition was compared to seed plant species composition of standing vegetation for each sample distance along the 150-m-long axis. For this purpose, we used presence–absence data for standing vegetation in 10×10 m quadrates, as provided by Oosterhoorn and Kappelle (2000).

23.4 Seedling Abundance and Diversity

During 6 months of monitoring, in all 4,940 seedlings were counted of which 97 % was identified to species level. A total of 1,203 emerged from forest soil, and 3,737 from pasture soil samples. Seedlings per sample ranged from eight (forest) to 462 (pasture). Average density was 491 seedlings/m^2 for forest samples and 1,271 for pasture samples. The number of seeds found in pasture samples was significantly higher than that in forest samples ($P<0.01$).

Seedlings belonged to 80 species (68 genera, 37 families), Asteraceae (16 species) and Solanaceae (eight species) being most important. *Physalis* and *Solanum* had three species each. Half of all 4,940 seedlings belonged to only two genera: *Hydrocotyle* (1,446 seedlings, 29.3 %) and *Gnaphalium* (1,030 seedlings, 20.9 %). Species were classified into nine pioneer trees (*Bocconia, Brunellia, Malva, Monnina, Oreopanax, Sida, Vaccinium, Viburnum* and *Wigandia*), 17 shrubs, 48 herbs and six climbers. Trees represented 0.3 % of all seedlings, whereas shrubs totaled 20.0 %, herbs 76.8 %, and climbers less than 0.01 %. Pastures were significantly richer in herb seedlings than were forests ($P<0.001$). Shannon-Wiener's diversity index H' ranged from 0.41 in pasture to 3.5 in forest. A lower seedling per species ratio was observed toward the end of the 6-month monitoring period.

23.5 Seed Dispersal Strategies

A total of 3,849 seedlings belonged to 36 species with an anemochorous (wind) dispersal strategy (655 seedlings in forest samples, and 3,194 in pasture), whereas 585 seedlings in 22 species were zoochorous (bird and/or mammal dispersed; forest: 467; pasture: 118), and another 506 seedlings in 22 species had an unknown or different dispersal mode, e.g., autochorous or hydrochorous (forest: 81; pasture: 425). Animal-dispersed seeds were significantly more abundant ($P<0.001$) in forest samples than in pasture, whereas wind-dispersed seeds were significantly more important ($P<0.001$) in pasture than in forest.

23.6 Changes Along the Forest Interior–Edge–Pasture Gradient

Seed density increased from the forest interior (490.8 germinated seeds per m^2) across the edge into the pasture (1,270.5 seeds per m^2) for all species, and also for wind-dispersed species alone (Fig. 23.1A–C). A *Hydrocotyle*-dominated seedling peak occurred in the pasture at 30 m from the edge. Forest soil samples ($n=20$; 55 species) and pasture soil samples ($n=24$; 55 species) did not show any significant difference in species richness (# of species per sample). Species richness did not change significantly along the forest interior–edge–pasture gradient (Fig. 23.1D–F). The average number of zoochorous seed plant species per sample, however, was significantly higher in the forest than in the pasture ($P<0.001$), and the average number of anemochorous species per sample was higher in pasture than in forest ($P<0.05$). Compared to forest samples, the seedling/species ratio for all species combined was significantly higher in pasture soil samples ($P<0.01$; Fig. 23.1G).

Similarity between pairs of forest soil seed bank samples was higher (0.48–0.66) than that between pairs of pasture samples (0.32–0.56; Table 23.1), indicating a greater variety in species compositions of pasture soil seed banks. Although soil seed banks (this study) did not reflect standing vegetation (Oosterhoorn and Kappelle 2000) very highly, soil seed bank composition was significantly more similar to the standing floristic composition in pastures than in forests (Table 23.2): mean Sørensen's similarity index $C_s \pm 1$ SE=0.0175±0.05 in forest, and 0.065±0.014 in pastures; unpaired t test: $t=-2.68$, df=8, $P=0.028$. The mean number (±1 SE) of species shared by standing vegetation in 100-m^2 quadrats and associated soil seed bank was 2.30±0.25.

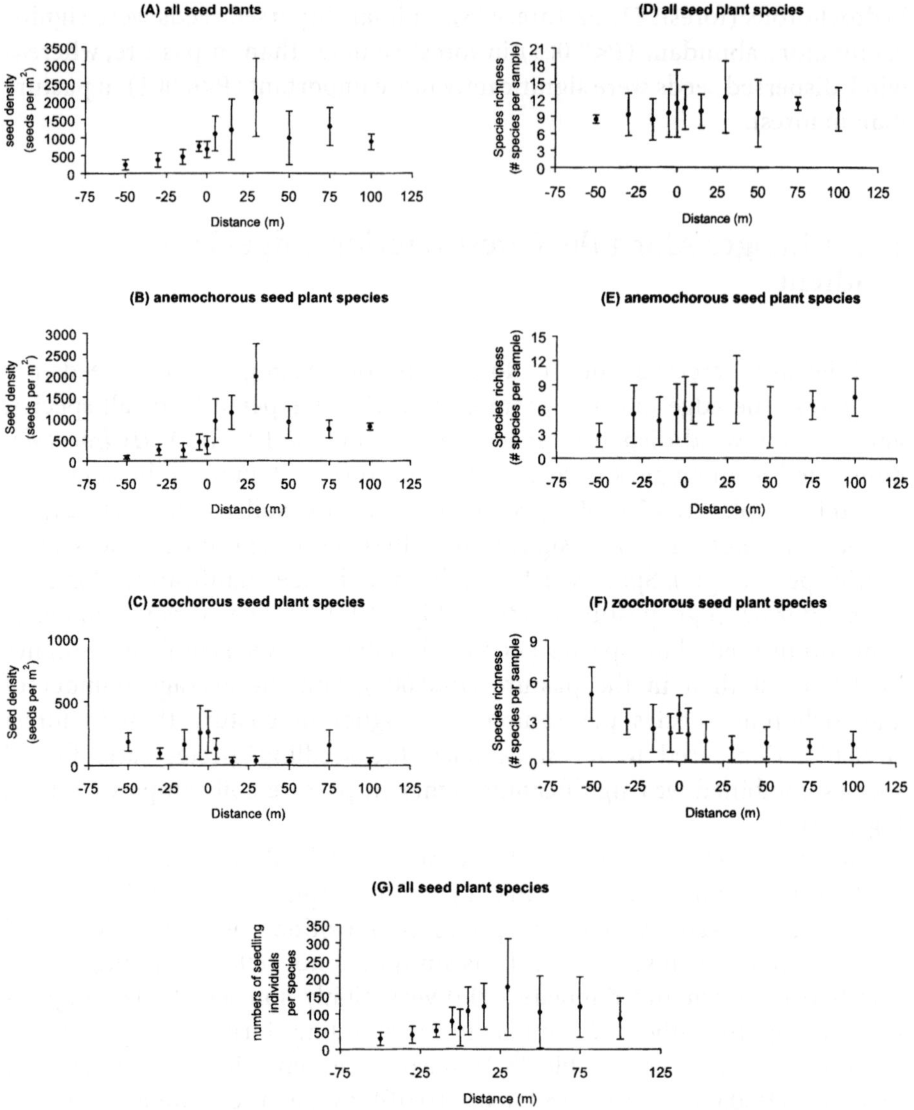

Fig. 23.1A–G. Seed density, species richness, and seedling number measured per distance unit along the forest interior–edge–pasture gradient for seed plant species that emerged from the soil seed bank samples. **A** Seed density (*n*=4; # of germinated seeds per m^2; mean and SD) for all species. **B** Seed density for anemochorous species only. **C** Seed density for zoochorous species only. **D** Species richness (*n*=4; # of species per sample; mean and SD) for all species. **E** Species richness for anemochorous species only. **F** Species richness for zoochorous species only. **G** Seedling numbers per species for all species (*n*=4; mean and 1 SE)

Table 23.1. Sørensen's similarity index (C_s) values for pairs of forest soil seed bank samples, and for pairs of pasture soil seed bank samples at four sampled transect sites in Costa Rica's montane oak forest–pasture zone (see main text for full names and locations of transect sites AR, LD, PR and SG)

	Forest soil samples				Pasture soil samples			
	AR	LD	PR	SG	AR	LD	PR	SG
AR	–	0.50	0.66	0.48	–	0.32	0.41	0.56
LD		–	0.56	0.59		–	0.42	0.38
PR			–	0.63			–	0.53
SG				–				–

Table 23.2. Sørensen's similarity index (C_s) values for pairs of sampled soil seed bank (this study) and standing vegetation (census data from Oosterhoorn and Kappelle 2000) per distance unit along the forest interior–edge–pasture gradient for four sampled transect sites in Costa Rica's montane oak forest–pasture zone. The edge sample at a distance of 0 m was excluded from this analysis

Dist.	Forest				Edge	Pasture					
(m)	50	30	15	5	0	5	15	30	50	75	100
AR	0.15	0.04	0.05	0.12	–	0.35	0.11	0.50	0.11	0.20	0.77
LD	0.04	0.04	0.03	0.03	–	0.07	0.22	0.13	0.00	0.13	0.33
PR	0.06	0.05	0.07	0.12	–	0.17	0.18	0.08	0.17	0.37	0.33
SG	0.00	0.00	0.00	0.11	–	0.13	0.11	0.25	0.18	0.24	0.24
Mean	0.06	0.03	0.04	0.10	–	0.18	0.16	0.24	0.12	0.24	0.42
1 SE	0.03	0.01	0.01	0.02	–	0.06	0.03	0.09	0.04	0.05	0.12

23.7 Conclusions

Mean soil seed densities for this study's forest and pasture are within ranges known from tropical lowland soil seed banks (150–900 seeds per m^2 for forest, and 370–7,623 for pastures; Garwood 1989), and similar to values for Colombian cloud forest (Giraldo and Uribe 1994). Soil seed bank species richness did not change significantly along the distance gradient, though it does along a Mexican lowland maize cropland–edge–forest gradient (Quintana-Ascencio et al. 1996). Richness was lower than in other studies (Williams-Linera 1993), possibly due to the relatively short monitoring period of the present study (see also Chap. 19). Species that need a longer period to successfully germinate may not have been recorded (Brown 1992). Furthermore, as it was not possible to spread the soil samples in very thin layers over the trays, an underestimation of the actual seed bank size and composition may have occurred

(Brown 1992; Dalling et al. 1997). Proportions of shrub and herb seedlings in forest soil samples were relatively high in comparison to known ranges (Garwood 1989).

In tropical lowland forests, seed dispersal by animals is much more important than wind dispersal (Howe and Smallwood 1982; Walker and Neris 1993; Chaps. 13, 18, 19 and 26). The contrasting importance of wind dispersal in this tropical montane study is probably due to the high number of herbs of temperate origin, such as *Arenaria*, *Conyza*, *Gnaphalium*, *Hieracium*, *Holcus*, *Plantago*, *Rumex*, *Senecio*, *Sonchus* and *Valeriana* (Kappelle et al. 1992). The dominance of wind-dispersed seeds we recorded in forest soils is known to occur in many tropical forests (Young et al. 1987; Aide and Cavalier 1994; Quintana-Ascencio et al. 1996; Dalling and Denslow 1998). The large proportion of wind-dispersed herb seeds in both forests and pastures is consistent with population dynamics theory stating that *r*-strategists tend to produce larger quantities of small, light seeds that are better protected against predation and can form long-lived soil seed banks (Garwood 1989; de Steven 1991). In the present study, soil seed banks did not closely reflect standing vegetation, particularly in forest samples. This implies that standing vegetation is not a good proxy for soil seed banks in these forests.

Dedication We dedicate this chapter to the late Menno Oosterhoorn, friend and colleague, who whole-heartedly supported our fieldwork and data analysis in the late 1990s.

Acknowledgements We are grateful to A.M. Cleef and F. Bouman for their conceptual support and critical comments. We thank L. Serrano and E. van Omme for field assistance, and Costa Rica's National Museum, universities (UNA, UCR) and INBio for herbarium facilities and species identification. The hospitality of the rural families in San Gerardo de Dota is much appreciated. Funding was provided by The Netherlands Organization for Scientific Research (NWO), the University of Amsterdam, the University of Groningen, and Bever Outdoor Sports. Research permission was provided by Costa Rica's Ministry of Environment and Energy.

References

Aide TM, Cavalier J (1994) Barriers to lowland tropical forest succession in the Sierra Nevada de Santa Marta, Colombia. Restor Ecol 2(4):219–229

Brown D (1992) Estimating the composition of a forest seed bank: a comparison of the seed extraction and seedling emergence methods. Can J Bot 70:1603–1612

Cubiña A, Aide TM (2001) The effect of distance from forest edge on seed rain and soil seed bank in a tropical pasture. Biotropica 33:260–267

Dalling JW, Denslow JS (1998) Soil seed bank composition along a forest chronosequence in seasonally moist tropical forest, Panama. J Veg Sci 9:669–678

Dalling JW, Swaine MD, Garwood NC (1997) Soil seed bank community dynamics in seasonally moist lowland tropical forest, Panama. J Trop Ecol 13:659–680

De Steven D (1991) Experiments on mechanisms of tree establishment in old-field succession: seedling emergence. Ecology 72:1066–1075
Ewel J (1980) Tropical succession: manifold routes to maturity. Biotropica 12 Suppl 2:2–7
Fraver S (1994) Vegetation responses along edge-to-interior gradients in the mixed hardwood forests of the Roanoke River basin, North Carolina. Conserv Biol 8:822–832
Garwood NC (1989) Tropical soil seed banks: a review. In: Leck MA, Parker VT, Simpson RL (eds) Ecology of soil seed banks. Academic Press, London, UK, pp 149–209
Giraldo R, Uribe A (1994) Evaluación del banco de semillas en bosque de niebla premontano en la Reserva Natural La Planada (Nariño). In: Cavalier J (ed) Resumenes del Simposio Nacional: sobre Diversidad Biológica, Conservación y Manejo de los Ecosistemas de Montaña en Colombia. Universidad de los Andes, Bogotá
Helmer EH (2000) The landscape ecology of tropical secondary forest in montane Costa Rica. Ecosystem 3:98–114
Herrera W (1986) Clima de Costa Rica. EUNED, San José
Holl KD, Kappelle M (1999) Tropical forest recovery and restoration. Trends Ecol Evol 14(10):378–379
Howe HF, Smallwood J (1982) Ecology of seed dispersal. Annu Rev Ecol Syst 13:201–228
José S, Gillespie AR, George SJ, Kumar BM (1996) Vegetation responses along edge-to-interior gradients in a high altitude tropical forest in peninsular India. For Ecol Manage 87:51–62
Kappelle M (1996) Los bosques de roble (*Quercus*) de la Cordillera de Talamanca, Costa Rica: biodiversidad, ecología, conservación y desarrollo. Instituto Nacional de Biodiversidad (INBio), Santo Domingo de Heredia, Costa Rica
Kappelle M, Juárez ME (1995) Agroecological zonation along an altitudinal gradient in the montane belt of the Los Santos Forest Reserve in Costa Rica. Mount Res Dev 15(1):19–37
Kappelle M, Cleef AM, Chaverri A (1992) Phytogeography of Talamanca montane *Quercus* forests, Costa Rica. J Biogeogr 19(3):299–315
Kappelle M, Kennis PAF, de Vries RAJ (1995) Changes in diversity along a successional gradient in a Costa Rican upper montane *Quercus* forest. Biodiv Conserv 4:10–34
Kappelle M, Geuze T, Leal ME, Cleef A (1996) Successional age and forest structure in a Costa Rican upper montane *Quercus* forest. J Trop Ecol 12:681–698
Laurance WF, Bierregaard Jr RO, Gascon C, Didham RK, Smith AP, Lynam AJ, Viana VM, Lovejoy TE, Sieving KE, Sites Jr JW, Andersen M, Tocher MD, Kramer EA, Restrepo C, Moritz C (1997) Tropical forest fragmentation: synthesis of a diverse and dynamic discipline. In: Laurance WF, Bierregaard RO Jr (eds) Tropical forest remnants: ecology, management and conservation of fragmented communities. Univ Chicago Press, Chicago, IL, pp 502–514
Lopez de Casenave J, Pelotto JP, Protomastro J (1995) Edge-interior differences in vegetation structure and composition in a Chaco semi-arid forest, Argentina. For Ecol Manage 72:61–69
Lovejoy TE, Bierregaard Jr RO, Rankin JM, Schubart HO (1983) Ecological dynamics of tropical forest fragments. In: Sutton L, Whitmore TC, Chadwick AC (eds) Tropical rain forest: ecology and management. Blackwell, Oxford, UK, pp 377–384
Magurran AE (1988) Ecological diversity and its measurement. Croom Helm, London, UK
Oosterhoorn M, Kappelle M (2000) Vegetation structure and composition along an interior-edge-exterior gradient in a Costa Rican montane cloud forest. For Ecol Manage 126:291–307
Parker VT, Simpson RL, Leck MA (1989) Patterns and process in the dynamics of seed banks. In: Leck MA, Parker VT, Simpson RL (eds) Ecology of soil seed banks. Academic Press, London, UK, pp 367–384

Quintana-Ascencio PF, González-Espinosa M, Ramírez-Marcial N, Dominguez-Vázquez G, Martínez-Icó M (1996) Soil seed banks and regeneration of tropical rain forest from milpa fields at the Selva Lacandona, Chiapas, Mexico. Biotropica 28:192–209

Teketay D, Granström A (1995) Soil seed banks in dry Afromontane forests of Ethiopia. J Veg Sci 6:777–786

Thompson K, Bakker JP, Bekker RM (1996) Soil seed banks of North-West Europe: methodology, density and longevity. Cambridge Univ Press, Cambridge, UK

Walker LR, Neris LE (1993) Post-hurricane seed rain dynamics in Puerto Rico. Biotropica 25(4):408–418

Wijtzes WH (1990) Dispersal strategies of upper montane primary forest and secondary vegetation on the Pacific side of the Cordillera de Talamanca, Costa Rica. MSc Thesis, University of Amsterdam, Amsterdam

Williams-Linera G (1990) Vegetation structure and environmental conditions of forest edges in Panama. J Ecol 78:356–373

Williams-Linera G (1993) Soil seed banks in four lower montane forests of Mexico. J Trop Ecol 9(2):321–337

Young KR, Ewel JJ, Brown BJ (1987) Seed dynamics during forest succession in Costa Rica. Vegetatio 71:157–173

24 Frugivorous Birds, Habitat Preference and Seed Dispersal in a Fragmented Costa Rican Montane Oak Forest Landscape

J.J.A.M. WILMS and M. KAPPELLE

24.1 Introduction

Frugivorous birds play an important role in seed dispersal (Levey et al. 2005), especially in the tropics (Stiles 1985; Guevara et al. 1986; Corlett 1998). Their role in lowland rainforest recovery on abandoned pastures has been extensively documented (Debussche et al. 1982; McDonnell and Stiles 1983; Guevara et al. 1986; Gorchov et al. 1993; Guevara and Laborde 1993; Martínez-Ramos and Soto-Castro 1993; McClanahan and Wolfe 1993; Robinson and Handel 1993; Galindo-González et al. 2000).

Numerous bird species thrive well in secondary growth tropical lowland habitats where croplands, shrublands and pastures alternate (Estrada et al. 1993; Gorchov et al. 1993). In such patchy landscapes, isolated remnant forest trees may offer protection to birds, and serve as stepping stones during their journey in search for closed old-growth forest fragments (Guevara et al. 1986; McClanahan and Wolfe 1987). In this way, isolated bird-dispersed forest trees may act as 'seed-trapping' centers of succession following lowland rainforest clearing (Debussche et al. 1982; McDonnell and Stiles 1983; Guevara et al. 1986; Guevara and Laborde 1993).

To date, however, little is known on the ecology of frugivorous birds in human-influenced tropical montane forest landscapes. Neither do we know much about their role in the recovery of cleared upland forest (Stiles 1985; Kappelle et al. 1994; Poulsen 1994; Long 1995; Shiels and Walker 2003). Even less is known on the interaction between frugivorous bird species and specific ornithochorous tree species in relation to montane forest succession (Chap. 25). To further gain insight into these processes, we studied the role of frugivorous, seed-dispersing birds in montane oak forest recovery following clearing, burning and grazing in Costa Rica's Talamanca highlands.

Ecological Studies, Vol. 185
M. Kappelle (Ed.) Ecology and Conservation of Neotropical Montane Oak Forests
© Springer-Verlag Berlin Heidelberg 2006

24.2 Study Area

This study was carried out in fragmented oak forests and grasslands in the upper watershed (2,200–2,800 m a.s.l.) of the Savegre River, near San Gerardo de Dota (9°35'40"N, 83°44'30"W) in Costa Rica's 62,000-ha Los Santos Forest Reserve (Kappelle and Juárez 1995; Chap. 30). Recently, these highlands have been recognized as an Endemic Bird Area (EBA) by BirdLife International (Long 1996). In this EBA, 52 endemic bird species are found, ten of which are restricted to montane cloud forests (Long 1995).

Average annual rainfall in this river valley is 2,000–3,000 mm, and annual temperature varies in the range 10–15 °C. The dry season lasts from January to April (Kappelle 1996). Soils are derived from volcanic ash, are acid (pH 3.5–5.5), and moderately fertile (Chap. 4). Natural vegetation is 30–50 m tall, evergreen broad-leaved old-growth oak forest (Chap. 10). The vegetation of the upper Savegre watershed is characterized by patches of mature ('primary') and 10–40 year old, recovering (secondary) forests, dense scrublands, fern brakes, blackberry fields, pasturelands with isolated trees, and fruit tree orchards. This diverse, multifaceted landscape mosaic is the result of intensive logging, burning, and subsequent changes in land use (Kappelle and Juárez 1995; Chaps. 10 and 30). During the last decade, information has become available on changes in forest structure and species composition along a successional gradient in these forests (Kappelle 1993, 1996; Kappelle et al. 1994, 1995, 1996; Chap. 17).

24.3 Habitat Selection and Plot Establishment

Three upper montane (2,300–3,200 m altitude) forest plant communities were selected for sampling: (1) undisturbed mature old-growth forest (MF), (2) successional (secondary) forest (SF), and (3) pasture with isolated trees (PI). The MF community is an evergreen, old-growth cloud forest dominated by 35–45 m tall *Quercus copeyensis* (*Q. bumelioides* – K.C. Nixon, personal communication), *Q. costaricensis* and *Q. seemannii*. The SF community is a 10–35 year old secondary forest with 7–15 m tall canopy trees, and thrives at sites that were cleared, burned, grazed, and subsequently abandoned. The PI community grows in moderately grazed pastures with scattered and isolated remnant forest trees of mature stature. Descriptions of structure and composition of these communities are presented elsewhere (Kappelle et al. 1994; Chap. 17).

We established nine randomly selected plots, three per plant community. Plot size was dependent on horizontal visibility, and ranged from 0.6 ha in PI plots, to 0.1 ha in SF plots and 0.3 ha in MF plots. Vertical visibility was excellent, even in MF plots, as alpine rope climbing techniques were applied to get access to old-growth forest canopies (Perry 1978; Whitacre 1981; Ter Steege and Cornelissen 1988). Tree climbing was also applied to obtain samples of fruiting tree branches for identification of species that are part of the birds' diets. Plot location was determined using aerial black/white photographs taken in 1992 (scale 1:15,000), and a global positioning system (GPS).

24.4 Vegetation Sampling

Vegetation structure and floristic composition were assessed in each plot using standard ecological census techniques (Jongman et al. 1987; Kent and Coker 1992). Relative vertical (aerial) crown cover or shoot cover projection was estimated for each terrestrial vascular plant species. Growth forms were also recorded. Plant specimens were collected for identification, and stored at Costa Rica's National Herbarium (CR) and the National Biodiversity Institute (INB).

Two-way indicator species analysis (TWINSPAN) was applied on a plant data matrix (presence/absence and relative aerial tree crown cover data) comprising nine plots×96 tree species occurring in at least two plots (Hill 1979; Kent and Coker 1992). For classification purposes, plant species cover percentages were converted into nine cover classes using an adapted form of the logarithmic octave-scaling technique proposed by Gauch (1982): 0 % (-), <1 % (1), 1 % (2), 2–3 % (3), 4–7 % (4), 8–15 % (5), 16–31 % (6), 32–63 % (7), and 64–100 % (8).

24.5 Bird Censusing

Bird identification and censusing using the approach of Stiles and Skutch (1994) took place in 1996. In each plot observations were made twice, once in the (late) dry season (February–April) and once in the (early) wet season (May–August). Observations were conducted on three consecutive days (dry season: 5:30–10:30 AM; wet season: 5:00–10:00 AM) for 5 h each morning when bird activity is greatest (Blake 1992). This resulted in 54 observation days, with 6 days spent in each of the nine plots (3 days per season, two seasons). Binoculars were used, and sightings were recorded with a portable voice tape recorder (Wheelwright 1991). The consumption of tree fruits by birds – as part of their diet – was censused, as dispersal of tree seeds seems to be the most important aspect inducing forest regeneration (Guevara et al. 1986; Kappelle 1993; Robinson and Handel 1993).

During each 5-h observation session, the number of bird species, and the position, behavior and diet of each species was recorded every 15 min for a period of 3 min, following Gibbons et al. (1996). In addition to the 54 observation days spent inside the plots (270 observation hours), some more observation days were used to assess the diet of noted frugivores in randomly chosen places at various times of the day. This record accounted for some 120 observation hours outside the plots. Bird size classes (small: 0–50 g; medium: 50–100 g; large: >100 g) were based upon body weight data, as provided by Stiles and Skutch (1994).

24.6 Quantitative Data Analysis

Data were first tested on normal distributions, and normally distributed data were subsequently tested using one-way ANOVA. For nonparametric data, we applied Kruskal Wallis tests and Mann-Whitney U tests with Bonferroni correction (Krebs 1985; Magurran 1988; Rice 1989; Sokal and Rohlf 1995). We defined 'frequently observed species' as those species that were observed on at least 6 days (N_{obs}>5, with N_{obs} being the number of observation days) of a total of 54 observation days (N_{max}=54, i.e., nine plots×six observation days). TWINSPAN software was also applied to a bird presence/absence matrix to determine the composition of bird communities and that of ecological species groups, indicative for specific phases of oak forest recovery. The level of similarity between species compositions was assessed using Sørensen's index of similarity (coefficient of community, CC) for pairs of communities and plots (Magurran 1988). Calculated similarity index values were tested on significance, using ANOVA after conducting an arcsine square root transformation (Sokal and Rohlf 1995).

24.7 Plant Communities

The TWINSPAN vegetation analysis (nine plots×96 tree species) showed a division between (1) mature, old-growth forest and secondary forest plant communities representing closed forest bird habitats, and (2) low-stature, pasture and open secondary growth plant communities representing open and non-forest bird habitats.

Community group 1 is characterized by the plant species *Alfaroa costaricensis*, *Anthurium concinnatum*, *Ardisia compressa*, *Chusquea tomentosa*, *Clusia* sp., *Cyclanthera langaei*, *Ilex pallida*, *Lycopodium thyoides*, *Meliosma glabrata*, *Myrsine coriacea*, *Nectandra cufodontisii*, *Oreopanax capitatus*, *Palicourea* sp., *Passiflora sexflora*, *Quercus seemannii* and *Rondeletia buddleoides*. Community group 2 is typified by *Ageratina subcordata*, *Buddleja cordata*, *Buddleja nitida*, *Cirsium subcoriaceum*, *Conyza bonariensis*, *Fuchsia microphylla, F. paniculata, Gnaphalium americanum, Holcus lanatus, Hydrocotyle bowlesioides, Lachemila standleyi, Miconia schnellii, Monochaetum floribundum, Oxalis spiralis, Pennisetum clandestinum, Plantago australis* and *Rhynchospora aristata.* Plant community data were consistent with previously published descriptions of species assemblages (Kappelle et al. 1994; Chap. 17).

24.8 Bird Diversity and Habitat Preference

Sixty bird species were observed in nine plots during 270 observation hours distributed over 54 days. Another 15 bird species were observed outside the plots, totaling 75 species for the study area. This richness corresponds to 53.2 % of 141 species known from the upper Savegre River watershed (2,000–3,400 m; Bader 1995). About 45.3 % (34 species) of the 75 species are frugivorous (Wheelwright 1983; Wheelwright et al. 1984; Hilty and Brown 1986; Stiles and Skutch 1994).

The TWINSPAN frugivorous bird community analysis (nine plots×34 species) showed a division into three ecological species groups:

- a group of ten bird species with a habitat preference for closed forest plant communities 1;
- a group of 11 bird species with a habitat preference for open, bushy pastures and young, secondary growth plant communities 2; and
- a group of 13 bird species with no specific habitat preference. This latter group includes the resplendent quetzal *Pharomachrus mocinno* (Table 24.1), a spectacular bird for which more diet information has recently become available (García 2004; Chap. 25).

The most frequently observed frugivorous birds are (1) in MF, *Chamaepetes unicolor, Melanerpes formicivorus, Catharus gracilirostris, Turdus plebejus, Myadestes melanops, Ptilogonys caudatus, Chlorospingus pileatus* and *Pselliphorus tibialis*; (2) in SF, *Chamaepetes unicolor, Empidonax atriceps, Elaenia frantzii, Turdus plebejus, Turdus nigrescens, Ptilogonys caudatus, Chlorospingus pileatus* and *Pezopetes capitalis*; and (3) in PI, *Columba subvinacea, Pharomachrus mocinno, Elaenia frantzii, Catharus gracilirostris, Turdus plebejus, Turdus nigrescens* and *Ptilogonys caudatus.*

Significance tests (Kruskal Wallis and Mann-Whitney U tests, $P<0.05$) applied to observation frequency data (mean±1 SE for 18 species with $N_{obs}>5$) for each frugivorous bird species in each plant community showed a significant plant community preference among nine bird species, and no significance among another set of nine species. *Melanerpes formicivorus, Turdus plebejus* and *Myadestes melanops* had a significant preference for MF, whereas *Columba fasciata, Elaenia frantzii* and *Turdus nigrescens* showed a significant

Table 24.1. List of 34 frugivorous bird species observed in fragmented tropical montane oak forest in Costa Rica. Species are assigned to groups in accordance with their habitat preference. For each species, its family, species and common names are given. Twenty-two fruit tree species on which bird species fed during observation are listed as well

Family	Species	Common name	Tree species[a]
Bird species with a habitat preference for closed forest (ten species)			
Corvidae	*Cyanolyca argentigula*	Silvery-Throated Jay	-
Corvidae	*Vireo leucophrys*	Brown-Capped Vireo	-
Cracidae	*Chamaepetes unicolor*	Black Guan	10, 11, 15
Emberizidae	*Pezopetes capitalis*	Large-Footed Finch	-
Emberizidae	*Pheucticus tibialis*	Black-Thighed Grosbeak	21, 22
Emberizidae	*Pselliphorus tibialis*	Yellow-Thighed Finch	3, 4
Icteridae	*Amblycercus holosericeus*	Yellow-Billed Cacique	-
Trogonidae	*Trogon collaris*	Collared Trogon	15, 21
Turdidae	*Catharus frantzii*	Ruddy-Capped Nightingale-Thrush	-
Turdidae	*Myadestes melanops*	Black-Faced Solitaire	3, 9, 10
Bird species with a habitat preference for open pasture or secondary growth (11 species)			
Accipitridae	*Elanoides forficatus*	American Swallow-Tailed Kite	-
Columbidae	*Columba fasciata*	Band-Tailed Dove	-
Emberizidae	*Junco vulcani*	Volcano Junco	-
Emberizidae	*Pheucticus ludovicianus*	Rose-Breasted Grosbeak	-
Fringillidae	*Carduelis xanthogastra*	Yellow-Bellied Siskin	3
Picidae	*Piculus rubiginosus*	Golden-Olive Woodpecker	-
Psittacidae	*Pyrrhura hoffmanni*	Sulfur-Winged Parakeet	9, 22
Thraupidae	*Piranga bidentata*	Flame-Colored Tanager	5
Thraupidae	*Thraupis episcopus*	Blue-Gray Tanager	-
Turdidae	*Turdus grayi*	Clay-Colored Robin	3, 9, 12
Turdidae	*Turdus nigrescens*	Sooty Robin	1, 3, 5, 9

Bird species with no specific habitat preference (13 species)			
Columbidae	*Columba subvinacea*	Ruddy Pigeon	3
Corvidae	*Parula gutturalis*	Flame-Throated Warbler	-
Corvidae	*Vireo carmioli*	Yellow-Winged Vireo	-
Picidae	*Elaenia frantzii*	Mountain Elaenia	1, 3, 7, 9
Picidae	*Melanerpes formicivorus*	Acorn Woodpecker	-
Ptilogonatidae	*Ptilogonys caudatus*	Long-Tailed Silky-Flycatcher	3, 5, 6, 7, 8, 9
Ramphastidae	*Aulacorhynchus prasinus*	Emerald Toucanet	9, 15, 16, 17, 18, 19, 20, 21
Thraupidae	*Chlorophonia callophrys*	Golden-Browed Chlorophonia	9
Thraupidae	*Chlorospingus pileatus*	Sooty-Capped Bush-Tanager	3, 4
Trogonidae	*Pharomachrus moccino*	Resplendent Quetzal	10, 12, 13, 14, 15, 16, 17, 18
Turdidae	*Catharus gracilirostris*	Black-Billed Nightingale-Thrush	1, 3
Turdidae	*Turdus plebejus*	Mountain Robin	1, 2, 3, 4, 7, 8, 9, 10, 11, 12, 16, 21
Tyrannidae	*Empidonax atriceps*	Black-Capped Flycatcher	3

[a] Tree species: 1, *Viburnum costaricanum*; 2, *Vaccinium consanguineum*; 3, *Fuchsia paniculata*; 4, *Palicourea salicifolia*; 5, *Miconia tonduzii*; 6, *Cleyera theaeoides*; 7, *Moninna xalapensis*; 8, *Solanum dotanum*; 9, *Freziera candicans*; 10, *Ilex pallida*; 11, *Cornus disciflora*; 12, *Billia hippocastanum*; 13, *Ocotea pharomachrosorum*; 14, *Myrcianthes rhopaloides*; 15, *Nectandra cufodontisii*; 16, *Ocotea insularis*; 17, *Ocotea austinii*; 18, *Ocotea pseudopalmana*; 19, *Buddleja cordata*; 20, *Myrica pubescens*; 21, *Sapium pachystachys*; 22, *Croton xalapensis*

preference for both SF and PI. *Pharomachrus mocinno* and *Catharus gracilirostris* appeared significantly less often in SF, and *Pezopetes capitalis* significantly less often in PI.

24.9 Bird Species Diet

During the fieldwork period, 20 of the 34 frugivorous bird species were observed while foraging on 22 different tree and shrub fruits, resulting in a total of 68 bird–plant frugivorous interactions (Table 24.1, Fig. 24.1). Small to medium-sized birds foraged mainly on fruits of successional trees as *Fuchsia paniculata, Miconia tonduzii, Monnina xalapensis* and *Viburnum costaricanum*, whereas medium to large-sized birds foraged largely on mature forest tree species in, e.g., Lauraceae, including *Ocotea pharomachrosorum* (the 'Quetzal-bearing' *Ocotea*). *Fuchsia paniculata* and *Freziera candicans* were

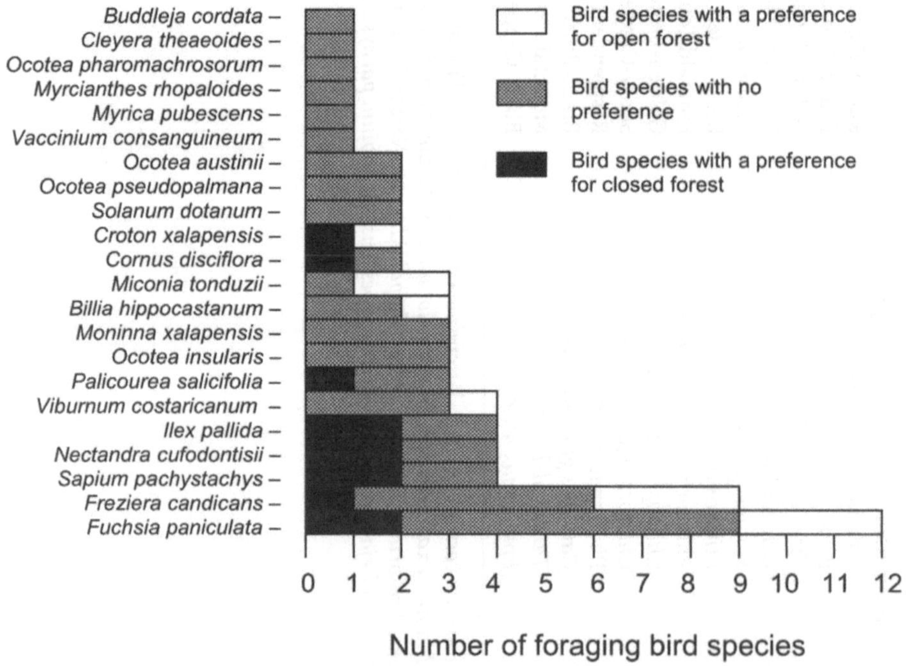

Fig. 24.1. Number of bird species per diet plant species for a total of 20 frugivorous birds feeding on 22 trees and shrubs. Data are from observations on foraging behavior made in 1996, and resulted in a total of 68 bird–plant frugivorous interactions. A distinction has been made for bird species with preferences for either open or closed forest, as well as for species without any specific habitat preference. For details on bird species, see Table 24.1

the two most visited ornithochorous tree species, with 12 and nine avian frugivores foraging on their fruits, respectively.

24.10 Birds, Plant Communities and Seasonality

Frugivorous bird species richness per plant community resulted in 21 species (61.8 %) in MF, 26 species in SF, and 22 species in PI. The largest number of bird species (23) was observed during the dry season in successional forest at 2,600 m elevation (Fig. 24.2). A comparison between bird species richness and plant community (Kruskal Wallis test, $P<0.05$), and bird species richness and season (Mann-Whitney U test with Bonferroni correction) showed no significant difference between plant communities. However, a significant season preference was found (Kruskal Wallis and Mann-Whitney U tests, $P<0.05$) for *Chamaepetes unicolor* in the dry season, and for *Catharus gracilirostris* and *Ptilogonys caudatus* in the wet season.

Table 24.2 shows Sørensen's similarity between bird species compositions in three plant communities for (1) the entire set of 34 frugivorous bird species, and (2) a subset of 18 frequently observed species. It appears that the

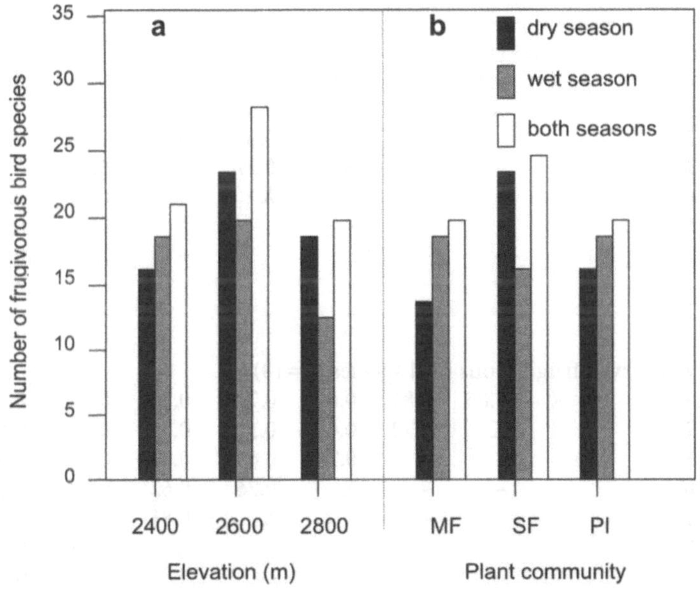

Fig. 24.2a, b. Number of frugivorous bird species observed during dry and wet seasons at three elevations (**a**) and in three successional plant communities (**b**) in the montane oak forest belt in Costa Rica. *MF* Undisturbed mature old-growth oak forest, *SF* successional (secondary) forest, *PI* pasture with isolated trees

Table 24.2. Sørensen similarity index values for pairs of plant communities in fragmented tropical montane oak forest in Costa Rica. Values are given taking into account all 34 frugivorous bird species, and only those 18 species that were frequently observed

Plant community[a]	All frugivorous bird species: Number of bird species	Plant community[a] MF	SF	PI	Frequently observed frugivorous bird species: Number of bird species	Plant community[a] MF	SF	PI
MF	21	X	0.68	0.65	16	X	0.94	0.91
SF	26	–	X	0.75	18	–	X	0.84
PI	22	–	–	X	15	–	–	X

[a] MF, undisturbed mature old-growth forest; SF, successional (secondary) forest; PI, pasture with isolated trees

Table 24.3. Sørensen similarity index values for pairs of plots in three different plant communities in fragmented tropical montane oak forest in Costa Rica. A total of three plots were sampled per plant community. Values are given taking into account all frugivorous bird species, and only those species that were frequently observed

Plot[a]	Number of species per plot	Plot[a] MF 1	MF 2	MF 3	SF 1	SF 2	SF 3	PI 1	PI 2	PI 3
All frugivorous bird species (*n*=34)										
MF 1	16	X	0.71	0.76	0.57	0.56	0.71	0.55	0.59	0.57
MF 2	15	–	X	0.64	0.37	0.39	0.61	0.43	0.56	0.52
MF 3	13	–	–	X	0.64	0.69	0.77	0.77	0.65	0.72
SF 1	12	–	–	–	X	0.57	0.53	0.72	0.61	0.67
SF 2	16	–	–	–	–	X	0.65	0.69	0.59	0.64
SF 3	18	–	–	–	–	–	X	0.77	0.67	0.67
PI 1	13	–	–	–	–	–	–	X	0.76	0.80
PI 2	21	–	–	–	–	–	–	–	X	0.67
PI 3	12	–	–	–	–	–	–	–	–	X
Frequently observed frugivorous bird species (*n*=18)										
MF 1	14	X	0.85	0.81	0.67	0.72	0.80	0.67	0.76	0.64
MF 2	12	–	X	0.72	0.45	0.52	0.71	0.56	0.67	0.61
MF 3	13	–	–	X	0.70	0.83	0.83	0.77	0.86	0.75
SF 1	10	–	–	–	X	0.76	0.62	0.78	0.80	0.76
SF 2	11	–	–	–	–	X	0.81	0.83	0.77	0.82
SF 3	16	–	–	–	–	–	X	0.83	0.84	0.74
PI 1	13	–	–	–	–	–	–	X	0.93	0.92
PI 2	15	–	–	–	–	–	–	–	X	0.85
PI 3	11	–	–	–	–	–	–	–	–	X

[a] MF, undisturbed mature old-growth forest, SF, successional (secondary) forest, PI, pasture with isolated trees

frugivorous bird species composition is similar in all three plant communities, whether all frugivores -- including rarely observed species – are considered, or only those species that were frequently observed.

Table 24.3 shows Sørensen's similarity between bird species compositions found in the nine plots, for (1) the entire set of 34 frugivorous bird species, and (2) a subset of 18 frequently observed species. ANOVA applied to both sets of bird species showed no significant differences in similarity within (three plots, three pairs) and between (six plots, nine pairs) each plant community. An exception was the similarity recorded for the 18 frequently observed species found within plant community PI, which turned out to be significantly higher (mean±1 SE=0.90±0.03; $P<0.05$) than that of other paired combinations (means varying in the range 0.69–0.80). Frugivorous bird species composition in open, successional habitats was significantly different (ANOVA, mean±1 SE=0.82±0.02; $P<0.05$) from the composition found for all habitats combined (ANOVA, mean±1 SE=0.70±0.03; $P<0.05$).

24.11 Seed-Dispersing Birds and Ornithochorous Trees

Seven bird species did not show any significant habitat preference, and were qualified as seed dispersers that may play a key role at sites where oak forest needs to be restored: *Columba subvinacea*, *Pharomachrus mocinno*, *Aulacorhynchus prasinus*, *Catharus gracilirostris*, *Turdus plebejus*, *Parula gutturalis* and *Chlorospingus pileatus*. They were observed in similar quantities in both closed forest and open successional habitats, and may disperse seeds from fruiting parent trees found in old-growth mature forest while venturing into open pastures and young secondary growth.

It has been reported that 30–50 % of tropical forest bird species may depend on fruits for their diet, and 50–90 % of tropical forest plant species may depend on birds for seed dispersal (Stiles 1985; Fleming et al. 1987; Estrada et al. 1993; Arango 1994). This is particularly the case in montane cloud forests where seed dispersal by monkeys and bats is very limited. Arango (1994) recorded that 51 % of the bird species in a Colombian montane forest (Alto Quindío) was frugivorous, and 34 % of the tree species in that forest was ornithochorously dispersed. Close to our study area near Cerro de la Muerte (3,491 m altitude), Stiles (1985) recorded 43 bird-dispersed trees and shrubs – a number corresponding to about 75 % of all woody species known from that area (Kappelle et al. 2000). In all, 37 % of the avifauna found at the nearby La Muerte peak fed on fruits. We recorded relative values within a similar range, and 45.3 % of the bird species we observed were frugivorous.

24.12 Are Forest-Dependent Birds More Threatened?

Species such as *Chamaepetes unicolor*, *Trogon collaris*, *Catharus frantzii*, *Vireo leucophrys* and *Pheucticus tibialis* that have a preference for mature forest habitat may suffer severely from continued deforestation, their populations being potentially threatened. Long (1995) classified the black guan *Chamaepetes unicolor* as a restricted range species confined mainly to tropical montane cloud forests in the Costa Rican and Panamanian Highlands Endemic Bird Area (EBA). Further studies on the behavior and natural history of the ten bird species that appear to be restricted to mature, closed montane oak forest are highly recommended (Table 24.1).

24.13 Acorn Dispersal by Jays

It is expected that birds such as the Costa Rican silvery-throated jay (*Cyanolyca argentigula*) will behave similarly to its North American temperate oak forest counterpart, the blue jay (*Cyanocitta cristata*). In the northern, cool oak forests, the blue jay is known to disperse acorns and speed up oak forest regeneration (Darley-Hill and Johnson 1981; Johnson and Webb 1989). Johnson and colleagues observed that acorns are a valuable, albeit inconsistent, source of food, and that more than 180 different kinds of birds and mammals use temperate oak acorns as food in the northern temperate autumn; among these are the afore-mentioned blue jays, as well as crows, red-headed woodpeckers, pigeons, pheasants, turkey, ducks, quail, deer, squirrels, mice, chipmunks, badgers, and even raccoons (see also Chap. 13). Jays, in particular, enjoy feeding on these oak nuts. They seem to form a symbiotic relation with oaks (Bosema 1979). If jays do not eat the acorns on the spot, they carry them away and hide them in the ground, forming a winter store (scatter-hoarding). However, not all the acorns buried will be found again. Many are left in the ground, and some of these germinate and grow into new trees.

We recommend conducting similar studies on the silvery-throated jay and oaks such as the locally endemic *Quercus copeyensis*, *Q. costaricensis* and *Q. seemannii*, which still abound in Costa Rica's montane oak forests. Ethological studies on the silvery-throated jay and other foraging Corvidae – in relation to research on the phenological behavior of oaks (Céspedes 1991) – may well offer new insights urgently needed to accelerate oak forest recovery and enhance oak forest restoration at abandoned sites in the future.

24.14 Conclusions

The role of frugivorous birds is essential for montane oak forest recovery, as seeds of most tree species are endozoochorously dispersed (Stiles 1985; Wijtzes 1990). More detailed data on the abundance of fruit trees important to key avian seed dispersers are urgently needed. Future ecological research should focus particularly on those birds that forage specifically on fruits of mature forest trees occurring also in pastures. This knowledge is indispensable for accelerating oak forest recovery.

We found that numerous bird species show significant preference for either closed or open habitats. Examples are *Chamaepetes unicolor* in closed forest, and *Piculus rubiginosus* in open pastures. Open, early successional habitats may locally become important to rare and endangered birds. This phenomenon has previously been observed in the Ecuadorian Andes (Welford 2000). However, a specific set of birds, including *Turdus plebejus*, do not show any particular preference for either open or closed habitats. Species belonging to this group are of key importance, as they potentially contribute to successful seed dispersal at recovering sites. Thus, they may serve as key dispersal agents as they transport tree seeds from mature closed forest into non-forested secondary scrub and pastures. These species may promote forest recovery at cleared and fragmented sites, especially when remnant isolated trees are present.

Similarly, isolated trees may function as attractive perching places, foraging spots, stepping stones, and even nesting places for frugivorous bird species venturing into pastures (McDonnell and Stiles 1983; Guevara et al. 1986; McClanahan and Wolfe 1987, 1993; Aide and Cavelier 1994; Stiles and Skutch 1994; Galindo-González et al. 2000; Chap. 25). Indeed, we observed numerous frugivorous birds in pastures where they were attracted by isolated fruiting trees, including the small-seeded pioneer tree *Fuchsia paniculata* and the large-seeded *Ocotea* spp. Therefore, we recommend planting these trees as well as *Freziera candicans* in reforestation programs that aim at speeding up tropical oak forest recovery in open pastures. Other montane forest tree genera such as *Guarea*, *Miconia* and *Schefflera* also seem to be promising in this respect, as they may produce fruit over prolonged time periods, thereby providing birds with a fairly constant fruit supply (Carlo et al. 2004).

Acknowledgements We are grateful to A.M. Cleef, F. Bouman and J. Wattel for providing helpful advice. M.B. van den Bergh and J. Vogel helped during fieldwork. J. Sánchez assisted in bird identification. Staff at INBio and the National Museum identified many of the collected plant specimens. R.A. Wesselingh assisted with statistical analysis. The hospitality of the rural families in San Gerardo de Dota is much appreciated. Funding was provided by The Netherlands Organization for Scientific Research (NWO, grant 895.100.003), the University of Amsterdam, the Amsterdam University Foundation, the

Alberta Mennega Foundation, the STIR Foundation, and Bever Outdoor Sports. Research permission was granted by Costa Rica's Ministry of Environment and Energy (MINAE).

References

Aide TM, Cavelier J (1994) Barriers to lowland tropical forest restoration in the Sierra Nevada de Santa Marta, Colombia. Restor Ecol 2:219–229

Arango S (1994) El papel de las aves dispersoras de semillas en la regeneración de pastizales en el Alto Quindío. In: Cavelier J, Uribe A (eds) Diversidad biológica, conservación y manejo de los ecosystemas de montaña en Colombia. Universidad de los Andes, Bogotá, p 17

Bader MT (1995) Guidebook to the Quetzal Education Research Center (QERC). Bader, Lakewood, UK

Blake JG (1992) Temporal variation in point counts of birds in a lowland wet forest in Costa Rica. The Condor 94:265–275

Bosema I (1979) Jays and oaks: an eco-ethological study of symbiosis. Behaviour 70:1–117

Carlo T, Collazo JA, Groom MJ (2004) Influences of fruit diversity and abundance on bird use of two shaded coffee plantations. Biotropica 36(4):602–614

Céspedes R (1991) Fenología de *Quercus seemannii* Lieb. (Fagaceae), en Cartago, Costa Rica. Rev Biol Trop 39(2):243–248

Corlett RT (1998) Frugivory and seed dispersal by vertebrates in the Oriental (Indomalayan) Region. Biol Rev Cambridge Philos Soc 73:413–448

Darley-Hill S, Johnson WC (1981) Acorn dispersal by the blue jay (*Cyanocitta cristata*). Oecologia 50:231–232

Debussche M, Escarré J, Lepart J (1982) Ornithochory and plant succession in Mediterranean abandoned orchards. Vegetatio 48:255–266

Estrada A, Coates-Estrada R, Meritt Jr D, Montiel S, Curiel D (1993) Patterns of frugivore species richness and abundance in forest islands and in agricultural habitats at Los Tuxtlas, Mexico. Vegetatio 107/108:245–257

Fleming TH, Breitwisch R, Whitesides GH (1987) Patterns of tropical vertebrate frugivore diversity. Annu Rev Ecol Syst 18:91–109

Galindo-González J, Guevara S, Sosa VJ (2000) Bat- and bird-generated seed rains at isolated trees in pastures in a tropical rainforest. Conserv Biol 14(6):1693–1703

García M (2004) Ecología y conservación del quetzal (*Pharomachrus mocinno costaricensis*) en la Reserva Forestal Los Santos (RFLS), Costa Rica. MSc Thesis, Programa Regional de Manejo de Vida Silvestre, Universidad Nacional, Heredia, Costa Rica

Gauch HG (1982) Multivariate analysis in community ecology. Cambridge Univ Press, Cambridge, UK

Gibbons DW, Hill D, Sutherland WJ (1996) Birds. In: Sutherland WJ (ed) Ecological census techniques. Cambridge Univ Press, Cambridge, UK, pp 227–259

Gorchov DL, Cornejo F, Ascorra C, Jaramillo M (1993) The role of seed dispersal in the natural regeneration of rain forest after strip-cutting in the Peruvian Amazon. Vegetatio 107/108:339–349

Guevara S, Laborde J (1993) Monitoring seed dispersal at isolated standing trees in tropical pastures: consequences for local species availability. Vegetatio 107/108:319–338

Guevara S, Purata SE, van der Maarel E (1986) The role of remnant forest trees in tropical secondary succession. Vegetatio 66:77–84

Hill MO (1979) TWINSPAN – a Fortran program for arranging multivariate data in an ordered two-way table by classification of individuals and attributes. Cornell University, Ithaca, NY

Hilty SL, Brown WL (1986) Birds of Colombia. Princeton Univ Press, Princeton, NJ

Johnson WC, Webb T (1989) The role of blue jays (*Cyanocitta cristata*) in the postglacial dispersal of fagaceous trees in eastern North America. J Biogeogr 16:561–571

Jongman RHG, Ter Braak CFJ, van Tongeren OFR (eds) (1987) Data analysis in community and landscape ecology. Pudoc, Wageningen

Kappelle M (1993) Recovery following clearing of an upper montane *Quercus* forest in Costa Rica. Rev Biol Trop 41(1):47–56

Kappelle M (1996) Los bosques de roble (*Quercus*) de la Cordillera de Talamanca, Costa Rica: biodiversidad, ecología, conservación y desarrollo. Instituto Nacional de Biodiversidad (INBio), Santo Domingo de Heredia

Kappelle M, Juárez ME (1995) Agroecological zonation along an altitudinal gradient in the montane belt of the Los Santos Forest Reserve in Costa Rica. Mount Res Dev 15(1):19–37

Kappelle M, van Velzen HP, Wijtzes WH (1994) Plant communities of montane secondary vegetation in the Cordillera de Talamanca, Costa Rica. Phytocoenology 22(4):449–484

Kappelle M, Kennis PAF, de Vries RAJ (1995) Changes in diversity along a successional gradient in a Costa Rican upper montane *Quercus* forest. Biodiv Conserv 4:10–34

Kappelle M, Geuze T, Leal ME, Cleef AM (1996) Successional age and forest structure in a Costa Rican upper montane *Quercus* forest. J Trop Ecol 12:681–698

Kappelle M, van Omme E, Juárez ME (2000) Lista de la flora vascular terrestre de la cuenca superior del Río Savegre, San Gerardo de Dota, Costa Rica. Acta Bot Mex 51:1–38

Kent M, Coker P (1992) Vegetation description and analysis: a practical approach. Belhaven Press, London

Krebs CJ (1985) Ecology: the experimental analysis of distribution and abundance, 3rd edn. Harper and Row, New York

Levey DJ, Bolker BM, Tewksbury JJ, Sargent S, Haddad NM (2005) Effects of landscape corridors on seed dispersal. Science 309:146–148

Long AJ (1995) The importance of tropical montane cloud forests for endemic and threatened birds. In: Hamilton LS, Juvik JO, Scatena FN (eds) Tropical montane cloud forests. Springer, Berlin Heidelberg New York, pp 79–106

Long AJ (1996) Establishing conservation priorities using endemic birds. In: Harcourt CS, Sayer JA (eds) The conservation atlas of tropical forests: the Americas. Simon and Schuster, New York, pp 35–46

Magurran AE (1988) Ecological diversity and its measurement. Croom Helm, London

Martínez-Ramos M, Soto-Castro A (1993) Seed rain and advanced regeneration in a tropical rain forest. Vegetatio 107/108:299–318

McClahanan TR, Wolfe RW (1987) Dispersal of ornithochorous seeds from forest edges in central Florida. Vegetatio 71:107–112

McClahanan TR, Wolfe RW (1993) Accelerating forest succession in a fragmented landscape: the role of birds and perches. Conserv Biol 7:279–288

McDonnell MJ, Stiles EW (1983) The structural complexity of old field vegetation and the recruitment of bird-dispersed plant species. Oecologia 56:109–116

Perry D (1978) A method of access into the crowns of emergent and canopy trees. Biotropica 10:155–157

Poulsen BO (1994) Movements of single birds and mixed-species flocks between isolated fragments of cloud forest in Ecuador. Stud Neotrop Fauna Environ 29:149–160

Rice WR (1989) Analyzing tables of statistical tests. Evolution 43:223–225

Robinson GR, Handel ST (1993) Forest restoration on a closed landfill: rapid addition of new species by bird dispersal. Conserv Biol 7:271-278

Shiels AB, Walker LR (2003) Bird perches increase forest seeds on Puerto Rican landslides. Restor Ecol 11(4):457–465

Sokal RR, Rohlf FJ (1995) Biometry. Freeman, New York

Stiles FG (1985) On the role of birds in the dynamics of Neotropical forests. In: Diamond AW, Lovejoy TE (eds) Conservation of tropical forest birds. International Council for Bird Preservation, ICPB, Cambridge, UK, Tech Publ 4, pp 49–59

Stiles FG, Skutch AF (1994) A guide to the birds of Costa Rica. Helm, London

Ter Steege H, Cornelissen JHC (1988) Collecting and studying bryophytes in the canopy of standing rain forest trees. In: Glime JM (ed) Methods in bryology. Proc Bryol Meth, pp 285–290

Welford MR (2000) The importance of early successional habitats to rare, restricted-range, and endangered birds in the Ecuadorian Andes. Bird Conserv Int 10(4):351–360

Wheelwright NT (1983) Fruits and the ecology of the resplendent quetzals. The Auk 100:286–301

Wheelwright NT (1991) How long do fruit-eating birds stay in the plants where they feed? Biotropica 23(1):29–40

Wheelwright NT, Haber WA, Murray KG, Guindon C (1984) Tropical fruit-eating birds and their food plants: a survey of a Costa Rican lower montane forest. Biotropica 16:173–192

Whitacre DF (1981) Additional techniques and safety hints for climbing tall trees and some equipment and information sources. Biotropica 13:286–291

Wijtzes WH (1990) Dispersal strategies of upper montane primary forest and secondary vegetation on the Pacific side of the Cordillera de Talamanca, Costa Rica. MSc Thesis, University of Amsterdam, Amsterdam

25 Diet and Habitat Preference of the Resplendent Quetzal (*Pharomachrus mocinno costaricensis*) in Costa Rican Montane Oak Forest

M. García-Rojas

25.1 Introduction

In order to set more efficient conservation measures, some scientists have proposed a research model based on vulnerable and highly mobile species (Wheelwright 1983; Powell and Bjork 1995). These organisms are sensitive to ecological disruptions, feed on scarce resources distributed in patches within the landscape, face high predation or competition, and use an array of habitats throughout their annual life cycle. The Resplendent Quetzal (*Pharomachrus mocinno*), inhabitant of the Central American montane oak forest ecosystem, fits this description well (Solórzano et al. 2004). It forms part of the ecological guild of altitudinal migrants – animals which move mainly through continuous forest, and become affected by habitat loss and fragmentation (Powell and Bjork 1995; Chaves 2001). This guild is formed by seed-dispersal agents and is recognized for its ecological function in maintaining montane forest dynamics. Its behaviour increases the probabilities of seed survival and promotes light gap colonization, enrichment of isolated forest patches, and maintenance of tree species populations with limited distribution ranges (Guindon 1997; Kappelle 2001; Chap. 24).

In Costa Rica, the Resplendent Quetzal, *P. mocinno costaricensis* (Fig. 25.1), is found in the highlands of the Tilarán, Central Volcanic, and Talamanca mountain ranges. Because of its habitat loss and illegal trade of adults, chicks, eggs and feathers throughout its distribution range, *P. mocinno* has been included in the CITES Appendix I (http://www.cites.org) and in the 'vulnerable' category of the IUCN Red Lists (http://www.iucn.org). The quetzals' habitat conservation implies not only its survival, but also the effective protection of associated organisms which use the same ecological gradient, and the processes which keep the ecosystem healthy (Loiselle and Blake 1991; Powell and Bjork 1994; Chaves 2001).

Ecological Studies, Vol. 185
M. Kappelle (Ed.) Ecology and Conservation of Neotropical Montane Oak Forests
© Springer-Verlag Berlin Heidelberg 2006

Fig. 25.1. Photograph of a male quetzal (*Pharomachrus mocinno costaricensis*), taken by Richard Laval in 2001 in Monteverde, Costa Rica

In the Talamanca mountain range, *P. mocinno* had not yet been studied with regard to the abundance of the resources traditionally considered critical for this species, amongst others, food, nests, resting and protection structures. Therefore, the goal of this study was to test the hypothesis that a positive relationship exists between quetzal abundance and the abundance of its potential food source and the structures potentially used for nesting. Understanding this dynamics between quetzal abundance and habitat quality will contribute to conservation planning and ecosystem management in the Talamanca Mountains, as well as in other regions where quetzals interact with human populations.

25.2 Study Site

This research was conducted between 1,100 and 3,060 m above sea level (a.s.l.) on the Pacific slope of the Talamanca mountain range (9°50'N, 83°90'W), in the Los Santos Forest Reserve (RFLS). This reserve is densely forested (INBio and MNCR 2001), and is composed of state and privately owned lands, the latter in communities such as Providencia, Jaboncillo, Lira, Copey, Siberia and San Gerardo de Dota. RFLS serves as a major buffer zone to the La Amistad Biosphere Reserve (Kappelle and Juárez 1995).

The average total annual rainfall is 2,643 mm. The annual average temperature at 3,490 m a.s.l. is 7 °C, and at 300 m a.s.l. it is 22 °C (ICE 2002). There is a dry season from January through March, and a rainy season during the rest of the year, with a reduction in rainfall between July and August ('veranillo').

Fragile and important ecosystems preserved in this region include the cloud forest association (Holdridge 1987), a key element in the natural hydrological system of the RFLS. The climatic conditions of cloud forests support the presence of vital springs, which supply water to several human communities in the lower basin of the Savegre and Parrita rivers. According to Holdridge's life zone system (Holdridge 1987), there are four types of habitat in the RFLS: montane rain forest (MRF), lower montane wet forest (LMWF), premontane wet forest (PWF), and premontane rain forest (PRF).

25.3 Methods

25.3.1 Quetzal Abundance

Research was conducted from October 2001 to June 2002. Sampling was done along pre-existing trails in the study area. A total of 19 km of transects was established and proportionally distributed in the study area, based on the availability of each habitat or life zone (Table 25.1). Along each transect, quetzal censuses were conducted once a month, between 07:00 and 10:30 AM, while walking at a speed of 15 m per minute. All birds seen or heard within a 25-m-broad belt along each transect side were included in the dataset used for subsequent calculations.

As the sampling effort differed between habitats, a relative abundance index was developed by dividing the total number of quetzals per month observed in each habitat by the total length (km) walked along these transects per habitat. Nevertheless, absolute quetzal abundance was compared to its expected distribution according to habitat offer, using a goodness of fit test with the $\lozenge^2$ statistic (Zar 1984), in order to determine the species' relative preference (Johnson 1980; Ojasti 2000).

25.3.2 Habitat Variables

Circular plots with a 25-m diameter were located and centred on each transect line, every 250 m (Retamosa 1999). In all, 84 plots were established, corresponding to a total area of 39,224 m^2 or 3.9 ha. With a global positioning system (GPS), elevation and location of the centroid of each plot were acquired with a location error of ±11 m. Data were associated to rainfall and temperature information available from the nearest climate station. In the plots, all trees and shrubs with stems >10 cm diameter at breast height (DBH) were

Table 25.1. Elevation, area, transect length and number of sample plots per Holdridge's life zone, in the Los Santos Forest Reserve, Costa Rica. Data are given for the period 2001–2002

Life zone	Mean elevation (m a.s.l.)	Area (ha)	Transect length (km)	Number of plots
Montane rain forest	2,857±195.93	9,597	6	27
Lower montane wet forest	2,219±114.08	24,609	8	36
Premontane wet forest	1,660±53.45	8,391	2	9
Premontane rain forest	1,391±173.94	12,875	3	12
Total		55,472	19	84

identified to genus level, and when possible down to species. Each tree or shrub was assigned to a height class through a visual estimation. Height classes were <5, 5–10, 10–15, 15–20, 20–25 and >25 m.

25.3.3 Habitat Indices

On the basis of own field observations and existing literature, suitable habitat variables were defined, capable to explain the abundance of Resplendent Quetzals in the RFLS. Raw data from plots were transformed into the following habitat indices (Retamosa 1999):

1. *potential food offer index* (PFOI): the number of trees and shrubs which offer fruits eaten by quetzals divided by the total number of trees in a plot; this index serves as an indicator of potential trophic resource supply;
2. *mixture quotient* (MQ): the number of tree and shrub genera divided by the total number of tree individuals in a plot; this index is considered an indicator for horizontal within-plot diversity;
3. *phytosociological value of superior strata* (*canopy*) (PVSS): the number of trees >20 m tall divided by the total number of trees in a plot; this index represents the plot's vertical diversity; and
4. *potential nests index* (PNI): number of dead standing trees >5 m tall in a plot divided by the total number of tree individuals.

25.4 The Quetzal's Habitat Preference

The absolute abundance of Resplendent Quetzals was not proportional to the availability of each habitat type. Quetzals showed a relative preference for MRF, as well as a relative rejection for PRF (X^2=30.31, df=3, P<0.001; Table 25.2). Higher relative abundance of quetzals was found between February and June (Fig. 25.2), coinciding with the breeding season (F=4.15, df=1, 76; P=0.011).

The habitats showing a higher relative abundance were, in descending order, MRF, LMRF and PWF (F=2.8, df=3, 73; P=0.046). Quetzal presence was recorded throughout the study period in MRF and LMWF (Fig. 25.2). However, there was a reduction in quetzal abundance in those habitats in April, influencing an increase in PWF. In PRF, there were no records of quetzals either by calls or traces throughout the study period, although it is known that they use this habitat in other regions of their distribution range, for instance, at Monteverde (Powell and Bjork 1995; Guindon 1997).

PWF was the habitat with a significantly higher potential food offer (F=4.48, df=3, 50; P=0.007) and a more diverse horizontal structure (F=5.96, df=3, 80; P=0.001), although this was the habitat with the lowest phytosocio-

Table 25.2. Absolute abundance and relative habitat preference of the Resplendent Quetzal (*Pharomachrus mocinno*). Data are given per Holdridge's life zone, in the Los Santos Forest Reserve, Costa Rica, for the period 2001–2002

Life zone	No. of observed individuals	Bonferroni confidence intervals	No. of expected individuals
Montane rain forest[a]	22	13–31	10
Lower montane wet forest	33	23–43	27
Premontane wet forest	5	0–10	9
Premontane rain forest[b]	0	0–1	14

[a] Relative preference ($P<0.001$)
[b] Relative rejection ($P<0.001$)

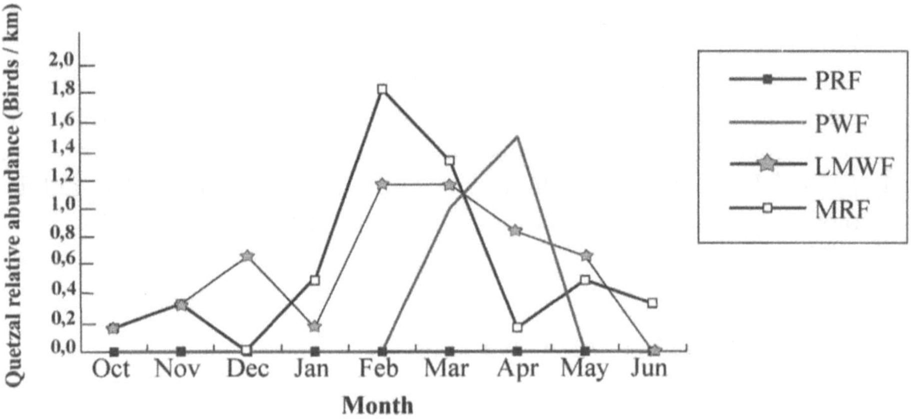

Fig. 25.2. Relative abundance (individuals per km) of Resplendent Quetzals (*Pharomachrus mocinno*) according to Holdridge's life zones, in the Los Santos Forest Reserve, Costa Rica, during the period October 2001–June 2002. *MRF* Montane rain forest, *LMWF* lower montane wet forest, *PWF* premontane wet forest, *PRF* premontane rain forest

logical value ($F=3.91$, df=3, 80; $P=0.018$). Trees with smaller stems were found in PRF ($F=5.98$, df=3, 80; $P=0.001$), but potential nest structure availability was found to be higher in this life zone ($F=1.35$, df=3, 57; $P=0.26$).

A stepwise multiple regression analysis revealed that elevation and potential food offer index were the most efficient explanatory variables for quetzal presence. The complete model (square root of quetzal abundance=0.0005* elevation+0.3278*$\log_{10}$ (PFOI)–0.0959) explained 51 % of the variation in the response variable ($R^2=0.51$, $n=84$, $P<0.01$). Elevation as the main component of this regression model was negatively correlated with temperature and positively correlated with stem diameter. These variables were associated with quetzal abundance in the one-by-one analysis. Thus, elevation can be interpreted as a proximate variable which elucidates underlying forces that affect *P. mocinno*'s natural history.

25.5 The Quetzal's Diet

In 84 plots, a total of 2,025 tree stems>10 cm DBH was identified. These belonged to 145 species, 68 genera and 49 families. Twenty-five tree species (17.24 %) formed part of RFLS' real and potential quetzal's diet, of which 13 (52 %) belonged to Lauraceae (Table 25.3). It is important to note that this author did not observe all listed diet species when fed upon by quetzals; such tree species were previously recognized as diet trees by other authors (Wheelwright 1983; Solórzano et al. 2000). Although consumption of species of wild avocados (Lauraceae) by *P. mocinno* was not quantified by this research, other species of trees can produce fruits which have a comparable rate of consumption. Among those, a type of dogwood called 'lloró' (*Cornus disciflora*, Cornaceae) seemed to be important, besides 'corral' (*Symplocos serrulata*, Symplocaceae).

Two interesting facts related to the quetzal's diet in the RFLS should be mentioned. First, blackberries (*Rubus* sp., Rosaceae) are frequently consumed by the bird. This plant is a vital socioeconomic resource for peasants in the area (Kappelle and Juárez 1995; Kappelle et al. 2000; Chap. 30), due to its high productivity and only few maintenance requirements. The second fact is the observation of a male quetzal sallying for a coffee bean (*Coffea arabica*, Rubiaceae) in a plantation in Providencia. This anecdotic but interesting scene occurred somewhere at about 1,600 m a.s.l. Though this event was observed only once, coffee was included in the list of the quetzal's diet plant species (Table 25.3).

Table 25.3. List of plants that are part of the potential and real diet of the Resplendent Quetzal (*Pharomachrus mocinno costaricensis*) in the Los Santos Forest Reserve in Costa Rica. Data are given for the period 2001–2002 (except where stated otherwise)

Common name	Family	Scientific name
Anonillo[b]	Annonaceae	*Guatteria* sp.
Zarnilla[a]	Asteraceae	*Clibadium surinamensis*
Lloró[a b]	Cornaceae	*Cornus disciflora*
	Flacourtiaceae	*Hasseltia* sp.
Aguacatillo[a d]	Lauraceae	*Aiouea costaricensis*
Aguacatillo[c d]	Lauraceae	*Beilschmiedia ovalis*
Aguacatillo, Ira[a d]	Lauraceae	*Nectandra cufodontisii*
Aguacatillo[d]	Lauraceae	*Ocotea austinii*
Ira zoncho[a d]	Lauraceae	*Ocotea calophylla*
Ira[d]	Lauraceae	*Ocotea holdridgeiana*
Ira[d]	Lauraceae	*Ocotea insularis*
Aguacatillo/Ira[a d]	Lauraceae	*Ocotea whitei*
Aguacatillo[a d]	Lauraceae	*Ocotea laetevirens*
Aguacatillo[a d]	Lauraceae	*Cinnamomum triplinerve* (=*Phoebe cinammonifolia*)
Aguacatillo[a d]	Lauraceae	*Cinnamomum hammelianum* (=*Phoebe hammeliana*)
Aguacatillo[d]	Lauraceae	*Ocotea pharomachrosurum*
Aguacatón[a d]	Lauraceae	*Persea* sp.
Santa María[b c]	Melastomataceae	*Conostegia* sp.
Higuerón[b c]	Moraceae	*Ficus* sp.
Tucuico[b]	Myrsinaceae	*Ardisia compressa*
Café[a]	Rubiaceae	*Coffea arabica*
Cafecillo[b]	Rubiaceae	*Guettarda poasana*
Corral[a b]	Symplocaceae	*Symplocos* sp.
Corral[a b]	Symplocaceae	*Symplocos serrulata*
Dama[a b c]	Verbenaceae	*Citharexylum* sp.

[a] Observed by the author
[b] Observed by Wheelwright (1983)
[c] Observed by Solórzano et al. (2000)
[d] Observed by staff at the Quetzal Education Research Center (QERC), Oklahoma Southern Nazarene University (Campus San Gerardo de Dota)

25.6 Discussion

The present study has found that in high-elevation habitats where *P. mocinno* was more abundant, food offered to quetzals was less and forest diversity lower. The explanatory variables (food and potential nest offer) did not fully explain the quetzal's temporal and spatial distribution in the RFLS. Results suggest that general fruit offer – not necessarily fruits of Lauraceae – slightly better explains quetzal abundance, although not in a satisfactory way. Nevertheless, similar conclusions have been obtained using complementary methodologies along the Caribbean slope of the Tilarán mountain range in Costa Rica (Chaves 2001), as well as in Mexico (Solórzano et al. 2000). There, marginal relationships were found between lauraceous fruit production and frugivorous abundance, including that of quetzals.

Results also show a movement within and outside the RFLS by at least a part of the local quetzal population, indicating a possible migratory pattern. This aspect must be considered with caution because radio-telemetry studies are required in order to confirm this hypothesis. However, the persistence of these movements suggests the current high quality and health of the quetzal habitat, due to the fact that quetzal distribution is limited by distance to extensive forest, and between forest fragments (Guindon 1997; personal observations).

The inherent geological, natural and socioeconomic history of the Talamanca mountain range should be considered while interpreting the research findings. In the Talamanca montane forest, northern flora is a significant element of composition (Kappelle et al. 1992; Gentry 2001), represented especially by oak trees (*Quercus* spp., Fagaceae), which have a larger basal area and higher frequency of occurrence (Blaser and Camacho 1991; Kappelle 1996; Chap. 10). The mature oak forests' structural and microclimatic conditions, influenced by the elevation, were indirectly associated with the quetzal habitat preference, and are key factors when trying to explain quetzal distribution and abundance in the RFLS.

In this regard, the biophysical microclimatic conditions might set the standard for the quetzal's physiological parameters, which limit the species performance more than suspected to date (Estudillo 2000). Within the context of environmental conditions, a dynamic habitat selection might be a non-random way to deal with the temporal and spatial dynamics of the microclimatic gradients and their effects (Karr and Freemark 1983). In the case of the quetzal, information has been generated under captivity conditions, which might be helpful for studies in the wild (Estudillo 2000). However, further in situ research is certainly required.

A special feature in the landscape in the RFLS is a friendly matrix which encourages quetzal survival. The presence of blackberry fields, which also help in soil conservation and local economy, is of relevance for the quetzal's diet. Blackberry production is an activity which reduces the need for forest conversion to pasture for cattle and other uses (Kappelle and Juárez 1994, 1995; Kappelle et al. 2000; Chap. 30), keeping a stable habitat (personal observations). In this regard, the reduced presence of cattle also allows natural forest regeneration and recovery of the original habitat (Powell and Bjork 1994; Kappelle 2001; Chaps. 17, 18 and 21).

25.7 Conclusions

The results of this study can be used by planners and landowners to set reforestation and habitat management programmes (Chaps. 30, 32 and 33). The feasibility of planting certain tree and shrub species must yet be determined by means of systematic research in plant nurseries. Also, energetic forests can be created in order to supply fuel wood for local demand, nevertheless discouraging the use of trees which are important for the quetzal and other frugivorous species (e.g., Chap. 30).

Finally, an economic and social incentive program should be conceived and implemented considering the value of forest (Chap. 33) for wildlife conservation. In this way, landowners protecting forest fragments can receive extra benefits by protecting the quetzal's habitat.

The RFLS' oak forest conservation implies the protection of an array of species and processes. However, we must be aware of the fact that conservation measures focusing on only one species, such as the quetzal, should be adapted to the inherent conditions of the region of interest. This study's findings, related to the natural history of the quetzal in the Talamancan oak forest, demonstrate that the RFLS is useful in protecting quetzal habitat while supporting human populations.

Acknowledgements I am very grateful to Maarten Kappelle for allowing me to express part of the results of my MSc. thesis in this book. This research was possible due to generous financial and logistical assistance provided by the Instituto Costarricense de Electricidad (ICE), Idea Wild, the Proyecto Desarrollo Sostenible de la Cuenca Hidrográfica del Río Savegre (MINAE, Agencia Cooperación Española), the Oklahoma Southern Nazarene University's (SNU) Quetzal Education Research Center (QERC), Hotel de Montaña Savegre, Mirador de Quetzales 'Eddy Serrano', the Museo Nacional, the Herbario Juvenal Valerio, and Programa Regional en Manejo y Conservación de Vida Silvestre (PRMVS) at the Universidad Nacional (UNA). I am indebted to L. Grandas, J. Fallas, E. Chaves, M. McCoy and M. Scally for advice and support to the final version of this paper. Richard Laval provided the photograph of the quetzal. Finally, I thank the wonderful people who made me part of their families in the communities of the Los Santos Forest Reserve.

References

Blaser J, Camacho M (1991) Estructura, composición y aspectos silviculturales de un bosque de roble (*Quercus* spp.) del piso montano de Costa Rica. Colección Silvicultura y Manejo de Bosques Naturales, Centro Agronómico Tropical de Investigación y Enseñanza, Turrialba, Costa Rica, Inf Téc 185

Chaves J (2001) Movimientos altitudinales de aves frugívoras grandes en la vertiente del Caribe de la Cordillera de Tilarán, Costa Rica: causas y consecuencias para la conservación. MSc Tesis, Escuela de Biología, Universidad de Costa Rica

Estudillo J (2000) Conservación en América Latina: Salvar al Quetzal. Natl Geogr 6(3)

Gentry AH (2001) Patrones de diversidad y composición florística en los bosques de las montañas neotropicales. In: Kappelle M, Brown AD (eds) Bosques Nublados del Neotrópico. Instituto Nacional de Biodiversidad, Costa Rica, pp 85–123

Guindon C (1997) The importance of forest fragments to the maintenance of regional biodiversity surrounding a tropical montane reserve, Costa Rica. PhD Dissertation, Faculty of the School of Forestry and Environmental Studies, Yale University, Connecticut

Holdridge LR (1987) Ecología basada en zonas de vida. Instituto Interamericano de Ciencias Agronómicas, Costa Rica

ICE (2002) Datos climatológicos de las estaciones meteorológicas de las cuencas de los ríos Pirris y Savegre, Costa Rica. Instituto Costarricense de Electricidad, San José

INBio and MNCR (2001) Caracterización de la vegetación de la cuenca hidrográfica del río Savegre. Instituto Nacional de Biodiversidad, Museo Nacional de Costa Rica, Informe Final Proyecto de Desarrollo Sostenible de la Cuenca Hidrográfica del Río Savegre, Costa Rica. Instituto Nacional de Biodiversidad, Santo Domingo de Heredia

Johnson DH (1980) The comparison of usage and availability measurements for evaluating resource preference. Ecology 61(1):65–71

Kappelle M (1996) Los bosques de roble (*Quercus*) de la Cordillera de Talamanca, Costa Rica: biodiversidad, ecologia, conservación y desarrollo. Instituto Nacional de Biodiversidad (INBio), Santo Domingo de Heredia, Costa Rica

Kappelle M (2001) Costa Rica. In: Kappelle M, Brown AD (eds) Bosques nublados del Neotrópico. Instituto Nacional de Biodiversidad, Costa Rica, pp 301–370

Kappelle M, Juárez ME (1994) The Los Santos Forest Reserve: a bufferzone vital for the Costa Rican La Amistad Biosphere Reserve. Environ Conserv 21(2):166–169

Kappelle M, Juárez ME (1995) Agroecological zonation along an altitudinal gradient in the montane belt of the Los Santos Forest Reserve in Costa Rica. Mount Res Develop 15(1):19–37

Kappelle M, Cleef AM, Chaverri A (1992) Phytogeography of Talamanca montane *Quercus* forests, Costa Rica. J Biogeogr 19(3):299–315

Kappelle M, Avertin G, Juárez ME, Zamora N (2000) Useful plants within a *campesino* community in a Costa Rican montane cloud forest. Mount Res Dev 20(2):162–171

Karr J, Freemark K (1983) Habitat selection and environmental gradient dynamics in the "stable" tropics. Ecology 64(6):1481–1494

Loiselle B, Blake J (1991) Temporal variation in birds and fruits along an elevational gradient in Costa Rica. Ecology 72(1):180–193

Ojasti J (2000) Manejo de fauna silvestre neotropical. In: Dallmeier F (ed) SIMAB series no 5. Smithsonian Institution/MAB Program, Washington, DC

Powell G, Bjork R (1994) Implications of altitudinal migration for conservation strategies to protect tropical biodiversity: a case study of the Resplendent Quetzal (*Pharomachrus mocinno*) at Monteverde, Costa Rica. Bird Conserv Int 4:161–174

Powell G, Bjork R (1995) Implications of intratropical migration on reserve design: a case study using *Pharomachrus mocinno*. Conserv Biol 9(2):354–362

Retamosa M (1999) Selección de hábitat y distribución potencial del pájaro sombrilla (*Cephalopterus glabricollis*) en la Cordillera de Tilarán y su vertiente Atlántica, Costa Rica: implicaciones para su conservación. MSc Tesis, Programa Regional de Manejo de Vida Silvestre, Universidad Nacional Autónoma de Costa Rica

Solórzano S, Ávila L, Castillo S, Valverde T (2000) Quetzal abundance in relation to fruit availability in a cloud forest in southeastern Mexico. Biotropica 32(3):523–532

Solórzano S, Baker AJ, Oyama K (2004) Conservation priorities for resplendent quetzals based on analysis of mitochondrial DNA control-region sequences. The Condor 106:449–456

Wheelwright NT (1983) Fruits and ecology of Resplendent Quetzals. Auk 100:286–301

Zar J (1984) Biostatistical analysis, 2nd edn. Prentice Hall, New Jersey

26 Small Terrestrial Rodents in Disturbed and Old-Growth Montane Oak Forest in Costa Rica

M.B. van den Bergh and M. Kappelle

26.1 Introduction

It is well known that rodents such as squirrels and mice often function as important seed dispersers and predators in temperate oak forests (Krajicek 1955; Barnett 1977; Fox 1982; Schweiger et al. 2004) and tropical lowland rainforests (Adler 1995; Ceballos 1995; Lambert and Adler 2000; Janzen and Forget 2001; Demattia et al. 2002). In tropical montane oak forests, for example, they may feed on acorns during mast seeding years, affecting tree seed germination patterns (Chap. 13). Often, ground-dwelling rodents play a prominent role in the food chain of species-rich, neotropical communities, as they serve as staple food for predators such as carnivorous birds and medium to large-sized mammals (e.g., Chinchilla-Romero 1997).

However, still little is known about the responses of small terrestrial rodents to changes in forest cover as a result of habitat fragmentation and loss of tropical forest (Kasene 1984; Adler 1994). Even less is known on the role terrestrial Rodentia play in forest succession, canopy closure, and biodiversity recovery following slash-and-burn practices or commercial logging. Data on presence and abundance of small ground-dwelling rodents in disturbed tropical montane environments are much scarcer than for lowland rainforests (Kappelle 1996). Only few studies on small-sized terrestrial rodent are available for neotropical montane oak forests (e.g., Lanzewizki 1991; Johnson and Vaughan 1993).

We conducted live-trapping sessions for censusing small terrestrial rodent species populations with the aim to focus on diversity and distribution of this functional group of mammals along a disturbance gradient, ranging from cattle-grazed pasturelands to undisturbed mature, old-growth forest (van den Bergh and Kappelle 1998). Our main purpose was to identify key indicator rodent species characteristic for certain levels of disturbance and/or recovery

Ecological Studies, Vol. 185
M. Kappelle (Ed.) Ecology and Conservation of Neotropical Montane Oak Forests
© Springer-Verlag Berlin Heidelberg 2006

in a neotropical montane oak forest, and as such, contribute to much needed knowledge for montane oak forest restoration in a fragmented landscape.

26.2 Study Area

The study was carried out in the core of the fragmented montane oak forest zone (2,200–2,800 m a.s.l.) in the upper watershed of the Savegre River, near the village of San Gerardo de Dota in the Los Santos Forest Reserve (Dota County), situated in the western part of Costa Rica's Talamanca Range. Average annual rainfall is 2,000–3,000 mm, and average annual temperature varies in the range 10–15 °C, depending on altitude. The dry season lasts from early January to late April (Kappelle 1996). Soils are derived from volcanic ash, are acid (pH 3.5–5.5), and moderately fertile (Chap. 4). Natural vegetation is 30–50 m tall, evergreen broad-leaved oak forest (Chaps. 4 and 10). Since the beginning of the 1950s, conversion of the region's oak forests into grasslands and croplands has led to a complex landscape mosaic (Kappelle and Juárez 1995). After abandonment of low-production pasturelands in the late 1970s and early 1980s, stands of successional forests have developed along edges of old-growth forests and pastures (Kappelle et al. 1995; Chaps. 17 and 30).

26.3 Habitat Selection

Five different habitats were selected along a man-induced disturbance gradient in the upper montane forest belt (2,300–2,800 m a.s.l.). These habitats are, from low to high levels of disturbance:

1. 30–40 m tall, dense, old-growth oak-dominated 'closed mature forest' (CMF);
2. oak-dominated 'open mature forest', with a relatively open, 30–35 m high canopy (OMF);
3. 3–7 m high secondary 'successional shrubland' (SSL);
4. 0.5–2 m tall secondary 'abandoned pastureland' (APL); and
5. <0.5 m high 'grazed pastureland' (GPL).

In each habitat, a 0.25-ha plot (50×50 m) was established, with exception of APL in which a 0.28-ha plot (37.5×75 m) had to be laid out due to this site's different patch shape (Table 26.1).

Table 26.1. Live-trapping and population data for five different habitats, sampled along a successional gradient in a fragmented neotropical montane oak forest landscape, Costa Rica

Plots	Habitat type[a]					
	CMF	OMF	SSL	APL	GPL	All
Plot size (ha)	0.25	0.25	0.25	0.28	0.25	1.28
Trap nights[b]	10.00	9.00	10.00	10.00	10.00	49.00
Species richness[c]	4.00	5.00	4.00	4.00	4.00	7.00
Species density[d]	92.00	180.00	160.00	196.00	80.00	142.00
Capture frequency[e]	3.20	10.70	11.60	8.00	5.80	7.90
Capture frequency per ha	12.80	42.80	46.40	28.60	23.20	30.80

[a] CMF, closed mature forest; OMF, open mature forest; SSL, successional shrubland; APL, abandoned pastureland; GPL, grazed pastureland
[b] Total number of trap nights
[c] Total number of species per plot
[d] Total number of individuals per hectare
[e] Mean number of captures per trap night

26.4 Rodent Trapping

In all but the APL plot, a total of 25 Sherman traps (trap size: 23×9×8 cm) was placed in a 5×5 grid with a 12.5-m distance between two neighboring traps, in order to trap small mammals alive. Similarly, in APL 28 traps were located in a 4×7 grid. A standardized capture-recapture method was used to estimate species diversity, distribution and abundance in each habitat type (Leslie 1952). Plots were studied from April to June, covering the transition from the dry to the wet season. Sherman traps were checked before noon during 10 consecutive days (Table 26.1). Only OMF traps were checked over 9 days. Bait consisted of a mixture of rolled oats and peanut butter with a touch of vanilla flavor. During afternoons, traps were checked for diurnal catches and bait was renewed for the next trap-night. Traps were covered with litter and/or leaves for camouflage and insulation.

26.5 Data Collection and Analysis

We recorded data on the presence, sex, weight, and length of trapped individuals. Weight was measured using a Pesola pocket scale (max. weight 300 g). Length measurements included the head–body length, measured from the tip of the nose to the inflection point of the tail, and the tail length, measured from the inflection point with the body to the fleshy tip of the tail. On-site species identification was done with the aid of Emmons' (1990) field guide, after studying rodent specimens in Costa Rica's national collections. Control specimens were collected and stored at the country's National Museum. Collected specimens were identified with help of B. Rodríguez (personal communication, at the National Museum in San José), and on basis of field guides (Mora and Moreira 1984; Emmons 1990). Trapped individuals were marked upon first capture with acrylic paint before being released.

26.6 Rodent Species Diversity

Seven rodent species distributed over two families (five Muridae, two Heteromyidae) were represented in a total of 389 captures (185 individuals; Table 26.2). The most abundant species was *Peromyscus mexicanus* (Mexican

Table 26.2. Data on abundance and captures of seven rodent species for five different habitats, sampled along a successional gradient in a fragmented neotropical montane oak forest landscape, Costa Rica

	Habitat type[a]					
	CMF	OMF	SSL	APL	GPL	All plots
Abundance[b]						
Peromyscus mexicanus	19	21	17	13	5	75
Scotinomys xerampelinus	3	7	20	24	7	61
Oryzomys albigularis	1	7	1	0	6	15
Reithrodontomys creper	0	6	0	9	0	15
Reithrodontomys cf. *sumichrasti*	0	0	0	9	2	11
Heteromys oresterus	2	4	0	0	0	6
Heteromys cf. *desmarestianus*	0	0	2	0	0	2
Total (all species)	25	45	40	55	20	185
Captures[c]						
Peromyscus mexicanus	26	67	58	26	30	207
Scotinomys xerampelinus	3	12	44	29	14	102
Oryzomys albigularis	1	12	1	0	9	23
Reithrodontomys creper	0	9	0	15	0	24
Reithrodontomys cf. *sumichrasti*	0	0	0	10	5	15
Heteromys oresterus	2	4	0	0	0	6
Heteromys cf. *desmarestianus*	0	0	12	0	0	12
Total (all species)	32	104	115	80	58	389

[a] CMF, closed mature forest; OMF, open mature forest; SSL, successional shrubland; APL, abandoned pastureland; GPL, grazed pastureland
[b] Total number of captured individuals
[c] Total number of captures, including recaptures

Deer Mouse, *ratón pata blanca*, closely related to *P. nudipes*, the Naked-footed Deer Mouse, which is common at Monteverde; Anderson 2000), with 75 individuals recorded in 207 captures (Table 26.2). The other six species were *Scotinomys xerampelinus* (Singing Mouse, *ratón cantante*), *Oryzomys albigularis* (White-throated Rice Rat, *ratón arrocero*), *Reithrodontomys creper* (Chiriquí Harvest Mouse, *ratón de las cosechas*), *R.* cf. *sumichrasti* (Sumichrasti's Harvest Mouse, *ratón de las cosechas*), and *Heteromys oresterus* and *H.* cf. *desmarestianus*, both being Spiny Pocket Mice (*ratas de campo*). Rodent species richness was very similar for all habitat sites, with 4–5 species per habitat regardless of its position along the disturbance axis.

26.7 Rodent Body Sizes and Abundance

Average head–body lengths of individuals ranged from 6.9 cm for *Scotinomys xerampelinus* to 14.8 cm for *Oryzomys albigularis*. Similarly, average tail lengths of individuals ranged from 5.9 cm for *Scotinomys xerampelinus* to 17.1 cm for *Oryzomys albigularis* (Table 26.3). The latter had the largest weight (average of 93.4 g), whereas *Scotinomys xerampelinus* had the lowest average weight (14.0 g).

Population density and capture frequency (absolute and per area values) were higher in habitats suffering intermediate levels of disturbance, being

Table 26.3. Data on sex, weight, and head-to-body and tail lengths of seven rodent species captured along a successional gradient in a fragmented neotropical montane oak forest landscape, Costa Rica

	Males	Females	Weight[a]	HB-length[b]	T-length[c]
Peromyscus mexicanus	43	32	51.6±4.6	11.9±0.7	12.3±0.6
Scotinomys xerampelinus	36	25	14.0±2.2	6.9±0.4	5.9±0.8
Oryzomys albigularis	8	7	93.4±14.2	14.8±0.9	17.1±2.8
Reithrodontomys creper	13	2	22.8±2.9	8.7±0.8	13.0±1.0
Reithrodontomys cf. *sumichrasti*	4	7	14.9±2.5	7.4±0.5	8.7±0.8
Heteromys oresterus	2	4	74.8±3.8	12.7±0.8	17.0±0.7
Heteromys cf. *desmarestianus*	1	1	24.0±5.0	8.8±0.3	9.8±1.3
Total (all species)	107	78	–	–	–

[a] 1Weight (mean±1SD) in g, based on measurements taken at each individual's first catch

[b] HB-length (mean±1SD) corresponds to the head–body length (length from the tip of the nose to the inflection point of the tail) of an individual; HB-length in cm, based on measurements taken at each individual's first catch

[c] T-length (mean±1SD) corresponds to the tail length (length from the inflection point with the body to the fleshy tip of the tail) of an individual; T-length in cm, based on measurements taken at each individual's first catch

twice as high in OMF, SSL, and APL. Abundance values differed strongly among species. Males (107 in total) appeared to prevail over females (78) in captures (Table 26.3).

26.8 Changes Along the Disturbance Gradient

This study presents results on small rodents that were live-trapped in mature and recovering montane oak forest in Costa Rica, in order to evaluate the relative distribution of species to each other and to microhabitat properties found along a successional gradient. Species diversity and abundance of small terrestrial rodents proved to be relatively high, with murid species being dominant. Observed abundance values were concordant with classical rank-abundance values (Magurran 1988). High abundances (60–75 individuals) were recorded for *P. mexicanus* and *S. xerampelinus*, whereas the other three Muridae, *O. albigularis*, *R. creper* and *R.* cf. *sumichrasti*, appeared to be four to five times less common. As López Barrera and Manson (Chap. 13) point out, mast seeding may produce a large pulse of food resources that allow populations of acorn (*Quercus* seed) predators and/or dispersers such as *Peromyscus* spp. to increase into the following year. However, so far it is unknown if mast seeding occurred in Costa Rican *Quercus* species the years (1994–1995) before we conducted live trapping (1996) at the study site.

Together, *P. mexicanus* and *S. xerampelinus* represented 79.4 % of all captures (including recaptured individuals), and 73.5 % of all captured individuals. Both Heteromyidae species, by contrast, seemed to be rather rare in the area (≤ 6 individuals), and appeared to avoid both abandoned and grazed pasturelands. *P. mexicanus* and *S. xerampelinus* occurred at all five sites, although they differed in habitat preference, *P. mexicanus* appearing with higher numbers in CMF and OMF, and *S. xerampelinus* showing higher abundance in SSL and APL. This difference is explained by the fact that the latter, more insectivorous species requires a rather dense vegetation cover at ground level, as its peak in activity is during morning hours (Hooper 1972).

26.9 Habitat Preferences

The species *O. albigularis* preferred OMF as well as the edge of GPL bordering mature forest. This observation is consistent with data presented earlier by Timm et al. (1989) for the Braulio Carrillo Park in northern-central Costa Rica. *R. creper* and *R.* cf. *sumichrasti* were most abundant in APL. These observations are concordant with data on habitat preferences of Nicaraguan *Reithrodontomys* species (Jones and Genoways 1970). *R. creper* was also trapped in OMF, but in lower numbers. Similarly, two individuals of *R.* cf. *sumichrasti* occurred in GPL. Low capture results of *H. oresterus* and *H.* cf. *desmarestianus* suggest these species' rarity in the area. Whereas *H. oresterus* preferred mature forest habitats, *H.* cf. *desmarestianus* was recorded only in SSL. *H. oresterus* may be qualified as an indicator species for low levels of disturbance.

In general, male individuals were more often trapped in CMF and OMF as well as APL, whereas males and females appeared equally in SSL and GPL. Whether these capture differences between sexes is due to behavior or population composition remains unknown. *O. albigularis* was the largest terrestrial rodent species captured, being over six times as heavy and twice as long as *R.* cf. *sumirachrasti.* Data on weight and head–body and tail lengths of captured individuals were in accordance with data in Emmons (1990), with the exception of values recorded for *H.* cf. *desmarestianus.* Individuals of this species were much smaller in this study than the data range given by these authors would suggest (49–103 g, 10.8–14.8 cm). The weight distribution for male and female individuals of the most abundant species, *P. mexicanus,* clearly illustrates within-species weight differences found between sexes in rodents.

26.10 Conclusions

Results are generally consistent with those presented by Lanzewizki (1991), and Johnson and Vaughan (1993), who recorded five myomorph species of which *P. mexicanus* (*P. nudipes,* in their study) and *S. xerampelinus* were most common. Our study revealed seven species of terrestrial myomorph rodents, including *O. albigularis* and *H. oresterus* as new local records. Both datasets indicate the importance of within-habitat micro-environmental heterogeneity for rodent populations in these forests.

Although most species were not confined to a specific successional forest phase, *Heteromys* species did seem to prefer older stages such as closed and open mature forest. However, the data did not suffice for a thorough analysis of co-occurrences of species and variability between local species assemblages. More grid-based plots (replicates) need to be sampled during longer trapping sessions, preferably over a series of consecutive years, in order to achieve statistically reliable results. The results presented only support a much-needed first inventory of species, and offer preliminary insights into species abundance and habitat preference. Large-scale, long-term species monitoring is required to assess the current status and trends of small terrestrial rodent species populations and assemblages in neotropical montane oak forests.

Acknowledgements We are grateful to A.M. Cleef and F. Bouman for providing scientific guidance. A. Velázquez gave helpful advice and practical support and B. Rodríguez identified rodents. The hospitality of the rural families in San Gerardo de Dota is much appreciated. Funding was provided by the Netherlands Organization for Scientific Research (NWO, grant 895.100.003), University of Amsterdam, the Hugo de Vries and Van Tienhoven Foundations, the European Science Foundation and Bever Outdoor Sports. The VZZ Association provided Sherman traps. Research permission was kindly provided by Costa Rica's Ministry of Environment and Energy (MINAE).

References

Adler GH (1994) Tropical forest fragmentation and isolation promote asynchrony among populations of a frugivorous rodent. J Anim Ecol 63:903–911

Adler GH (1995) Fruit and seed exploitation by Central American spiny rats. *Proechimys semispinosus*. Stud Neotrop Fauna Environ 30:237–244

Anderson SD (2000) Reproduction and dynamics of deer mice. In: Nadkarni NM, Wheelwright NT (eds) Monteverde: ecology and conservation of a tropical cloud forest. Oxford Univ Press, New York, pp 238–239

Barnett RJ (1977) The effect of burial by squirrels on germination and survival of oak and hickory nuts. Am Midl Nat 98:319–330

Ceballos G (1995) Vertebrate diversity, ecology and conservation in neotropical dry forests. In: Bullock SH, Mooney HA, Medina E (eds) Seasonally dry tropical forests. Cambridge Univ Press, Cambridge, UK, pp 195–220

Chinchilla-Romero FA (1997) Diet of *Panthera onca*, *Felis concolor* and *Felis pardalis* (Carnivora: Felidae) in Parque Nacional Corcovado, Costa Rica. In: Anon (ed) Abstr Vol Association of Tropical Biology Annu Meet, San José, p 47

Demattia EA, Curran LM, Rathcke BJ (2002) Effects of small rodent seed predators on forest recruitment in Corcovado National Park, Costa Rica. In: Anon (ed) Abstr Vol Ecological Society of America Annu Meet, Tucson, AR, pp 111–112

Emmons LH (1990) Neotropical rainforest mammals: a field guide. Univ Chicago Press, Chicago, IL

Fox JF (1982) Adaptation of gray squirrel behavior to autumn germination by white oak acorns. Evolution 36:800–809

Hooper ET (1972) A synopsis of the rodent genus *Scotinomys*. Museum of Zoology, University of Michigan, Occ Pap 665:1–32

Janzen PA, Forget PM (2001) Scatter-hoarding rodents and tree regeneration. In: Bongers F, Charles-Dominique P, Forget PM, Thery M (eds) Nouragues: dynamics and plant-animal interactions in a Neotropical rainforest. Kluwer, Dordrecht, pp 275–288

Johnson WE, Vaughan C (1993) Habitat use of small terrestrial rodents in the Costa Rican highlands. Rev Biol Trop 41:185–191

Jones JK Jr, Genoways HH (1970) Harvest mice (genus *Reithrodontomys*) of Nicaragua. W Found Vert Zool Occ Pap, vol 2

Kappelle M (1996) Los bosques de roble (*Quercus*) de la Cordillera de Talamanca, Costa Rica: biodiversidad, ecología, conservación y desarrollo. Instituto Nacional de Biodiversidad (INBio), Santo Domingo de Heredia

Kappelle M, Juárez ME (1995) Agroecological zonation along an altitudinal gradient in the montane belt of the Los Santos Forest Reserve in Costa Rica. Mount Res Dev 15(1):19–37

Kappelle M, Kennis PAF, de Vries RAJ (1995) Changes in diversity along a successional gradient in a Costa Rican upper montane *Quercus* forest. Biodiv Conserv 4:10–34

Kasene JM (1984) The influence of selective logging on rodent populations and the regeneration of selected tree species in the Kibale forest, Uganda. Trop Ecol 25:179–195

Krajicek JE (1955) Rodents influence red oak regeneration. USDA Forest Service, Centr Stat For Exp Sta Note no 91:1–2

Lambert TD, Adler GH (2000) Microhabitat use by a tropical forest rodent, *Proechimys semispinosus*, in Central Panama. J Mammol 18(1):70–76

Lanzewizki T (1991) Populationsökologische Untersuchungen an Kleinsäugern in einem Eichen-Wolkenwald (*Quercus* spp.) der Montanstufe Costa Ricas. MSc Thesis, Philips-Universität, Marburg, Germany

Leslie PH (1952) The estimation of population parameters from data obtained by means of the capture-recapture method. II. The estimation of total numbers. Biometrika 39:362–88

Magurran AE (1988) Ecological diversity and its measurement. Croom Helm, London, UK

Mora JM, Moreira I (1984) Mamíferos de Costa Rica. Universidad Estatal a Distancia (UNED), San José, Costa Rica

Schweiger EW, Holt RD, Pierotti R, Diffendorfer J (2004) The relative importance of small-scale and landscape-level heterogeneity in structuring small mammal distributions. In: Barrett GW, Peles JD (eds) Landscape ecology of small mammals. Springer, Berlin Heidelberg New York, pp 175–211

Timm RM, Wilson DE, Clauson BL, Laval RK, Vaughan CS (1989) Mammals of the La Selva–Braulio Carrillo Complex, Costa Rica. US Fish and Wildlife Service, North American Fauna, vol 75

Van den Bergh MB, Kappelle M (1998) Diversity and distribution of small terrestrial rodents along a disturbance gradient in montane Costa Rica. Rev Biol Trop 46(2):331–338

27 Habitat Preference, Feeding Habits and Conservation of Baird's Tapir in Neotropical Montane Oak Forests

M.W. Tobler, E.J. Naranjo, and I. Lira-Torres

27.1 Introduction

The Baird's tapir (*Tapirus bairdii*) is the largest terrestrial mammal in Mesoamerica and it is distributed from Veracruz and Oaxaca in southern Mexico to north-western Colombia (Reid 1997; Lawton 2000; Kappelle and Brown 2001; Naranjo 2002). It occurs in a wide variety of habitats, from sea level to over 3,600 m, including marshes, mangroves, swamps, moist tropical forests, riparian woodlands, monsoonal deciduous forests, dry deciduous forests, montane cloud forests, and above the treeline in paramo (Matola et al. 1997; Naranjo and Vaughan 2000; Kappelle and Horn 2005). *T. bairdii* has been categorized as endangered according to the IUCN Red List of Threatened Species (IUCN 2004), and populations in many areas are threatened by both forest fragmentation and overhunting.

Large herbivores such as tapirs are known to have an important impact on the structure and plant diversity of their habitat. Tapirs play an important role as long-distance seed dispersers, ingesting whole seeds and dropping them intact with their faeces (Bodmer 1991; Rodrigues et al. 1993; Fragoso 1997; Olmos 1997; Lawton 2000; Fragoso et al. 2003). On the other hand, tapirs are also effective seed predators for many plant species (Janzen 1981; Olmos 1997). Downer (2001) suggests that the mountain tapir (*T. pinchaque*) is one of the most important seed dispersers in montane cloud forests and the paramo of the high Andes Mountains.

Tapirs also have an impact on the composition and regeneration of vegetation by selectively browsing on young shoots and samplings of many plant species. After only 1.5 years of excluding large mammals in a montane cloud forest in Colombia, Lizcano (2004) found significant differences in plant composition, compared to sites accessible to these mammals. This suggests that

Ecological Studies, Vol. 185
M. Kappelle (Ed.) Ecology and Conservation of Neotropical Montane Oak Forests
© Springer-Verlag Berlin Heidelberg 2006

tapirs play an important role in shaping and maintaining the composition and structure of montane cloud forests, and it is therefore important to learn more about their ecology in this ecosystem.

Most earlier studies on Baird's tapir have been conducted in lowland tropical rainforest (Terwilliger 1978; Fragoso 1987; Naranjo 1995; Foerster and Vaughan 2002). Williams (1984) studied the Baird's tapir in the deciduous forest of Santa Rosa, Costa Rica, and Naranjo and Cruz (1998) investigated its distribution and ecology along an altitudinal gradient in Chiapas, Mexico. This chapter summarizes and compares results of two studies of the Baird's tapir in montane cloud forest, one in the Sierra Madre of Chiapas, Mexico (Lira et al. 2004), and one in the Cordillera de Talamanca, Costa Rica (Tobler 2002).

27.2 Study Areas

27.2.1 Cordillera de Talamanca, Costa Rica

The Cordillera de Talamanca stretches from central Costa Rica south-eastwards into Panama, separating the Atlantic coast from the Pacific coast. Its highest peak is Cerro Chirripó at 3,819 m elevation. Much of the area is protected by the Amistad Biosphere Reserve, the Chirripó National Park, and the Rio Macho Forest Reserve. A detailed floristic description can be found in Kappelle (1996; Chap. 10). The study area is located in the north-western part of the Cordillera de Talamanca, near the village of Villa Mills at an elevation of 2,600–3,200 m. The mean annual rainfall at 3,000 m is 2,642 mm, with a dry season from January to April and a wet season from May to December. Mean annual air temperature is 10.9 °C, with a maximum of 22 °C and a minimum below 0 °C (Blaser 1987). For this study, three sites were chosen:

1. Villa Mills (VM) – this area is closest to the village of Villa Mills. Elevation varies in the range 2,500–2,800 m, and vegetation is all old-growth oak forest. The area is part of the Rio Macho Forest Reserve, established in 1964 (Kappelle 1996). The trails in this area are used a few times per week by local people.
2. Parcelas (PA) – this area is part of an ongoing project on sustainable forestry carried out by the Centro Agronómico Tropical de Investigación y Enseñanza (CATIE) in collaboration with Costa Rica's Ministerio de Ambiente y Energía (MINAE). Elevation range is 2,650–2,750 m. The area consists of 11 different research plots where 20–30 % of the basal area of all trees was harvested in 1991 (Aus der Beek and Sáenz 1996; Chap. 18). To extract the timber, a 5-m-wide access road was built in 1990. Smaller logging tracks lead from this road to the various plots. During this study, there

were no logging activities. The roads are not accessible to the public by car, but are frequently used by local people and visitors travelling by foot. A more detailed description of the project can be found in Aus der Beek and Sáenz (1996), and Chapter 18.

3. Cerro Cuericí (CU) – this area lies on a mountain ridge of the continental divide and has an elevational range of 2,800–3,200 m. Much of the area is within the Chirripó National Park, which was founded in 1975 (Kappelle 1996; Chaps. 4, 10 and 30). This old-growth oak forest contains large, old stands of bamboo (*Chusquea* spp.) in its understorey, and is only rarely crossed by pedestrians on the trail leading to Cuericí.

27.2.2 El Triunfo Biosphere Reserve, Chiapas, Mexico

El Triunfo Biosphere Reserve, established in 1990, covers a large part of the central Sierra Madre in Chiapas, Mexico. The reserve is divided into two different zones: a core zone (25,763 ha), under strict protection but open for research, and a buffer zone (93,458 ha) where also sustainable agriculture is permitted. This study was carried out in the central part of the reserve in the polygon called 'El Triunfo', at an elevation of 1,000–2,450 m. The mean annual temperature is 18–22 °C, and rainfall varies in the range 2,500–3,500 m.

There are seven major vegetation associations for the area: *Gaultheria-Ugni-Vaccinium*, *Quercus-Matudaea-Hedyosmum-Dendropanax*, *Liquidambar-Quercus-Pinus*, *Cupressus-Pinus*, *Ficus-Coccoloba-Dipholis-Sapium*, *Garcinia-Inga-Desmopsis* and *Quercus salicifolia* (Long and Heath 1991). A comparison of several neotropical montane cloud forests showed that the forest at El Triunfo is very similar to that in the Cordillera de Talamanca, both in species diversity as well as composition at plant genus level (Vázquez-García 1995).

The study was conducted in both the core and the buffer zones of the reserve, at the following sites:

1. core zone with ecotourism (CZE) – the area has a well-established trail system with a relatively high frequency of visiting tourists,
2. core zone without ecotourism (CZ) – the area has fewer trails, and trails are not accessible to tourists, and
3. buffer zone with coffee production (BZC) – this area of the buffer zone is used mainly for coffee production.

27.3 Methods

27.3.1 Relative Abundance and Habitat Use

To assess the relative abundance of tapirs in the different parts of the study areas, we used tracks and faeces encountered along transects. In Costa Rica, eight transects each 200 m long were installed orthogonally to existing trails at each of the specific study sites. These transects were surveyed 22 (VM) and 21 (PA, CU) times between March and July 1999. Transects were usually checked between 5 and 11 AM, each transect being revisited at intervals of at least 4 days.

In Mexico, the existing trail system was used for transects. Twelve transects were located in the core zone (six in CZE, and six in CZ), and three in the buffer zone. Length of transects varied between 0.5 and 10 km. Transects were surveyed between June and September 2000. Using a map of vegetation types (National Forest Inventory 2000), we made sure that each transect was within only one of the three chosen vegetation types: old-growth montane oak forest, secondary montane oak forest, and coffee plantation.

An χ^2-test was used to compare expected with observed numbers of tracks for different parts of the study area and different vegetation types. The expected number of tracks was calculated by multiplying the total number of tracks found during the study by the total length of transects located in each specific part divided by the total length of all transects.

27.3.2 Feeding Habits

Faeces are a good indicator of the composition of the diet of tapirs. In his study on lowland tapirs (*T. terrestris*), Bodmer (1990) found that the composition of faeces corresponds well with that of samples taken from stomach contents. Results from Foerster (1998) obtained by direct observations of Baird's tapirs in Corcovado in southern coastal Costa Rica are consistent with those based on faecal analysis by Naranjo (1995) in the same area.

Faeces were collected along transects and in other parts of the study area. Samples were dried in the sun and stored in plastic bags for subsequent analysis. Composition was analysed using a frequency method first described by Chamrad and Box (1964), and modified by Naranjo and Cruz (1998). A sample of the faeces was spread out evenly on paper. A comb with 10 needles was then used to sample 100 points. At each point, the type of plant remains was recorded. In Costa Rica, we distinguished between leaves, stems, fibres, seeds and fruits, and fine material. In Mexico, we distinguished between leaves, stems, seeds and fruits. For the analysis in this chapter, we used data only on leaves, stems and fruits for comparison between Mexico and Costa Rica.

In addition to the frequency method, composition of the samples from Mexico was determined using weight. A subsample of 3 g was divided into three components (leaves, stems and fruits), each weighed at a precision of 0.1 g. Results from both methods were compared using a Pearson's R test. Larger parts of leaves and stems found in the faeces were identified to obtain specific information on the plant species eaten by tapirs. Furthermore, plants observed in the field as browsed upon by tapirs were also identified. Only plants found close to fresh tracks were taken into account.

27.3.3 Hunting

In an effort to accumulate information on tapir hunting in the past and present as well as on the knowledge of local people on tapirs, structured interviews were conducted.

In Mexico, 24 people from the following communities were interviewed: Santa Rita, Nueva Colombia, Siete de Octubre, Municipio de Ángel Albino Corzo, Vega del Palmar and Municipio de la Concordia. In Costa Rica, interviews were held with 15 people from Villa Mills.

People who were known to spend lengthier periods of time in the forest were asked if they had seen tapirs in the study area and, if so, when. They were also asked how many tapirs were hunted in the past and present, what parts of the animal were used, and if they had noticed a decline or increase of animal and/or track sightings.

27.4 Relative Abundance and Habitat Use

Near Villa Mills in Costa Rica, a total of 95.7 transect km was walked, along which 36 tracks were found. As the different transects within one area were relatively close to each other, we made a more conservative estimate using only the number of days tracks found in an area, thus avoiding counting the same animal twice. CU had a significantly higher number of tracks than did the other two areas (VM, PA), both when looking at number of tracks found (X^2=8.652, df=2, P=0.013) and the number of days when tracks were found (X^2=7.14, df=2, P=0.028). There was no difference between VM and PA (X^2=0.06, df=1, P=0.806). CU had the highest track frequency (0.39–0.63/km), followed by VM (0.2–0.23/km) and PA (0.15–0.26/km; Table 27.1).

At El Triunfo in Mexico, a total of 456.9 transect km was walked, along which 281 individual tracks were found. There were significant differences in the number of tracks found in the different areas (X^2=90.581, df=2, P<0.001). CZ had the highest track frequency (1.1/km), followed by CZE (0.4/km). No tracks were found in BZC (Table 27.1).

Table 27.1. Baird's tapir tracks found in Villa Mills, Costa Rica, and El Triunfo, Mexico. Significant differences are given in the main text

Area	Length of transects (km)	Days controlled (*N*)	Total length (km)	Days with tracks (*N*)	Tracks[a] (*N*)	Tracks expected (*N*)	Tracks/km[b]
Villa Mills, Costa Rica							
PA	1.28	21	26.9	4	7	10.1	0.15–0.26
VM	1.60	22	35.2	7	8	13.2	0.20–0.23
CU	1.60	21	33.6	13	21	12.6	0.39–0.63
El Triunfo, Mexico							
BZC	2.50	10	25.0	–	0	15.4	0.00
CZE	27.85	10	278.5	–	113	171.3	0.41
CZ	15.34	10	153.4	–	168	94.3	1.10

[a] Several tracks along one transect on the same day were counted only once
[b] The lower value is day with tracks divided by total length, the higher value tracks divided by total length

For the data collected at El Triunfo, we compared track density for different forest types: old-growth montane oak forest, secondary montane oak forest, and coffee plantation. Pooling the data from all areas and transects shows that secondary montane oak forests are used more intensively than expected, the reverse being the case for the other two types (X^2=31.467, df=2, P<0.001). When looking at the data for CZ and CZE separately, we find that in CZ old-growth forest is preferred (X^2=10.267, df=1, P=0.001), whereas in CZE secondary forest is preferred (X^2=76.815, df=1, P<0.001; Table 27.2).

Results from both studies show that the Baird's tapir prefers undisturbed areas, being found less frequently in those impacted by human activity. Indeed, at both sites, relative tapir abundance was more than twice as high in undisturbed areas. Probably, tapirs do not completely avoid disturbed areas but do keep away from trails used by local people and tourists. In Villa Mills where hunting took place in the past, tapirs still seem to avoid the area close to the village.

Track densities found in our study areas (Villa Mills: 0.15–0.63/km; El Triunfo: 0.41–1.10/km) are within the range reported in other studies. Naranjo (1995) found 0.6 tracks/km in Corcovado, Costa Rica, whereas Naranjo and Cruz (1998) reported 0.1–0.7 tracks/km for different forest types in the La Sepultura Biosphere Reserve, Mexico. Compared to El Triunfo, the lower track density in Villa Mills is possibly caused by the fact that tapirs were intensively hunted here for several years, and the population at this site may still be recovering, as former densities have not been re-established yet.

Table 27.2. Expected and observed numbers of Baird's tapir tracks in different vegetation types at El Triunfo, Mexico. *CZ* Core zone without ecotourism, *CZE* core zone with ecotourism

	All areas obs.[a]	All areas exp.	CZ obs.[b]	CZ exp.	CZE obs.[c]	CZE exp.
Old-growth montane oak forest	90	116.8	83	62.9	7	53.8
Secondary montane oak forest	191	148.8	85	105.1	106	59.2
Coffee plantation	0	15.4	0	0	0	0

[a] X^2=31.467, df=2, P<0.001
[b] X^2=10.267, df=1, P=0.001
[c] X^2=76.815, df=1, P<0.001

Results from the habitat analysis at El Triunfo suggest that tapirs completely avoid areas where agriculture is the main land use. Conversion of forest to agricultural land destroys most tapir habitat. Extensive agricultural areas are probably efficient barriers to tapir dispersal, leaving fragmented populations isolated from each other.

The question whether tapirs prefer old-growth, rather than secondary forests is not fully answered yet. Overall, secondary forest seems to be preferred, though at CZ a preference for old-growth forest is apparent. At CZE, old-growth forest was almost entirely avoided. The preferred use of secondary forests would agree with results from other studies (Fragoso 1991; Naranjo 1995; Foerster and Vaughan 2002). Secondary forests often have a higher amount of young plants with fresh leaves and shoots, a preferred component of the tapir's diet.

27.5 Feeding Habits

We analysed the composition of 90 faecal samples from El Triunfo, and 13 from Villa Mills. The composition of the faeces was very similar for both sites, the most frequent plant parts encountered being leaves and stems (Table 27.3). Fruits made up for only 3.9 % of the composition at El Triunfo, and were almost absent at Villa Mills. There was no seasonal difference in composition in El Triunfo (Mann-Whitney's U=1,974.0, df=1, P=0.2115).

Results from the frequency and weight methods were highly correlated (Pearson's R=0.008, df=270, P<0.0001), indicating that the frequency method is valid for composition analysis.

Table 27.3. Composition of faecal samples of the Baird's tapir collected at El Triunfo, Mexico, and at Villa Mills, Costa Rica. Significant positive correlation data are given in the main text

Study area	Frequency (%)			Weight (%)		
	Leaves	Stems	Fruits	Leaves	Stems	Fruits
El Triunfo (rainy season, *N*=40)	49.4	45.8	4.8	60.0	35.2	4.8
El Triunfo (dry season, *N*=50)	42.3	54.5	3.2	55.7	40.4	3.9
El Triunfo (both seasons, *N*=90)	45.5	50.6	3.9	57.6	38.1	4.3
Villa Mills (*N*=13)	48.3	51.4	0.3			

Twenty plant species eaten by tapirs were identified at browsing sites near Villa Mills. An additional seven plant species were identified from leaves and stem parts found in faeces. These species represented 24 families and 24 genera. Remains of *Chusquea* spp. (Poaceae) were found in all faecal samples. Other species frequently encountered were *Quercus costaricenses* (Fagaceae), and several ferns which could not be identified further. In the field, *Anthurium* spp. (Araceae), *Chusquea* spp., *Buddleja* spp. (Buddlejaceae), *Columnea* sp. (Gesneriaceae) and various ferns were often browsed upon.

At El Triunfo, we found 45 species of plants consumed by tapirs, belonging to 27 families and 35 genera. The most common families were Solanaceae (13 %), Rubiaceae (12 %) and Asteraceae (11 %). Complete lists of all recorded diet species can be found in Tobler (2002) and Lira et al. (2004).

Usually, the leaves and new shoots of younger plants and herbs (<1 m) were eaten. At least some leaves commonly remained on the plant. Often, only one or two species were eaten at a given spot, even though there were other plants available which were eaten on other occasions. At many spots, only a small number of plants were eaten before the animal likely moved on.

In both study areas, tapirs appeared to have a very broad diet, feeding on a large number of different plants. This is probably one of the reasons why they can easily adapt to a wide range of habitats along a strong elevational gradient, from lowland rainforest and montane cloud forest to alpine paramo. As shown by other studies (Janzen 1982; Bodmer 1990; Downer 1996; Foerster 1998), however, tapirs prefer some diet plant species over others, and appear to selectively feed upon these.

The proportion of fruit found in the diet of tapirs in El Triunfo, and especially in Villa Mills, is lower than that documented in other studies. Naranjo and Cruz (1998) found that fruits can make up to 6.2–8.1 % of the tapir's diet in several forest types, ranging from lowland moist forest to montane forests in Chiapas, Mexico.

In the lowland moist forest of Corcovado, Costa Rica, the percentage of fruits varied in the range 3.2–12 % based on faecal analysis (Naranjo 1995), and 12.4–22.3% based on direct observations (Foerster 1998). Fruits may account for about 33% of the diet of *T. terrestris* in a lowland tropical rainforest (Bodmer 1990). The higher proportion of fruits in the lowlands is probably related to a higher availability of large fleshy dicot and palm fruits preferred by tapirs, compared to relatively small fruits of most montane oak forest trees. Interestingly, however, Downer (1996, 2001) reported *T. pinchaque* to disperse as many as 86–264 plant species in montane cloud forest and paramo in the Andes of Ecuador.

There was no seasonal variation in the tapir's diet at El Triunfo. This is different from most lowland regions where there are clear differences between dry and wet seasons (Naranjo 1995; Foerster 1998; Naranjo and Cruz 1998). Again, this is most likely due to the small proportion of fruits in the tapir's diet at El Triunfo. Fruit availability in lowland moist forest, however, varies strongly with season (Altrichter et al. 2001).

27.6 Hunting

All local people interviewed in Villa Mills had been living there for at least 15 years. Two-thirds of the 15 persons interviewed had seen tapirs at least once in their lives, and all had seen tapir tracks. Seven had hunted tapirs in the past, and 11 had eaten tapir meat at least once in their life.

Intensive hunting of tapirs in the area of Villa Mills started about 25 years ago and ended 10–15 years ago. The number of animals killed during that time period fluctuated between eight and 20 per year. Subsistence hunting was most frequent, and all animal parts were used. Rarely were tapirs killed for sport or adventure. Tapirs were usually hunted with firearms and the help of dogs. Hunting pressure declined in the late 1980s as a result of stricter law enforcement and a change in attitude due to national education programs. Today, it is estimated that about one animal per year is still killed by poachers. Very few inhabitants of Villa Mills would still hunt tapirs if this were not prohibited by law, and most of these people are now in favour of the protection of tapirs and the forests in general.

Many people noticed a decrease of sightings during the period of intensive hunting. Some reported a slight increase during the last few years, and a couple of persons mentioned that they had seen more tracks and faeces further inside the forest in almost inaccessible areas.

In El Triunfo, ten of the 24 people interviewed had seen tapirs at least once in their life, and some three during the last 5 years. Only five persons said that they had hunted tapirs in the past for their meat and skin. All the people interviewed worked in agriculture, most of them in coffee production. Only one person reported an incident in which tapirs caused damage to crops.

Interviews with guards from the El Triunfo reserve revealed that, during the last ca. 21 years (1980–2001), seven tapirs were hunted in the area. In this case, all hunting was for subsistence purposes.

27.7 Conclusions

Deforestation is the main threat to tapir populations in Mesoamerica. At El Triunfo, 80.2 % of the area was covered by old-growth vegetation in 1975. In 1995, this was only 68.9 % (Arreola et al. 1997). During those 20 years, 13,526 ha of forest had been converted to coffee plantations.

In Costa Rica, the least-fragmented forest types are the lower and upper montane forests (Kappelle 2001). The former occur at elevations of 1,400–2,300 m (Kappelle 2004), and represent 21 % of the remaining forests, 80 % or more being canopy closure (Sánchez-Azofeifa et al. 2001). The latter occur at elevations of 2,300–3,200 m (Kappelle 2004), and make up 8 % of all the remaining forests in Costa Rica (Sánchez-Azofeifa et al. 2001). Whereas forests within protected areas are relatively little impacted, adjacent areas are becoming increasingly more fragmented, prohibiting connectivity between protected areas and therefore isolating populations (Sánchez-Azofeifa et al. 2003).

Montane oak forests not only serve as an important habitat for tapirs, they also connect populations of large mammals at lower elevations along both sides of the mountains. It is therefore crucial not only to protect montane oak forests but also to assure their connectivity with lower-elevation forest habitats. If we loose mid-altitude forests such as lower montane forests, tapir populations in the montane oak forests will remain isolated, and mountain ranges may become barriers separating populations in the lowlands of both main slopes.

Following habitat loss, hunting is the second most important threat to tapir populations. Tapirs have a very low reproduction rate, and even a low hunting pressure may quickly lead to a population decline (Bodmer et al. 1997; Carrillo et al. 2000). Tapirs are protected throughout their range, and hunting happens only occasionally for subsistence reasons. The example from Villa Mills shows that education programs can help change people's attitude towards conservation and hunting. Although education is a good strategy, a stricter control of protected areas by park guards (law enforcement) is another important prerequisite.

As shown above, human presence seems to have a strong impact on tapir abundance. This conflicts directly with the concept of ecotourism as a strategy to conserve an area while generating income for local communities. Based on our findings, we recommend that the trail system in any area be kept to a minimum, making sure that there are sufficiently large sectors with no access for tourists. In addition, the number of people visiting an area should be limited, and regular studies should evaluate the impact of tourism on local native fauna.

The impact of selective logging in Villa Mills seems to be low (see also Chap. 18). There was no significant difference in track frequency between PA and VM. At all places in PA where tracks were found, many browsing signs were observed, indicating that despite the timber extraction, this habitat is still suitable for tapirs. Forest activities are temporally concentrated in this region; felling cycles are estimated to be about 28 years for forest in this study area (Aus der Beek and Sáenz 1996). This may cause a difficulty for tapirs only during a relatively short period of time. Provided that there are sufficiently large, undisturbed areas where tapirs can retreat during this time, they will probably start using the area again shortly after the disturbance.

Montane oak forests are an important habitat for the Baird's tapir but our knowledge on their status and ecology is still very scanty. Further detailed studies on home range size, habitat use and population dynamics are necessary to develop adequate management plans and to design protected areas to ensure the long-term survival of this species.

Acknowledgements We thank Proyecto Tapir (Instituto de Historia Natural y Ecología, Chiapas, México), Epigmenio Cruz, El Colegio de la Frontera Sur and CONACYT for their technical support during the fieldwork at El Triunfo. The fieldwork in Costa Rica was supported by CATIE, and special thanks go to Grace Sáenz, Ligia Quirós, Geoffrey Venegas, Marlen Camacho, Oscar Araya and Álvaro Abarca.

References

Altrichter M, Carrillo E, Fuller T (2001) White lipped peccary (*Tayassu pecari*) diet and fruit availability in a Costa Rican rain forest. Rev Biol Trop 49:1105–1114

Arreola MA, Noble LC, Altamirano MA, Vergara ME, Zúniga LFG, Becerril RA, Cuevas GG, Guillén CV (1997) Evaluación de los cambios de uso del suelo y la cobertura vegetal en la Reserva de la Biosfera El Triunfo, Chiapas, México. Instituto para el Desarrollo Sustentable en Mesoamérica, A.C., WWF, IHN, INE, San Cristóbal de Las Casas, Chiapas, Mexico

Aus der Beek R, Sáenz G (1996) Impacto de las intervenciones silviculturales en los robledales de alturaé estudio de caso en la Cordillera de Talamanca, Costa Rica. Rev For Centroam 17:30–37

Blaser J (1987) Standörtliche und waldkundliche Analyse eines Eichen-Wolkenwaldes (*Quercus* spp.) der Montanstufe in Costa Rica. PhD Thesis, Georg-August Universität, Göttingen, Germany

Bodmer RE (1990) Fruit patch size and frugivory in the lowland tapir (*Tapirus terrestris*). J Zool 222:121–128

Bodmer RE (1991) Strategies of seed dispersal and seed predation in Amazonian ungulates. Biotropica 23:255–261

Bodmer RE, Eisenberg JF, Redford KH (1997) Hunting and the likelihood of extinction of Amazonian mammals. Conserv Biol 11:460–466

Carrillo E, Wong G, Cuaron AD (2000) Monitoring mammal populations in Costa Rican protected areas under different hunting restrictions. Conserv Biol 14:1580–1591

Chamrad AD, Box TW (1964) A point frame for sampling rumen contents. J Wildlife Manage 28:473–477

Downer CC (1996) The mountain tapir, endangered 'flagship' species of the high Andes. Oryx 30:45–58

Downer CC (2001) Observations on the diet and habitat of the mountain tapir (*Tapirus pinchaque*). J Zool 254:279–291

Foerster CR (1998) Ecología de la danta Centroamericana *Tapirus bairdii* en un bosque húmedo tropical de Costa Rica. MSc Thesis, Universidad Nacional, Heredia, Costa Rica

Foerster CR, Vaughan C (2002) Home range, habitat use, and activity of Baird's tapir in Costa Rica. Biotropica 34:423–437

Fragoso JM (1987) The habitat preferences and social structure of tapirs. MSc Thesis, University of Toronto, Ontario, Canada

Fragoso JM (1991) The effects of selective logging on Baird's tapir. In Mares MA, Schmidly DJ (eds) Latin-American mammalogy: history, biodiversity and conservation. University of Oklahoma Press, Norman, OK, pp 295–304

Fragoso JM (1997) Tapir-generated seed shadows: scale-dependent patchiness in the Amazon rain forest. J Ecol 85:519–529

Fragoso JM, Silvius KM, Correa JA (2003) Long-distance seed dispersal by tapirs increases seed survival and aggregates tropical trees. Ecology 84:1998–2006

IUCN (2004) IUCN Red List of Threatened Species. World Conservation Union. http://www.iucnredlist.org

Janzen DH (1981) Digestive seed predation by a Costa Rican Bairds' tapir. Biotropica 13:59–63

Janzen DH (1982) Wild plant acceptability to a captive Costa Rican Baird's tapir. Brenesia 19/20:99–128

Kappelle M (1996) Los bosques de roble (*Quercus*) de la Cordillera de Talamanca, Costa Rica: biodiversidad, ecología, conservación y desarrollo. Instituto Nacional de Biodiversidad (INBio), Santo Domingo de Heredia, Costa Rica

Kappelle M (2001) Costa Rica. In: Kappelle M, Brown AD (eds) Bosques Nublados del Neotrópico. Instituto Nacional de Biodiversidad, Costa Rica, pp 301–370

Kappelle M (2004) Tropical montane forests. In: Burley J, Evans J, Youngquist JA (eds) Encyclopedia of forest sciences, vol 4. Elsevier, Oxford, UK, pp 1782–1793

Kappelle M, Brown AD (eds) (2001) Bosques nublados del Neotrópico. Instituto Nacional de Biodiversidad, Costa Rica

Kappelle M, Horn SP (eds) (2005) Páramos de Costa Rica. Instituto Nacional de Biodiversidad (INBio), Santo Domingo de Heredia, Costa Rica

Lawton RO (2000) Baird's tapir. In: Nadkarni NM, Wheelwright NT (eds) Monteverde: ecology and conservation of a tropical cloud forest. Oxford Univ Press, New York, pp 2342–2343

Lira I, Naranjo EJ, Güris DM, Cruz E (2004) Ecología de *Tapirus bairdii* (Perissodactyla: Tapiridae) en la Reserva de la Biosfera El Triunfo (Polígono I), Chiapas, México. Acta Zool Mex 20:1–21

Lizcano DJ (2004) Efecto de los grandes mamíferos en el bosque alto andino. In: Proc 6th Int Congr Manejo de Fauna Silvestre en la Amazonía y Latinoamérica, Iquitos, Peru

Long A, Heath M (1991) Flora of El Triunfo Biosphere Reserve, Chiapas, Mexico. An Inst Biol, Universidad Nacional Autónoma de México (UNAM), Ser Bot, pp 133–172

Matola S, Cuaron AD, Rubio-Torgler H (1997) Status and action plan of the Baird's tapir (*Tapirus bairdii*). In: Brooks DM, Bodmer ER, Matola S (eds) Tapirs: status survey and conservation action plan. IUCN, Gland, Switzerland, pp 29–45

Naranjo EJ (1995) Habitos de alimentación del tapir (*Tapirus bairdii*) en un bosque tropical húmedo de Costa Rica. Vida Silv Neotrop 4:32–37

Naranjo EJ (2002) Population ecology and conservation of ungulates in the Lacandon Forest, Mexico. PhD Dissertation, University of Florida, Gainesville, FL

Naranjo EJ, Cruz E (1998) Ecologia del tapir (*Tapirus bairdii*) en la Reserva de la Biosfera La Sepultura, Chiapas, Mexico. Acta Zool Mex 73:111–123

Naranjo EJ, Vaughan C (2000) Ampliación del ámbito altitudinal del tapir centroamericano (*Tapirus bairdii*). Rev Biol Trop 48:724

National Forest Inventory (2000) Inventario Nacional Forestal 2000. Secretaria de Medio Ambiente y Recursos Naturales, Universidad Nacional Autónoma de México and Instituto Nacional de Estadistica Geografia e Informática, México, DF, México. http://infoteca.semarnat.gob.mx/Metadatos/inventario2000.html

Olmos F (1997) Tapirs as seed dispersers and predators. In: Brooks DM, Bodmer ER, Matola S (eds) Tapirs: status survey and conservation action plan. IUCN, Gland, Switzerland, pp 46–66

Reid FA (1997) A field guide to the mammals of Central America and Southeast Mexico. Oxford Univ Press, New York

Rodrigues M, Olmos F, Galetti M (1993) Seed dispersal by tapir in Southeastern Brazil. Mammalia 57:460–461

Sánchez-Azofeifa GA, Harriss RC, Skole DL (2001) Deforestation in Costa Rica: a quantitative analysis using remote sensing imagery. Biotropica 33:378–384

Sánchez-Azofeifa GA, Daily GC, Pfaff ASP, Busch C (2003) Integrity and isolation of Costa Rica's national parks and biological reserves: examining the dynamics of land-cover change. Biol Conserv 109:123–135

Terwilliger VJ (1978) Natural history of Baird's tapir on Barro Colorado Island, Panama Canal Zone. Biotropica 10:211–220

Tobler MW (2002) Habitat use and diet of Baird's tapirs (*Tapirus bairdii*) in a montane cloud forest of the Cordillera de Talamanca, Costa Rica. Biotropica 34:468–474

Vázques-García JA (1995) Cloud forest archipelagos: preservation of fragmented montane ecosystems in tropical America. In: Hamilton LS, Juvik JO, Scatena FN (eds) Tropical montane cloud forests. Springer, Berlin Heidelberg New York, pp 315–332

Williams KD (1984) The Central American tapir (*Tapirus bairdii*) in northwestern Costa Rica. PhD Thesis, Michigan State University, East Lansing, MI

Part VI
Conservation and Sustainable Use

28 Dynamics and Silviculture of Montane Mixed Oak Forests in Western Mexico

M. Olvera-Vargas, B.L. Figueroa-Rangel, J.M. Vázquez-López, and N. Brown

28.1 Introduction

Whilst it is widely accepted that silvicultural systems for natural forest must be grounded in a sound understanding of natural forest dynamics, recent developments in conservation biology suggest that this understanding is also a requirement for the successful management of protected forest areas. Many valued forest communities are early- or mid-successional assemblages. Managers may need to protect minimum dynamic areas, manipulate the magnitude and frequency of disturbance, and facilitate natural regeneration. In this chapter, we outline key aspects of the dynamics of subtropical montane *Quercus* forests of the Sierra de Manantlán, a Biosphere Reserve situated in western Mexico (19°24'32"–19°31'02" N; 103°57'44"–104°01'09" W).

This is a *Quercus* species-rich area, with 31 species so far recognised for the region (Vázquez-García et al. 1995), co-occurring in large congeneric associations but also intermingled with non-*Quercus* broadleaved and pine species. There are two main *Quercus* assemblages found at different altitudes across the Sierra de Manantlán (Cuevas-Guzmán et al. 1997):

1. a dry oak forest ecosystem (400–1,500 m above sea level, a.s.l.), characterised by deciduous trees up to 15 m tall; typical oaks in this system are *Quercus castanea*, *Q. glaucescens*, *Q. magnoliifolia* and *Q. rugosa*;
2. a sub-deciduous oak forest ecosystem (above 1,500 m a.s.l.), characterised by semi-deciduous, 20–35 m tall trees, including *Quercus crassipes*, *Q. candicans*, *Q. acutifolia* and *Q. laurina*.

In this chapter, we review the dynamics of sub-deciduous oak forests in the Sierra de Manantlán. We examine whether there are distinct assemblages of *Quercus* and other species, and whether successional changes occurred in

Ecological Studies, Vol. 185
M. Kappelle (Ed.) Ecology and Conservation of Neotropical Montane Oak Forests
© Springer-Verlag Berlin Heidelberg 2006

their structure and composition over an 11-year period. We review the role which regeneration dynamics may play in driving change. Our investigations have enabled us to formulate management guidelines for conservation or silvicultural purposes. Our analysis is based on long-term data from a network of 105 permanent sample plots (psps, 500 m^2 each), located in the south-eastern sector of the Sierra de Manantlán Biosphere Reserve.

The plots are scattered over an area of approximately 4,000 ha, at altitudes of 2,020–2,250 m. Thirty-two plots were established in 1991, 28 in 1994, and 45 in 1998. These have been re-enumerated approximately every 4 years since 1991. We have followed a standardised inventory protocol (Olvera-Vargas et al. 1996) within the plots, measuring all individuals in three size classes: seedlings (individuals <1.30 m tall), saplings (individuals >1.30 m tall and having <5 cm diameter at breast height, DBH), and adult trees (≥5 cm DBH and ≥1.30 tall). In addition, a detailed characterisation of the forest environment in the vicinity of the plots was conducted.

28.2 Spatial Variation in Floristic Composition

Over the last decade, one major focus in community ecology research has been the study of processes which control the distribution and abundance of species. Niche differences including trade-offs between dispersal and competitive abilities have been used to account for high-diversity assemblages (Tilman 1988, 1994) but many of the patterns generated by such processes can equally well be explained by neutral theories (Bell 2001; Hubbell 2001). High-altitude Mexican oak forests provide a valuable testing ground for these ideas. The oaks found here form a large congeneric set of species with similar life-history traits and limited dispersal ability. We analysed whether tree species composition was strongly related to habitat type or simply reflected stochastic drift from place to place.

In our inventory, we identified 38 tree species (including nine species of *Quercus*) from 28 genera and 22 families. The richest genus was *Quercus*, with relative frequencies (at plot level) varying in the range 0.94–100 %.

We carried out an ordination of plot data using non-metric multidimensional scaling (NMDS), a rank-order multivariate technique (Prentice 1977) based on Bray-Curtis (Sørensen) dissimilarity. This analysis revealed two groups of plots (Fig. 28.1). In one group, species such as *Quercus crassipes*, *Prunus serotina*, *Pinus leiophylla* and *Alnus jorullensis* are the most abundant taxa. The second group is dominated by species such as *Q. candicans*, *Q. laurina*, *Q. gentryi* and *Q. scytophylla*, intermingled with other broadleaved

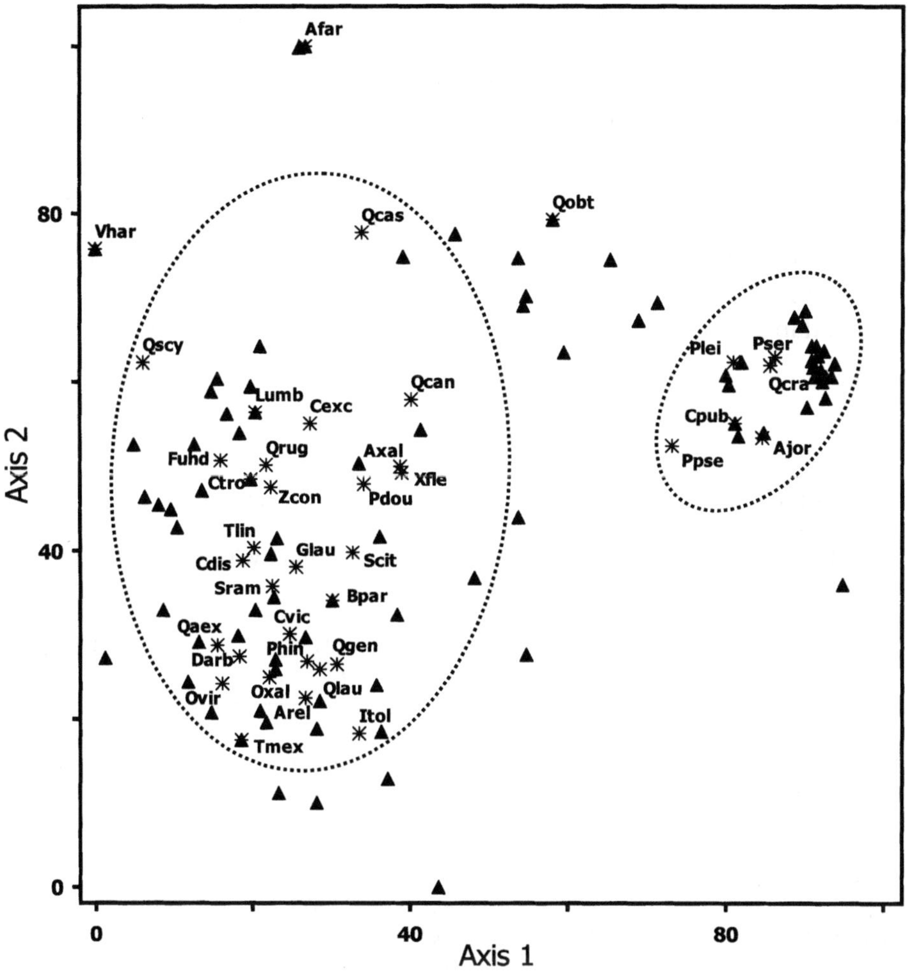

Fig. 28.1. First two axes of non-metric multidimensional scaling (stress 14.9; $P = 0.019$) for mixed oak forest in Manantlán, showing groups of plots and species which share similarities (Sørensen index). *Asterisks* Relative position of species on the axes, *triangles* relative position of permanent sample plots on the same axes in terms of relative species dominance, *lines encircling groups of plots and species* are approximate boundaries between zones. Geographical range distance between psps 0.0010-9.5 km. Abbreviations of species names: Arel = *Abies religiosa var. religiosa;* Afar = *Acacia farnesiana;* Ajor = *Alnus jorullensis subsp. lutea;* Axal = *Arbutus xalapensis;* Bpar = *Buddleja parviflora;* Ctro = *Carpinus tropicalis;* Cvic = *Clethra vicentina;* Cdis = *Comarostaphylis discolor subsp. discolor;* Cexc = *Cornus excelsa;* Cpub= *Crataegus pubescens;* Darb = *Dendropanax arboreus;* Fuhd = *Fraxinus uhdei;* Glau = *Garrya laurifolia;* Itol = *Ilex tolucana;* Lumb = *Lippia umbellata;* Oxal = *Oreopanax xalapensis;* Ovir = *Ostrya virginiana;* Phin = *Persea hintonii;* Pdou = *Pinus douglasiana;* Plei = *Pinus leiophylla;* Ppse = *Pinus pseudostrobu;* Pser = *Prunus serotina;* Qaex = *Quercus aff. excelsa;* Qcan = *Quercus candicans;* Qcas = *Quercus castanea;* Qcra = *Quercus crassipes;* Qgen = *Quercus gentryi;* Qlau = *Quercus laurina;* Qobt =*Quercus obtusata;* Qrug = *Quercus rugosa;* Qscy = *Quercus scytophylla;* Sram = *Styrax ramirezii;* Scit = *Symplocos citrea;* Tlin = *Ternstroemia lineata subsp. lineata;* Tmex = *Tilia mexicana;* Vhar = *Viburnum hartwegii;* Xfle = *Xilosma flexuosum;* Zcon = *Zinowiewia concinna*

species such as *Persea hintonii, Ilex tolucana, Oreopanax xalapensis* and *Styrax ramirezii.*

We used canonical correspondence analysis (CCA) in order to identify the most important environmental gradients underlying compositional differences between our plots. Unconstrained CCA (Ter Braak 1987; Table 28.1) showed that differences in elevation created the strongest gradient in species composition. Elevation is unlikely to act directly on species composition over such a small altitudinal range. We contend that it is correlated with an intermediate factor such as rainfall. The south-western slopes of the Sierra de Manantlán are directly exposed to moisture-saturated winds from the Pacific,

Table 28.1. Canonical correspondence analysis (CCA) for adult species composition in montane mixed oak forest in Manantlán, Mexico. Correlation coefficients (*r*) of environmental variables and *t*-values (*t*) are presented for axes 1 and 2 with eigenvalues between brackets; critical value* (df≥18; ×=0.05)=2.1

Variable	Code	Axis 1 (0.444)		Axis 2 (0.273)	
		r	*t*	*r*	*t*
Topography					
Elevation	ELEV	0.241	**2.743**	−0.437	**−5.618**
Regular topography	Top1	−0.183	**−2.489**	−0.070	−1.119
Terrain					
Slope	Slo	−0.157	−1.854	−0.102	−1.402
Aspect	ASP	0.195	**2.574**	0.141	**2.166**
Flat catena	Cat1	−0.111	−0.891	−0.089	−0.830
Ridge top	Cat2	0.152	2.009	0.061	0.941
Upper slope	Cat3	0.149	0.964	0.069	0.520
Middle slope	Cat4	−0.120	−0.600	−0.038	−0.222
Lower slope	Cat5	−0.066	−0.340	−0.013	−0.078
Base of slope	Cat6	−0.021	−0.210	−0.036	−0.414
Vegetation					
Old-growth	Mat1	−0.030	−0.458	0.071	1.236
Mature canopy	Mat2	0.143	2.003	0.128	2.080
Young reproductive canopy	Mat3	0.185	**2.633**	0.233	**3.847**
One horizontal stratum	CaLa1	0.036	0.499	−0.084	−1.332
Soil					
Organic matter content	OMC	0.019	0.263	0.000	0.006
Cation exchange capacity	CEC	−0.106	−1.267	0.026	0.362
Texture	Tex3	−0.120	−1.157	−0.090	−1.013
Clay	Tex2	−0.156	−0.644	−0.079	−0.382
Sand	Tex1	−0.244	−1.000	−0.135	−0.641
Phosphorous	P	−0.142	−1.962	−0.054	−0.877

and receive orographic rainfall, whereas the central part of the plateau is relatively dry. This hypothesis is supported by our knowledge of the ecology of species associated with each group. The boundary between the two floristic associations is not sharp but rather constitutes a broad xeric-mesic ecotone (Fig. 28.1), accompanied by a gradual change in stand physiognomy. Species composition is also significantly correlated with our measure of forest structure, implying that spatial variation is also partly attributable to successional history.

A simple Mantel test of the correlation between matrices of differences in species composition and geographical distance between plots showed a strong and significant relationship (Table 28.2). This implies that spatial variation in oak forest composition is highly influenced by dispersal limitation. However, it is difficult to distinguish the effects of dispersal limitation from those of niche specificity in natural assemblages, because environmental conditions often change with distance, at least in temperate forests (Gilbert and Lechowicz 2004). If dispersal limitation alone is important in determining spatial variation in species composition, we would expect any correlation between environmental variability and community dissimilarity to be masked once effects of geographical distance on environmental dissimilarity are accounted for. In order to test this, we used partial Mantel tests to compare matrices of environmental dissimilarity and assemblage dissimilarity between plots, controlling for effects of geographical separation. The data show that differences between plots in terms of elevation, topography and canopy maturity continued to account for a significant proportion of variation in species composition, even when the effects of geographic distance had been statistically eliminated. We conclude that floristic variation over an area of 4,000 ha in Manantlán is caused by local environmental differences acting

Table 28.2. Mantel correlation coefficients (simple and partial tests) between differences in floristic distance, differences in environmental distances, and spatial distances. Statistical significances of each correlation coefficient were calculated using 10,000 permutations

Floristic distance with	*r*	*p*
Spatial distance	0.287	<0.001
Spatial distance (elevation eliminated)	0.295	<0.001
Spatial distance (topography eliminated)	0.287	<0.001
Spatial distance (aspect eliminated)	0.285	<0.001
Spatial distance (Mat3 eliminated)	0.286	<0.001
Elevation (spatial distance eliminated)	0.133	<0.001
Topography (spatial distance eliminated)	0.061	<0.001
Aspect (spatial distance eliminated)	0.010	n.s.
Mat3 (spatial distance eliminated)	0.076	<0.001

together with dispersal limitation. In previous studies of this same forest, Figueroa-Rangel and Olvera-Vargas (2000) and Olvera-Vargas et al. (2000) found that the most xeric zones, characterised by larger physiographic units and gently sloping terrain, and more exposed to solar radiation tend to be less species-rich, compared with zones of more complex physiography located in more humid environments.

28.3 Patterns of Change over Time

Because trees are such long-lived organisms, compositional change in such forests occurs at a rate revealed only by very long-term studies, such as this one. Shorter-term species abundance data do not reveal subtle changes in dominance caused by variations in the relative size of individuals. We have used a detrended correspondence analysis (DCA; Hill and Gauch 1980) of species dominance data to reveal the direction and rate of any successional trends. We chose the method described by Whittaker (1989) to track the movement of individual plots in the ordination space over two or more consecutive enumerations.

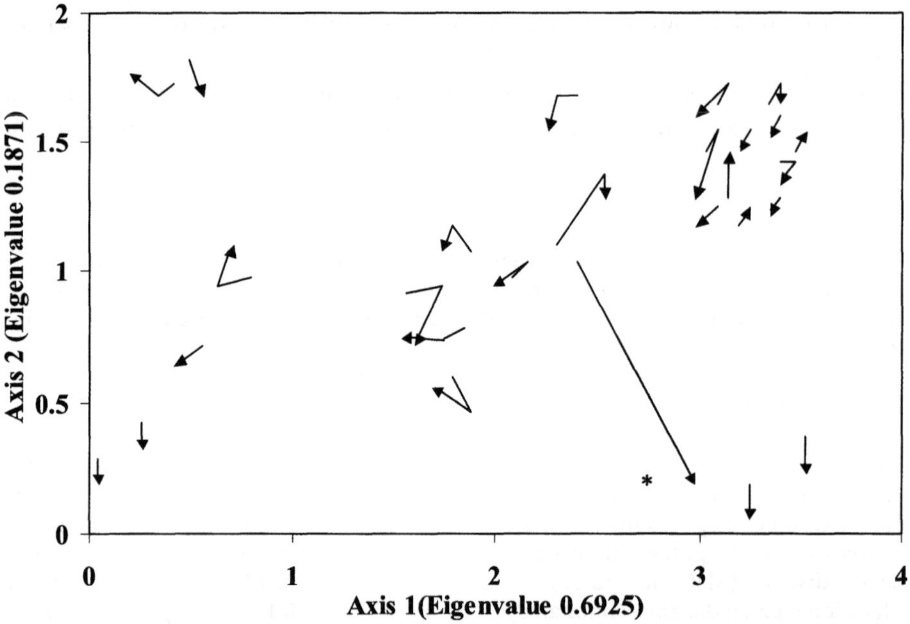

Fig. 28.2. Detrended correspondence analyses of mixed oak forests in Manantlán based on 11 years of re-measurements. *Arrow directions and lengths* Plot migrations and the proportion in rate of changes respectively, *asterisks* plots which did not change over time

The results showed little congruence between plots in the direction or rate of change (Fig. 28.2). Fulton and Harcombe (2002) argued that if community changes are predictable, then vegetation patches of similar species composition should have analogous successional trajectories and rates of change. In the present case, the lack of consistent trends in community composition implies that changes at our study plots are stochastic fluctuations driven by fine-scale competitive interactions, rather than successional processes. Similar results were reported by Yoshida and Kamitani (2000) in a study carried out in Niigata, Central Japan, investigating the effect of interspecific competition in a *Quercus-Fagus-Magnolia* canopy-dominated forest. They observed that plots of similar species composition followed very different successional trajectories, depending on initial dominance by species differing in shade tolerance.

28.4 The Regeneration Dynamics

We identified 30 species of seedlings (23 genera, 20 families) and 21 sapling species (20 genera, 14 families) regenerating under mixed oak canopies in Manantlán. Two *Quercus* species, *Q. gentryi* and *Q. obtusata*, did not occur as seedlings or saplings. There were considerably more seedlings (n=3,434) than saplings (n=634) in 2002. High mortality at this life stage may exert strong selective pressures on assemblage composition. Natural regeneration of oaks is well-known to be problematic, and many species are believed to show preferential recruitment at some distance from conspecific adults (the 'escape hypothesis' sensu Howe and Smallwood 1982). This has been documented in Mexican *Quercus* forests (Quintana-Ascencio et al. 1992; Chaps. 14 and 16). This density-dependent recruitment may be an important process driving fluctuations in species composition over time.

We compared the species composition of seedlings, saplings and adult trees in each plot, using the quantitative version of the Sørensen similarity coefficient (Magurran 1988). In general, both seedling and sapling populations were dissimilar to adult tree assemblages. The mean similarity between seedlings and adults was 0.16, and between saplings and adults only 0.11. More than half the plots showed greater similarity between seedlings and adults than between saplings and adults. This suggests that few seedlings survive to grow into saplings beneath conspecific adults and, as a consequence, plots may show compositional fluctuations over time.

Nevertheless, results from partial Mantel tests revealed a significant correlation (Table 28.3) between ecological (cf. seedling and sapling frequencies) and geographical distances which was independent of environmental conditions. This discrepancy is interpreted to be a consequence of physiographic patchiness in Manantlán, and this may partially explain the different floristic

Table 28.3. Results from a simple (values in *upper right corner*) and partial (values in *lower left corner*) Mantel statistical analysis for seedling abundance (Sørensen percentage dissimilarity), environmental heterogeneity (18 variables) and geographical distances (0.10–9.5 km for seedlings and saplings) after testing the null hypothesis that there is no association between each pair of distance data matrices

	Simple mantel statistical analysis		
	Seedlings abundance	Environmental heterogeneity	Geographical distance
Partial mantel statistical analysis			
Seedlings abundance	–	r=–0.0515	r=0.2657
	–	p=0.0949	p=0.0002
Environment heterogeneity	r=0.0618	–	r=0.0302
	p=0.0657	–	p=0.1182
Geographical distance	r=0.2678	r=0.0456	–
	p=0.0002	p=0.0415	–

patterns. Battaglia et al. (2000) claim that micro-physiographic variability acts as a strong filter which may have an independent or interacting effect on patterns of regeneration, particularly those influencing seedling establishment. Still, it is difficult to decouple the effects of competition from those derived from environmental heterogeneity and which eventually may represent a constraint for seedling establishment. Intuitively, however, the evidence hints at environmental filters, and subsequent competition with already established vegetation (adults). Hofer et al. (2004) hypothesised that if competitive interactions strongly determine the distributional pattern of the assemblage, then it is to be expected that close neighbours in niche space, e.g. species pairs with high overlap in resource use, be segregated either spatially or temporally. Riley and Jones (2003) claim that environmental factors influencing woody plant regeneration also vary spatially across forest stands as well as within stands at scales as small as 1 m or less.

28.5 Implications for Silvicultural Management

There is a widespread belief that natural regeneration of *Quercus* forest ecosystems is problematic (see, for instance, Loftis and McGee 1993 and references therein). This notion has led to much speculation on how regeneration may be enhanced through improved silviculture (Brose et al. 1999). We have found that many *Quercus* species are prolific seed producers; in a typical crop year, it is common to observe seeds profusely covering the forest floor. Seedling survival is poor under conspecific adults but, beneath a contrasting canopy, regeneration is more than adequate to sustain most *Quercus* species populations. Our investigations in Manantlán (Olvera-Vargas et al. 1998; Olvera-Vargas and Figueroa-Rangel 1998) have also shown that species typically found in a xeric habitat, such as *Q. crassipes*, *Q. rugosa* and *Prunus serotina*, have an exceptional capacity to sprout, from either roots or stumps, or both. For example, when a tree of this group of species is felled, it usually re-sprouts copiously around the stump after a short period of stagnation. These shoots invariably grow faster and more vigorously than those established from acorns. The amount of canopy disturbance caused by harvesting can also have a significant influence on the survival and growth of species belonging to the two main floristic formations which we have identified. Large canopy gaps create cold, dry and open environments, thereby favouring the establishment of xeric species, whereas smaller openings resulting from selective felling maintain mild, humid and shaded environments which promote the establishment of more mesic vegetation.

Considering the forest dynamics outlined above, we suggest that the choice of optimum silvicultural system for mixed oak forests in Manantlán should depend on the floristic formation. Xeric-type forest assemblages, dominated by *Quercus crassipes*, should be managed using even-aged methods such as a shelter-wood technique (Smith 1986). This involves cutting in three phases. In the first phase, the oldest trees, or those with unwanted characteristics, are removed to create space for more highly preferred trees, thereby also halting the development of undesirable specimens or entire species. In the second phase, favourable conditions allowing oak regeneration (basically controlling canopy opening) are promoted. In the third phase, the remaining mature trees are harvested, once the new crop has been established.

In some stands it is possible to apply partial cuttings (low-intensity thinning), which might be conducted on the most abundant and conspicuous species such as *Ternstroemia lineata* and *Prunus serotina*. However, we consider that it is important to maintain species mixtures, and subsequently implement intermediate cuttings on these species before the final oak harvest is carried out. Apparently, these species have little influence on oak diameter growth but they seem to provide favourable stand conditions for oak regeneration. Commercial thinning of oaks can reduce competition either in the canopy or in the intermediate stratum (Olvera-Vargas and Figueroa-Rangel 1998).

In mesic formations, dominated mainly by *Q. candicans*, *Q. laurina* and *Q. rugosa*, the application of uneven-aged methods such as selective felling would be more suitable, given the ecology of the constituent species. This silvicultural method might be more compatible with multiple-purpose management, and it affords greater flexibility in dealing with varying light-tolerance ranges in mixed species stands (Lüpke 1998). With this method, only mature oak trees above a specified minimum diameter should be removed, although other commercial tree species (e.g. *Symplocos citrea* and *Ilex tolucana*) could be considered for felling, too. The application of this method requires more technical skills during harvesting, due to differential species performance, particularly in growth. In turn, this necessarily implies the development of species-specific minimum diameter cutting limits.

28.6 Conclusions

Montane forest assemblages of the Sierra de Manantlán contain a large number of sympatric *Quercus* species which coexist through a combination of niche differentiation, dispersal limitation, and density-dependent seedling recruitment. Species composition is very variable from site to site, and oak species dominating the canopy vary over relatively short distances. We propose that canopy composition will vary strongly with time as a result of density-dependent recruitment. The Sierra de Manantlán is an important site for the study of oak biogeography on the American continent, and may offer important insights into speciation patterns.

Although the Sierra de Manantlán is a protected area, human impacts are now a significant force shaping the structure and composition of the forest. There are substantial challenges ahead in ensuring that exploitation of valuable forest resources, including *Quercus* timber, does not compromise the productive capacity or biodiversity of these forests. We contend that the true threat to oak survival in Manantlán is that associated to human impacts. With this in mind, future research should consider the effect of man-induced disturbances on oak forest dynamics and forest recovery after silvicultural interventions.

Acknowledgements We are grateful to Saul Moreno Gómez who started this project with us 13 years ago. We also owe our gratitude to the people of El Terrero and Toxin Ejidos, especially to Oscar Sánchez Jiménez and Abel Ceja Gutiérrez, who acted as exceptional field guides during all those years. We thank Ramon Cuevas Guzmán and Luis Guzmán for help with taxonomic identification, Mario González Espinosa, Boris Zeide and John Palmer for useful advice to improve the methodology, and all those people who have contributed to the establishment and re-measurement of the plots since 1991. We are grateful for financial support from CONABIO, CONACyT-Mexico, the University of Guadalajara, DFID-United Kingdom, and WWF.

References

Battaglia LL, Fore SA, Sharitz RR (2000) Seedling emergence, survival and size in relation to light and water availability in two bottomland hardwood species. J Ecol 88:1041–1050
Bell G (2001) Neutral macroecology. Science 293:2413–2418
Brose P, van Lear D, Cooper R (1999) Using shelterwood method harvests and prescribed fire to regenerate oak stands on productive upland sites. For Ecol Manage 113:125–141
Cuevas-Guzmán R, Benz BF, Jardel PE (1997) Sierra de Manantlán region and Biosphere Reserve. In: Heywood DS, Herrera-MacBryde O, Villalobos J, Hamilton AC (eds) Centres of plant diversity: a guide and strategy for their conservation, 1st edn. World Conservation Union-World Wildlife Fund, Washington, DC, pp 158–161
Figueroa-Rangel BL, Olvera-Vargas M (2000) Regeneration patterns in relation to canopy species composition and site variables in mixed oak forests in the Sierra de Manantlán Biosphere Reserve, Mexico. Ecol Res 15:249–265
Fulton MR, Harcombe PA (2002) Fine-scale predictability of forest community dynamics. Ecology 83:1204–1208
Gilbert B, Lechowicz M (2004) Neutrality, niches, and dispersal in a temperate forest understory. Proc Natl Acad Sci USA 101:7651–7656
Hill MO, Gauch JHG (1980) Detrended correspondence analysis: an improved ordination technique. Vegetatio 47:47–58
Hofer U, Bersier L-F, Borcard D (2004) Relating niche and spatial overlap at the community level. Oikos 106:366–376
Howe HF, Smallwood J (1982) Ecology of seed dispersal. Annu Rev Ecol Syst 13:201–228
Hubbell SP (2001) The unified neutral theory of biodiversity and biogeography, 1st edn. Princeton Univ Press, Princeton, NJ
Loftis D, McGee CE (1993) Oak regeneration: Serious problems, practical recommendations. USDA Forest Service, Southeastern Forest Experiment Station, Asheville, NC, Gen Tech Rep SE-84, p 319
Lüpke BV (1998) Silvicultural methods of oak regeneration with special respect to shade tolerant mixed species. For Ecol Manage 106:19–26
Magurran AE (1988) Ecological diversity and its measurement. Chapman and Hall, London, UK
Olvera-Vargas M, Figueroa-Rangel BL (1998) Ecology and silviculture of oak and mixed-oak forests in the Sierra de Manantlán, Mexico: seeking sustainable forest management in a biosphere reserve. In: Guariguata MR, Finegan B (eds) Ecology and management of tropical secondary forest: science, people and policy, 1st edn. Centro Agronómico de Investigación y Enseñanza (CATIE), Turrialba, Costa Rica, pp 121–135
Olvera-Vargas M, Moreno Gómez S, Figueroa-Rangel BL (1996) Sitios permanentes de investigación silvícola: manual para su establecimiento, 1st edn. Universidad de Guadalajara, Guadalajara, Jalisco, Mexico
Olvera-Vargas M, Figueroa-Rangel BL, Moreno Gómez S, Solís-Magallanes A (1998) Resultados preliminares de la fenología de cuatro especies de encino (*Quercus*) en Cerro Grande, Reserva de la Biosfera Sierra de Manantlán. Biotam 9:7–18
Olvera-Vargas M, Figueroa-Rangel BL, Bongers F (2000) Zonation and management of mountain forests in the Sierra de Manantlán, Mexico. In: van der Maarel E (ed) Zonation and management of mountain forests particularly on volcanoes. Opulus Press, Uppsala, Sweden, pp 17–22
Prentice IC (1977) Non-metric ordination methods in ecology. J Ecol 65:85–94

Quintana-Ascencio PF, Ramírez-Marcial N, González-Espinoza M (1992) Acorn removal, seedling survivorship, and seedling growth of *Quercus crisipipilis* in successional forests of the highlands of Chiapas, Mexico. Bull Torrey Bot Club 119:6–18

Riley J, Jones R (2003) Factors limiting regeneration of *Quercus alba* and *Cornus florida* in formerly cultivated coastal plain sites, South Carolina. For Ecol Manage 177:571–585

Smith DM (1986) The practice of silviculture , 8th edn. Wiley, New York

Ter Braak CJF (1987) The analysis of vegetation-environment relationships by canonical correspondence analysis. Vegetatio 69:69–77

Tilman D (1988) Plant strategies and the dynamics and structure of plant communities. Princeton Univ Press, Princeton, NJ

Tilman D (1994) Competition and biodiversity in spatially structured habitats. Ecology 75:2–15

Vázquez-García JA, Cuevas-Guzmán R, Cochrane TS, Iltis HH, Santana-Michel FJ, Guzmán-Hernández L (1995) Flora de Manantlán, 1st edn. Botanical Research Institute of Texas, Forth Worth, TX, Sida Bot Misc no 13

Whittaker RJ (1989) The vegetation of the Storbeen Gletschervorfeld, Jotunheimen, Norway. III. Vegetation-environment relationships. J Biogeogr 16:413–433

Yoshida T, Kamitani T (2000) Interspecific competition among three canopy-tree species in a mixed-species even-aged forest of central Japan. For Ecol Manage 137:221–230

29 Vascular Epiphytes and Their Potential as a Conservation Tool in Pine-Oak Forests of Chiapas, Mexico

J.H.D. Wolf and A. Flamenco-S.

29.1 Introduction

Approximately 10 % of all vascular plant species worldwide grow predominantly as epiphytes on terrestrially rooted plants (Benzing 1990). Epiphytes occur mainly in tropical areas, and most species are found at mid-elevation on mountains (reviewed by Wolf and Flamenco-S 2003). In tropical America, from Mexico as far south as Colombia, oaks often dominate the mid-elevation forest (Kappelle and Brown 2001). Hence, it is not surprising that epiphytes are considered an important component of oak forests in terms of plant diversity and number of individuals (e.g. González-Espinosa et al. 1991; Kappelle and Brown 2001; Chaps. 6, 7 and 16). The abundance of epiphytes in oak forest is, moreover, facilitated by the suitable bark characteristics of oaks for epiphytes when, for example, compared with pine trees.

Despite the significant proliferation of epiphytes in oak forests, epiphytes are generally not included in systematic forest inventories. This paper is the first to pay special attention to the epiphytes supported by oak trees. The emphasis is on the diversity and biogeography of epiphytes in oak forests, particularly of forests which differ in altitude and amount of rainfall. In addition, we focus on the human impact on the epiphyte vegetation, and investigate if ornamental epiphytes may be employed as a conservation tool.

Our studies took place in the state of Chiapas. Chiapas is one of the botanically best-explored regions in the tropics, and its great physiographic variability results in explicit biotic distribution patterns. Moreover, Chiapas is of interest because the use and disturbance of its pine-oak forest is well-studied (Chaps. 13, 14, 16 and 27).

Ecological Studies, Vol. 185
M. Kappelle (Ed.) Ecology and Conservation of Neotropical Montane Oak Forests
© Springer-Verlag Berlin Heidelberg 2006

29.2 Physiography, Forest Formations and Anthropogenic Disturbance

The state of Chiapas in southern Mexico is situated at 14.5–18.0° N and 90.3–94.5° W, and comprises approximately 75,000 km^2. The climate varies strongly from semi-deserts to wet areas where annual rainfall exceeds 3,500 mm, and from tropical lowlands to temperate mountains.

Oak trees prevail in areas with a pronounced dry season of 3–6 months above 1,000 m elevation where they may form pure stands (Breedlove 1978). Typically, however, oaks grow mixed with pine trees: the pine-oak forest. In mountain areas experiencing a significant amount of precipitation every month of the year, the pine-oak forest is replaced by more diverse montane rain forest and evergreen cloud forest.

The structure of the pine-oak forest relates to the type and amount of anthropogenic disturbance. Oaks are logged only to provide fuel, whereas the associated pine trees provide timber. Forest regeneration is left to natural processes. Oak tree stumps frequently sprout, but pines rely on seed dispersal. The net result of frequent anthropogenic disturbances is often that the dominance of pines over oaks is increased (González-Espinosa et al. 1991; Ramírez-Marcial et al. 2001; Chaps. 14 and 16).

29.3 Epiphyte Diversity, Composition and Distribution

29.3.1 Sampling and Analysis

To assess the epiphyte distribution patterns in the state of Chiapas, we compiled label data from epiphyte specimens in six herbaria known to have relatively large collections from the state of Chiapas. Hemi-epiphytes were included, but the ecologically different heterotrophic Loranthaceae were not. The spatial distribution of species was analysed in a geographical information system (ArcInfo), where the latitudinal and longitudinal positions of species were superimposed on digitised topographic, physiographic, rainfall and vegetation maps which had been prepared at ECOSUR (for more details, see Wolf and Flamenco-S 2003).

In addition to species counts, we used Chao's non-parametric diversity estimator to estimate total species richness in different regions, with help of the statistical program EstimateS (Colwell 1997).

29.3.2 The Chiapas Epiphyte Database

The database of the state of Chiapas comprises 12,276 records, of which 3,923 are in the pine-oak forests and 3,215 in the montane rain forest. Despite the large number of records, the assessment of diversity and distribution patterns of epiphytes in the state would benefit from additional sampling. First, approximately 20 % of all species in the database are known from single collections, suggesting that more species may be found in these forests. Second, it would be helpful if a bias in the botanical collections could be reduced, which shows a preference of botanists for road sides, waterways and nature reserves. In addition, there appears to be a bias for certain taxonomic groups, too (Table 29.1). For example, botanists avoid Cactaceae – at least, this is what the low collection efficiency suggests. Acknowledging that there is room for improvement of the database, we nevertheless consider that the large number of records make it possible to obtain a meaningful assessment of the distribution of pine-oak forest epiphytes in Chiapas.

29.3.3 Epiphytes of the Pine-Oak Forest

In all, we confirm the presence of 720 species of vascular epiphytes in the pine-oak forest, of an estimated total number (S_{Chao}) of 946 species. This value is higher than that previously reported (608 species, Wolf and Flamenco-S 2003) because, in this case, also the records from areas of potential pine-oak forest cover are included, raising the number of records from 2,269 to 3,923. A complete species list is available on request from the first author. The high number of epiphyte species is surprising, compared with the value of 2,110 species for the whole of Peru (Ibisch et al. 1996). Since Mexico is situated at the northern limits of the American tropics, it predictably harbours fewer species of vascular epiphytes than do countries near the Equator (Gentry and Dodson 1987).

Orchids are by far the most diverse taxonomic group, comprising 45 % (322 species) of all species (Table 29.1). Other important groups are ferns and their allies (156 species, 22 %), Bromeliaceae (71 species, 10 %), Araceae (46 species, 6 %) and Piperaceae (40 species, 6 %). The familiar makeup of the epiphyte flora of the pine-oak forest is not notably different from that in other neotropical mountain areas (e.g. Hartshorn and Hammel 1994; Webster and Rhode 2001). However, orchid dominance in Chiapas is less pronounced than in Peru, where nearly two thirds (63 %) of all epiphytes are orchids (Ibisch et al. 1996). The most species-rich orchid genera in the pine-oak forest are *Epidendrum* (52), *Encyclia* (29), *Pleurothallis* (26), *Oncidium* (18), *Maxillaria* (13), *Spiranthes* (12) and *Stelis* (10). In the Bromeliaceae, most species are in *Tillandsia* (50). In contrast to South American florulas, *Guzmania* has only

Table 29.1. Representation of vascular epiphyte families in the pine-oak forest and in the montane rain forest, which includes the pine-oak-*Liquidamber* forest and the evergreen cloud forest[a]

Dry season length	Pine-oak forest 3–6 months					Montane rain forest <3 months				
Forest formation	*N*	$S_{obs.}$	(%)	S_{Chao}	CE	*N*	$S_{obs.}$	(%)	S_{Chao}	CE
Orchidaceae	1,715	322	44.7	441.8	18.8	1,316	334	46.7	438.6	25.4
Pterodophyta	985	156	21.7	181.5	15.8	992	180	25.1	221.1	18.2
Bromeliaceae	397	71	9.9	93.6	17.9	162	50	6.9	94.1	30.9
Araceae	168	46	6.4	64.0	27.4	246	43	6.0	47.0	17.5
Piperaceae	261	40	5.6	64.0	15.3	144	33	4.6	57.5	22.9
Cactaceae	34	16	2.2	28.5	47.1	16	8	1.1	–	50.0
Begoniaceae	58	13	1.8	15.7	22.4	62	14	2.0	18.0	22.6
Clusiaceae	39	11	1.5	–	28.2	45	8	1.1	12.5	17.8
Araliaceae	111	10	1.4	12.0	9.0	72	10	1.4	11.5	13.9
Crassulaceae	31	10	1.4	18.0	32.3	2	1	0.1	–	–
Ericaceae	52	9	1.3	–	17.3	64	10	1.4	10.3	17.8
Marcgraviaceae	13	4	0.6	–	30.8	20	4	0.6	–	25.0
Gesneriaceae	5	3	0.4	–	–	24	6	0.8	–	25.0
Asteraceae	9	2	0.3	–	22.2	20	3	0.4	–	15.0
Burmanniaceae	3	2	0.3	–	–	0	0	0	–	–
Bignoniaceae	2	1	0.1	–	–	0	0	0	–	–
Dioscoreaceae	2	1	0.1	–	–	1	1	0.1	–	–
Moraceae	2	1	0.1	–	–	8	2	0.3	–	–
Onagraceae	28	1	0.1	–	3.6	7	1	0.1	–	–
Solanaceae	8	1	0.1	–	12.5	3	1	0.1	–	–
Cyclanthaceae	0	0	0	–	–	5	4	0.6	–	–
Lentibulariaceae	0	0	0	–	–	2	1	0.1	–	–
Liliaceae	0	0	0	–	–	1	1	0.1	–	–
Rubiaceae	0	0	0	–	–	3	1	0.1	–	–
Total	3,923	720	100	946.2	18.4	3,215	716	100	933.9	22.3

[a] Given are the number of collections (*N*), the number of observed epiphyte species ($S_{obs.}$), the relative contribution to total diversity (%), the total estimated number of species (S_{Chao}) and the collection efficiency (CE), i.e. the number of species encountered per 100 collections (– denotes insufficient data). The montane rain forest includes the evergreen cloud forest and the pine-oak *Liquidamber* forest

one species. In the Pteridophyta, *Polypodium* (32) and *Asplenium* (30) are the most rich in species, followed by *Elaphoglossum* (18), *Hymenophyllum* (15) and *Blechnum* (11). Of the remaining genera, only *Anthurium* (19) and *Peperomia* (37) are present with more than 15 species each.

29.3.4 Epiphyte Distribution Patterns

In the pine-oak forest, epiphyte richness is not measurably different from that of the less seasonal and more humid montane rain forest, which has 716 confirmed species (Table 29.1). This is surprising because wetter forests are believed to support more species of epiphytes (Gentry and Dodson 1987). Apparently, in Chiapas the loss of high-humidity species is compensated by a gain in epiphytes adapted to the seasonal climatic conditions of the pine-oak forest. Accordingly, the pine-oak forest supports a unique epiphyte assemblage with species which are clearly drought-adapted. Examples are the common cactus *Epiphyllum crenatum*, and several *Tillandsia* species which show no, or only weakly developed phytotelma and absorptive foliar trichomes over the entire shoot, i.e. Benzing's (2000) ecological type V, which typically uses the crassulacean acid metabolic (CAM) pathway (Table 29.2).

The annual rainfall within the pine-oak forests varies in the range 800–5,000 mm, but most collection records (3,501 of 3,923) fell in the range 1,000–2,000 mm. In this interval, again there is little variation in richness but there certainly is differentiation in the taxonomic composition of the epiphyte vegetation (Fig. 29.1). Several species show a clear preference for wetter areas (Table 29.2). Drought-adapted *Tillandsia* species show a preference for

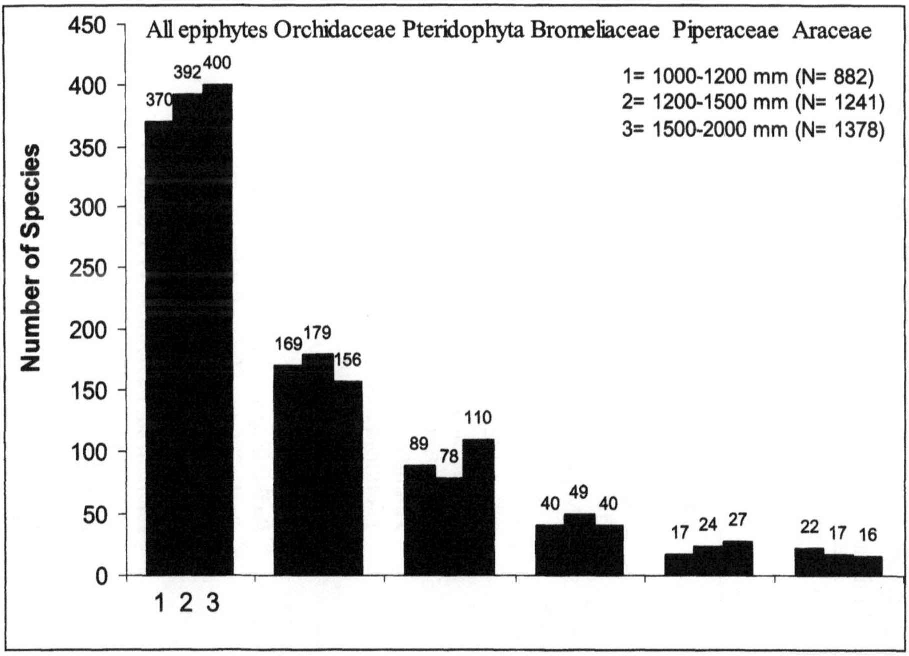

Fig. 29.1. Number of epiphyte species ($S_{obs.}$) in main plant groups of the pine-oak forest per rainfall-cohort

Table 29.2. Relative contribution of species in the database between the pine-oak forest (POF) and the montane rain forest (MRF), and between two annual rainfall-cohorts[a]

		Forest type		Annual rainfall-cohort (mm)	
		POF	MRF	1,000–1,500	1,500–2,000
		(*N*=3,923)	(*N*=3,215)	(*N*=2,123)	(*N*=1,378)
Araceae	*Anthurium chiapasense* Standley	7.6	46.7	–	–
	Anthurium fraternum Schott	35.7	6.2	–	–
	Anthurium huixtlense Matuda	2.5	31.1	–	–
	Philodendron smithii Engler	5.1	31.1	–	–
Araliaceae	*Oreopanax obtusifolius* L. O. Williams	–	–	12.1	94.3
Begoniaceae	*Begonia oaxacana* A. DC.	–	–	16.1	101.6
Bromeliaceae	*Tillandsia caput-medusae* E. Morr.	33.1	3.1		
	Tillandsia carlsoniae L. B. Smith	28.0	0.0	–	–
	Tillandsia lampropoda L. B. Smith	–	–	59.9	7.3
	Tillandsia polystachia (L.) L.	28.0	3.1	–	–
	Tillandsia punctulata Schldl. et Cham.	–	–	57.5	7.3
	Tillandsia seleriana Mez	51.0	6.2	75.2	14.5
	Tillandsia vicentina Standley	30.6	3.1	–	–
Cactaceae	*Epiphyllum crenatum* (Lindley) G. Don	33.1	0.0	–	–
Onagraceae	*Fuchsia splendens* Zucc.	–	–	12.1	166.9
Orchidaceae	*Barkeria spectabilis* Bateman ex Lindley	73.9	0.0	132.5	0.0
	Coelia guatemalensis Rchb. f.	40.8	0.0	–	–
	Elleanthus cynarocephalus (Rchb. f.) Rchb. f.	5.1	65.3	–	–
	Encyclia aromatica (Bateman) Schltr.	35.7	0.0	66.3	0.0
	Encyclia chondylobulbon (A. Rich. & Galeotti) Dressler & Pollard	28.0	0.0	–	–
	Epidendrum cnemidophorum Lindley	–	–	12.1	87.1
	Epidendrum eximium L. O. Williams	–	–	8.1	58.1
	Epidendrum parkinsonianum Hook.	33.1	3.1	–	–
	Epidendrum propinquum A. Rich. & Galeotti	53.5	9.3	12.1	123.4
	Jacquiniella cobanensis (Ames & Schltr.) Dressler		–	40.3	7.3
	Laelia superbiens Lindley	25.5	3.1	–	–

	Lycaste cruenta Lindley	63.7	12.4	102.5	7.3
	Maxillaria variabilis Bateman ex Lindley	–	–	126.1	21.8
	Nageliella purpurea (Lindley) L. O. Williams	89.2	18.7	–	–
	Oncidium leucochilum Bateman ex Lindley	158.0	15.6	262.3	14.5
	Pleurothallis dolichopus Schltr.	5.1	37.3	–	–
	Pleurothallis pubescens Lindley	25.5	3.1	–	–
	Ponera glomerata Correll	30.6	3.1	47.6	7.3
	Rhyncholaelia glauca (Lindley) Schltr.	30.6	0.0	–	–
	Rhynchostele stellata Soto Arenas & Salazar		–	12.1	101.6
	Spiranthes cinnabarina Llave & Lex.	33.1	3.1	49.9	7.3
	Stelis guatemalensis Schltr.	5.1	68.4	–	–
	Stelis microchila Schltr.	0.0	77.8	–	–
	Stelis ovatilabia Schltr.	2.5	43.5	–	–
Piperaceae	*Peperomia campylotropa* A. W. Hill	38.2	0.0	–	–
	Peperomia liebmannii C. DC.	–	–	23.5	181.4
	Peperomia molithrix Trel. & Standley	28.0	3.1	–	–
	Peperomia obtusifolia (L.) A. Dietr.	–	–	0.0	79.8
Pteridophyta	*Asplenium alatum* H. & B. ex Willd.	5.1	31.1	–	–
	Asplenium cuspidatum Lam.	20.4	115.1	–	–
	Asplenium praemorsum Sw.	53.5	3.1	–	–
	Asplenium serra Langsd. & Fisch.	–	–	9.7	72.6
	Asplenium sessifolium Desv.	–	–	4.1	87.1
	Campyloneuron amphostenon (Kunze ex Klotzsch) Fée		–	8.1	101.6
	Elaphoglossum muscosum (Sw.) Moore	–	–	12.1	79.8
	Pleopeltis astrolepis (Liebm.) Fourn.	5.1	31.1	–	–
	Polypodium fissidens Maxon	38.2	6.2	12.1	79.8
	Polypodium lindenianum Kunze	–	–	9.7	72.6
	Polypodium plesiosorum Kunze ex Mett.	–	–	8.1	79.8

[a] The relative contribution is the number of records/total number of records in that vegetation type or rainfall-cohort multiplied by 10,000. The montane rain forest includes the pine-oak-*Liquidamber* forest and the evergreen cloud forest. Only species with a preference for either forest formation or rainfall-cohort are shown, arbitrarily defined as being collected there at least five times more often (or with a difference of 20 points, in the case of a zero value in the other type). Rare species with ten or fewer records are omitted

the 1,000–1,500 mm cohort, in contrast to seven fern species with a bias for the 1,500–2,000 mm cohort.

The environmental variable *altitude* integrates variations in humidity, rainfall, temperature and light, among others, along elevation gradients. In the present case, after an initial increase, the number of epiphyte species decreases at elevations above 2,000 m (Fig. 29.2). Epiphytes showed the same pattern throughout the state of Chiapas, which is in agreement with epiphyte

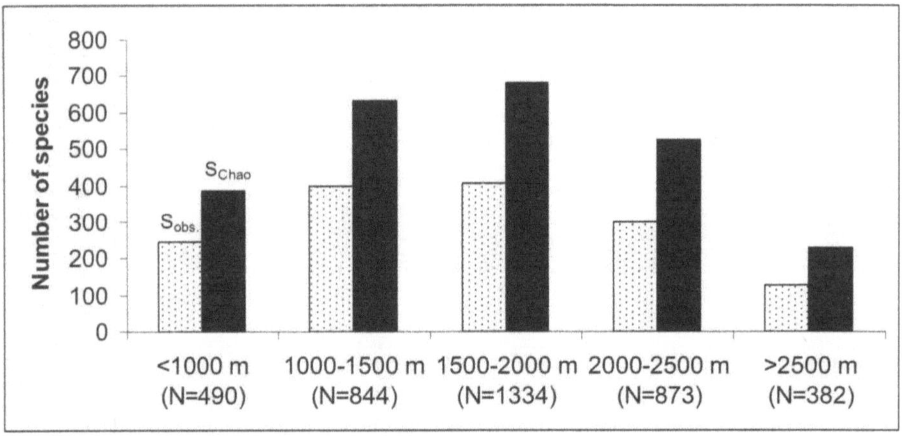

Fig. 29.2. Number of observed epiphyte species ($S_{obs.}$) and number of estimated species (S_{Chao}) in the pine-oak forest per altitudinal interval

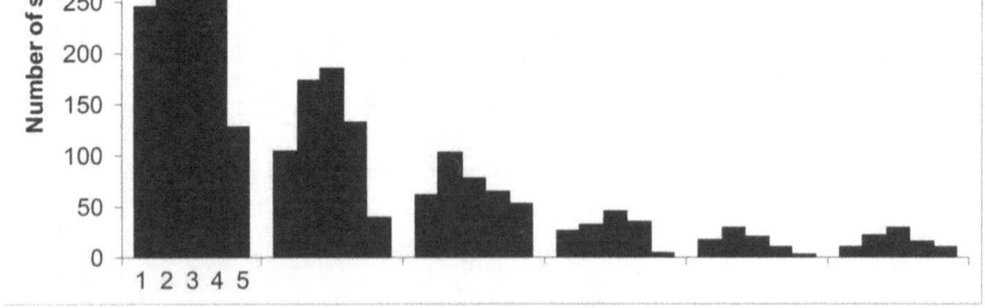

Fig. 29.3. Number of epiphyte species ($S_{obs.}$) in main plant groups of the pine-oak forest per altitudinal interval

diversity patterns on mountains elsewhere in tropical America, reviewed by Wolf and Flamenco-S (2003).

The decrease in species with altitude is due to a decrease in all major groups/families (Fig. 29.3). However, the decrease in Pteridophyta is less than that in other groups, and ferns and related groups thus contribute more to total diversity at higher elevations.

29.4 Pine-Oak Epiphytes and Man

29.4.1 Epiphyte Response to Anthropogenic Disturbance in Pine-Oak Forest

Epiphyte Biomass and Diversity

Oaks and pines are an important source for fuel and timber in Chiapas, where pristine pine-oak forests are today nearly absent. Wolf (2005) extensively studied the response of epiphytes to anthropogenic disturbance. Over a period of several years, approximately 40,000 individual epiphytes were sampled on 560 oak trees in 16 stands of pine-oak forest along a long disturbance gradient in the San Cristóbal de Las Casas area (Central Plateau). Pines generally supported few epiphytes and were therefore not included in this study. In total, 74 species of epiphytes were found. Bromeliads, comprising ten species, were by far the most abundant taxonomic group, contributing more than 95 % to the total epiphyte biomass. Wolf (2005) demonstrated that in the more disturbed forests, the number of epiphyte species and their total biomass were substantially lower. Compared with less disturbed stands, in heavily disturbed forest stands an average of 16.6 versus 27.5 species, and values of 215.5 versus 1,085.0 kg dry weight per hectare were recorded. In part, the lower tree densities and tree sizes in the more disturbed stands explain this pattern. The pattern, however, persisted after correction for differences in the size of oak trees between stands. Thus, epiphyte diversity and biomass also decreased on existing host trees which had not been logged or which had since established.

Epiphyte Species Composition

Based on Wolf's (2005) data, here we report on the qualitative response of the epiphytes to disturbance in more detail. For this purpose, we performed a constrained multivariate gradient analysis using CANOCO software, version 4.0 (Ter Braak 1988). As explanatory variables, we choose only those disturbance variables which explained a significant amount of variation in the epiphyte data (Monte Carlo, $P<0.05$), these being (1) the basal area of trees (diameter at breast height DBH>5 cm), (2) the presence–absence of tree stumps and (3) the relative number of sprouted oak trees. The results are pre-

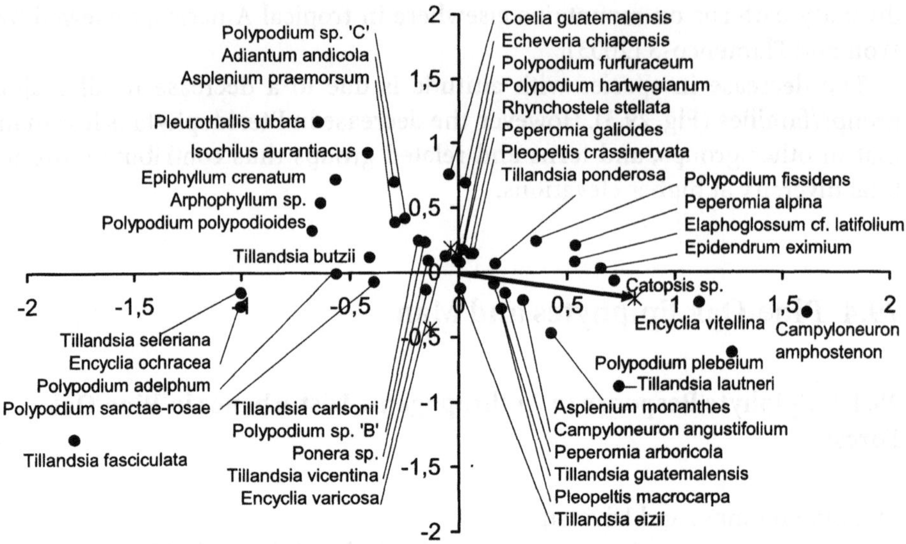

Fig. 29.4. Ordination diagram depicting species distribution on first and second axes generated by a canonical correspondence analysis, CANOCO (Ter Braak 1988). Species biomass values were log-transformed and the axes are scaled according to Hill's scaling. The axes are constrained by three environmental variables which indicate forest disturbance. The *arrow* corresponds to the basal area of the trees (DBH>5 cm) in the forest. The *asterisks* denote the presence of tree stumps (–0.04, 0.19) and the relative number of sprouted oak trees (–0.13, –0.41). All species (74) were used in the analysis, but only those with an effective number of occurrences (N_2 in CANOCO) exceeding 2.0 are shown. The first axis is significantly different from randomly extracted axes (Monte Carlo, $P<0.05$), and this axis explains 15.5 % of the variation between the 16 plots

sented in an ordination diagram in Fig. 29.4. The first ordination axis correlates best with the basal area of the trees (inter-set correlation 0.86). The more drought-tolerant epiphyte species are found in the more disturbed forests at the left side of the diagram, confirming earlier observations (Hietz and Hietz-Seifert 1995; Barthlott et al. 2001). Disturbance-tolerant species include the widespread poikilohydric fern *Polypodium polypodioides*, and several bromeliads which use CAM, e.g. *Tillandsia fasciculata* and *T. seleriana*. By contrast, high basal area forests are characterized by species susceptible to desiccation, such as the phytotelm bromeliad *Tillandsia guatemalensis* (Castro-Hernández et al. 1999).

Epiphyte Response to Disturbance Explained from a Dispersal Assembly Perspective

Interestingly, epiphytes show higher resilience to disturbance in terms of biomass and diversity when trees were selectively logged, as opposed to clear-cutting in a 20–30 year cycle (Wolf 2005). After approximately 25 years, stands at Milpoleta and La Florecilla-3 which had previously been clear-cut, evidenced by the high proportion of sprouted stems and by trunk diameter frequency distributions, had oak basal areas similar to those in less disturbed forest (Table 29.3). Nevertheless, these forests were still poor in epiphytes, suggesting that the arrival of epiphyte propagules from outside the patch is a slow process. In general, tree plantations support only few epiphytes (e.g. Barthlott et al. 2001; Merwin et al. 2003).

Dispersal limitation may also explain why sparing some of the bigger trees in the forest apparently enhances resilience to disturbance as well. This is shown by comparing the stands at Chilil-2 and Mitzitón – these have similar proportions of sprouted trees (20 versus 34 % respectively), and a comparably low oak basal area (8.0 versus 7.2 m^2/ha respectively), indicating a similar disturbance history. Nevertheless, Mitzitón has many more epiphyte species than does Chilil-2 – 24 versus 16 species. The difference is even more obvious for epiphyte biomass. The 35 oak trees at Mitzitón support 108.1 kg dry weight whereas those at Chilil-2 barely 0.8 kg. Wolf (2005) suggested that the difference in type of disturbance explains these contrasting patterns. It seems that, in contrast to Chilil-2, at Mitzitón the bigger trees were spared. At Mitzitón and Chilil-2, the oak basal area of trees with a 1.30-m trunk diameter (DBH) exceeding 45 cm was 2.4 and 0.6 m^2/ha respectively. Large host trees may serve a nearby epiphyte seed source to facilitate epiphyte colonization in successional forests. This 'rescue effect' has obvious implications for wise forest management aiming to conserve the rich and abundant epiphyte vegetation.

The multivariate analysis of the distribution of species in this landscape provides additional evidence for dispersal limitation. Here, the constrained first axis explained only 15.5 % of total variation in the data. Thus, disturbance has only limited value as a predictor of epiphyte species composition. Apart

Table 29.3. Structural characteristics of pine-oak forest stands of comparable oak basal areas at various sites, and the number of epiphytes and their biomass found on 35 oak host trees (extracted from Wolf 2005)

Site	Altitude (m)	Sprouted oaks (%)	Oak basal area (m^2 ha^{-1})	Number of epiphyte species	Epiphyte biomass (kg dry weight)
Milpoleta	2,425	94.3	41.6	13	16.0
La Florecilla-3	2,350	82.9	22.6	14	16.6
San Antonio	2,370	31.4	25.9	29	98.9
El Chivero	2,360	11.4	17.7	34	97.6
Basom-1	2,490	14.3	45.3	23	103.3
Chilil-1	2,300	11.4	20.3	35	74.7

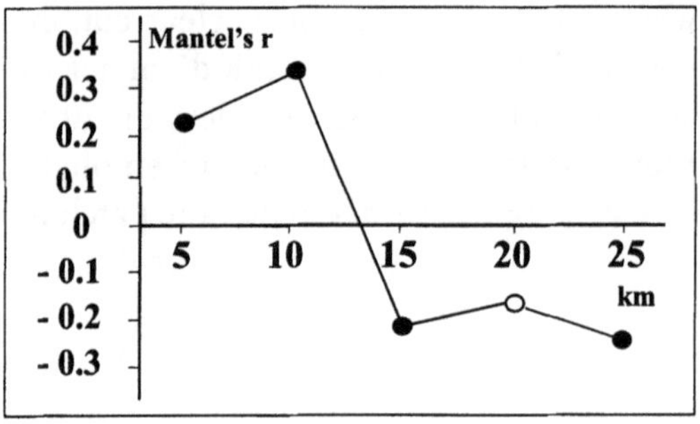

Fig. 29.5. Mantel's correlogram showing epiphyte spatial dependence between sites. The correlations (Mantel's *r*) are calculated from a species similarity matrix, using the Raup and Crick probabilistic coefficient as a measure of species similarity, and an Euclidian distance matrix computed from the geographical coordinates of the 16 sites. Biomass values were log-transformed. *Closed symbols* indicate a significant value of Mantel's *r* ($P<0.05$). The R Package was used for computations (Casgrain and Legendre 2001)

from random variation, historical differences and unknown environmental factors, this may be explained by the relative position of sites in the landscape (Wolf 2005). The nature of this spatial dependence can be visualized in a correlogram (Fig. 29.5). Stands close together, up to a distance of ca. 10 km, are more similar in epiphytes than is the case for sites further apart, possibly due to similar seed rain (cf. the dispersal assembly perspective, Hubbell 2001). For epiphytes which are dispersed mainly by wind, this seems a sensible approach. At the smaller spatial scale of tree crowns, too, epiphytes grow clumped around mother plants (e.g. Bader et al. 2000).

In summary, dispersal and colonization could well explain the poor proliferation of epiphytes in plantation coppices, the rescue effect of big trees on epiphyte community structure, and the similarity of epiphytes in adjacent sites. Whether this is true or false, the important conclusion is that if we want to understand epiphyte distribution patterns within the landscape, attention should be paid to the spatial structures at this scale as well.

29.4.2 Epiphytes as a Tool for Pine-Oak Forest Conservation

Introduction

Conservation efforts in tropical areas have often failed, apparently because they do not provide a direct economic benefit for local stakeholders (e.g. Gullison et al. 2000). Consequently, it has been advocated to exploit the potential of the canopy for the harvesting of non-timber forest products as a means of financial compensation for conservation, a concept known as 'canopy farming' (Neugebauer et al. 1996; term coined by R.A.A. Oldeman at the 1993 Global Forestry Conf Beyond UNCED – Response to Agenda 21, Bandung, Indonesia). Vascular epiphytes are one of the potential forest products in the canopy. Here, we focus on bromeliads, a widely treasured group of ornamental epiphytes which is by far most abundant in the pine-oak forests.

Bromeliads are traditionally used by indigenous tribes of Maya origin for ceremonial purposes and for the decoration of sacred sites. It is only recently that the western world has discovered the potential of bromeliads for (home-) decoration. Indeed, *Tillandsia* is now regarded as a valuable cash crop in several countries. Many plants, however, are being collected from the wild, and the observation that locally many bromeliad populations are declining casts doubts upon the sustainability of this harvesting practice (Holbrook 1991). In an attempt to obtain clarity and to control export, seven species of *Tillandsia* have been added to Appendix II of CITES (Luther 1994). None of these occur in the pine-oak forests. In general, the most severe threat to bromeliads and other epiphytes is probably the loss and degradation of forests. On the Central Plateau, annual deforestation rates for 1974–1984 and 1984–1990 were 1.58 and 2.13% respectively (Ochoa-Gaona and González-Espinosa 2000).

Sound management plans for sustainable exploitation are needed if bromeliads are to be used as a tool for overall forest conservation. Different approaches may be adopted for species which attain high or low densities, the latter being in the majority. Common species may be harvested from the wild if sustainability of yield can be guaranteed, whereas rarer species would rather need to be cultivated. If harvesting of bromeliads is to be used as a tool for biodiversity conservation, it should also be compatible with socio-economic and ecological sustainability, topics which are not addressed in this chapter.

Towards Sustainable Harvesting of Bromeliads from Pine-Oak Forests

Specifically for the bromeliads in the pine-oak forest, Wolf and Konings (2001) have suggested an empirical approach to guarantee sustainability of yield. Working at La Florecilla near San Cristóbal de Las Casas, these authors propose that harvesting should be permitted only from populations with (1) a high population density, which (2) are evenly distributed in space and (3) whose reproductive potential will by and large not be affected by removal.

Harvesting from a small population might negatively affect the local survival of a species (e.g. Young et al. 1996). Wolf and Konings (2001) suggest a minimum population density limit for exploitation amounting to 9,000 large rosettes/ha, which is ten times higher than an apparently stable *Tillandsia circinnata* population in Florida (Benzing 2000).

Regarding the spatial distribution requisite, Wolf and Konings (2001) consider that populations evenly dispersed within a homogeneous habitat are at carrying capacity, the extinction risk being smaller in this case. Presumably also for epiphytic populations at capacity, this implies that essentially the abundance of epiphytes on trees of larger inhabitable size is proportionally greater than that on smaller trees. Predictably, in more disturbed forest the good correlation between tree size and epiphyte abundance should deteriorate. This was indeed observed at La Florecilla, and subsequently confirmed in a more extensive study on epiphytes in disturbed pine-oak forests (Wolf 2005).

To ensure that the removal of rosettes does not affect the reproductive capacity of the population, Wolf and Konings (2001) propose to exploit only that part of the bromeliad population which grows in the lower stratum of the forest. Plants growing near the forest floor are unlikely to play an essential role as providers of progeny for populations of bromeliads in the canopy, the environment to which they seem best adapted to survive (Benzing 1990).

In the 160-ha forest at La Florecilla, *T. vicentina* showed both a satisfactory average population density of ca. 24,000 rosettes>20 cm tall/ha on oaks, and an even spatial distribution. Less than 30 % of the population occurred in the lower stratum of the forest, up to a height of 6 m. Of those, about 2,700 rosettes per ha grew as solitary rosettes (genets), which are more attractive because of their symmetric growth. Based on a 4-year rotation cycle, Wolf and Konings (2001) estimate that at La Florecilla, it is possible to sustainably harvest ca. 700 of such solitary rosettes of *T. vicentina* per ha per year from the understory. This amounts to an annual sustainable yield of 112,000 rosettes, i.e. ca. 3 % of the total number of rosettes from the forest at La Florecilla.

Continuous monitoring should always be part of a management plan, since these thresholds – although strict – are nonetheless arbitrary, and often presumptions need to be tested in the field. For example, plants harvested from the lower stratum are presumably replaced naturally but the rate at which this occurs, and impacts on other forest organisms have to date not been investigated. Nevertheless, the estimated high number of plants which may be harvested with a sustainable yield raises hope that some natural bromeliad populations may be employed to help protect the forest by generating income to local stakeholders, without affecting forest integrity.

Cultivation of Bromeliads in Pine-Oak Forests

Most bromeliad species develop small populations for which cultivation is the only option if these species are to be exploited in a sustainable manner. The cultivation of ornamental bromeliads has 'exploded' over the last 10–15 years, especially in the temperate western world. In The Netherlands, one bromeliad grower alone produces over 20 million plants annually, mostly for export, even to some tropical countries (Manzanares 2002). However, most of the canopy bromeliads show physiological dependence on CAM, and those species take too long to reach maturity to be commercially viable in temperate regions. One way bromeliad growers in The Netherlands overcome this constraint is through the development of new varieties, thereby creating an own market. Energy restrictions in tropical areas are less severe, and several native CAM bromeliads are now being produced in countries such as Colombia, Ecuador and Guatemala (personal observations). This includes many species from the pine-oak forest in Chiapas, for example, *Tillandsia butzii*, *T. fasciculata*, *T. flabellate* and *T. juncea*.

To counteract the slow growth in the wild exhibited by *Tillandsia eizii*, another attractive epiphytic bromeliad from the pine-oak forest, in vitro culture techniques have been applied to propagate and enhance the early stages of development (Pickens et al. 2003). The cultivation of *T. eizii* is urgently needed, since this 'sacred plant' is heavily collected from the forest by indigenous people. The protocols of Pickens (2003) have enhanced seed germination and increased growth rate tenfold. Tissue-cultured plants were successfully transplanted and have exhibited normal morphology until flowering after 4 years. In the wild, this large phytotelm semelparous species is estimated to require at least 15 years to flower (Wolf, unpublished data).

29.5 Conclusions

The pine-oak forests of Chiapas are rich in epiphytes, harbouring 720 confirmed species. The epiphyte vegetation is uunique and adapted to the marked seasonal climate. Anthropogenic disturbance of the pine-oak forest reduces the richness and abundance of epiphytes, not only in terms of surface area but also on the remaining or re-growing trees, and causes a shift towards more drought-resilient species. Epiphytes show resilience to disturbance when trees are selectively logged, rather than being periodically clear-cut. Notably, the epiphyte community is less affected when some of the larger trees are spared which may serve as nearby epiphyte seed source in the successional forest. Nearby sites are more similar in epiphytes than are sites further apart. Hence, epiphyte protective reserves should be well distributed over a physiographically uniform region. Finally, we conclude that in some areas it may well be possible to harvest up to 700 solitary high-quality rosettes per hectare in a sustainable way. This prognosis raises hope for a successful employment of epiphytic bromeliad populations as a tool to help conserve pine-oak forests.

Acknowledgements We thank Juan Castillo, José L. Godónez-A., Teresa Santiago-V. and, in particular, Guadalupe Olalde who compiled most of the database information. Miguel Ángel Castillo and Dario Navarrete at the ECOSUR-GIS laboratory are acknowledged for providing digitised information on the distribution of rainfall, forest formations and physiographic regions. We also thank R.A.A. Oldeman for permitting free use of the term 'canopy farming'. Financial support was provided by El Colegio de la Frontera Sur, ECOSUR, the Comisión Nacional para el uso y Conocimiento de la Biodiversidad, CONABIO grants B060 and L050, and the Stichting Het Kronendak, which also provided valuable scientific and other advise.

References

Bader M, van Dunne HJF, Stuiver HJ (2000) Epiphyte distribution in a secondary cloud forest vegetation; a case study of the application of GIS in epiphyte ecology. Ecotropica 6:181–195

Barthlott W, Schmit-Neuerburg V, Nieder J, Engwald S (2001) Diversity and abundance of vascular epiphytes: a comparison of secondary vegetation and primary montane rain forest in the Venezuelan Andes. Plant Ecol 152:145–156

Benzing DH (1990) Vascular epiphytes. Cambridge Univ Press, Cambridge, UK

Benzing DH (ed) (2000) Bromeliaceae: profile of an adaptive radiation. Cambridge Univ Press, Cambridge, UK

Breedlove DE (1978) The phytogeography and vegetation of Chiapas (Mexico). In: Graham A (ed) Vegetation and vegetational history of northern Latin America. California Academy of Sciences (CAS), San Francisco, CA, pp 149–165

Casgrain P, Legendre P (2001) The R package for multivariate and spatial analysis, version 4.0 d.5: user's manual. Département de Sciences Biologiques, Université de Montréal, Montréal, Canada (http://www.fas.umontreal.ca/BIOL/legendre/)

Castro-Hernández JC, Wolf JHD, García-Franco JG, González-Espinosa M (1999) The influence of humidity, nutrients and light on the establishment of the epiphytic bromeliad *Tillandsia guatemalensis* in the highlands of Chiapas, Mexico. Rev Biol Trop 47:763–773

Colwell RK (1997) EstimateS: statistical estimation of species richness and shared species from samples. User's guide and application, version 5 (http://viceroy.eeb.uconn.edu/estimates)

Gentry AH, Dodson CH (1987) Diversity and biogeography of neotropical vascular epiphytes. Ann Missouri Bot Gard 74:205–233

González-Espinosa M, Quintana-Ascencio PF, Ramírez-Marcial N, Gayatán-Guzmán P (1991) Secondary succession in disturbed *Pinus-Quercus* forests in the highlands of Chiapas, Mexico. J Veg Sci 2:351–360

Gullison RE, Rice RE, Blundell AG (2000) 'Marketing' species conservation. Nature 404:923–924

Hartshorn GS, Hammel BE (1994) Vegetation types and floristic patterns. In: McDade LA, Bawa KS, Hespenheide HA, Hartshorn GS (eds) La Selva: ecology and natural history of a neotropical rain forest. Univ Chicago Press, Chicago, pp 73–89

Hietz P, Hietz-Seifert U (1995) Composition and ecology of vascular epiphyte communities along an altitudinal gradient in Central Veracruz, Mexico. J Veg Sci 6:487–498

Holbrook NM (1991) Small plants in high places: the conservation and biology of epiphytes. Tree 6(10):314

Hubbell SP (2001) The unified neutral theory of biodiversity and biogeography. Princeton Univ Press, Princeton, NJ

Ibisch PL, Boegner A, Nieder J, Barthlott W (1996) How diverse are neotropical epiphytes? An analysis based on the 'Catalogue of the flowering plants and gymnosperms of Peru'. Ecotropica 2:13–28

Kappelle M, Brown AD (eds) (2001) Bosques nublados del Neotrópico. Instituto Nacional de Biodiversidad (INBio), Santo Domingo de Heredia, Costa Rica

Luther HE (1994) A guide to the species of *Tillandsia* regulated by Appendix II of CITES. Selbyana 15:112–131

Manzanares JM (2002) Jewels of the jungle. Bromeliaceae of Ecuador, part I. Bromelioideae, Imprenta Marscal, Quito, Ecuador

Merwin MC, Rentmeester SA, Nadkarni NM (2003) The influence of host tree species on the distribution of epiphytic bromeliads in experimental monospecific plantations, La Selva, Costa Rica. Biotropica 35:37–47

Neugebauer B, Oldeman RAA, Valverde P (1996) Key principles in ecological silviculture. In: Ostergaard TV (ed) Fundamentals of organic agriculture. Proc 11th IFOAM Int Sci Conf, Copenhagen, Denmark

Ochoa-Gaona S, González-Espinosa M (2000) Land use and deforestation in the highlands of Chiapas, Mexico. Appl Geogr 20:17–42

Pickens KA, Affolter JM, Wetzstein HY, Wolf JHD (2003) Enhanced seed germination and seedling growth of *Tillandsia eizii* in vitro. Hortscience 18:101–104

Ramírez-Marcial N, Gonzalez-Espinosa M, Williams-Linera G (2001) Anthropogenic disturbance and tree diversity in montane rain forests in Chiapas, Mexico. For Ecol Manage 154:311–326

Ter Braak CJF (1988) CANOCO: an extension of DECORANA to analyze species-environment relationships. Vegetatio 75:159–160

Webster GL, Rhode RM (2001) Plant diversity of an Andean cloud forest - checklist of the vascular flora of Maquipucuna, Ecuador. Univ California Press, Berkeley, CA

Wolf JHD (2005) The response of epiphytes to anthropogenic disturbance of pine-oak forests in the highlands of Chiapas, Mexico. For Ecol Manage 212:376–393

Wolf JHD, Flamenco-S A (2003) Patterns in species richness and distribution of vascular epiphytes in Chiapas, Mexico. J Biogeogr 30:1689–1707

Wolf JHD, Konings CJF (2001) Toward the sustainable harvesting of epiphytic bromeliads: a pilot study from the highlands of Chiapas, Mexico. Biol Conserv 101:23–31

Young A, Boyle T, Brown T (1996) The population genetic consequences of habitat fragmentation in plants. Tree 11:413–418

30 Land Use, Ethnobotany and Conservation in Costa Rican Montane Oak Forests

M. Kappelle and M.E. Juárez

30.1 Introduction

A large number of studies on human impact on tropical mountains underline the vast destructive and often irreversible effects that settlements and inappropriate land use practices may have on local forest resources (e.g., Baker and Little 1976; Budowski 1982; Stadel 1986; Churchill et al. 1995; Kappelle 1996, 2004; Kappelle and Brown 2001; Bewket 2002; Benítez 2003). However, still little is known about the impact of man on tropical highland oak forests, the often unsustainable use, and the utilization of native plants by indigenous peoples and locally dwelling peasants (*ladino* colonists). At many places in Mexico, Central America and Colombia, oak trees have been cut for timber, fuelwood and other uses, and entire oak forests have been converted to pastures and croplands (Kappelle and Juárez 1995; Helmer 2000; Chaps. 16, 17, 21 and 31).

In order to gain a better insight into the past and present use of these forests and their vascular plants, we conducted a case study and assessed land use history, changing trends in agricultural practices, and current ethnobotanical knowledge in the montane oak forest zone of Costa Rica's Talamanca Mountains, with emphasis on the largely cleared Los Santos Forest Reserve, a Human Inhabited Protected Area (HIPA).

30.2 Colonization, Deforestation and Land Use History

The Costa Rican montane oak forest zone has suffered from human intervention since the arrival of the first indigenous peoples over 10,000 years ago. Today's valley of Santa Maria de Dota at 1,500 m elevation is believed to have

Ecological Studies, Vol. 185
M. Kappelle (Ed.) Ecology and Conservation of Neotropical Montane Oak Forests
© Springer-Verlag Berlin Heidelberg 2006

been inhabited in recent pre-Columbian times by a Huetar indigenous tribe known as 'Ota' – hence, the county name Dota from 'De Ota', in the central sector of the Los Santos Forest Reserve (Chinchilla 1987; Ureña 1990). However, it was not until the mid 19th Century that this upland oak forest region became colonized by settlers, with dramatic consequences. During that historic period, landless sons of traditional coffee farmers in search of new land migrated from the Central Valley of Costa Rica, from rural towns around San José and Cartago, to the mountains of the high Talamancas (Rodríguez and Vargas 1988; Ureña 1990; Carrière 1991). They settled especially in what is presently known as the Los Santos Forest Reserve, by that time an area where land was still available without any legal constraints (Kappelle and Juárez 2000). In this way, new rural population centers were established at 1,000–1,500 m elevation, such as Santa Maria de Dota (since 1863), San Pablo de León Cortés, and San Marcos de Tarrazú. During the early and mid 20th Century, new generations of farmers continued to migrate even further southeast, establishing small hamlets known as *caserios*, e.g., Providencia de Dota, San Gerardo de Dota, and Villa Mills. During the 1940s and 1950s, these hamlets started to grow and develop into villages, as the construction of the Panamerican highway (*Carretera Interamericana Sur*) opened up new markets (Schubel 1980; Siles de Guerrero 1980).

The farmers that settled at elevations above 1,800 m – the altitudinal limit of coffee growth in Costa Rica – initially practiced slash-and-burn techniques, extracted timber for fence posts, fuelwood and charcoal, and cultivated crops (maize, legumes) for domestic use. They also gathered blackberries and edible palm hearts locally known as *pejibaye*, and raised dairy cattle and pigs (Kappelle and Juárez 1995; Kappelle et al. 2000). In the upper Savegre Valley, near the village of San Gerardo de Dota, deforestation rates reached their peak between 1950 and 1980, according to a study based on retrospective monitoring and interpretation of historical sets of aerial photographs covering almost 60 years (1941, 1956, 1969, 1984, 1992, and 1998), validated by ground data collected in 1996 and 2001 (van Omme et al. 1997; Acevedo et al. 2002). Deforestation rates increased from an initial 6.6 ha year^{-2} in 1954–1956, to as much as 21.3 ha year^{-2} over the period 1956–1969, the value being 20.2 ha year^{-2} for the period 1969–1984. Between 1984 and 1996, the deforestation rate dropped significantly (0.4 ha year^{-2}), and finally leveled off to almost zero in 2001 (Cháves et al. 2001). As a result, today the valley of San Gerardo de Dota comprises a landscape mosaic of old-growth montane oak forest, successional stages of recovering forests, blackberry fields, apple orchards, and pastures (Fig. 30.1).

The current lower deforestation rate is largely explained by a significant change in land use practices that has taken place since the mid 1970s when the Los Santos Forest Reserve was established. At that time, activities such as traditional logging for timber and charcoal, and forest conversion to pasture were progressively replaced by fruit tree cultivation on former pastures,

Fig. 30.1. Panorama of the western slope (with east orientation) of the Savegre River valley at 2,000–2,300 m elevation near San Gerardo de Dota, Costa Rica. The view shows a landscape mosaic with old-growth montane *Quercus copeyensis*-dominated oak forest along the crests, successional stages of recovering forests at its lower edges, pastures with isolated dark-leaved *Quercus* and gray-leaved *Buddleja* trees, fences of live cypress trees (*Cupressus lusitanica*), and young apple orchards (*Malus pumila*) with uncovered soil at the bottom. Photograph taken by M. Kappelle in 1992

hatching of introduced rainbow trout (*Oncorhynchus mykiss*), and exploitation of ecotourism as non-traditional sources of income. This account is probably one of the first reporting a significant change in human behavior toward a more responsible attitude in conserving human-inhabited neotropical montane oak forest.

30.3 Altitudinal Zonation of Agroecological Belts

The mid-montane, oak-dominated highlands of Costa Rica's Talamanca Range (1,000–3,000 m) show a sequential altitudinal zonation of different agroecological belts (Kappelle and Juárez 1995), similarly to those of the Andes (Zimmerer 1999; Fig. 30.2). Coffee plantations have replaced most of the premontane and lower montane forest zones, and dominate at 1,000–1,800 m – the upper limit of coffee growth in the country. Examples are found in the Pacific counties of León Cortés, Tarrazú, Dota, Pérez Zeledón, and Coto Brus.

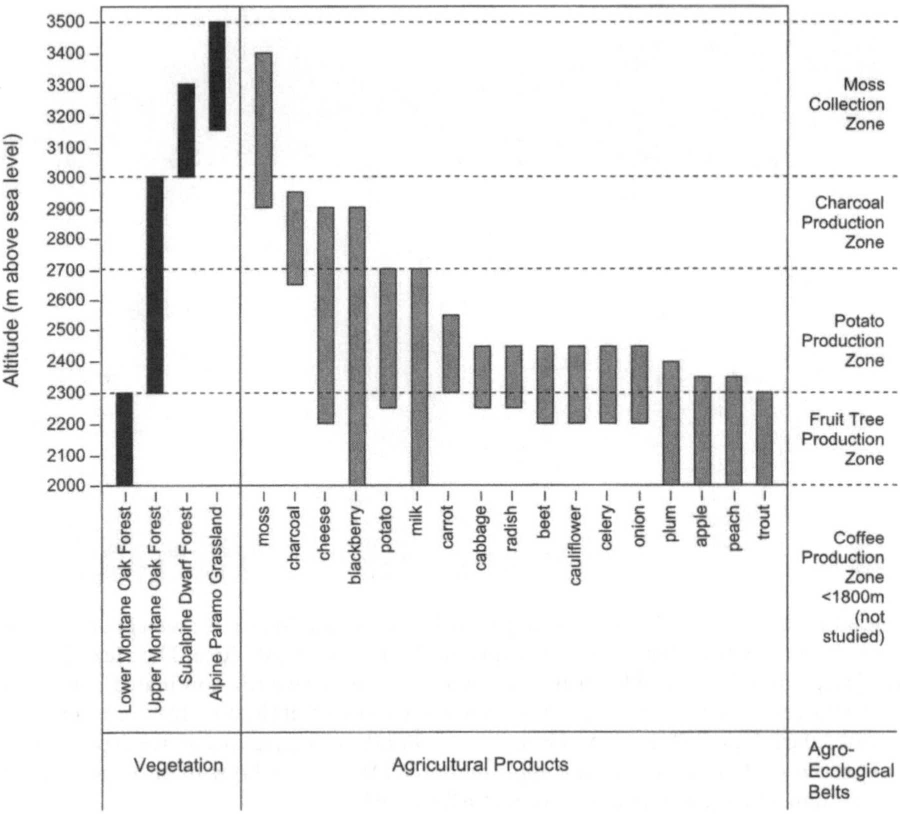

Fig. 30.2. Altitudinal zonation of mountain vegetation, agricultural products, and agroecological belts above 2,000 m elevation along the slopes of the Savegre River valley, Costa Rica. The highest altitude corresponds to the summit of the La Muerte peak at 3,491 m

Climbing up the mountains, starting at the upper limit of the coffee belt, one passes through a lower montane oak forest zone (1,800–2,300 m) that has largely been cleared during the mid 19th Century. After forest removal, these lands were initially used as pastures. However, during the mid 1970s, they were converted into fruit orchards of introduced species such as apple, peach, and plum (Kappelle and Juárez 2000). Other lands were kept as blackberry orchards, and planted with native *Rubus* species. The agricultural lands around the Dota County villages of Trinidad, Copey, Providencia and San Gerardo, in the heart of the Los Santos Forest Reserve, offer a good example of how today's fruit tree plantations thrive.

Higher up the mountain, at 2,300–3,000 m, is found the upper montane forest belt. In its pristine state, this belt is dominated by oak species such as *Quercus seemannii*, *Q. copeyensis* (now known as *Q. bumelioides* – K.C. Nixon, personal communication, Chap. 1) and *Q. costaricensis* (Chaps. 4 and 10).

However, many oak forest patches at 2,300–2,700 m elevation have been removed over the last century, and have been replaced by annual crops (vegetables) such as potato (two harvests a year), beet, carrot, onion, cauliflower, cabbage, celery, lettuce and radish (Fig. 30.2). Many of these vegetable crops are grown applying high inputs of fertilizer and pesticides (herbicides, fungicides, nematicides, and insecticides). They are artificially irrigated by means of overhead sprinklers during the dry season (January–May). Vegetables are largely sold at markets in San José, Cartago, and San Isidro del General (Kappelle and Juárez 1995).

Even higher, at elevations of 2,700–3,000 m, oak forest has been cleared for charcoal production (Schubel 1980). This was still a major source of income for farmers during the early 1990s (Chap. 31), and beyond (M.E. Juárez, personal observation). Charcoal is produced mainly on the basis of oak (*Quercus seemannii*, *Q. copeyensis* and *Q. costaricensis*), and to a lesser extent, alder (*Alnus acuminata*), *Buddleja nitida*, *Drimys granadensis*, *Nectandra* spp., *Ocotea* spp., Podocarpaceae and *Weinmannia pinnata* (Kappelle and Juárez 1995). Today, the production of charcoal from living trees is prohibited by national legislation. Therefore, charcoal producers currently uncover decaying oak logs still scattered around in pastures (Chap. 31). They represent evidence from historic clearing campaigns that often took place in the 1950s, 1960s and early 1970s.

In the subalpine dwarf forest and alpine paramo grassland environments, poor peasants gather mosses and hepatics for ornamental arrangements, especially in the period before the Christmas season. These non-vascular plants are locally important as non-timber forest products (NTFPs) of considerable commercial value (Siles de Guerrero 1980; Romero 2002). They are sold to truck drivers who take them to urban markets ('*ferias*') in Costa Rica's Central Valley.

As clearing of forested land became illegal in the Los Santos Forest Reserve, after its establishment in 1975, loggers (*madereros*), charcoal producers (*carboneros*) and cattle-farmers (*ganaderos*) had to change their non-sustainable land use practices. However, only few farmers and their rural families were able to successfully develop alternative socioeconomic activities by growing fruit trees, establishing artificial ponds for hatching of introduced rainbow trout (*Oncorhynchus mykiss*), or initiating activities in the field of ecotourism (e.g., bird-watching tours). A large number of farmers, however, lacked technical knowledge and skills to implement other land use systems, neither did they have the capital needed (including bank loans) for initial investments, nor could they count on appropriate social networks in order to achieve their production goals. Today, many of them still produce charcoal, or grow blackberries in the upper parts of the watersheds and try to seek new, economically remunerative agricultural activities in order to improve their relatively low living standards (Kappelle and Juárez 2000).

30.4 Ethnobotany

On all (sub)tropical continents, montane forest products and services have extensively been used for many centuries and probably millenniums, to maintain local populations. Traditionally, but also in modern times, rural communities have gathered numerous timber and non-timber forest products from these forests, including food, fodder, fiber, fuel, medicines, dyes, gums, oils, antioxidants, spices, poisons, ornamental plants, and pets. The use of plants for a variety of purposes is still a common phenomenon in neotropical montane oak forests, such as those that we studied in Costa Rica.

An ethnobotanical survey among farmers (*campesinos*) in the Costa Rican village of San Gerardo de Dota in Savegre River valley (Los Santos Forest Reserve) demonstrated that 32.0% of the vascular plant flora (189 of 590 species) known from the area is perceived by farmers as useful (Kappelle et al. 2000). The number of useful plant species per farm varied strongly in the range 22–117, depending largely on the origin of the farmer family, the age of the interviewees, and the time of arrival in the Savegre Valley. In general, elder farmers originating from neighboring valleys and living several decades in the area had a greater knowledge of useful plants than was the case for colonists of younger age who recently immigrated from regions further away. Similarly, about 57% of the useful plant species was known by only one or two *campesino* families, whereas widely known species were few and corresponded mainly to introduced fruit and timber trees.

Plant families that were found to be richest in useful species were Poaceae (13 species), Asteraceae (12), Rosaceae (9), Lauraceae (8), Solanaceae (8), Apiaceae (6), Cucurbitaceae (6), Verbenaceae (6), Brassicaceae (5), and Fabaceae (5). Of the 189 useful plant species (100%), 23.8% was used for medicines, 39.7% for food, and 24.3% for construction (timber) or as combustible (fuelwood, charcoal; Fig. 30.3). Types of less important use included dye, ornamental, fodder, gum, oil, and poison. A total of 61.9% of all plants had only one kind of use. The introduced and exotic trees *Cupressus lusitanica* (cypress) and *Eucalyptus globulus* showed the highest diversity in use types (7), together with the native tree *Alnus acuminata* (alder). Trunks (53%) and fruits (47%) were the main plant species' organs used, followed by leaves (33%) and branches (30%). Over 27.5% of all plants were used on a daily basis, and 34.9% only occasionally. About 11.6% was very rarely used (Kappelle et al. 2000).

At present, use of traditional and native species is becoming less common in San Gerardo de Dota. Trends in use frequency are in favor of introduced and economically important species. Therefore, it is questionable if folk knowledge of useful native plants – especially medicinal plants – will remain common good or disappear on a short or medium term. Today, the farmers' village of San Gerardo rapidly transforms from a rural hamlet depending on

Fig. 30.3a–c. Relative number (in percentages) of vascular plant species (n=189) distributed as **a** use type category, **b** type of useful plant organ, and **c** number of use types per plant species

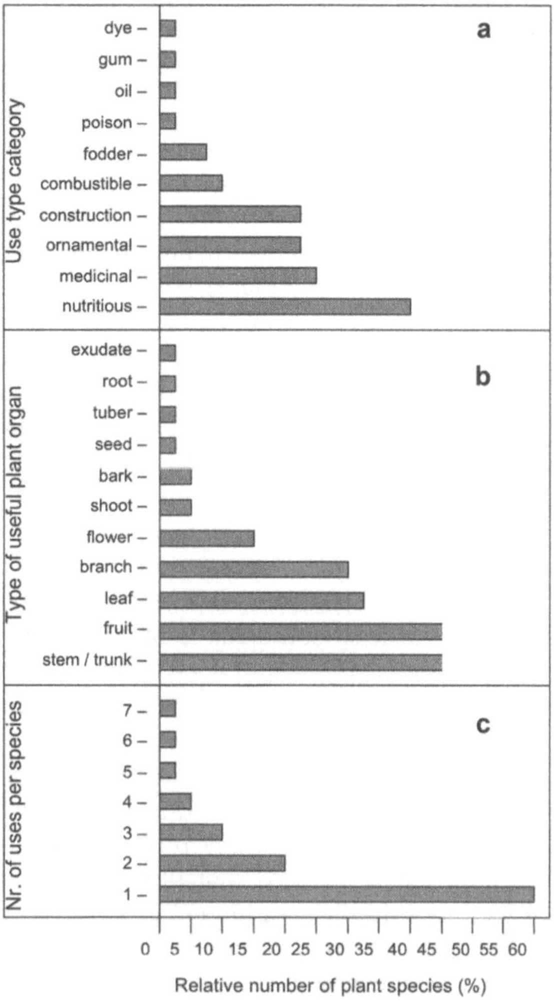

agricultural subsistence into a booming town where fruit and trout export and ecotouristic enterprises dominate.

30.5 Protected Areas Preserving Montane Oak Forests

In 1982, the UNESCO designated the transboundary La Amistad Biosphere Reserve in mountainous eastern Costa Rica and western Panama (Talamanca highlands). The 612,570-ha Costa Rican sector of this oak forest-dominated Biosphere Reserve included the 50,920-ha Chirripó National Park established in 1975, the 58,495-ha Tapantí–Macizo de la Muerte National Park established in 1999, the Costa Rican part of the 193,929-ha La Amistad International Park (PILA) established in 1982, the Las Tablas Protected Zone, as well as a number of Indigenous Reserves. The 62,000-ha Los Santos Forest Reserve (LSFR), established in 1975, was not included in this biosphere reserve, as it was considered to be in an advanced state of degradation. However, during the 1990s the LSFR has been considered a vital buffer zone for the La Amistad Biosphere Reserve, due to its strategic location between the urban centers of the Central Valley to the north and west, and the densely forested Biosphere Reserve to the east (Kappelle and Juárez 1994).

In 1983, a large part of the La Amistad Biosphere Reserve was designated as a World Heritage Site. Ten years later it was categorized as a Center of Plant Diversity and as a key Endemic Bird Area (Chaverri et al. 1994; Kappelle and Juárez 1994). It also serves as one of the main cores of the Mesoamerican Hotspot of Biodiversity (Myers et al. 2000). Since 2003, Amistad includes Costa Rica's eleventh Ramsar site – the unique high-altitude peat lands locally known as the *Turberas de Talamanca* (Ramsar site no. 1286).

Due to these national and international conservation action strategies, over 60 % (i.e., 6,489 km^2) of Costa Rica's upland territory (>1,000 m elevation) had at least some status of protection in 1999 (Kappelle and Juárez 2000). Knowing that about 80–90 % of these protected highlands had a dense forest cover during the mid 1990s (Kappelle and Juárez 2000), we may conclude that Costa Rica's montane oak forests are currently among the best protected in the Neotropics. However, lower montane oak forest patches below 2,000 m, immediately east and south of the highly populated metropolitan Central Valley, still lack any protected status (Castro and Kappelle 2000). Therefore, it is stressed that these threatened areas should receive major attention from conservation policy makers over the coming years.

At the same time, our assessment demonstrates that deterioration of the ecological integrity of these tropical montane oak forests may positively be counteracted by new socioeconomic alternatives implemented by a rural, visionary population of mountain people, who are well aware of the strong relation between the health of the ecosystem they inhabit and the future perspectives of their population (see also Messerli and Ives 1997). This trend is a hopeful sign for numerous other tropical montane forest sites that still suffer from continued deforestation, biodiversity loss, and degradation of ecosystem goods and services, which doubtlessly affect the long-term future of mankind (Aldrich et al. 1997; Kappelle and Brown 2001; Kappelle 2004).

30.6 Involving Local People in Conservation Action

In an even broader perspective, we may conclude that the management of protected areas in tropical montane forest environments has undergone significant changes over the last two decades (Aldrich et al. 1997; Kappelle and Brown 2001; Bubb et al. 2004; Kappelle 2004). The most important aspect of this transformation concerns the involvement of local people in protected area management. Whereas in the 1960s and 1970s national parks, wildlife reserves, and other protected nature areas were places managed only by guards under national command – as was the case in the Costa Rican Los Santos Forest Reserve (LSFR) discussed in this chapter – today, the participation of local populations in conserving protected areas is considered a prerequisite for successful management in many regions.

However, not only protected areas have benefited from the active involvement of local people in management. In many places, the people themselves have profited, while conserving natural resources and reducing poverty in an integrated manner. A striking example of this new and successful approach is the development of ecotouristic activities, like in the LSFR. Here, for instance, local people have been able to make a living out of protected areas while preserving endangered species such as the resplendent quetzal (*Pharomachrus mocinno*), a bird of mythical significance for the locals and for foreign birdwatchers, too (Chap. 25). Today, local people treasure the presence of magnificent trees and mammals as an asset in attracting North American and European nature visitors. Colonists came to understand that a huge tree has much more economic value when it can be admired over decades by international ecotourists, rather than being chopped down and 'once and for all' sold as timber at the local market. This example demonstrates the ability of protected areas to contribute to poverty alleviation locally as well as regionally, as people's willingness to protect biodiversity in situ is increasing (Adams et al. 2004).

If we understand that both local people as well as civil society as a whole can directly benefit from environmental goods (e.g., food, fodder, fiber, water, soil) and services (e.g., soil erosion control, flood prevention, climate regulation) provided by biodiversity conserved in situ, then we are on the right path toward successful conservation action (Daily 1997; MA 2005; Chap. 33). Indeed, it is the integration of biodiversity conservation with sustainable development that helps to directly alleviate poverty among the poorest and most marginalized on Earth (Adams et al. 2004).

30.7 Linking Biodiversity Conservation to Poverty Alleviation

The key to successful poverty alleviation is participatory conservation, particularly in developing and transitional economies (Adams et al. 2004). Here, ecotourism directly focused at supporting the poor can play an important role. Policy makers and other stakeholders involved in ecotourism should be specifically working on developing ways for tourism to genuinely alleviate poverty in and near economically poor but biodiverse, human-inhabited protected areas (HIPAs). Thus, a viable and attractive tourism product is necessary if people are to visit, and benefits to be felt by all involved – poor local people, national and international visitors, the tourism industry, policy makers, and conservationists alike.

Today, protected area systems in developing and transitional countries aim to protect not only biodiversity but also the environmental goods and services this provides (Daily 1997; MA 2005). As such, local people wish – and often need – to have access to these natural goods and services. Unfortunately, strikingly different interests among stakeholders often cause conflicts, especially at the borders of protected areas. Protected area managers frequently have goals and objectives that contrast strongly with those of local dwellers. Therefore, over the years, state departments as well as private organizations have included different levels of buffer zones around core areas, generally following the UNESCO Man and Biosphere Principles. The LSFR in Costa Rica is an example of this approach, as it serves the Amistad Biosphere Reserve as a key buffer zone (Kappelle and Juárez 1994).

In the buffer zones where protection and extraction of resources are often combined, integrated conservation-and-development projects should be carried out. The Spanish Government-funded ARAUCARIA project in LSFR's Savegre River valley is a good example of such an approach (Acevedo et al. 2002; Chap. 25). It is here, at the edge of the strictly protected area and the less protected multipurpose area, that the greatest successes in addressing the dual issues of biodiversity protection and poverty alleviation may be achieved. Particularly, the involvement of local people, such as former hunters and gatherers, in the preservation of 'their' protected area is crucial. Therefore, empowerment of local communities should be promoted and enhanced wherever possible, particularly as poor local people often depend for their survival on the use of natural resources such as found in Costa Rica's upper montane oak forests (Kappelle and Juárez 1995, 2000).

If we were to empower local people as 'owners' of protected land they use and concurrently conserve, the elimination of poverty as well as the strengthening and conservation of biodiversity may go hand in hand. Stimulating participatory management of biodiversity in situ is the key, and should therefore be promoted by policy makers and decision makers alike. However, capacity building, and particularly the transfer of skills, knowledge and technology to local people – the future custodians of biodiversity – is a prerequisite. At the same time, local people will not need to abandon their villages and migrate to cities in search of work and income – as is happening today in Costa Rica's LSFR – as local natural resources will not necessarily diminish but rather remain intact as a result of more sustainable management. Women and youngsters can play a particularly important role in this new approach, and should be encouraged to do so. Gender issues should receive special attention from conservation planners and decision-makers alike.

30.8 Macroeconomic Trends, Conventions and Conservation Implications

Other points of interest are macroeconomic processes and international trade relations. All these issues may affect biodiversity while being intended to alleviate poverty in first instance. They deserve our special attention. The liberation of trade through free-trade agreements (e.g., the Central American Free Trade Agreement, CAFTA), and the prevention of new trade barriers may help to positively impact poverty reduction and biodiversity conservation as well, but should be put forward within the framework of UN conventions, especially the Convention on Biological Diversity (CBD), targeted for the year 2010 (Balmford et al. 2005), and the associated CoP-7 procedure related to national gap assessments for protected area networks in CBD-signatory countries. Socioeconomic inequity among countries, peoples, regions, and localities should be minimized while preserving biodiversity. The participative management of protected areas may form a sustainable basis for achieving these larger goals, especially as we are to reduce biodiversity losses significantly by the year 2010 – one of the CBD's main targets – while offering a sustainable livelihood to local people, the custodians of biodiversity, today and in the future.

30.9 Conclusions

We conclude that, if we are to preserve a large part of the remaining neotropical montane oak forest and its variety of life as expressed in its genes, species and ecosystem types in the long term, we will need to elaborate a conservation strategy in which not only networks of protected core areas, buffer zones, and corridors form a fundamental component, but also participatory planning strategies in which different local and regional stakeholder groups and decision-makers are entirely involved, in order to establish a broad-based, consensus-oriented conservation framework (Calderón et al. 2004; Kappelle 2004).

Active restoration will be one of the key conservation strategies in recovering the neotropical montane forest landscape matrix over time (Holl and Kappelle 1999; Peterson and Haines 2000; Wijdeven and Kuzee 2000). The ecoregional and site conservation planning strategies applied by The Nature Conservancy and its partners are a good example of such an approach (Benítez 2003; Groves 2003; Calderón et al. 2004). Such an actor-oriented conservation planning process is particularly vital as a prerequisite for long-term conservation and sustainable use, for it is the recognition and valuation of the whole set of environmental goods and services offered by these forests to local and regional peoples (Chap. 33) – and strategies including compensation payments to forest owners for these goods and services ('easements' or *servidumbres*) – that will make its conservation economically successful (Daily 1997; Kappelle 2004; Balmford et al. 2005).

Acknowledgements We are very grateful to the farmers and their families in the Los Santos Forest Reserve who collaborated unconditionally in the course of this study, and whose hospitality is much appreciated. A.M. Cleef continuously supported this work during the last 15 years. We thank Guillaume Avertin (Université Pierre et Marie Curie, Paris, France) for field assistance, and staff at Costa Rica's National Museum, universities (UNA, UCR), and INBio for herbarium facilities and species identification. Main funding was provided by The Netherlands Organization for Scientific Research (NWO-WOTRO), the University of Amsterdam, and the European Commission (Erasmus Programme). Research permission was granted by Costa Rica's Ministry of Environment and Energy (MINAE).

References

Acevedo H, González J, Bustamante J, Paniagua L, Cháves R (2002) Ecosistemas de la cuenca hidrográfica del Río Savegre, Costa Rica. In: Kappelle M, Castro MV (eds) Serie Ecosistemas de Costa Rica, vol 1. Instituto Nacional de Biodiversidad (INBio), Santo Domingo de Heredia, Costa Rica

Adams WM, Aveling R, Brockington D, Dickson B, Elliott J, Hutton J, Roe D, Vira B, Wolmer W (2004) Biodiversity conservation and the eradication of poverty. Science 306:1146–1149

Aldrich M, Billington C, Edwards M, Laidlaw R (1997) Tropical montane cloud forests: an urgent priority for conservation. UNEP-WCMC, Cambridge, UK, WCMC Biodiversity Bull 2

Baker PT, Little MA (1976) Man in the Andes: a multidisciplinary study of high altitude Quechua. Dowden, Hutchinson and Ross, Stroudsburg, PA

Balmford A, Bennun L, Ten Brink B, Cooper D, Côté IM, Crane P, Dobson A, Dudley N, Dutton I, Green RE, Gregory RD, Harrison J, Kennedy ET, Kremen C, Leader-Williams N, Lovejoy TE, Mace G, May R, Mayaux P, Morling P, Phillips J, Redford K, Ricketts TH,

Rodríguez JP, Sanjayan M, Schei PJ, van Jaarsveld AS, Walther BA (2005) The convention on biological diversity's 2010 target. Science 307:212–213
Benítez S (2003) The Condor Biosphere Reserve in Ecuador: use of the functional landscape approach to conservation of montane ecosystems. Mount Res Dev 23(3):212–214
Bewket W (2002) Land cover dynamics since the 1950s in the Chemoga watershed, Blue Nile Basin, Ethiopia. Mount Res Dev 22(3):263–269
Bubb P, May I, Miles L, Sayer J (2004) Cloud forest agenda. UNEP-WCMC, Cambridge, UK
Budowski G (1982) The socioeconomic effects of forest management on the lives of people living in the area: the case of Central America and some Caribbean countries. In: Hallsworth EG (ed) Socioeconomic effects and constraints in tropical forest management. Wiley, New York, NY, pp 87–102
Calderón R, Boucher T, Bryer M, Sotomayor L, Kappelle M (2004) Setting biodiversity conservation priorities in Central America: action site selection for the development of a first portfolio. The Nature Conservancy (TNC), San José, Costa Rica
Carrière J (1991) The political economy of land degradation in Costa Rica: Latin American development, rethinking the social theory. Int J Polit Econ 21(1):10–30
Castro MV, Kappelle M (2000) Mapping and monitoring mountain development in Costa Rica. In: Price M, Butt N (eds) Forests in sustainable mountain development: a state of knowledge report for 2000. CABI, Oxon, UK, IUFRO Res Ser 5, pp 12–13
Chaverri A, Herrera B, Herrera-McBryde O (1994) La Amistad Biosphere Reserve, Costa Rica – Panamá. In: Davis SD, Heywood VH, Herrera-McBryde O, Villa-Lobos J, Hamilton AC (eds) Centers of Plant Diversity: a Guide and Strategy for Their Conservation, vol 3. The Americas. World Conservation Union (IUCN), World Wildlife Fund (WWF) and Smithsonian Institution (SI), Washington, DC, pp 209–214
Cháves R, Bustamante J, Paniagua L, González J, Acevedo H, Castro MV, Kappelle M (2001) Caracterización de la vegetación en la cuenca hidrográfica del Río Savegre, Costa Rica. Instituto Nacional de Biodiversidad (INBio), Ministerio de Ambiente y Energía (MINAE), Museo Nacional, y Proyecto Araucaria (Cooperación Española), Santo Domingo de Heredia, Costa Rica, four CD-ROMs
Chinchilla E (1987) Atlas cantonal de Costa Rica. Instituto de Fomento y Asesoría Municipal (IFAM), Imprenta Nacional, San José
Churchill SP, Balslev H, Forero E, Luteyn JL (eds) (1995) Biodiversity and conservation of Neotropical montane forests. New York Botanical Garden Press, Bronx, NY
Daily GC (1997) Nature's services: societal dependence on natural ecosystems. Island Press, Washington, DC
Groves C (2003) Drafting a conservation blueprint: a practitioners guide to planning for biodiversity. Island Press, Washington, DC
Helmer EH (2000) The landscape ecology of tropical secondary forest in montane Costa Rica. Ecosystems 3(1):98–114
Holl KD, Kappelle M (1999) Tropical forest recovery and restoration. Trends Ecol Evol 14(10):378–379
Kappelle M (1996) Los bosques de roble (*Quercus*) de la Cordillera de Talamanca, Costa Rica: biodiversidad, ecología, conservación y desarrollo. Instituto Nacional de Biodiversidad (INBio), Santo Domingo de Heredia
Kappelle M (2004) Tropical montane forests. In: Burley J, Evans J, Youngquist JA (eds) Encyclopedia of Forest Sciences, vol 4. Elsevier, Oxford, UK, pp 1782–1793
Kappelle M, Brown AD (eds) (2001) Bosques nublados del Neotrópico. Instituto Nacional de Biodiversidad (INBio), Santo Domingo de Heredia
Kappelle M, Juárez ME (1994) The Los Santos Forest Reserve: a bufferzone vital for the Costa Rican La Amistad Biosphere Reserve. Environ Conserv 21(2):166–169

Kappelle M, Juárez ME (1995) Agroecological zonation along an altitudinal gradient in the montane belt of the Los Santos Forest Reserve in Costa Rica. Mount Res Dev 15(1):19–37

Kappelle M, Juárez ME (2000) Mountain forests, biodiversity and people in Costa Rica. In: Price M, Butt N (eds) Forests in sustainable mountain development: a state of knowledge report for 2000. CABI, Oxon, UK, IUFRO Res Ser 5, pp 38–46

Kappelle M, Avertin G, Juárez ME, Zamora N (2000) Useful plants within a *campesino* community in a Costa Rican montane cloud forest. Mount Res Dev 20(2):162–171

MA (2005) Living beyond our means: natural assets and human well-being. Millennium Ecosystem Assessment (MA), World Fish Center, United Nations Environment Programme (UNEP), Kuala Lumpur, Malaysia

Messerli B, Ives JD (1997) Mountains of the world: a global priority. Parthenon, London

Myers N, Mittermeier RA, Mittermeier CG, da Fonseca GAB, Kent J (2000) Biodiversity hotspots for conservation priorities. Nature 403:853–858

Peterson CJ, Haines BL (2000) Early successional patterns and potential facilitation of woody plant colonization by rotting logs in premontane Costa Rican pastures. Restor Ecol 8(4):361–369

Rodríguez S, Vargas E (1988) El recurso forestal en Costa Rica: políticas públicas y sociedad. Editorial Universidad Nacional (EUNA), Heredia, Costa Rica

Romero C (2002) Commercially harvested non-vascular epiphytes in a montane tropical forest: ecology, effects of logging and management. MSc Thesis, University of Florida, Gainesville, FL

Schubel RJ (1980) The human impact on a montane oak forest, Costa Rica. MSc Thesis, University of California at Los Angeles (UCLA), Los Angeles, CA

Siles de Guerrero G (1980) Estudio socioeconómico y técnico de productos de carbón, recolectores de mora y lana en las Reservas de Río Macho y Los Santos. Ministerio de Agricultura y Ganadería (MAG), San José, Inf Téc 10

Stadel C (1986) Altitudinal patterns of agricultural activities in the Patate-Pelileo area of Ecuador. Mount Res Dev 6(1):53–62

Ureña A (1990) Reseña histórica del Cantón de Dota. Editorial Serrano Elizondo, San José

Van Omme E, Kappelle M, Juárez ME (1997) Land cover/use changes and deforestation trends over 55 years (1941-1996) in a Costa Rican montane cloud forest watershed area. In: Abstr Vol Conf Geo-Information for Sustainable Land Management, ITC, Enschede, The Netherlands, p 5.14

Wijdeven SMJ, Kuzee ME (20000 Seed availability as a limiting factor in forest recovery processes in Costa Rica. Restor Ecol 8(4):414–424

Zimmerer K (1999) Overlapping patchworks of mountain agriculture in Peru and Bolivia: toward a regional-global landscape model. Human Ecol 27(1):135–165

31 Charcoal Production in a Costa Rican Montane Oak Forest

R. AUS DER BEEK, G. VENEGAS, and L. PEDRONI

31.1 Introduction

31.1.1 Charcoal as an Alternative Energy Source

Wood waste from silvicultural treatments and timber harvesting may account for nearly half of all biomass harvested in managed forests (Flores 1984). Waste from such forests has an enormous potential for energetic purposes. Especially the production of charcoal seems promising in the Neotropics. However, in many countries the use of wood waste derived from timber harvesting has not optimally been explored.

In Costa Rica, for instance, the use of wood waste for energy has been relatively low during the 1970s and early 1980s (MIEM 1986). However, in the late 1980s use of bio-energy reached up to 32 % of the national energy balance (DSE 1989). At that time, charcoal made up for only 0.44 % of all bio-energy sources. That tiny amount corresponded to an annual production of some 6,000 ton (t) of charcoal (Salazar 1986; see also Eckardt 1982), mainly for consumption in restaurants and households (e.g., for barbecue parties).

Costa Rica's National Plan for Energy for the period 1986–2005 clearly identified the need for other energy sources as alternatives for more expensive oil imports (MIEM 1986). In that plan, charcoal was recognized as a major source that could help meet the national energy demand. The plan noted that the availability of raw waste materials remaining after timber-oriented harvesting in Costa Rican forests represented a significant potential, and could well be used for charcoal production in a more sustainable way.

Ecological Studies, Vol. 185
M. Kappelle (Ed.) Ecology and Conservation of Neotropical Montane Oak Forests
© Springer-Verlag Berlin Heidelberg 2006

31.1.2 Charcoal Production History in the High Talamancas

In the mid-1990s, charcoal production was of primary importance to the poorest farmers living in villages such as Villa Mills, which lie scattered along the Panamerican Highway at the borders of the Los Santos and Rio Macho forest reserves – the latter now being part of the Tapantí–Macizo de la Muerte National Park. These two protected areas are located in Costa Rica's high Talamanca Mountains, the backbone of southern Central America (Chap. 30).

At that time, these charcoal-producing farmers, or so-called carboneros, prepared approximately 4,000 t charcoal per year. That amount corresponded to two thirds of the national supply (Salazar 1986). In fact, about half of the income of these rural people came from charcoal production (Siles 1980). Charcoal was produced mainly from oaks growing between 2,700 and 3,000 m in montane forest (Kappelle and Juárez 1995; Chap. 30). About half of all charcoal was produced from dead trees in mature oak forests and, to a lesser extent, secondary forests. Schubel (1980) estimated that about 374,000 trees had been cut for charcoal production during the 30 years following the opening of the Panamerican Highway in the early 1940s. During that period (1941–1984), around 13.3 % of the montane oak forests in the neighboring Savegre River valley in the Los Santos Forest Reserve was lost due to clearing for timber and pastures (van Omme et al. 1997; Kappelle et al. 2000). Fortunately, during the 1990s farmers became more aware of the negative impact of traditional charcoal production on the forest environment and on their own health (see last section of this chapter).

Observations in this study area revealed that the carboneros used to maintain several earth pits at different stages of burning simultaneously, in order to ensure continuous production for weekly pickup by buyers. Apparently, the carboneros did not distinguish between logs with low and high timber values, and even used first-class logs with a high potential for other, more enduring purposes (e.g., construction).

After the creation of the Rio Macho and Los Santos forest reserves in the late 1960s and early 1970s, charcoal production decreased dramatically (Chap. 30). Most of the population depending on forest cutting migrated out of the area. As a consequence, between 1972 and 1976 the local population decreased by two thirds. At the beginning of the 1980s, only 890 people were reported to live in these two forest reserves (Siles 1980). However, since the late 1990s and early 2000s, this trend seems to have reversed as numerous foreigners from North America and Europe have bought land and moved into the area to start ecotourism businesses or simply retire (M. Kappelle, personal communication).

During the mid-1990s, we conducted a participatory study to test alternative technologies for the production of charcoal for the benefit of the farmers' health and living conditions, as well as for the environment (Pedroni 1991; Aus der Beek and Navas 1993). The need for such a study was quite clear in view of (1) the potential of charcoal as a bio-energetic alternative; (2) the importance of charcoal production for the income of the rural poor; (3) the need for reduced negative impacts on mature oak forests (see also Chap. 18); and (4) the need for cleaner and healthier production processes (MIEM 1986).

31.1.3 Scope of this Study

The aim of this study was to analyze the manufacturing process by comparing the traditional production method (the earth pit) with an alternative, transportable metal kiln using oak wood from the montane forests near the Talamancan village of Villa Mills (2,600–2,900 m elevation). The transportable metal kiln has repeatedly been reported to be successful (Earl 1975; Paddon and Harker 1979; Whitehead 1980; Briane et al. 1985). This chapter summarizes the findings of this study.

31.2 Charcoal Production Process

31.2.1 General Aspects of the Production Process

Three wood properties appear to be important when considering the effectiveness and efficiency of charcoal production methods: woody species traits, wood dimension, and wood moisture content (FAO 1985). Broad-leaved species deliver the best charcoal due to their high wood density and high lignin content. For example, oak (*Quercus*) meets processing requirements quite well. In Costa Rica's Talamancan highlands, it offers the preferred raw material for charcoal production (Schubel 1980; Eckardt 1982).

When producing charcoal, it is recommended to reduce the quantity of oak bark as it has a high ash content that negatively affects product quality. Similarly, wood moisture content should be reduced to a minimum to aim for high quality charcoal. Evaporation from wood, however, consumes a huge amount of energy. If this step has to be made during the carbonization process itself, a large amount of wood destined to be converted into charcoal needs to be burned to ensure that sufficient energy is produced. In this case, production efficiency can be increased considerably if the wood is previously dried under the sun for approximately 2–3 weeks. Additionally, lower moisture contents will reduce the time required for carbonization, and therefore lower production costs. This is of particular relevance in montane oak forests where heavy rains may have strongly negative impacts (see below).

Furthermore, carbonization rates are generally influenced by the size of the wood pieces used. Larger pieces require more time to pass through this process due to a slower heat transfer within the wood structure. Smaller and thinner pieces of wood, on the other hand, may show an excessively fast heat transfer, thus reducing market value. FAO (1985) concluded that good-quality charcoal is produced from wood pieces of 45–60 cm length and diameters of some 20 cm.

Temperature levels during the carbonization process will also determine charcoal quality. Low temperatures may result in a higher productivity but are associated with lower charcoal quality. Carbonization temperatures of 500 °C will ensure the production of high-quality charcoal containing at least 75 % of fixed carbon.

In view of these requirements, it is important to select the most adequate technology to ensure production of high-quality charcoal. The next two sections provide short descriptions of the two production methods assessed and compared in this study: the traditional earth pit, and the transportable metal kiln (TPI type Mark V).

31.2.2 The Traditional Earth Pit

The method known as the 'traditional earth pit' is the most common processing tool applied in the Talamanca Mountains. This is probably the oldest method to produce charcoal in the world (FAO 1985). Normally, a large pit is dug in the forest ground (often in pastures) and properly sealed with soil material (Fig. 31.1). The size of the pit differs from place to place, depending on the availability of raw material and the practices of the villagers. For the purpose of the current study, the cavity (earth pit) assessed was designed to fit the same amount of wood (4 m^3 wood) as that consumed by the transportable metal kiln (see below). In order to avoid excessive penetration of air in the pit (which would burn the wood, rather than converting it into charcoal), the air entrance was kept as small as possible. The carbonization process started at the pit's end where the air entered, and slowly advanced toward the opposite end where the air exited.

The main advantage of the 'traditional earth pit' method is its low cost and low investment requirements. However, this system wastes a lot of wood. It does not control air circulation well, and may cause burning of a large portion of the loaded wood. Moreover, charcoal quality differs throughout the pit. At the pit's air entrance end, charcoal is heated for a longer period than at the other end, and therefore may contain less volatile material in the former case. Production quality using the earth pit is further reduced by soil contamination when the pit is discharged. An additional weakness is the increased reabsorption of pyro-ligneous acid, which produces a lot of smoke when charcoal is consumed, thus making it less attractive for the end consumer.

The main ecological impact of this method is the negative effect on surrounding natural regeneration, the vegetation in the vicinity often being cut

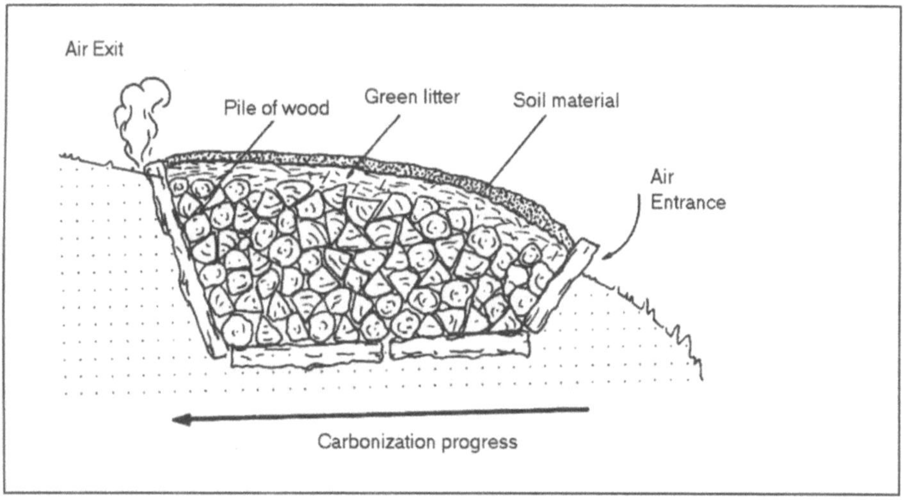

Fig. 31.1. A model of the earth pit in which charcoal is produced in a traditional way

and used to cover the pit. Villagers are accustomed to use the same pit during several production cycles, and therefore repeatedly remove vegetation around the pit. Finally, the carboneros' health is negatively affected, the smoke damaging their eyes and lungs.

31.2.3 The Transportable Metal Kiln

The use of transportable metal kilns for charcoal production spread through Europe during the early 1930s. During the Second World War (1939–1945), this technology was further developed in the UK, where several designs were tested by the Forest Products Research Laboratory (FPRL). The improved technology was transferred to developing countries toward the end of the 1960s, particularly being adopted in Uganda.

Over decades, the UK's Tropical Products Institute (TPI) gained much experience in the use and adaptation of several types of metal kilns. Based on this knowledge, it developed the 'TPI model' characterized by an optimum balance in terms of manufacturing cost, endurance, easy functioning, and high efficiency and productivity, being particularly suitable for developing countries.

As highlighted in the literature (Risi 1942; Briane et al. 1985; ICAP 1988), some of the model's major advantages when compared to the traditional earth pit are (1) better control over air circulation during carbonization; (2) limited need of supervision; (3) higher productivity; (4) little loss and contamination of charcoal while downloading the kiln; (5) ability to function

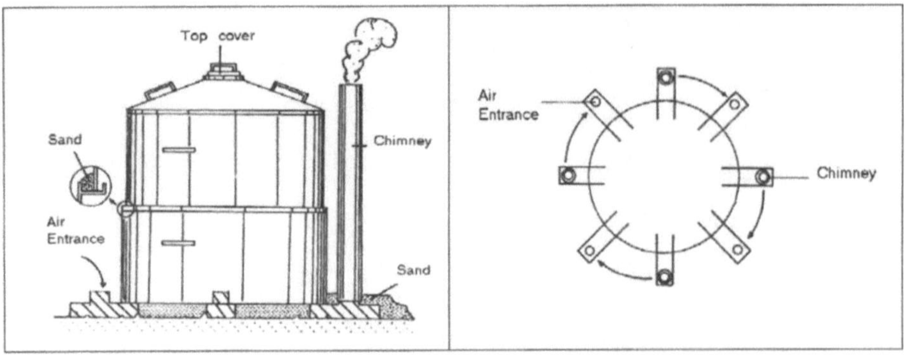

Fig. 31.2. A model of the transportable metal kiln in which charcoal is produced in a modern way

under any weather condition; and (6) short duration of the carbonization process.

The disadvantages, on the other hand, are (1) the need for initial investments; (2) the need for more accuracy while preparing and loading raw materials (wood); (3) the difficulty in moving the kiln from one processing site to another; and (4) the kiln's short lifespan (3–4 years).

The transportable metal kiln used in this study (Fig. 31.2) is similar to the type known as 'Mark V' (Earl 1975; Briane et al. 1985) and the original TPI (Paddon and Harker 1979). It has cylindrical components, and a conic cover with an opening at its top (vapor exit). It is arranged on eight conducts for access and exit of air, laid out in circular order. During carbonization, four pipes (smoke outlets) are rotated periodically.

31.3 Study Design

In order to compare the two production methods described above, we based our work on a participatory approach with two carboneros from the Villa Mills area (Mr. C.H. Araya Mena and Mr. G.V. Mena Granados), highly skilled in producing charcoal from traditional earth pits. Additionally, they were trained in using the transportable metal kiln. Both men contributed to simultaneously implementing ten carbonization sessions, using the earth pit as well as the transportable metal kiln.

Each system had a wood capacity of 4 m^3, in which 1 m^3 was equivalent to 0.65 m^3 of piled wood. As raw material, wood from two oak species was used: *Quercus copeyensis*, now known as *Q. bumelioides* (Chap. 1), with two carbonization sessions; and *Q. costaricensis*, with eight carbonization sessions. Wood was derived from the montane oak forests near the CATIE study area at

Villa Mills (see also Chap. 18). Three months before processing, the collected wood was cut in 0.85-m-long pieces and piled up. The piles were covered with plastic to protect them from rain. Shortly before carbonization took place, samples were taken from each pile and delivered to the Instituto Tecnólogico de Costa Rica (ITCR) for wood moisture content measurements. Both carbonization processes were compared in terms of duration, productivity (in kg of produced charcoal), and quality of produced charcoal. Quality analysis was also carried out by ITCR.

After each process was completed, we randomly took eight samples of 2.5 kg each from the charcoal obtained by each method. This corresponded to a sampling intensity of approximately 15–20 %. The quality of the samples was assessed chemically following the Standard Test Method for Chemical Analysis of Wood Charcoal, stipulated by Norm D-1762 of the American Society for Testing and Materials (ASTM 1964). Quality analysis included assessment of (1) moisture content, (2) content of volatile matter, (3) ash content, (4) fixed carbon content, (5) caloric value, and (6) bulk density.

31.4 Charcoal Production Processing Time

The time periods required for each production step per method are summarized in Table 31.1. Results show that approximately 230 h (roughly 10 days) are required for each process when using the traditional earth pit method. The portable metal kiln method required only 163 h (roughly 7 days). Average effective working time was about 16 h at the earth pit, and 15 h when applying the metal kiln.

Although the process duration when using the metal kiln is shorter than that for the earth pit, the values obtained in the study are considerably higher

Table 31.1. Time periods required for each of the charcoal production steps for two processing methods (earth pit vs. metal kiln). Average time (with exception of totals) is given in minutes (±confidence interval of 95 %)

Production step	Earth pit	Metal kiln
Collection of green wood	70±11	0±0
Loading	374±44	299±40
Setting fire	39±20	401±60
Carbonization	9,062±725	7,015±1,182
Cooling	3,074±866	1,952±463
Unloading	496±26	138±28
Total	~230 h	~163 h

than the 30–40 h reported in the literature (Earl 1975; Paddon and Harker 1979), and the period of 5 days obtained in previous trials from the same area (Pedroni 1991). The differences may be explained by the specific climate of the area, and the moisture content of the wood used during the process implementation in the current study.

31.5 Productivity Levels

Productivity results for both processes expressed in charcoal weight (kg) and amount of remaining raw material show a higher productivity when using *Quercus copeyensis*, rather than *Q. costaricensis* (Table 31.2). Carbonization efficiency ranges from 72 kg (metal kiln) to 84 kg (earth pit) of charcoal produced from 1 m^3 of *Q. copeyensis* wood, and from 47 kg (metal kiln) to 65 kg (earth pit) of charcoal produced from 1 m^3 of *Q. costaricensis* wood. Results also indicate considerably higher productivity achieved when using earth pits (261 kg of charcoal from *Q. costaricensis*, and 338 kg from *Q. copeyensis*) than when applying the metal kiln method (189 kg of charcoal for *Q. costaricensis*, and 290 kg for *Q. copeyensis*).

However, considering productivity in relation to process duration over a one-year period, the earth pit would produce approximately 9,500 kg of charcoal whereas the metal kiln would produce some 9,800 kg. Hence, the difference between the two methods is not well expressed in terms of annual production as such, but rather in the workload and the amount of wood consumed during the process, being almost 40 % greater when the metal kiln is used.

In general, the values obtained in this study are far below the findings from similar trials with earth pits conducted in the same area by Pedroni (1991). This author reported an average efficiency of approximately 156 kg of charcoal obtained from 1 m^3 of *Q. copeyensis* wood. One of the causes is probably directly related to the temperatures reached during the carbonization process. The correlation that exists between the content of fixed carbon in produced charcoal and maximum temperatures of the carbonization process

Table 31.2. Amount of charcoal and remaining raw material obtained during the carbonization process for two methods and two oak species

	Earth pit		Metal kiln	
	Quercus costaricensis	*Quercus copeyensis*	*Quercus costaricensis*	*Quercus copeyensis*
Amount of charcoal (kg)	261	338	189	290
Amount of raw material (kg)	161	246	31	140

(Briane et al. 1985) suggests that process temperatures in our study probably did not exceed 280 °C. This value is far below the level of 500 °C required for high-quality carbonization.

These low temperatures are most probably due to the wet and cold weather that occurred during the study period. High annual rainfall (2,500–6,500 mm $year^{-1}$), wind-blown mist, and frequent presence of clouds are the key characteristics of these montane oak forests (Kappelle 1996). This was consequently reflected in the high average moisture content (42 %) in the wood used during the carbonization process.

Carbonization was faster for the metal kiln, probably because of the protection against the infiltration of moisture. Here, carbonization occurred at lower temperatures. This may be due to the limited thermal resistance of the kiln. These circumstances caused a relatively low productivity level. On the other hand, carbonization temperatures were lower for the earth pit process, in which carbonization rate was much slower. In this case, productivity was also low but still higher than for the metal kiln system.

31.6 Quality Levels

Most of the specifications used to control charcoal quality have originated in the steel and chemical industries (e.g., ASTM 1964). When charcoal is exported, buyers tend to make use of these industrial specifications even though the main outlet of the imported charcoal may well be domestic cooking, including barbecue fires. Charcoal quality is defined by properties that are measured and appraised separately. For the current study, the following main specifications were considered: moisture content, volatile matter other than water, fixed carbon content, ash content, caloric power, and density.

Within the scope of the present chapter, it is not possible to provide quantitative detail for each of the qualitative properties of charcoal. However, a qualitative comparison of our results for both oak species with specifications and standards (from the USA and UK) reported elsewhere (Doat and Petroff 1975; Paddon and Harker 1979) did reveal some interesting patterns (Table 31.3).

Notably, the data show that the quality of the charcoal produced in this study does not meet international export standards, especially because of the high content of volatile matter. This drawback could be successfully mastered by achieving higher temperatures during the carbonization process, and by reducing the wood moisture content. Despite these limitations in the international arena, oak charcoal produced in these high-altitude forests is still preferred for the domestic market, which focuses on household consumption. As expected, the quality of charcoal from *Q. copeyensis* is slightly higher than from *Q. costaricensis* due to the former having a higher wood density.

Table 31.3. Comparison of charcoal quality levels (dry basis) for two oak species and two methods as reported in this and two similar studies. USA and UK import standards are also given

Criteria	This study		Similar studies		Import standards	
	Metal kiln	Earth pit	Paddon and Harker (1979)	Doat and Petroff (1975)	USA	UK
Quercus costaricensis						
Moisture (%)	8.0	7.9	3.5	–	5.0–8.0	5.0
Volatiles (%)	24.5	20.4	13.8	11.8–16.0	–	13.7
Ash (%)	3.5	3.3	2.2	10.7–16.7	–	3.2
Fixed carbon (%)	72.0	76.1	84.1	81.0–86.0	–	84.2
Caloric power	6,860	7,220	7,800	7,700–8,200	–	7,200
Wood density	0.27	0.3	–	0.2–0.3	0.3	–
Quercus copeyensis						
Moisture (%)	6.3	6.0	3.5	–	5.0–8.0	5.0
Volatiles (%)	24.1	20.2	13.8	11.8–16.0	–	13.7
Ash (%)	1.4	1.1	2.2	10.7–16.0	–	3.2
Fixed carbon (%)	74.4	78.7	84.1	81.0–86.0	–	84.2
Caloric power	7,097	7,469	7,800	7,700–8,200	–	7,200
Wood density	0.36	0.38	–	0.2–0.3	0.3	–

Although the results obtained with the earth pit appear to be better than those from the metal kiln, the observed differences are weak.

31.7 Ownership and the Future of the 'Carboneros'

One of the main causes for the low net return of charcoal manufacturing is the long chain of intermediary traders of this product, extending from the producers at the selling site along the Panamerican Highway to the end consumers in the main cities of Costa Rica (own observations, and M.E. Juárez, personal communication). Transporters, wholesalers and supermarkets are those making most profit along the charcoal commercialization chain.

For this reason, in 1993 local carboneros established the Villa Mills Producers Association (ASOPROFOR). It involved charcoal producers – men and women, including local transporters – who had taken over the role of these intermediaries by packing, transporting and storing charcoal themselves, thus increasing the benefit to the local communities. These traditional manufacturers produced charcoal from dead trees and byproducts from sustainable forest management, the women packed charcoal in attractive 2.5-kg bags, and local transporters carried and sold the product directly to the main supermarkets in San José and Cartago.

The association helped not only to increase profit for the producers, but also to give more power to women in a traditionally male-dominated society, and by joining women and men in collaborative efforts for rural development. The association's main focus was not only to create direct income based on charcoal production and commercialization, but also to contribute to conservation of biodiversity, water, air quality and scenic beauty of the highland ecosystems. The aim was also to promote local activities to improve the quality of life of the local communities, and the conservation of natural resources by taking advantage of new niches in Costa Rica's prospering ecotourism sector. Based on their own experiences and tradition, these carboneros helped to increase people's understanding about man–nature relationships, and the importance of maintaining a balance between the communities and their natural environment (see also Eco-Index 2003).

Recently, the ASOPROFOR stopped commercializing charcoal due to heavy price competition and insufficient product supply. At present, only three farmers are still involved in charcoal production, mainly during their spare time. Thus, the charcoal production activity is slowly disappearing from the high altitude oak forests of Costa Rica. There is no doubt about the health benefits for the former carboneros, now that they no longer produce charcoal. In fact, toward the mid-1990s they were seeking new remunerative agricultural activities in order to improve their living standards (Kappelle and Juárez 1995; Chap. 30).

31.8 Conclusion

Today, local communities are undergoing an abrupt change from traditional subsistence agriculture toward a more modern economy based on fruit culture, trout export and ecotourism (Kappelle and Juárez 2000; Kappelle et al. 2000; Murillo 2003). Therefore, it is quite likely that charcoal production soon will be part of history in the high Talamancas of southern Costa Rica.

References

ASTM (1964) Procedures for testing. Committee on Soils for Engineering Purposes, American Society for Testing and Materials, Philadelphia, PA

Aus der Beek R, Navas S (1993) Técnicas de producción y calidad del carbón vegetal en los robledales de altura de Costa Rica. CATIE, Turrialba, Costa Rica. Col Silvic Man Bosq Nat 8:1–41

Briane D, Doat J, Riedhacker A (1985) Guide technique de la carbonisation de bois. EDISUD, Paris

Doat J, Petroff G (1975) La carbonisation des bois tropicaux: essais de laboratoire et perspectives industrielles. Rev Bois Foret Trop 159:55–72

DSE (1989) Balance energético nacional. Dirección Sectorial de Energía, San José, Costa Rica

Earl DE (1975) Informe sobre el carbón vegetal. FAO, Rome

Eckardt N (1982) Charcoal production in the high altitude oak forest in the Talamanca Mountains, Costa Rica. Undergraduate Thesis, Associated Colleges of the Midwest (ACM), Universidad Nacional, Heredia

Eco-Index (2003) Production and commercialization of vegetal charcoal from trees dead of natural causes or from sustainably managed forests. http://www.eco-index.org

FAO (1985) Industrial charcoal making. UN Food and Agricultural Organization, Rome, FAO Forestry Pap 63

Flores JG (1984) Diagnóstico del sector industrial forestal y alternativas de solución. San José, Costa Rica

ICAP (1988) Producción de carbón vegetal en Costa Rica: situación y perspectivas. Instituto Centroamericano de Administración Pública, San Jose, Costa Rica

Kappelle M (1996) Los bosques de roble (*Quercus*) de la Cordillera de Talamanca, Costa Rica: biodiversidad, ecología, conservación y desarrollo. Instituto Nacional de Biodiversidad (INBio), Santo Domingo de Heredia, Costa Rica

Kappelle M, Juárez ME (1995) Agroecological zonation along an altitudinal gradient in the montane belt of the Los santos Forest Reserve in Costa Rica. Mount Res Dev 15:19–37

Kappelle M, Juárez ME (2000) Mountain forests, biodiversity and people in Costa Rica. In: Price M, Butt N (eds) Forests in sustainable mountain development: a state of knowledge report for 2000. IUFRO Res Ser 5. CABI, Oxon, UK, pp 38–46

Kappelle M, Avertin G, Juárez ME, Zamora N (2000) Useful plants within a campesino community in Costa Rican montane cloud forest. Mount Res Dev 20(2):162–171

MIEM (1986) Plan nacional de energía (1986–2005). Ministerio de Industria, Energía y Minas, San José, Costa Rica

Murillo K (2003) Comunidades de Talamanca mejoran su calidad de vida a través del ecoturismo. Rainforest Alliance. http://www.rainforest-alliance.org

Paddon AR, Harker AP (1979) The production of charcoal in a portable metal kiln. Tropical Products Institute, London

Pedroni L (1991) Sobre la producción de carbón en los robledales de altura en Costa Rica. CATIE, Turrialba, Costa Rica. Col Silvic Man Bosq Nat 4:1–29

Risi MJ (1942) L'Industrie de la carbonization du bois dans la province du Quebec. Ministère des Terres et des Forets du Québec, Canada

Salazar R (1986) Producción y mercado de carbón vegetal en Costa Rica. ICAITI, San José, Costa Rica

Schubel RJ (1980) The human impact on a montane oak forest in Costa Rica. PhD Thesis, University of California, Los Angeles, CA

Siles G (1980) Estudio socioeconomico y técnico de productores de cabon, recolectores de mora y lana en las Reservas de Rio Macho y Los Santos. Dirección General Forestal, San José, Costa Rica

Van Omme E, Kappelle M, Juárez ME (1997) Land cover/use changes and deforestation trends over 55 years (1941–1996) in a Costa Rican montane cloud forest watershed area. In: Abstr Vol Conf Geo-Information for Sustainable Land Management, ITC, Enschede, The Netherlands, p 5.14

Whitehead W (1980) The construction of a transportable charcoal kiln. Tropical Products Institute, London

32 Criteria and Indicators for Sustainable Management of Central American Montane Oak Forests

B. HERRERA and A. CHAVERRI †

32.1 Introduction

The evolution of the concept of forest management, from its traditional focus on timber and wood harvesting toward more sustainable practices, has been attributed to changes in the social perception of nature (Sheil et al. 2004). Over the last decades, modern society has come to value and recognize that forests are not only sources of timber, but also critical ecosystems for water production, as sources of medicinal products, as carbon sinks and reservoirs, areas for recreation, and landscapes of great scenic beauty (Dawkins and Philip 1998). The multifunctional characteristic of forests has been recognized in several national and international forums, and is reflected by current definitions of sustainable forest management (Castañeda 2000). One of the most widely accepted definition was purported by the International Tropical Timber Organization (ITTO):

"*Sustainable forest management is the process of managing permanent forest land to achieve one or more clearly specified objectives of management with regard to the production of a continuous flow of desired forest products and services without undue reduction of its inherent values and future productivity and without undue undesirable effects on the biophysical and social environment*" (ITTO 1992).

Within the framework of a number of international efforts and agreements, initiated after the 1992 United Nations Conference for Environment and Development (UNCED) held in Rio de Janeiro, a significant number of countries (from the developed and developing world) have generated principles, criteria and indicators (C&I) to assess and monitor their progress in implementing sustainable forest management (Castañeda 2000). Noteworthy, global-scale events that have engaged in this growing area of interest and

Ecological Studies, Vol. 185
M. Kappelle (Ed.) Ecology and Conservation of Neotropical Montane Oak Forests
© Springer-Verlag Berlin Heidelberg 2006

work include the FAO/ITTO Expert Meeting on the Harmonization of Criteria and Indicators for Sustainable Forest Management, held in Rome in February 1995, and the Intergovernmental Seminar on Criteria and Indicators, organized by the Government of Finland in Helsinki in August 1996 (Granholm et al. 1996).

In Central America, the issue of sustainable forest management was first formally addressed at a regional level through the Lepaterique Process, which was initiated in January 1997, following the recommendations of an Expert Meeting on C&I for Sustainable Forest Management organized by the Central American Council of Forests and Protected Areas (CCAB-AP) in Lepaterique, Honduras. Experts from all Central American countries identified eight national-level criteria and 53 related indicators. At the regional level, four criteria and 40 indicators were accepted by the gathered experts (FAO 1997). This meeting was followed by sub-regional workshops and national seminars, which reviewed the degree of applicability of the standards and made specific recommendations for future implementation. A set of seven agreed, national-level criteria emerged from the workshops (FAO 1997).

At the regional and national levels, the monitoring of the management of each forest type is imperative. Therefore, specific standards should be developed and tested in order to improve the implementation of the Central American process on C&I (FAO 1997). The C&I identified at the regional and national levels should be considered mutually compatible, although there are some differences in purpose and scope. The set of C&I developed at the national level can assist in identifying those for use at the Forest Management Unit (FMU) level (Castañeda 2000). National-level indicators contribute toward the development and regular updating of policy instruments (laws, policies, regulations), whereas indicators at the FMU level can support the setting and rectification of forest management prescriptions over time to meet established national goals (Castañeda 2004), and can also serve as certification tools.

Because indicators at the FMU level will be influenced by factors such as forest type and biophysical conditions, as well as socioeconomic considerations (Castañeda 2000), C&I at this level may be different among individual forest areas. For this reason, and considering the prevailing ecological and socioeconomic conditions of tropical montane oak forests (see Chaps. 18, 28, and 31), it is necessary to define a specific set of C&I for this type of forest in order to guide and secure its sustainable management. Other international initiatives, such as the Montreal Process on C&I for the conservation and sustainable management of temperate and boreal forest, have proven useful in developing standards for specific forest types (Castañeda 2000). In this chapter, a set of standards for the management of montane oak forests in Central America at three different scales of application (regional, national, and FMU) is presented. For the first two levels, only the criteria are shown. Future papers will include the corresponding indicators for these two levels. For the FMU level, the complete standard is presented and discussed. Furthermore, implications for forest policy and future improvements in the standard are discussed.

32.2 Ecological Factors Determining Montane Oak Forest Management

The ecological and topographic conditions under which Neotropical montane oak forest grows determine specific aspects that must be considered in the forest planning process to ensure sustainable management (Dawkins and Philip 1998). Therefore, corresponding C&I must reflect these particular factors in the planning process.

Such ecological factors are related to the topography of the terrain on which the forest is situated, as well as specific ecological and socioeconomic functions that it fulfills, such as water production, soil protection, and biodiversity conservation (Jiménez et al. 1988; Chaverri and Hernández 1995). This ecological specificity must be reflected in the forest management standards at any scale of application, since these are characteristics that will ensure the spatial and temporal persistence of these montane oak forest ecosystems (Kappelle and Brown 2001).

Along the Central American montane oak forest belt, most of the productive stands are located on steep slopes, in many cases exhibiting values of 100 % (Aus der Beek and Sáenz 1992; Chaps. 18 and 31). This particular condition requires the application of specific forest management prescriptions that should be reflected in the indicators (see below).

Other key forest functions associated with this type of ecosystem must also be taken into consideration during the forest management planning process (Aus der Beek and Sáenz 1992). For example, the montane oak forest plays a critical role in conserving populations of endemic species (Chaverri et al. 1997). Therefore, it is important to know which species this ecosystem harbors, and to understand the ecological characteristics of these species and their interaction with other components of the system. This knowledge serves as the scientific baseline utilized to design silvicultural practices that will ensure the conservation and protection of target organisms (Aus der Beek and Sáenz 1992). Of particular interest are red-listed species such as the resplendent quetzal (*Pharomachrus moccino*; Chaps. 24 and 25), an emblematic bird species whose feeding preferences are associated with the presence of species in the Lauraceae (*Ocotea* spp., *Nectandra* spp.), big mammals such as the tapir (*Tapirus bairdii*, Chap. 27), peccary (*Pecari tajacu*), and jaguar (*Pantera onca*), some tree genera such as *Podocarpus* and *Magnolia*, and several orchids (Orozco 1991; Kappelle 1996; Chaverri et al. 1997; Chap. 10).

Another key function that montane forest ecosystems play is watershed protection (Stadtmüller 1994; Chaverri et al. 1997; Kappelle and Brown 2001; Chap. 21). It is well recognized that soil protection and water production are functions inherent to any forest ecosystem (Stadtmüller 1994). However, the topographic conditions (i.e., steep slopes) and high precipitation rates associated with the montane oak forest in Central America (Chaverri et al. 1997) make the region more prone to erosion, and these functions of particular relevance for the sustainable management of this ecosystem (Chaverri and Hernández 1995; Kappelle 2004; Chap. 21). The watersheds along the montane belt are known to produce high-quality water (Chaverri et al. 1997), due to the capacity of the forests to protect the soil surface and allow high rates of infiltration, maintaining a stable annual distribution of water (Stadtmüller 1994; Chap. 21). In the view of many experts, water production is the main function of montane oak forests. Therefore, during its management, mitigation actions that reduce or avoid negative impacts on this function must explicitly be considered (Chaverri et al. 1997; Kappelle and Brown 2001; Chap. 21).

32.3 Socioeconomic Factors and Montane Oak Forest Management

In Central America, montane oak forests provide a wide range of products to local people (Kappelle and Brown 2001, Chaps. 30 and 31). In Costa Rica, for example, vascular plants occurring in this ecosystem supply local communities with timber, charcoal (Chap. 31), fuelwood as well as orchids and ferns for ornamental purposes, among others (Kappelle and Juárez 2000). Due to deforestation, the area covered by this ecosystem type has been reduced significantly since the 1950s, although deforestation rates have decreased in the last two decades (Kappelle and Juárez 2000; Kappelle and Brown 2001). The main factors associated with the loss of this type of ecosystem are forest clearing for timber, fuelwood, and charcoal production, forest conversion to grassland for cattle ranching, and land degradation (Kappelle and Juárez 2000; Samudio 2001). According to Morales and Brown (1995), this negative trend can be attributed to productivity loss on marginal lands (which produce a migratory agriculture process), an increase in the lowland to highland immigration rate, and to the high population density observed in the highlands. The aforementioned pressures on montane oak forest make it imperative to consider local communities in the development of sustainable forest management tools, such as proposed in this chapter.

32.4 Development of Management Standards

32.4.1 Defining a Conceptual Framework and Attributes for C&I

In order to define the corresponding set of C&I, a hierarchical approach was used (Lammerts van Bueren and Blom 1997; Herrera and Corrales 2004). A prerequisite for a coherent and consistent standard is an ambiguous and well-explained hierarchical framework, which consists of different levels (i.e., principles, criteria and indicators) that facilitate the formulation of a set of parameters in a logical way and, at the same time, describes the function of each level (Lammerts van Bueren and Blom 1997; Fig. 32.1). The proposed standard was developed by maintaining its horizontal and vertical consistency. Horizontal consistency implies that parameters appearing at the same level (e.g., criterion) do not show any overlap. Vertical consistency refers to the logical relation between parameters at adjacent levels, which means that the parameter must appear at the correct hierarchical level and must be expressed in correct terms (Lammerts van Bueren and Blom 1997; Fig. 32.1).

Principles, criteria and indicators are "information tools at the service of forest management" (Prabhu et al. 1999; Herrera and Corrales 2004). They are used to conceptualize, evaluate, implement, monitor, and report sustainable forest management (Lammerts van Bueren and Blom 1997; McGinley and Finegan 2003). While a *principle* is defined as "a fundamental law or rule, serving as a basis for reasoning and action" (Lammerts van Bueren and Blom 1997), *criteria* "are the intermediate points to which the information provided by indicators can be integrated and where an interpretable assessment crys-

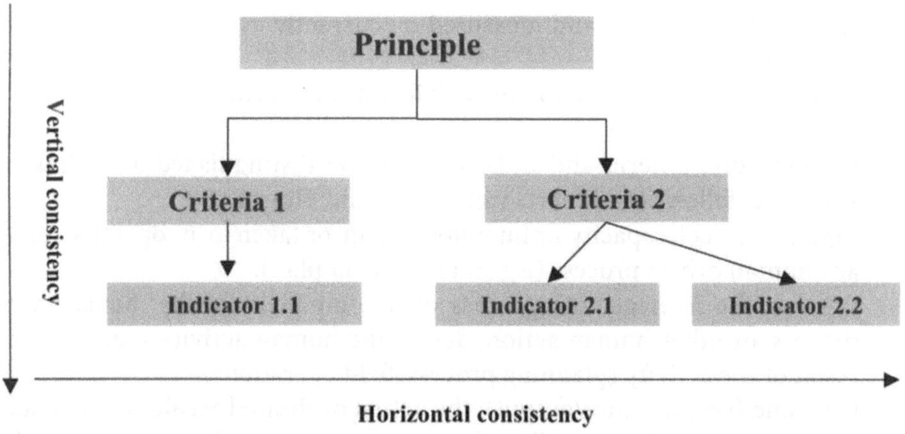

Fig. 32.1. Hierarchical structure demonstrating the various levels of organization for principles, criteria and indicators. See the main text for further explanation

tallizes" (Prabhu et al. 1999). *Indicators* "are any variables or components of a forest ecosystem or management system that are used to infer the status of a particular criterion" (Prabhu et al. 1999). The actual assessment of management performance should be based on a comparison between the actual value of the indicator and its reference value or norm (Lammerts van Bueren and Blom 1997). According to Lammerts van Bueren and Blom (1997), a *norm* is "the reference value of the indicator and is established for use as a rule or a basis for comparison". By comparing the norm with the actual measured value, the result demonstrates the degree of fulfillment of a criterion. For the purpose of this study, only C&I were developed and, where possible, norms were also identified. Thus, criteria for sustainable forest management of montane oak forest in Central America should exhibit the following characteristics:

- Have the capacity to measure progress toward sustainable development;
- Represent a regional consensus on sustainable forest management practices;
- Be comprehensive and clearly defined;
- Be developed on the basis of sound science;
- Take into consideration local capacity to implement standards; and
- Be limited in number and sensitive to regional characteristics of the forests.

Regarding the corresponding indicators, the set should have at least the following characteristics (Prabhu et al. 1999; Lammerts van Bueren and Blom 1997):

- Clarity: based on a language affordable to the general public, as well as technical and political audiences;
- Sound science: based on state-of-the-art knowledge and scientific research;
- Applicability: utilized and measured by currently available, cost-effective procedures; and
- Flexibility: useful in all montane oak forests in Central America.

Furthermore, criteria and indicators can be distinguished according to their type, as follows (Lammerts van Bueren and Blom 1997):

- Input: an object, capacity, or intention, put in, or taken in, or operated on by any human driven process (e.g., management plan).
- Process: the management process or a component of the management process, or other human action, describing human activities and not the result of the activity (planning process, field operations).
- Outcome (performance/output): the actual or desired result of a management process, which describes the state or capacity of the ecosystem, the state of a physical component or the state of the related social system or its components.

32.4.2 Defining the Geographic Area for Standards Development

It is important to define the geographic boundaries of the area for which the standard is developed, since its application is restricted to the particular ecological and socioeconomic conditions of the forest. In order to define such a geographic area, ecological criteria can be used for this purpose. The presence and distribution of native coniferous and oak-dominated forests can be used to characterize the highlands (1,800 (2,200)–2,800 (3,200) m above sea level, a.s.l.) in Central America (Lauer 1968; Chaverri and Herrera 1996; Kappelle and Brown 2001). In this regard, two major sectors can be identified. The first includes the countries from Guatemala to Nicaragua, whereas the second includes Costa Rica and Panama, where native coniferous and mixed forests are replaced by pure oak forest stands (Kappelle 1996; Kappelle and Brown 2001). The proposed standards apply only to the highland forests where there is no predominance of coniferous forests.

32.4.3 Selecting Criteria and Indicators

The proposed set of C&I was developed in two stages. During the first phase, a preliminary list of C&I was produced by the authors. In order to compile this list, data generated in various international forums (see Granholm et al. 1996; Prabhu et al. 1999; Castañeda 2000) regarding the development of C&I, as well as other pertinent technical literature were reviewed. During the second phase, the aforementioned preliminary list was circulated for review to a group of scientists and relevant institutions in Central America, in order to include a broad range of expertise and to attempt reaching regional-level consensus in the development of the final standard (Chaverri and Herrera 1996).

An important drawback for developing area-specific C&I sets is the lack of scientific information on the exact prerequisites for sustainable forest management (Lammerts van Bueren and Blom 1997). In order to bridge this gap of information, indirect (i.e., process and input) indicators, rather than outcome indicators (which require clear, 'hard' norms) were preferred. Process and input indicators were used to characterize policy and management. For the proposed outcome indicators, estimates based on the authors' experience in this type of forests served as the basis for the definition of a comparative value.

32.5 Criteria and Indicators at Different Scales of Application

32.5.1 Regional and National Levels

Below, the proposed criteria for sustainable montane oak forest management and monitoring at regional and national levels in Central America are summarized.

- Legal, political, institutional, technical, economical, and social frameworks are respected;
- Protection and conservation functions of forests are maintained;
- Forest productivity capacity is maintained;
- Social, economic and cultural benefits of forests are maintained and improved;
- Forest cover is conserved;
- Health of forest is maintained;
- Forest biodiversity is conserved;
- Current contribution to global ecological cycles of forests is maintained.

It is expected that if the C&I at the national level (not within the scope of this paper) are fulfilled, then the C&I at the regional level will also be met. Therefore, the same set of criteria - albeit not the same indicators - can be used for monitoring at both scales of application. Although there are some differences in semantics, this C&I set coincides with the criteria agreed upon during Central America's Lepaterique Process (FAO, no date).

32.5.2 Forest Management Unit (FMU)

The set of C&I - with a total of five criteria and 35 indicators - proposed for Central American montane oak forest management at the FMU level is as follows (with indicators underlined when defined as outcome indicators):

Criterion I. The biodiversity of the forest ecosystem is maintained
Indicators:

I.a. Areas considered ecologically relevant are protected;
I.b. After forest harvesting, all original species are present;
I.c. Threatened, key species, and endangered species are protected;
I.d. Natural regeneration is assured, and floristic composition and forest structure are similar to those of the undisturbed stand;
I.e. Key microhabitats for wildlife (e.g., dead trees, canopy structure, levels of vegetation density, and decaying logs) are conserved.

Criterion II. A continuous production of environmental goods and services is maintained

Indicators:

II.a. Forest management objectives are clear, based on relevant information and prioritized according to forest functions;

II.b. Forest management units are well defined at the field level, and annual harvest units are defined as well;

II.c. Forest production is regulated on basis of area, volume and stem diameter increment;

II.d. A pre-harvest inventory is conducted and used as the basis for further activities;

II.e. Forest interventions (e.g., tree felling) are planned and undertaken in such a way that the impact on the ecosystem is kept at a minimum;

II.f. The extracted basal area is <25 % of the total basal area per hectare;

II.g. Before forest activities are begun, involved personnel is trained;

II.h. Forest operations are continuously monitored;

II.i. Canopy openness is kept at a minimum;

II.j. Floristic composition, forest structure, and forest growth are monitored in permanent plots (one per 100 ha);

II.k. No new land use change is taking place at the FMU level, and affected areas are ecologically restored;

II.l. Mitigation actions that prevent forest fragmentation are defined;

II.m. Hunting and extraction of non-forest products are regulated.

Criterion III. Soil and water quality is protected

Indicators:

III.a. The area corresponding to forest roads and trails is <15 % of the FMU, and roads and trails are mapped;

III.b. Canopy openness is kept at a minimum;

III.c. Gallery forests are respected and protected;

III.d. Water and soil conservation actions are applied after forest harvesting;

III.e. Forest harvesting is conducted in areas with a slope<25°;

III.f. Environmental contamination is kept at a minimum, and the corresponding national health institutions approve chemical products to be used.

Criterion IV. Socioeconomic benefits of forest are secured

Indicators:

IV.a. Communities receive economic benefits derived from forest harvesting;

IV.b. Local people are hired and trained in sustainable management operations;

IV.c. Forest harvest intensity is determined on the basis of natural forest productivity;

IV.d. Monitoring of community participation is implemented;

IV.e. Woman participation is improved and assured in forest management operations;
IV.f. Local communities are informed, and participate in the preparation of forest management plans.

Criterion V. Legal, political, and institutional frameworks are respected and improved
Indicators:
V.a. A signed statement by the forest owner, stating that (s)he is willing to respect these frameworks, is required;
V.b. A forest management plan is prepared and approved by the corresponding national authorities;
V.c. Successful achievement of management plan objectives is monitored;
V.d. During the planning phase, specific studies that address themes such as forest growth rates, regeneration rates, composition and structural changes, social impact of management, and costs and efficiency are included and conducted;
V.e. FMU is protected from illegal human invasion after forest harvesting.

The criteria of this set encompass the ecological, social, economic and political characteristics adopted through other international C&I development processes (FAO 2001). However, considering the ecological characteristics of montane forests, it is important to include specific indicators that address the aforementioned ecological characteristics of this ecosystem, as stated above. For this reason, the standard includes indicators related to the percentage of extracted basal area, percentage of forest roads, and the minimum slope allowed to undertake forest harvest prescriptions, since these ecosystems are considered fragile, and after a silvicultural intervention, their recovery is slow (Stadtmüller 1994; Chap. 18). The former indicators can also be found in other standards (e.g., Campos et al. 1998), and therefore they are not exclusive of montane oak forest. However, the norms that belong to these indicators were defined more narrowly for this forest type. Therefore, the specific characteristic of the present C&I cannot be found only at the level of indicators, but also at the level of its associated norms.

Furthermore, it is important to highlight that the management objectives of the Neotropical montane oak forests can be multiple (Chaverri and Herrera 1996), and therefore, it may be a priority to secure one function over another, such as water, rather than timber production function. If this were the case, each objective should be monitored individually (Chaverri and Herrera 1996), as proposed in the present standard.

32.6 Conclusions

The specific ecological conditions of Neotropical montane oak forests require particular C&I if this type of ecosystem is to be sustainably managed. Like any other standard, the proposed C&I for montane oak forests address the socioeconomic, political, and ecological dimensions included in the definition of sustainable forest management (e.g., ITTO 1992). It should be highlighted, however, that some of the proposed C&I, both at national and FMU levels, could be applied to different types of tropical forest ecosystem (e.g., lowland forests). Nevertheless, it should be noted that the norms that belong to these indicators were defined specifically for the montane oak forests. Thus, if a set of C&I is locally developed, only adjustments at the indicator level are required. Furthermore, it is highly recommended that similar efforts in future define a set of complementary principles to the set of C&I for Central American montane oak forests.

It seems clear that criteria at the regional, national and FMU levels should be complementary. At the national level, C&I can be used to define forest policies, and to coordinate the forest and biodiversity management agendas among the relevant institutions. At the FMU level, however, different efforts have been executed in Central America (e.g., Campos et al. 1998), and the validation and definition of acceptable variation thresholds for each indicator are among the main challenges that need to be addressed to make the standards operative.

In order to improve the efficiency of indicators at different levels, especially those at the FMU level, specific values that define acceptable limits of variation (i.e., norms), protocols, and research are needed in order to test if the indicators indeed achieve the sustainability objectives. The range of natural variation needed to understand natural limits associated with disturbance can be used for this purpose. Therefore, the proposed norms for the montane oak forests should be considered as preliminary, since specific research is needed to define a more precise basis of comparison. Furthermore, it is still necessary to define what is conceived as an acceptable amount of divergence from a natural reference condition. Therefore, any C&I framework should be conceptualized as part of adaptive management schemes. Continuous integration of research results, lessons learned from practitioners, and information produced by monitoring approaches can result in a continuous evolvement of the proposed C&I over time. Pilot studies that cover the complete range of socioeconomic and ecological variation in the Central American montane oak forest belt should be conducted in order to validate the proposed standards. However, it must be recognized that developing specific sets of C&I could increase the complexity and costs associated with the standards preparation, since specific norms or thresholds are required. Lastly, in this validation process, it would be appropriate to explore the integration of indicators of similar scope, in order to simplify the standard formulation. For example, it may be necessary to study the linkages between terrestrial and aquatic indicators, the relation between ecosystem productivity and soil indicators, and that between ecosystem function and species diversity.

Acknowledgements This chapter is dedicated to the memory of the second author, the late Adelaida Chaverri, an outstanding tropical ecologist and a never-ending source of inspiration for many forest managers, conservationists, and scientists. I thank M. Libby, J. Mateo, and M. Kappelle, and two anonymous reviewers for their helpful comments on earlier drafts. Financial support was provided by CCAB-AP and FAO.

References

Aus der Beek R, Sáenz G (1992) Manejo forestal basado en la regeneración natural del bosque: estudio de caso en los robledales de altura de la Cordillera de Talamanca. CATIE, Turrialba, Costa Rica, Col Silvic Manejo Bosq Nat 6

Campos JJ, Lobo S, Müller E (1998) Development of criteria and indicators for sustainable forest management and forest certification in Costa Rica. In: Proc Int Conf Indicators for Sustainable Forest Management: Fostering Stakeholder Input to Advance Development of Scientifically Based Indicators. IUFRO, Melbourne, Australia, pp 91–93

Castañeda F (2000) Criteria and indicators for sustainable forest management: international processes, current status and the way ahead. Unasylva 51(4):34–40

Castañeda F (2004) Tendencias y perspectivas para las iniciativas de criterios e indicadores para la ordenación forestal sostenible. Rec Nat Amb 42:51–59

Chaverri A, Hernández O (1995) Ecology and management in montane oak forests: an option for conserving biodiversity. In: Churchill SP, Balslev H, Forero E, Luteyn JL (eds) Biodiversity and conservation of Neotropical montane forests. New York Botanical Garden Press, Bronx, NY, pp 609–617

Chaverri A, Herrera B (1996) Criterios e indicadores para el manejo forestal sostenible de los bosques de altura en Centroamérica. FAO/CCAD/CCB-AP. FAO, San José, Costa Rica

Chaverri A, Herrera B (1997) Criterios e indicadores para el manejo forestal sostenible en Centroamérica, con énfasis en bosques de altura. In: Morales E, Cartín F (eds) Congreso Forestal Centroaméricano: resúmenes de ponencias. FAO, San José, Costa Rica, pp 6–8

Chaverri A, Herrera B, Herrera-McBryde O (1997) La Amistad Biosphere Reserve, Costa Rica – Panamá. In: Davis SD, Heywood VH, Herrera-McBryde O, Villa-Lobos J, Hamilton AC (eds) Centers of Plant Diversity: a Guide and Strategy for their Conservation, vol 3. The Americas. World Conservation Union (IUCN), World Wildlife Fund (WWF) and Smithsonian Institution (SI), Washington, DC, pp 209–214

Dawkins HC, Philip MS (1998) Tropical moist forest silviculture and management: a history of success and failure. CABI, Oxon, UK

FAO (1997) Informe de la reunión de expertos sobre criterios e indicadores para la ordenación forestal sostenible en Centromérica. UN Food and Agricultural Organization, FAO/CCAD/CCAB-AP, Tegucigalpa, Honduras

FAO (2001) Criteria and indicators for sustainable forest management. UN Food and Agricultural Organization, Rome, Forest Management Working Papers, Work Pap 5 (http://www.fao.org/DOCREP/004/AC135E/AC135E00.hmt)

FAO (n.d) Sustainable forest management in Central America: proposal of criteria and indicators at the Forest Management Unit (FMU). UN Food and Agricultural Organization, FAO/CCAD/CCAB-AP/IUCN, Moravia, Costa Rica

Granholm H, Vähänen T, Sahlber S (eds) (1996) Seminario intergubernamental sobre criterios e indicadores para el manejo forestal sostenible. Ministry of Agricultura and Forestry, Helsinski, Finland

Herrera B, Corrales L (2004) Metodología para la selección de criterios e indicadores y análisis de verificadores para la evaluación del manejo forestal a escala de paisaje. Universidad Rafael Landivar, Ciudad de Guatemala, Guatemala, Ser Doc Téc 14

ITTO (1992) Criterios para la evaluación de la ordenación sostenible de los bosques tropicales. International Tropical Timber Organization, Yokohama, Serie ITTO/Desarrollo de Políticas, no 3

Jiménez W, Chaverri A, Miranda R, Rojas I (1988) Aproximaciones silviculturales al manejo de un robledal en San Gerardo de Dota, Costa Rica. Turrialba 38(3):208–214

Kappelle M (1996) Los bosques de roble (*Quercus*) de la Cordillera de Talamanca, Costa Rica: biodiversidad, ecología, conservación y desarrollo. Instituto Nacional de Biodiversidad (INBio), Santo Domingo de Heredia, Costa Rica

Kappelle M (2004) Tropical montane forests. In: Burley J, Evans J, Youngquist JA (eds) Encyclopedia of Forest Sciences, vol 4. Elsevier, Oxford, UK, pp 1782–1793

Kappelle M, Brown AD (eds) (2001) Bosques nublados del Neotrópico. Instituto Nacional de Biodiversidad (INBio), Santo Domingo de Heredia, Costa Rica

Kappelle M, Juárez ME (2000) Mountain forests, biodiversity and people in Costa Rica. In: Price M, Butt N (eds) Forests in sustainable mountain development: a state of knowledge report for 2000. CABI Publ, Oxon, UK, IUFRO Res Ser 5, pp 38–46

Lammerts van Bueren EM, Blom EM (1997) Hierarchical framework for the formulation of sustainable forest management standards. Tropenbos, Backhuys, Wageningen, The Netherlands

Lauer W (1968) Problemas de la division fitogeográfica en América Central. In: Troll C (ed) Geoecología de las regiones montañosas de las Américas Tropicales. Coll Geogr B 9:139–156

McGinley K, Finegan B (2003) The ecological sustainability of tropical forest management: evaluation of the national forest management standards of Costa Rica and Nicaragua, with emphasis on the need for adaptive management. For Pol Econ 5:421–431

Morales JM, Brown AD (1995) Entrega de los resultados del simposio sobre bosques nublados tropicales realizado en Puerto Rico en junio de 1993. Yungas 5(1):2–4

Orozco L (1991) Estudio ecológico y estructural de seis comunidades boscosas de la parte noreste de la Cordillera de Talamanca, Costa Rica. MSc Thesis, Universidad Nacional, Heredia

Prabhu R, Colfer CJP, Dudley RG (1999) Guidelines for developing, testing, and selecting criteria and indicators for sustainable forest management. The criteria and indicators toolbox series no 1. Center for International Forestry Research (CIFOR), Bogor, Indonesia

Samudio R (2001) Panamá. In: Kappelle M, Brown AD (eds) Bosques nublados del Neotrópico. Instituto Nacional de Biodiversidad (INBio), Heredia, pp 371–398

Sheil D, Nasi R, Johnson B (2004) Ecological criteria and indicators for tropical forest landscapes: challenges in the search for progress. Ecol Soc 9(1):7 (http://www.ecologyandsociety.org/vol9/iss1/art7)

Stadtmüller T (1994) Impacto hidrológico del manejo forestal de bosques naturales tropicales: medidas para mitigarlo. CATIE, Turrialba, Col Silvic Manejo Bosq Nat 10

33 Economic Valuation of Water Supply as a Key Environmental Service Provided by Montane Oak Forest Watershed Areas in Costa Rica

G. Barrantes Moreno

33.1 Introduction

Three basic reasons for conserving natural ecosystems can be considered. The first one is ecological: we have to preserve ecosystems for maintaining vital functions for all life on Earth, including human beings. A second reason is financial and socioeconomic. It is based on the concept that ecosystems provide raw materials (goods) and services for production and consumption (Daily 1997; Otárola et al. 2000; MA 2003). The third reason, not less important, is ethical and based on the idea that we have to respect all forms of life, and that life has an intrinsic value. These reasons imply a responsibility that we have to conserve ecosystems in order to be able to meet the basic needs for survival of our own and other species.

The socioeconomical valuation of ecosystems is increasingly based on flows of environmental goods and services they provide. We define 'environmental service' as the socially useful flow that ecosystems provide, which is used without being affected or lost during its production or consumption. Examples are water regulation, weather regulation, and carbon fixation (MA 2003). Similarly, environmental goods are here defined as the products that natural ecosystems provide to cover different societal needs, especially for improvement of living conditions, without been affected or lost during their production or consumption. Examples are wood, wicker, edible spices, meat, and water.

Flows of environmental goods and services are perceived as social benefits provided by natural ecosystems such as forests (Daily 1997; MA 2003). These social benefits are directly related to the quality and extension of ecosystems that provide benefits, as well as the integrity of their ecological functions. The more deteriorated these functions are, the more the well being of the human

Ecological Studies, Vol. 185
M. Kappelle (Ed.) Ecology and Conservation of Neotropical Montane Oak Forests
© Springer-Verlag Berlin Heidelberg 2006

population tends to deteriorate. This is why the conservation and sustainable use of natural resources should be guaranteed through integral maintenance of ecological functions that support ecosystem stability, and assure the permanent flow of environmental services that sustain present and future populations.

Recently, significant advances have been made in the field of promoting the application of environmental service valuation, especially in terms of financial approaches. This has facilitated the implementation of rates and measures concerning the promotion of a mechanism known as 'payment of environmental services'. In this regard, the country of Costa Rica has accomplished important advances during the last 10 years (Otárola et al. 2000; Barrantes and Vega 2002). In fact, water supply is one of the most important environmental services that have been addressed by policy makers in Costa Rica.

We offer a methodological approach to conduct economic appraisal of this environmental service (fresh-water supply), and apply it to the Savegre River watershed. In this region, montane oak and non-oak forests predominate (Kappelle 1996). Additionally, we present the case of a provincial utility company, the Empresa de Servicios Publicos de Heredia (ESPH), which has been at the forefront of implementing mechanisms for payments of environmental services such as water supply (Gámez 2004).

33.2 A Transformed Vision for Use of Environmental Services

Thinking of forests as mere sources of lumber represents an undervaluation of this precious resource. We know that a wide variety of environmental goods and services exists in forests that benefit society and add value to these ecosystems (Daily 1997; MA 2003). For instance, services such as scenic beauty (ecotourism!), water supply for different economic and domestic sectors, carbon fixation by impacting greenhouse gases (GHGs), soil conservation for agricultural production, genetic material for research, and the supply of food and medicinal products tremendously benefit national and international communities.

It needs to be pointed out that deforestation and land use change constitute the worst threat for future provision of such goods and environmental services. Despite the economic importance of environmental services from forest ecosystems, they are exploited as an environmental subsidy to the economy. They are neither properly analyzed nor well incorporated in the production costs of numerous economic activities.

In other words, there is an economic profit obtained by indiscriminately using environmental goods and services from forest ecosystems. However, the costs that this use represents to the society for maintaining this flow of goods and ser ices are not (yet) taken into account. This explains the lack of financial support for maintenance, protection, and conservation of these vital forest ecosystems.

33.3 Importance of Forests for Providing Water to Society

Fresh-water availability is a result of the retention capacity of forest ecosystems (Chap. 21). This retention and supply function is considered a key environmental service that benefits society both in terms of use for economic production and direct consumption. More specifically, 'fresh-water supply' is a so-called *provisioning service*, whereas 'water regulation' and 'water purification' are so-called *regulating services* (MA 2003).

Water supply allows the development of other goods and services on which society depends. The loss of forest ecosystems directly affects water regulation and, consequently, a diverse set of activities related to agriculture, aquaculture, industry, tourism, hydroelectricity, and drinking water supply. Other ecosystems related to water resources will be affected, too (Rudas 1995). More dramatically, the quality and quantity of available fresh-water resources determine the potential economical growth of a region or country (Azqueta 1994; Reynolds 1997).

It is expected that forest removal diminishes infiltration. This, in turn, produces an increment in water runoff during the rainy season, and has a negative effect on water storage (Álvarez 1995). Generally, there is a direct relationship between vegetation cover and natural water flows in tropical forests: the greater the vegetation extent, the larger the available water sources. In this regard, conservation, protection, and recuperation of watersheds is strongly recommended, as a larger forest will provide a better regulation of water resources, and diminish the amount of sediments that harm production structures (Calvo 1990; CCT-CINTERPEDS 1995).

Forest is often the most efficient type of cover, compared to other vegetation cover types, in terms of regulating water resources of good quality and large quantity. As a matter of fact, forest cover promotes water retention (Chap. 21), since root systems favor greater and better infiltration and reduce superficial runoff speeds (Ander 1991; Chap. 22). A study by CCT and CINTERPEDS (1995) determined that the runoff speed is less under forest cover than under pasture cover. This result supports the hypothesis that forests have a larger infiltration capacity (Table 33.1). Water quality was also evaluated, and a positive quality of 81.44 % was determined under a forest cover, whereas only 31.37 % was estimated for pastures.

The conversion of a forest into a pasture or other cover types may strongly reduce the soil's infiltration capacity, since the subsoil's absorption volume is favored in watershed areas with higher forest cover (Heuveldop et al. 1986).

Table 33.1. Average annual runoff (m^3 ha^{-1} $year^{-1}$) in Holdridge's life zones occurring in the Costa Rican study area. Positive and negative quality levels are evaluated as a function of the amount of sediment present in the water under two different Costa Rican vegetation cover types (after CCT and CINTERPEDS 1995)

Holdridge's life zone	Forest cover			Pasture cover		
	Total runoff	Positive quality	Negative quality	Total runoff	Positive quality	Negative quality
Tropical wet (T-w)	36,740	30,610	6,130	40,060	15,010	25,050
Premontane wet (P-w)	18,610	16,280	2,330	21,460	7,150	14,310
Premontane rainforest (P-r)	42,490	31,870	10,620	44,360	11,090	33,270
Lower montane wet (LM-w)	16,870	15,330	1,540	18,900	6,880	12,020
Montane wet (M-w)	9,120	8,070	1,050	10,340	3,580	6,760
Montane rainforest (M-r)	20,550	15,420	5,130	20,660	5,160	15,500
Absolute totals	144,380	117,580	26,800	155,780	48,870	106,910
Relative totals (%)	100.00	81.44	18.56	100.00	31.37	68.63

Table 33.2. Water infiltration into the ground under three different tropical vegetation cover types (after Heuveldop et al. 1986)

Time	Vegetation cover type Forest cover		Pasture cover		Bare soil		Total
(min)	(cm^3)	(%)	(cm^3)	(%)	(cm^3)	(%)	(cm^3)
5	60	69.52	21	24.33	5	6.14	86.30
10	119	67.70	46	26.05	11	6.26	175.80
30	360	68.90	127	24.31	36	6.79	522.50
60	715	69.55	250	24.32	63	6.13	1,028.00
Mean	–	68.92	–	24.75	–	6.33	–

According to Heuveldop et al. (1986), in an infiltration scenario with three types of cover (forests, pastures and bare soil), forest has the greatest infiltration efficiency (68.92 %), compared to pasture and bare soil (24.75 and 6.33 %, respectively; Table 33.2).

33.4 Economic-Ecological Valuation of Water

33.4.1 The Need for Economic-Ecological Valuation

Water supply is one of the main environmental services that should be valuated in economic and ecological terms in order to correctly adjust the rates established for water use. The purpose is to promote its rational use and the options for water conservation. Thus, economic-ecological valuation of water provides the monetary resources that help the start of a process to adjust water fees. In this way, a price mechanism is applied that will seek the optimum use of water in different social sectors. The idea behind this approach is that through a payment, water users recognize and pay tribute to water providers (owners or administrators) that maintain ecosystems that offer the water environmental service to society. Water resource owners and administrators, in turn, assume their responsibility to conserve the water-providing ecosystem – the forest – and, in fact, have a financial opportunity to do so.

The economic-ecologic valuation of water supply responds to the need for maintenance of forest ecosystems that offer large quantities of high-quality water resources. In this regard, there are three components to evaluate from a financial point of view: the forest's water productivity, the restoration of deforested areas, and water as production input. Once economically appraised, these aspects can be incorporated into the rate systems in order to environmentally adjust the actual fees. The economic evaluation of water resources aims to estimate water supply and demand (water budget). This information is key to the implementation of rate systems related to water utilization. Furthermore, the data obtained can be used to evaluate developmental possibilities, and to formulate measures oriented to the conservation and sustained use of water. The implementation of an adjusted water use fee presents the opportunity to generate additional income urgently needed for recovery, protection, and conservation of forested watersheds in tropical montane areas.

33.4.2 Capture Value of Forest Water Productivity

In order to assess the value of water as a key environmental service provided by forests, its production value should be viewed as a function of water retention (direct use value), in addition to other services provided (carbon fixation, scenic beauty, biodiversity, etc.). The increase of forest cover entails an opportunity cost, due to the decline in potential income that any economic activity could generate. This requires compensation to landowners, by an amount equal to, or higher than the opportunity cost, so that the owners can protect their lands and conserve the watersheds. This compensation should take place as a transfer of financial resources generated on the basis of provided goods and services.

Such a financial transfer is justified, because the forest's conservation, protection, and recovery are all aspects that generate positive externalities for economic activities, through a continuous and permanent flow of environmental services. Likewise, the costs of operation of productive systems could diminish over time, by having to spend less on systems maintenance. Similarly, there will be no need to exploit costly environmental services further away, as cheaper services are maintained nearby.

A forest's water productivity is determined by the water quantity it captures annually. Its financial value is related to the economic activity that makes use of this water. Conversion of forest to non-forest may be justified only if the annual revenue from non-forest use is greater than that generated by the forest, taking into account the environmental services it offers to society. In this respect, a hectare of forest needs to be protected when the value of its environmental services equals the opportunity cost of other land uses. In this way, the recovery and conservation of existing forests is based, in part, on its financial revenue from its environmental services.

Hence, the cost of opportunity is considered a valid method to financially appraise the water capture function of a forest, and valuate other environmental services for specific economic activities. This appraisal complies with the need to have an economic indicator available that valuates a forest's productivity. Forest maintenance by landowners should be compensated by society, as it will enable them to consider forest maintenance as a financially profitable activity (Castro and Barrantes 1998). To estimate the water capture value as a key component that determines the water productivity of a forest, it is necessary to:

- Estimate the annual water volume captured and stored by forests in watershed areas;
- Calculate the opportunity cost of land use in these areas; and
- Assess the forest's importance in terms of water productivity when compared to other environmental goods and services (e.g., biodiversity).

In addition, it is necessary to consider the positive effect that forests have on the quality of surface runoff water. The following equation (Eq. 33.1) estimates the capture value of a forest:

$$VC = \sum_{i=1}^{n} \frac{\alpha_i B_i Ab_i}{Oc_1} \qquad (33.1)$$

where VC is the water capture value of the forest in ¢ m^{-3}, in which ¢ is the Costa Rican currency known as *colones* (in July 2005, US $ 1 was equivalent to ¢ 475); B_i is the opportunity cost of the financial activity that competes with the forest for other land use in the watershed i (¢ ha^{-1} $year^{-1}$); Ab_i is the forest area in the watershed i (ha); Oc_i is the captured water volume in the watershed i (m^3 $year^{-1}$); α_i is the importance of the watershed's forest i in function of quantity and quality of the water resource ($0 \leq \alpha \leq 1$); and n is the number of watersheds involved.

Barrantes and Vega (2001) estimated that a total of 41,427 ha of forest existed in the year 2000 in the montane oak forest region and adjacent areas of the Savegre River watershed region (0–3,491 m a.s.l.) in the Los Santos Forest Reserve on the Pacific slope of Costa Rica's Cordillera de Talamanca (Kappelle 1996; Chaps. 10, 17, 21, 22, 23, 24, 25, 26, 30 and 31). Forest cover represents about 70 % of the watershed's total area. Twenty-four percent of the watershed is currently covered by pasture, making cattle raising the main activity that competes for land (Kappelle and Juárez 1995; Kappelle 1996; Chap. 30). The opportunity cost considered in the study was US \$ 119 ha $^{-1}$ year $^{-1}$. The projected opportunity cost was based on the net profits of cattle ranching, since this is the main competing activity.

Annual rainfall in the watershed varied in the range 2,293–6,108 mm year $^{-1}$ and average annual rainfall was 4,201 mm (Barrantes and Vega 2001). Actual evapotranspiration was 906 mm annually, implying a water supply of 3,295 mm per year, i.e., about $1{,}943 \times 10^6$ m ^{3} year $^{-1}$.

The importance of forest as a water resource was considered at 41.4 %. This percentage represents the portion of the opportunity cost that should be compensated by water consumers and paid to landowners involved in forest protection. Applying Eq. (33.1), a capture value of US \$ 0.001 m $^{-3}$ is obtained.

33.4.3 Restoration Value of Forest Ecosystems

During the last few decades, Neotropical montane forests including cloud forests have been severely degraded (Kappelle and Brown 2001). Natural recovery of tropical highland forests such as montane oak forests appears to be extremely slow (Ewel 1980; Kappelle et al. 1996). At many sites, active forest restoration seems to be the only conservation strategy that effectively addresses tropical montane forest recovery following clearing (Holl and Kappelle 1999). Restoration of degraded forests at the watershed level is a useful mechanism that enhances conservation of surface and ground water, and that prevents further soil erosion (Ramakrishna 1997; Chap. 21).

Forest restoration will have an implicit cost that needs to be considered via economic-ecological valuation of water, as this procedure will enable payment of funds needed for protection, recovery, restoration, and conservation of watersheds at high elevation (Castro and Barrantes 1998). Costs incurred in forest restoration should be determined by salary expenses, health insurances, employee benefits, fuel expenses, transportation, infrastructure, and other operational expenses as well as incentives needed for environmental protection.

Restoration payments should at least equal the costs needed for a degraded ecosystem to return to the natural conditions it had before human intervention took place (UN 1993). Therefore, an assessment should take place taking into account all costs associated with water resource protection. This exercise requires:

- Calculation of the amount of hectares that should be restored;
- Calculation of the restoration cost considering the natural forest condition before clearing;

- Assessment on the forest's importance in terms of water productivity; and
- Estimation of the water capture volume in the watershed.

Thus, in operational terms, the necessary resources for establishing a measure of recovery, protection, and maintenance of watersheds can be assessed using Eq. (33.2):

$$VP = \sum_{i=1}^{n} \sum_{j=1}^{m} \frac{\delta_{ij} C_{ij} Ar_i}{Oc_1} \qquad (33.2)$$

where VP represents forest restoration costs for the watersheds; C_{ij} represents the costs of activity j, needed for forest restoration in watershed i (¢ ha^{-1} $year^{-1}$); Ar_i is the watershed's restoration area i (ha); δ_{ij} is the fraction of the cost j needed for forest restoration in function of the watershed's water resource i (%); m is the number of inputs used; and n is the number of watersheds involved.

33.4.4 The Savegre River Watershed Area

In 2000 the Savegre River watershed in the Los Santos Forest Reserve presented conflicts of interest in land use in 24% (14,070 ha) of its area (Barrantes and Vega 2001). These conflicts are directly related to unsustainable activities affecting the terrain and soil that suffer from erosion.

The cost to cover forest reforestation in this watershed during a 5-year period was estimated at ¢ 297,316 ha^{-1} $year^{-1}$, with 43.31 % of this amount to be invested during the first year of restoration (CATIE 1996). After the fourth year, this amount is reduced to a fairly fixed amount (10 % of the total cost), as this is related only to maintenance costs. If we consider reforestation to be the only suitable restoration technique for this watershed, the restoration cost during the first year will equal US \$ 397 ha^{-1}, and water importance is estimated at 41.40 % and the water capture volume at some $1{,}943 \times 10^6$ m^3 annually. Then, Eq. (33.2) implies a restoration value of US \$ 0.0012 m^{-3}.

33.5 The ESPH Case: Environmental Service Payments in Practice

33.5.1 Legal Framework for Environmental Service Payments

Costa Rica's *Empresa de Servicios Públicos de Heredia* S.A. (ESPH) is a governmental utility company that provides drinking water from upstream montane oak and non-oak forests to the counties of Heredia, San Rafael, and San Isidro in the province of Heredia. Recently, ESPH decided to implement a scheme of *environmental service payments* (ESPs; in Spanish, *Pago de Servicios Ambientales*, PSA), in compliance with governmental Law #7789 and taking advantage of the country's legal and institutional structure related to ESP. The legal framework that permits such an initiative is provided by national laws # 7554 (Environment), # 7788 (Biodiversity), # 7593 (Regulating Authority of Public Services, ARESEP) and # 7575 (Forestry); for details, the reader can refer to the corresponding legal texts in the official, state-owned bulletin La Gaceta (1995, 1996 and 1998).

The ESP scheme recognizes the importance of forest conservation for the socioeconomic well-being of people, and acknowledges the need to invoice this service, which in turn ensures water supply to the population on the long term. Following the principle of social equity, this legal framework ensures that payments are made to forest owners, both public and private, who have assumed conservation costs. In this way, they are monetarily compensated for the opportunity costs they lose when rejecting traditional land use (Gámez 2004). As Gámez (2004) synthesizes, Forestry Law # 7575 defines the concept 'environmental service' as:

„*Those services which forests and plantations provide and which relate directly to environmental protection. They include mitigation of greenhouse gas (GHG) emissions (fixation, reduction, capture, storage and absorption), water protection for drinking and hydroelectric energy, and biodiversity protection for conservation, research and genetic improvement, and scenic beauty for tourism*".

Gámez (2004) identifies four environmental services that can be subject to a direct monetary compensation, amongst which is the protection of water resources (La Gaceta 1996). The legal innovation offered by the EPS scheme refers principally to forestry criteria (Chap. 32), and to the function of carbon fixation. On the other hand, the 1998 Biodiversity Law puts greater emphasis on water as a key environmental service.

33.5.2 Paying for Water Conservation

Recently, the economic value of the high-quality water-capturing, forested upper watersheds of the Segundo, Ciruelas, Bermúdez, Tibas and Pará rivers in the highlands of Heredia has been fully recognized by Heredian residents willing to pay, through the ESPH water use fee system.

More specifically, ESPH introduced adjustments of water use rates to make a sustainable and non-destructive water use really work. This resulted in the fact that end-consumers of water contribute directly to funding the costs of protecting the environment (forest conservation), and the service this provides to society. Payments are in particular made to the state-administered Braulio Carrillo National Park, and to private landowners taking care of forests on strategic grounds in the watershed.

The procedure began with an additional fee of ¢ 1.90 per cubic meter consumed (in 2004, this was already ¢ 3.80 m^{-3}). Forest owners who participated under this scheme received an amount of ¢ 23,000 ha^{-1} $year^{-1}$ from ESPH for protecting the water capturing zone. Currently, 850 ha is under protection and included in the ESP system. The goal is to expand to more than 1,600 ha over the next few years. The financial contribution made by the end-users responds to the principles of social equity and of 'who yields profit from this resource must pay'.

As pointed out by Gámez (2004), ESPH's initiative to implement an ESP system reflects a pragmatic vision to locally contribute to the solution of real problems and future threats that water resources may face in this region. The modest, yet determined financial self-sufficiency of this so-called PROCUENCAS program builds upon local organization and support to take direct, concrete action at the sub-watershed scale. This is a clear, practical, and healthy example of decentralization, in which it is possible to confront a problem from the beginning, and achieve results without having to either recur to legal processes, blame the lack of proactive institutional action, or justify the attitude of 'not doing anything'.

The 'water fee' reflects the progressive advancement of a consolidated public service and a consistent culture that pays for a regular utility service (high-quality drinking water). The investment in environmental protection evidences the gradual adjustment of a traditional water use culture toward a more integrated type of ecosystem management.

33.5.3 Investing in Maintaining Environmental Services

Well aware of the imminent risk to lose natural forest ecosystems ensuring water supply, and the costs of restoring those precious ecosystems, ESPH actively negotiates prices in favor of a more healthy future for sub-watershed forest. Until today, only public utility companies like ESPH and Costa Rica's National Company for Power and Light (CNFL) have made investments in the environment as part of their operations. In this regard, Heredian people and utility companies have responded as custodians of a common water resource heritage of key value. In this manner, Heredia's unique wealth of water resources is, through its PROCUENCAS program, correctly considered a natural asset and free endowment fund of exceptional value.

33.6 Conclusion

A methodological approach has been applied to conduct an economic appraisal of fresh-water supply in the montane forest zone of Costa Rica's Savegre River watershed. Additionally, the case of a provincial utility company is presented in which mechanisms for payments of environmental services (water supply) are involved. It is concluded that these novel approaches to economically valuating key environmental services offered by these montane oak and non-oak forests can greatly contribute to their long-term conservation.

References

Álvarez E (1995) Impacto hidrológico de la (de) (re) forestación en las regiones tropicales. Dirección de Ecología y de Recursos Naturales, Medellín, Colombia

Ander E (1991) El desafío ecológico. EUNED, San José, Costa Rica

Azqueta J (1994) Valoración económica de la calidad ambiental. Universidad de Alcalá de Henares, Madrid

Barrantes G, Vega M (2001) Evaluación del servicio ambiental hídrico en la cuenca del río Savegre con fines de ordenamiento territorial, Costa Rica. Fundación Instituto de Políticas para la Sostenibilidad (IPS), Heredia, Costa Rica

Barrantes G, Vega M (2002) Análisis del impacto social, económico, ambiental y organizacional de los incentivos a la conservación y del pago de servicios ambientales en Costa Rica. Fundación Instituto de Políticas para la Sostenibilidad (IPS), Heredia, Costa Rica

Calvo J (1990) Water resource development in Costa Rica (1970–2000). Hydrol Sci J 35(2):4

Castro E, Barrantes G (1998) Valoración económico ecológico del recurso hídrico en la cuenca de Arenal: el agua un flujo permanente de ingreso. Fundación Instituto de Políticas para la Sostenibilidad (IPS), Heredia, Costa Rica

CATIE (1996) Costos de establecimiento y manejo de plantaciones forestales y sistemas agroforestales en Costa Rica. Centro Agronómico de Investigación y Enseñanza, Turrialba, Costa Rica

CCT and CINTERPEDS (1995) Valoración económico ecológica del agua: primera aproximación para la interiorización de costos. Centro Científico Tropical and Centro Internacional en Política Económica para el Desarrollo Sostenible. CCT, San José, Costa Rica

Daily GC (1997) Nature's services: societal dependence on natural ecosystems. Island Press, Washington, DC

Ewel J (1980) Tropical succession: manifold routes to maturity. Biotropica 12 Suppl 2:2–7

Gámez L (2004) Los recursos hídricos como servicio ambiental y aplicaciones prácticas de su valoración: el caso de la Empresa de Servicios Públicos de Heredia (ESPH), Costa Rica. ESPH, Heredia, Costa Rica

Heuveldop J, Tasis JP, Quirós Conejo S, Espinoza Prieto L (1986) Agroclimatología tropical, 1st edn. EUNED, San José, Costa Rica

Holl KD, Kappelle M (1999) Tropical forest recovery and restoration. Trends Ecol Evol 14(10):378–379

Kappelle M (1996) Los bosques de roble (*Quercus*) de la Cordillera de Talamanca, Costa Rica: biodiversidad, ecología, conservación y desarrollo. Instituto Nacional de Biodiversidad (INBio), Santo Domingo de Heredia, Costa Rica

Kappelle M, Brown AD (eds) (2001) Bosques nublados del Neotrópico. Instituto Nacional de Biodiversidad (INBio) and World Conservation Union (IUCN), Santo Domingo de Heredia, Costa Rica

Kappelle M, Juárez ME (1995) Agroecological zonation along an altitudinal gradient in the montane belt of the Los Santos Forest Reserve in Costa Rica. Mount Res Dev 15(1):19–37

Kappelle M, Geuze T, Leal ME, Cleef AM (1996) Successional age and forest structure in a Costa Rican upper montane *Quercus* forest. J Trop Ecol 12:681–698

La Gaceta (1995) Leyes y Decretos: Ley Orgánica del Ambiente. San José, Costa Rica, Asamblea Legislativa de la República de Costa Rica en Diario Oficial La Gaceta, 13 de noviembre 1995, no 215

La Gaceta (1996) Leyes y Decretos: Asamblea Legislativa de la República y Gobierno de Costa Rica, Martes 16 de abril 1996, Ley Forestal no 7575. San José, Costa Rica, Alcance no 21 en Diario Oficial La Gaceta no 72, pp 1–8

La Gaceta (1998) Leyes y Decretos: Ley de Biodiversidad no 7788. Asamblea Legislativa de la República de Costa Rica, San José, Costa Rica

MA (2003) Ecosystems and human well-being: a framework for assessment. Millenium Ecosystem Assessment. Island Press, Washington, DC

Otárola M, Venegas I, Alfaro M (2000) Sistema de compensación de servicios ambientales para los robledales de la Cordillera de Talamanca, Costa Rica. Universidad Nacional, Heredia, Cienc Amb 18:37–59

Ramakrishna B (1997) Estrategias de extensión para el manejo integrado de cuencas hidrográficas: conceptos y experiencias. Instituto Interamericano de Cooperación para la Agricultura (IICA), Coronado, Costa Rica

Reynolds J (1997) Evaluación de los recursos hídricos en Costa Rica: disponibilidad y utilización. Proyecto de Cuentas Ambiéntales. CINPE, Universidad Nacional (UNA), Heredia, Costa Rica

Rudas G (1995) Uso del agua e incentivos económicos para la conservación de cuencas hidrográficas. Universidad Nacional, Bogotá, Colombia

UN (1993) The handbook of national accounting: integrated environmental and economic accounting. United Nations, New York, Ser F, no 61

Part VII
Synthesis

34 Neotropical Montane Oak Forests: Overview and Outlook

M. Kappelle

34.1 Introduction

The preceding chapters of this book discuss scientific research results on natural and managed oak forests growing in the highlands of the American Tropics. Chapter authors highlight evolutionary, ecological and socioeconomic aspects of specific oak forests and identify existing gaps in our understanding. Their work and studies published elsewhere form the basis for this chapter's overview of the current state of knowledge on ecological patterns and processes which determine the structure and functioning of these magnificent forests. The increasingly important human use and much needed conservation of these biodiverse and threatened habitats are dealt with as well. Cross-cutting trends and issues are discussed and conclusions drawn.

Neotropical montane oak forests are a special type of tropical montane forests (Kappelle 2004), in which tall oak trees (*Quercus*) dominate a 30- to 60-m-high forest canopy, often in pure stands or mixed with pines (*Pinus*) or other predominant trees. Though Neotropical oak forest history, structure, species diversity, species interactions, ecosystem functioning, post-disturbance recovery, and human use may vary considerably over short geographical ranges, regional trends are now becoming evident. Research presented in this book clearly illustrates these large-scale trends. The following sections in this chapter synthesize our knowledge for each of these trends and themes.

Ecological Studies, Vol. 185
M. Kappelle (Ed.) Ecology and Conservation of Neotropical Montane Oak Forests
© Springer-Verlag Berlin Heidelberg 2006

34.2 Modern Distribution and Biogeographical History

Highland oak forests in the Tropics of the American continent are latitudinally and longitudinally distributed from central Mexico (i.e. south of the line Mazatlán–Fresnillo–Tampico) through Central America (Guatemala, Belize's Maya Mountains, El Salvador, Honduras, Nicaragua, Costa Rica and Panama) into Andean Colombia (e.g. Antioquia, Boyacá, Cauca, Cundinamarca, Huila and Nariño) just north from Ecuador, where the southernmost native oaks grow at the Pasto Airport in Nariño (Pulido et al. in Chap. 11, Fig. 34.1). Oaks are also found in the insular Caribbean, in the western region of Cuba between Pinar del Río and Matanzas (Nixon in Chap. 1).

Altitudinally, Neotropical highland oak forests range from 500 m elevation up to 3,300 m in the Colombian Andes (Hooghiemstra in Chap. 2; Chap. 11) and 3,400 m in Costa Rica's Chirripó National Park (Kappelle and Van Uffelen in Chap. 4). *Quercus*, however, is also found at altitudes below 500 m, in lowland forests such as those thriving in Costa Rica's north-western Pacific, dry Guanacaste region (Kappelle in Chap. 10).

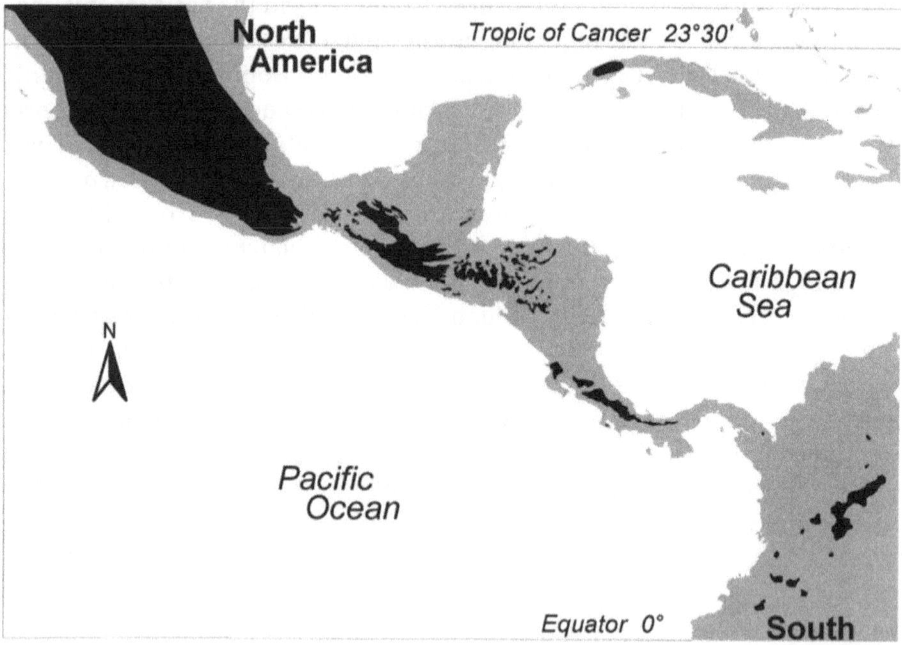

Fig. 34.1. Map showing the distribution of the genus *Quercus* (oak) in the Neotropics. Tropical oak in the Americas is found only in the Northern Hemisphere, at all latitudes between the Tropic of Cancer and the Equator. Most oaks are restricted to elevations over 500 m, such as found in the Mexican and Cuban sierras, the Belizean mountains, the Guatemalan volcanoes, the Honduran and Nicaraguan highlands, the Costa Rican and Panamanian cordilleras, and the Colombian Andes. Map preparation by Marco V. Castro

A common phenomenon observed on Neotropical mountains is the turnover of *Quercus* species along altitudinal gradients (Kappelle and Brown 2001). Oak species appear to be adapted to specific climatic and soil conditions prevailing at determined elevations. In Costa Rica, some 14 *Quercus* species are altitudinally distributed from sea level up to 3,400 m elevation. In this country, highest oak species diversity is found between 1,000 and 2,000 m where some 10–11 species may co-occur (Chap. 10). Similarly, in the small La Chinantla area of Oaxaca, Mexico, at least six species co-occur between 1,000 and 3,000 m elevation (Meave in Chap. 9).

The current latitudinal, longitudinal and altitudinal distribution of the genus *Quercus* and oak forests in the American Tropics is largely defined by the geological and climatic history of the continent, and the evolution of its flora (Kappelle et al. 1992). During the Pleistocene, the distribution of Neotropical montane oak forests was strongly affected by glacial cycles which included recurrent periods of cooling and warming. These climate fluctuations, exemplified by the Younger Dryas cooling event at the Pleistocene–Holocene boundary, made the oak forest altitudinal belt move up and down over considerable elevational distances (Islebe and Hooghiemstra in Chap. 3). As a result of this dynamic geoclimatic history, the modern flora of Neotropical montane oak forests is made up of a blend of tropical (60–75 %), temperate (18–35 %) and cosmopolitan (5–10 %) plant genera (Kappelle et al. 1992). Whereas at higher latitudes northern temperate (Holarctic) genera are more important, closer to the Equator southern temperate (Austral-Antarctic) and Neotropical genera become increasingly abundant (Kappelle in Chap. 10).

Evidence from long marine and terrestrial pollen records show how the distribution of the wind-pollinated Holarctic genus *Quercus* extended southwards from today's North America, through Central America, into northwestern South America (Hooghiemstra in Chap. 2). *Quercus* arrived in western Central Mexico at the start of the late Miocene, about 10 million years ago (Fournier 1982). The closure of the inter-oceanic channel which connected the tropical seas of the modern Atlantic and Pacific Oceans, and the subsequent formation of the Panamanian Isthmus took place some 4–5 million years ago (Keighwin 1982). However, detailed analysis of a long fossil pollen record from Colombia showed that *Quercus* migrated into today's basin of Bogotá not earlier than some 470,000 years before present (Van't Veer and Hooghiemstra 2000). As Hooghiemstra suggests (Chap. 2), the low migration rate of oak from Mexico into Colombia, even after the consolidation of Central America's Isthmus, may be largely due to its dependence on certain animals responsible for the dispersal of the tree's heavy seeds, known as acorns (López Barrera and Manson in Chap. 13, and Van Den Bergh and Kappelle in Chap. 26). Another reason for its slow migration may be the dependence of oak on the presence of specialized fungi with which the tree forms mycorrhizal associations at soil level (Mueller et al. in Chap. 5).

34.3 Forest Structure

Highland oak forests in tropical America vary considerably in structural features. Mature canopy trees may reach heights of 20–30 m in Mexico (Luna et al. in Chap. 8 and Meave et al. in Chap. 9), 35–45 m and occasionally up to 60 m in Costa Rica (Kappelle in Chap. 10), and 40 m in Colombia (Cuatrecasas 1934; Kappelle 1987; Pulido et al. in Chap. 11). In general, tree height decreases with elevation and may be greater on cloudy slopes, compared to drier slopes at the same altitude (Kappelle 1991; Kappelle and Van Uffelen in Chap. 4). Layering of horizontal forest strata is evident in most of these forests, and includes upper and lower canopy tree layers, a shrub layer, and a herb layer.

The upper canopy layer is often dominated by oak (*Quercus*) and pine (*Pinus*) in Mexico and northern Central America (Chaps. 2, 8, 9, 14 and 16). Pine, however, is not native to Costa Rica, Panama and South America (Hooghiemstra in Chap. 2). In the latter countries, highland oak forest tree canopies are dominated by oak and a variety of other tree families such as Araliaceae, Cunoniaceae, Magnoliaceae and Podocarpaceae (Chaps. 10, 17, 18, 21 and 22). As Pulido et al. (Chap. 11) point out, in some rare cases in Colombia oak forests are actually dominated by *Colombobalanus* (syn.: *Trigonobalanus*), a related and physiognomically similar tree, also in the Fagaceae. The lower canopy tree layer includes smaller individuals of tall trees and medium-sized adult trees in a number of families, e.g. Lauraceae, Melastomataceae, Myrsinaceae, Styracaceae and Theaceae (Chaps. 8–11).

The shrub layer is particularly rich in species, especially in Costa Rica and Colombia (Chaps. 10 and 11) where Ericaceae, Piperaceae, Rosaceae, Rubiaceae and Solanaceae predominate. The Mexican pine-oak forests are less rich in shrubs, though Ericaceae and Asteraceae may locally abound (Chaps. 8 and 9). In wetter montane oak forests closer to the Equator, bamboos (principally *Chusquea*) often accompany the bushes of the shrub layer (Widmer 1993, 1998; Chaps. 4, 10, 11, 21 and 22). Understory specialists such as dwarf palms (*Chamaedorea*, *Geonoma*) may locally predominate over bamboos, especially at elevations of 2,000–2,600 m (Groot et al. in Chap. 15). Some understory shrubs may develop very specific plant–pollinator (insect) relations, e.g. with bumblebees which may be vital to successful reproduction (Wesselingh et al. 1999, 2000).

The herb layer is less well developed and may include Acanthaceae, Araceae, Campanulaceae, Gesneriaceae and Scrophulariaceae, among other families (Chaps. 10 and 11). In the Araceae, the genus *Anthurium* – with terrestrial and epiphytic growth forms – may be particularly species rich and well adapted to locally differing light regimes and substrates (Chap. 15).

Mature oak forest densities of stems≥10 cm DBH (diameter at breast height) range from 500–750 individuals per hectare in Costa Rica (Blaser 1987; Jiménez et al. 1988) to 750–1,500 individuals per hectare in Mexico

(Meave et al. in Chap. 9). Occasionally, diameters of giant oaks may reach values over 120 cm (Chap. 10). With a basal area of about 50 m^2 per ha for stems≥10 cm DBH (Blaser 1987; Jiménez et al. 1988), values for mature old-growth montane oak forests in Costa Rica are among the highest found in tropical forests (Chap. 10). In addition, Hertel et al. (Chap. 22) revealed a very large fine root biomass (>1,300 g m^{-2}) for old-growth oak forest in Costa Rica, compared to other mature, humid tropical forests. It should be noted here that vascular epiphytes may also contribute considerably to oak forest biomass, as has been shown by Wolf and Flamenco-S (Chap. 29). These scholars reported a total of about 1,000 kg dry weight of vascular epiphytes per hectare in a little-disturbed old-growth pine-oak forest stand in Chiapas, Mexico.

34.4 Water and Nutrient Fluxes

Montane oak forests in the humid Tropics of the Americas often experience an almost diurnal presence of clouds (Kappelle in Chap. 10). That is why they are frequently called cloud forests (Spanish: *robledales nubosos, encinares nublados*), especially in Costa Rica, Panama and Colombia. Although knowledge of the overall effect of clouds through fog or horizontal precipitation on the hydrological input in tropical montane forests is still scanty, it has been widely recognized that, compared to other tropical forests, the specific atmospheric humidity regime of these forests represents one of the main factors causing the large array of differences in forest structure and functioning (Bruijnzeel 2001).

Köhler et al. (Chap. 21) measured incident rainfall (gross precipitation) in Costa Rican montane oak forests, and recorded 2,800–2,900 mm $year^{-1}$, of which 70–75 % corresponded to throughfall, 2–17 % to stemflow, and 10–25 % to canopy interception – depending on the successional stage of the forest. These authors found that nutrient concentrations in throughfall water exceeded those measured in incident rainfall. In upper canopy trees, they recorded a pH of stemflow water ranging from 4.2 to 5.7, and noted significantly higher nutrient concentrations in stemflow in these trees than in lower canopy trees. Total annual litter production in mature old-growth oak forest was 12,870 kg ha^{-1} $year^{-1}$. Leaves dominated the litter fraction, which contributed to some 56 % of total litter (Chap. 21).

In Costa Rican mature old-growth oak forests, soils generally have dark-brown humus profiles composed of fine organic material, free of litter fragments, and with only little mineral material. Between 2,000 and 3,000 m elevation, the thickness of soil humus profiles ranges from 10 to 20 cm at the Pacific slope, and up to 40 cm at the Atlantic slope (Kappelle et al. 1995b; Kappelle and Van Uffelen in Chap. 4). In these forests, the soil carbon pool size ranges from about 500 mol m^{-2} in the organic layer to 12.5 mol m^{-2} in the mineral topsoil, the C/N ratios (molar) in both soil layers fluctuating in the range 25–28. Similarly, N concentrations range from 100 to 150 mol m^{-3}, and P concentrations from 2.5 to 12 mol m^{-3} (Hertel et al. in Chap. 22).

34.5 Fungi and Lichens

Fungi are a key component of oak forests worldwide. Neotropical montane oak forests are no exception to that rule. Oaks have evolved often obligatory, highly specialized mutualistic relationships with certain macrofungi, termed ectomycorrhizae, to promote processes such as nutrient cycling, nutrient uptake, and decomposition of organic matter (Mueller and Bills 2004). In Costa Rican highland oak forests, at least 22 woody or tough macrofungi (polypore fungi) have been recorded (Mueller et al. in Chap. 5). According to Mueller et al. (Chap. 5), these species are adapted to significant daily fluctuations in temperature and to high humidity levels throughout the year. To date, an additional 400 species of Agaricales (mushrooms and boletes, in the Basidiomycetes) have been identified from these Costa Rican montane *Quercus*-dominated forests. About half are ectomycorrhizal, the others being putatively saprotrophic (Chap. 5).

Like fungi, lichens form a very important element of Neotropical oak forest biodiversity. Lichen growth in tropical oak forests is often abundant, probably due to the presence of suitable oak substrate and a favourable climate with high precipitation, frequent fog and moderate temperatures (Sipman in Chap. 6). Today, at least 460 lichen species have been recorded by Sipman and colleagues in montane oak forests of Mexico, Guatemala, El Salvador, Costa Rica and Colombia, suggesting that species richness is at least comparable with temperate oak forests, though actual diversity is probably twice as large (Chap. 6).

Epiphytic lichens inhabit a great number of microhabitats on oak trees. Holz (Chap. 7) reported some 60 species of macrolichens dwelling on standing mature oak trees in upper montane Costa Rica. Many demonstrate a specific host preference. Species richness is highly variable, with a mean of 2.7 lichen species per 600 cm^2 of random substrate area. Generally, species richness increases with tree height. Many lichens on the trunks and in the inner canopy tend to grow in pure patches, apparently as a mechanism of adaptation to promote successful interspecific competition for space and light (Chap. 7).

34.6 Plant Species Diversity

Recent work by Nixon (Chap. 1) demonstrates the huge species diversity within the Holarctic genus *Quercus* (oak) in the New World. Probably some 220 species of oak are found on the American continent, distributed from Canada southwards into Colombia. Mexico is the richest country, with some 160–165 species of native *Quercus* (Valencia-A 2004; Valencia-A and Nixon 2004), followed by the USA with 91 recognized species; Cuba and Colombia each harbour only a single native oak species (Nixon 1993 and in Chap. 1). The highlands of central and eastern Mexico have been identified as the major centre of oak species diversity in the Western Hemisphere (Nixon 1993). The best example of Mexico's incredibly large diversity within *Quercus* is found in the state of Oaxaca where some 70 species thrive (48 % of Mexico's total; Valencia-A 2004). This is particularly interesting, considering that only 25 % of this Mexican state is covered by pure oak forests and mixed pine-oak forests (Meave et al. in Chap. 9). In addition, it should be noted that the extraordinary diversity in Mexican oaks is also reflected in the enormous richness of non-plant species associated with oak, such as gall-forming insects, particularly gall wasps (Oyama et al. in Chap. 20). Major lineages of gall wasps are thought to have diverged from Mexico and Guatemala (Kinsey 1936), probably as a result of local oak diversity. This co-evolutionary history may be the main reason why morphological variation in galls induced by insects in oak species in Mexico is particularly high (Chap. 20).

Thousands of vascular plant species are believed to inhabit the montane oak forests of the Neotropics. In Costa Rican oak forests, at least 1,300 species have been formally recorded so far (Van Velzen et al. 1993; Kappelle 1996 and in Chap. 10, Kappelle et al. 2000). Over half of these are dicot species. Almost 500 terrestrial woody plants and some 130 pteridophytes have been documented along a single montane oak forest transect in Costa Rica's Chirripó National Park (Kappelle and Gómez 1992; Kappelle et al. 1995b; Kappelle and Van Uffelen in Chap. 4). Further to the northwest, Wolf and Flamenco-S (Chap. 29) confirm the presence of at least 720 species of vascular epiphytes in the highland pine-oak forests of Chiapas. Apparently, the vascular epiphyte diversity of these Mexican oak forests is strongly dependent on past levels of anthropogenic disturbance (Wolf 2005, and Chap. 29).

Plant diversity in other parts of the oak-inhabited Neotropical highlands may be as great or even greater. However, Pulido et al. (Chap. 11) are convinced that plant species richness of modern Colombian oak forest (577 recorded vascular plant species) is smaller than for similar forest in Costa Rica (Kappelle et al. 1996). This pattern may well be due to the long-lasting human activities which have strongly impacted the oak forests of the Northern Andes, for centuries, if not millennia (Van der Hammen and Gonzalez 1963; Herrera et al. 1989; Hooghiemstra in Chap. 2). Further analysis of large floristic databases, such as currently assembled for the highlands of Chiapas, may reveal specific and detailed patterns in, for example, montane oak forest tree species diversity (González-Espinosa et al. 2004 and in Chap. 16) and vascular epiphyte species richness (Wolf and Flamenco-S 2003 and in Chap. 19).

Not only vascular plant diversity is high in Neotropical montane oak forests; non-vascular plant species richness – especially that of bryophytes – is also extraordinarily large (Holz in Chap. 7). In a recent inventory conducted in Costa Rican highland oak forests, 251 bryophyte species (128 hepatics, one hornwort, and 122 mosses) were recorded (Holz et al. 2002; Holz and Gradstein 2005). The same study revealed the occurrence of 108 epiphytic bryophytes (67 hepatics and 41 mosses) dwelling on stems and branches of a total of 15 oak trees (Chap. 7). Many of these species are considered ecological indicator species which can be used to measure and monitor changes in microclimatic and substrate conditions of disturbed and recovering oak forest (Chaps. 7 and 29).

34.7 Animal Habitat Preferences and Diets

The largest mammal inhabiting the montane oak forests of Mesoamerica is the endangered Baird's tapir, *Tapirus bairdii* (Naranjo and Vaughan 2000). This herbivore mammal has an important impact on the structure and plant diversity of its habitat, as it disperses seeds which are ingested wholly and dropped intact (Tobler 2002; Tobler et al. in Chap. 27). Other large mammals observed in montane oak forest include feline carnivores such as the jaguar, *Panthera onca*, and the puma, *Puma concolor* (Aranda and Sánchez-Cordero 1996), still little studied in the Neotropical highlands (Almeida 2000).

Baird's tapir seems to prefer undisturbed oak forests, tending to avoid patches impacted by human activity. In Mexico and Costa Rica, relative tapir abundance in undisturbed areas is more than twice as high as that in disturbed zones (Chap. 27). Agricultural areas are considered important barriers to tapir movement, and may result in significant fragmentation of tapir populations. Following habitat loss, hunting is the second most important threat to tapir populations. Analysis of tapir faeces demonstrated the importance of leaves and stems as key components of this herbivore's diet, whereas fruits contribute to less than 10% of its food. In Costa Rica, remains of *Quercus costaricensis* were frequently encountered in tapir faecal samples (Chap. 27).

Rats and mice (Muridae and Heteromyidae) may abound in tropical montane oak forests – especially in years following mast seeding (López-Barrera Manson in Chap. 13; Van den Bergh and Kappelle in Chap. 26). Species of some myomorph genera prefer closed, mature oak forest (e.g. *Heteromys* and *Oryzomys*) whereas others are more numerous in open, shrubby or grassy habitats such as abandoned pastures (*Reithrodontomys*; Chap. 26). They are among the principal consumers and dispersers of acorns and other seeds (Chap. 13).

During a short field study in Costa Rican montane oak forests, some 20 birds have been observed foraging on over 20 different tree and shrub fruits, resulting in a total of 68 bird–plant frugivorous interactions (Wilms and Kappelle in Chap. 24). Lauraceous tree species within the genera *Ocotea* and *Nectandra* were most frequently fed upon. Small to medium-sized birds foraged mainly on fruits of fast-growing, light-dependent trees, whereas medium to large-sized birds preferred the fruits and seeds of slow-growing, mature forest tree species. Most of the birds observed displayed a preference for intact oak forest over open, disturbed habitats. Resplendent Quetzals, for instance, were much more abundant in mature old-growth forest than in pastures (Chap. 24). These birds prefer higher-elevation forests as their principal habitat, though they are known to altitudinally migrate as seasonal fruit availability changes (Wheelwright 1983). In Costa Rica, higher relative abundance of quetzals was found between February and June, coinciding with the breeding season (García-Rojas in Chap. 25).

34.8 Seed Predation and Dispersal

Many forest vertebrates depend on the availability of fruits and seeds which serve as main sources of food. Similarly, numerous plants depend on the presence and abundance of fruit- and seed-consuming vertebrates responsible for the dispersal of their seeds. It has been widely recognized that frugivorous and granivorous birds and mammals play a key role in the dispersal of seeds of trees, shrubs, herbs and climbers in tropical forests (e.g. Guevara et al. 1986; Adler 1995; Ceballos 1995; Janzen and Forget 2001; Levey et al. 2005; Wilms and Kappelle in Chap. 24; García-Rojas in Chap. 25).

Seedling emergence from soil seed bank material collected in Costa Rican montane oak forests and pastures demonstrated that at least one third of all plant species which emerged under greenhouse conditions had been dispersed zoochorously (Ten Hoopen and Kappelle in Chap. 23). Under natural conditions, emergence may be affected by litter (López-Barrera and González Espinoza 2000). In the particular case of oak (e.g. *Quercus rugosa*), animal-dispersed seeds (acorns) are likely to germinate soon after they are shed (Guariguata and Sáenz 2002; Bonfil in Chap. 12). Acorns cannot retain their viability long enough to survive until the next rainy season while forming part of the soil seed bank. The combination of these characteristics may be the main reason why acorns were not present in the collected soil seed bank material used during the seedling emergence observation study (Chap. 23).

Important fruit and seed predating and dispersing mammals at tropical latitudes include tapirs, monkeys, peccaries, bats, squirrels, rats and mice (Barnett 1977; Janzen 1983; Price and Jenkins 1986; Van den Bergh and Kappelle 1998 and in Chap.26; Demattia et al. 2002; López-Barrera and Manson in Chap. 13; García-Rojas in Chap. 25; Tobler et al. in Chap. 27).

Tropical montane oak seeds (acorns) are predated upon and/or dispersed by small mammals, particularly rodents (e.g. *Peromyscus* spp. and *Sciurus granatensis*; Chaps. 13 and 26). The huge amounts of tropical acorns available during mast seeding years may considerably affect local rodent populations, as has been demonstrated by a study on *Quercus laurina*, *Q. crassifolia* and *Q. rugosa* in the Chiapas highlands of Mexico (Chap. 13). It is well known that acorns are dispersed by jays in temperate oak forests. Jays seem to form a symbiotic relation with oaks. The same may apply for Neotropical montane oak forests which are inhabited by, e.g. the Silvery-Throated Jay (*Cyanolyca argentigula*, Chap. 24).

Certainly, a large percentage of tropical forest bird species strongly depend on fruits for their diet (Stiles 1985). Probably, frugivorous birds form the most important group of seed dispersers in the high-elevation oak forests of the tropical Americas, particularly in view of the low abundance of monkeys and bats in this cool 'temperate' highland ecosystem (Kappelle 1996). About half (34) of all bird species recorded in a Costa Rican montane oak forest showed frugivorous behaviour (Chap. 24). During a short survey, these avian dispersers were observed to forage on at least 22 fruit-bearing, endozoochorous tree species.

Many of these birds are considered key restoration agents, as they transport ornithochorous tree seeds from mature closed forest into neighbouring non-forested secondary growth and pastures on cleared forest lands (Chap. 24). In pastures, birds often drop seeds around isolated remnant forest trees, which offer them shelter on their way to distant forest fragments. Such scattered trees may act as seed traps and trigger forest succession at cleared sites (Guevara and Laborde 1993, and Chap. 24). A good example of a tropical oak forest bird accelerating forest recovery is that of the Resplendent Quetzal (*Pharomachrus mocinno*). This spectacular bird is recognized for its key ecological function in maintaining montane forest dynamics (Guindon 1997). Its behaviour increases the probabilities of seed survival and allows for light gap colonization, enrichment of isolated forest patches, and maintenance of tree species populations with limited distribution ranges (Chap. 25).

34.9 Responses to Disturbance

During the last 50 years, the clearing of large tracts of Neotropical montane forests, including oak forests, has led to severe land degradation in general, and to biodiversity depletion in particular (e.g. Rzedowski 1981; Budowski 1982; Monasterio et al. 1987; Bonfil 1991; González-Espinosa et al. 1995; Kappelle 1996). Since the early 1980s, vegetation and species response to tropical montane forest disturbance has received increased attention from scientists (Ewel 1980; Sugden et al. 1985; Brown and Lugo 1990; González-Espinosa et al. 1991; Walker et al. 1991; Kappelle 1993). A major conclusion of these studies concerns the decrease of the recovery rate with increasing elevation (e.g. Ewel 1980; Kappelle et al. 1995a, 1996). Regrowth following disturbance of tropical oak forests, for instance, appears to be extremely slow in the upper montane forest belt (2,400–3,200 m elevation), compared to faster recovery rates occurring at mid-elevation (1,500–2,400 m; Ewel 1980; Kappelle et al. 1994).

Recovery of oak species following disturbance may strongly depend on the availability of acorns and the activity of seed-dispersing animals (Chaps. 12 and 13). A study addressing seedling and population dynamics of *Quercus rugosa* on the slopes of the Ajusco Volcano, south of Mexico City, demonstrated that oak seedlings, once they emerged from dispersed acorns, were very fragile and had to survive a dry period of variable length in order to become vigorous saplings (Bonfil in Chap. 12). Therefore, only in years with higher than average winter rains (i.e. a reduced dry season) were conditions met for abundant seedling establishment. In average years, newly germinated seedlings died during their first dry season (Chap. 12).

Moisture, nutrient availability and light regime are among the most important factors affecting successful establishment of oak seedlings (Quintana-Ascencio et al. 1992). Most Neotropical oaks may be considered early successional species, as they require relatively large forest gaps or clearings for recruitment (Quintana-Ascencio et al. 2004; González-Espinosa et al. in Chap. 16). Some oaks may also regenerate along oak forest edges (Oosterhoorn and Kappelle 2000; López-Barrera et al. 2005, and Chaps. 13 and 16). Along such edges, microclimatic differences expressed in locally dry environments may affect oak seedling growth and physiological response, depending on the requirements of the species concerned (López-Barrera 2003, and Chap. 13; Asbjornsen et al. in Chap. 19). Mortality of oak seedlings (*Quercus acutifolia* and *Q. castanea*) in the Mixteca Alta region of Oaxaca, Mexico, is indeed significantly greater at open microsites – where moisture stress is higher – than at understory microsites (Asbjornsen et al. 2004, and Chap. 19). Compared to other broad-leaved trees and conifers in Chiapas, Mexico, seedling survival under field conditions (successional habitats) was highest for oaks (Ramírez-Marcial et al. 2005 and in Chap. 14). Survival rates of oak seedlings under greenhouse conditions, however, do not differ much from rates established for conifers (Chap. 14).

A recent analysis of the effects of reduced impact logging on oak forest regeneration in Costa Rica demonstrates that, at low harvesting intensities, the annual mortality rate for seedlings exceeds that recorded at moderate harvesting intensity (Sáenz and Guariguata 2001). This difference is probably related to variations in post-logging overhead light conditions. Comparing individual species, results for *Quercus costaricensis* and *Q. bumelioides* showed the largest, statistically significant differences in diameter growth rates among logging treatments, demonstrated by Guariguata et al. (Chap. 18). Guariguata's results revealed that, at the applied harvesting intensities, juveniles of *Quercus* (particularly *Q. costaricensis*) performed better in terms of radial growth than the other, non-oak study species.

During the last 15 years, community-level recovery of Neotropical montane oak forest has received increased attention as well (e.g. González-Espinosa et al. 1991; Kappelle et al. 1995a, 1996). The available studies on oak forest succession have mainly used chronosequences, particularly after agricultural field (e.g. *milpa*) and pasture abandonment. Research conducted in Chiapas, Mexico, and in the high Talamancas, Costa Rica, has revealed a series of successional pathways and sequential plant associations which occur after abandonment of old fields and pastures (Kappelle et al. 1994; González-Espinosa et al. in Chap. 16; Kappelle in Chap. 17). As expected, species diversity, structural complexity, stand height, basal area and accumulated litter increased with time (Chaps. 16, 17 and 22). Hertel et al. (Chap. 22), for instance, detected a significant linear increase in fine root mass with increasing age of secondary oak forests in highland Costa Rica.

However, recovery may take place at differing rates as a result of specific local stand or patch conditions. To gain a better insight into recovery rates, Kappelle et al. (1995a, 1996) estimated the theoretically minimum time required for abandoned montane pastures to return – via a successful natural recovery pathway – to a state in which their stand structure and floristic richness were within the ranges known for undisturbed mature oak forest. Analysis of beta diversity of terrestrial vascular plants led to an estimation of a minimum floristic recovery time of approximately 65 years. Similarly, a recovery period of 80–90 years was calculated for an abandoned pasture to become structurally similar, in terms of canopy height and basal area, to intact old-growth forest (Chap. 17).

To conclude, as González-Espinosa and colleagues correctly pointed out, floristic replacement probably delineates the major successional trends towards closely related sets of Neotropical old-growth oak forest types within certain ranges of altitudinal, climatic and edaphic conditions (Chaps. 16 and 17). The specific pathways followed by successional processes taking place on abandoned lands, and their characteristic levels of floristic and structural recovery will without doubt depend on the types and intensities of previous land use practices applied to Neotropical highland sites (González-Espinosa et al. 1991; Kappelle and Juárez 1995; Chaps. 16, 17, 18, 19, 21, 22 and 30).

34.10 Conservation and Sustainable Use

In the American Tropics, mountain oak forests have been used by people for centuries, if not millennia (Kappelle and Brown 2001). Probably, the indigenous peoples of the Mexican sierras, Guatemalan and Costa Rican highlands, and Colombian Andes were the first to use and manage Neotropical oak forests (e.g. Ramírez-Marcial et al. 2001; González-Espinosa et al. in Chap. 16). In pre-Columbian times, it is likely that these ethnic peoples relied heavily upon highland oak forests for obtaining food, fodder, fibre, fuel, medicines, dyes, gums, oils, antioxidants, spices, poisons, ornamental plants and pets (e.g. Kappelle et al. 2000). However, it is only recently that the Mesoamerican and Colombian montane oak forests have been significantly fragmented and degraded, as commercial logging and large-scale forest conversion for agricultural land and pastures became common practice (e.g. González-Espinosa et al. 1995; Kappelle 1996; Van Omme et al. 1997; Etter and Van Wijngaarden 2000; Helmer 2000).

For instance, since the early 1800s large oak forest areas in Mexican, Central American and northern Andean mountains have made way for coffee plantations and pastures for cattle ranching (Kappelle and Juárez 1995; Etter and Van Wijngaarden 2000). Due to increased population density, the last remaining tracts are now significantly threatened, since they are still being indiscriminately cut for timber, fuelwood and charcoal (Aus der Beek et al. in Chap. 31) throughout large parts of the region. This is especially alarming because these mountainous oak forests provide a large proportion of the drinking water and hydro-energy needed in large urban centres such as Guatemala City, San José and suburbs in Costa Rica, and Bogotá, Colombia.

Fortunately, over the last decades, society has come to value and recognize that tropical montane oak forests are not only sources of timber, but also critical ecosystems for water production (Barrantes Moreno in Chap. 33), sources of medicines (Kappelle and Juárez in Chap. 30), carbon sinks and reservoirs (Helmer and Brown 2000), areas for recreation (Chap. 30), and landscapes of great scenic beauty (see also Daily 1997, and Dawkins and Philip 1998). The now recognized multi-functionality of forests is reflected by current definitions of sustainable forest management which attempt to balance forest use and conservation (Castañeda 2000). In this context, Herrera and Chaverri (Chap. 32) stress the need for a clear definition of principles, criteria and indicators (C&I) necessary to assess and monitor sustainable forest management in highland oak forests. Such an approach, in combination with a sound understanding of natural oak forest dynamics (e.g. regeneration), is vital for the successful management of both silvicultural forest systems and protected forest areas (Olvera-Vargas et al. in Chap. 28). Only when C&I-based management guidelines for preservation and silvicultural purposes are in place will a sustainable balance between use and conservation be possible (Chaps. 28 and

32). The proposed framework for sustainable harvesting of bromeliads in the oak forests of Chiapas, Mexico, may serve as a good example of how a sound management plan may considerably enhance overall oak forest conservation while sustainably providing profit to local people (Wolf and Konings 2001; Wolf and Flamenco-S in Chap. 29).

If we are to preserve a large part of the remaining Neotropical montane oak forest and its variety of life as expressed in its genes, species and ecosystem types in the long term, we will need to elaborate a conservation strategy in which not only networks of protected core areas, buffer zones, and corridors form a fundamental component, but also participatory planning strategies in which different local and regional stakeholder groups and decision-makers are involved, in order to establish a broad-based, consensus-oriented conservation framework (Calderón et al. 2004; Kappelle 2004).

This new approach is well illustrated by the creation of the 'Alliance for the Conservation of the Pine-Oak Forest Ecoregion and its Birds in Meso-America' in 2003 (E. Secaira, The Nature Conservancy (TNC) and S. Pérez, Defensores de la Naturaleza, personal communications). The aim of this broad-based conservation partnership is to develop a regional conservation plan to save the montane oak forests of the highlands of southern Mexico (Chiapas), Guatemala, El Salvador, Honduras and Nicaragua, and in this way ensure the preservation of the winter habitat of Neotropical migratory birds threatened with extinction. Special focus is on an umbrella species known as the Golden Cheeked Warbler (*Dendroica chrysoparia*; Spanish: chipe mejilla dorada), which migrates during the boreal fall from the USA to its winter habitat, the montane pine-oak forests of southern Mexico and north-western Central America (Kroll 1980; King and Rappole 2000; Rappole et al. 2000).

The alliance is formed by a number of country-based non-governmental organizations (NGOs) including Pronatura-Chiapas, Instituto Montebello and Instituto de Historia Natural y Ecología (INHE) in Mexico, Fundación Defensores de la Naturaleza and Asociación de Reservas Naturales Privadas (Proyecto Cuchumatanes) in Guatemala, Salvanatura in El Salvador, Fundación EDUCA in Honduras, and the Alianza para las Áreas Silvestres (ALAS) in Nicaragua, as well as several international conservation organizations including TNC, the Texas Park and Wildlife Department (TPWD), and the Zoo Conservation Outreach Group.

It is hoped that such multi-country, multi-scale, multi-stakeholder initiatives will generate the ecosystem-based tools and knowledge urgently needed to help decision-makers find a way to sustainably manage and restore (Holl and Kappelle 1999) the threatened, species-rich highland oak forest ecoregions of the New World, for human well-being.

Acknowledgements I owe a special debt to my wife Marta E. Juárez who believed in this project and encouraged me to make my dream a reality.

References

Adler GH (1995) Fruit and seed exploitation by Central American spiny rats. *Proechimys semispinosus*. Stud Neotrop Fauna Environ 30:237–244

Almeida RT (2000) Evaluando y valorizando las áreas silvestres protegidas para la conservación del jaguar en Costa Rica: Área de Conservación La Amistad Pacífico. MINAE, San José, Tech Rep

Aranda M, Sánchez-Cordero V (1996) Prey spectra of jaguar (*Panthera onca*) and puma (*Puma concolor*) in tropical forests of Mexico. Stud Neotrop Fauna Environ 31:65–67

Asbjornsen H, Vogt KA, Ashton PMS (2004) Synergistic responses of oak, pine and shrub seedlings to edge environments and drought in a fragmented tropical oak forest, Oaxaca, Mexico. For Ecol Manage 192:313–334

Barnett RJ (1977) The effect of burial by squirrels on germination and survival of oak and hickory nuts. Am Midl Nat 98:319–330

Blaser J (1987) Standörtliche und waldkundliche Analyse eines Eichen-Wolkenwaldes (*Quercus* spp.) der Montanstufe in Costa Rica. PhD Thesis, Georg-August Universität, Göttingen

Bonfil C (1991) Los encinos y la Ciudad. Universidad Nacional Autónoma de México (UNAM), México DF, Mexico, Oikos, Boletín del Centro de Ecología

Brown S, Lugo AE (1990) Tropical secondary forests. J Trop Ecol 6:1–32

Bruijnzeel LA (2001) Hydrology of tropical montane cloud forests: a reassessment. Land Use Water Resources Res 1:1.1–1.8

Budowski G (1982) The socioeconomic effects of forest management on the lives of people living in the area: the case of Central America and some Caribbean countries. In: Hallsworth EG (ed) Socioeconomic effects and constraints in tropical forest management. Wiley, New York, NY, pp 87–102

Calderón R, Boucher T, Bryer M, Sotomayor L, Kappelle M (2004) Setting biodiversity conservation priorities in Central America: action site selection for the development of a first portfolio. The Nature Conservancy (TNC), San José, Costa Rica

Castañeda F (2000) Criteria and indicators for sustainable forest management: international processes, current status and the way ahead. Unasylva 51(4):34–40

Ceballos G (1995) Vertebrate diversity, ecology and conservation in neotropical dry forests. In: Bullock SH, Mooney HA, Medina E (eds) Seasonally dry tropical forests. Cambridge Univ Press, Cambridge, UK, pp 195–220

Cuatrecasas J (1934) Observaciones geobotánicas en Colombia. Trab Mus Nac Cienc Nat Ser Bot 27(45/48):1–222

Daily GC (1997) Nature's services: societal dependence on natural ecosystems. Island Press, Washington, DC

Dawkins HC, Philip MS (1998) Tropical moist forest silviculture and management: a history of success and failure. CABI, Oxon, UK

Demattia EA, Curran LM, Rathcke BJ (2002) Effects of small rodent seed predators on forest recruitment in Corcovado National Park, Costa Rica. In: Anon (ed) Abstr Vol Ecological Society of America Annual Meet, Tucson, AR, pp 111–112

Etter A, Van Wijngaarden W (2000) Patterns of landscape transformation in Colombia, with emphasis in the Andean region. Ambio 29(7):432–439

Ewel J (1980) Tropical succession: manifold routes to maturity. Biotropica 12(2):2–7

Fournier GR (1982) Palynostratigraphic analysis of cores from site 493. Proc Deep Sea Drilling Proj Leg 66:661–670

González-Espinosa M, Quintana-Ascencio PF, Ramírez-Marcial N, Gaytán-Guzmán P (1991) Secondary succession in disturbed *Pinus-Quercus* forests of the highlands of Chiapas, Mexico. J Veg Sci 2:351–360

González-Espinosa M, Ochoa-Gaona S, Ramírez-Marcial N, Quintana-Ascencio PF (1995) Current land-use trends and conservation of old-growth forest habitats in the highlands of Chiapas, Mexico. In: Wilson MH, Sader SA (eds) Conservation of Neotropical migratory birds in Mexico. Orono, Maine, Maine Agric For Exp Sta Misc Publ 727:190–198

González-Espinosa M, Rey-Benayas JM, Ramírez Marcial N, Huston MA, Golicher D (2004) Tree diversity in the northern Neotropics: regional patterns in highly diverse Chiapas, Mexico. Ecography 27:741–756

Guariguata MR, Sáenz GP (2002) Post-logging acorn production and oak regeneration in a tropical montane forest, Costa Rica. For Ecol Manage 167:285–293

Guevara S, Laborde J (1993) Monitoring seed dispersal at isolated standing trees in tropical pastures: consequences for local species availability. Vegetatio 107/108:319–338

Guevara S, Purata SE, Van der Maarel E (1986) The role of remnant forest trees in tropical secondary succession. Vegetatio 66:77–84

Guindon C (1997) The importance of forest fragments to the maintenance of regional biodiversity surrounding a tropical montane reserve, Costa Rica. PhD Dissertation, Faculty of the School of Forestry and Environmental Studies, Yale University, Connecticut

Helmer EH (2000) The landscape ecology of tropical secondary forest in montane Costa Rica. Ecosystems 3(1):98–114

Helmer EH, Brown S (2000) Gradient analysis of biomass in Costa Rica and a first estimate of trace gas emissions from biomass burning. In: Hall CAS (ed) Quantifying sustainable development. Academic Press, New York, pp 503–526

Herrera LF, Drenan R, Uribe C (eds) (1989) Cacicazgos prehispánicos del Valle de la Plata: El contexto medioambiental de la ocupación humana. Univ Pittsburgh Mem Lat Am Archaeol 2:95–118

Holl KD, Kappelle M (1999) Tropical forest recovery and restoration. Trends Ecol Evol 14(10):378–379

Holz I, Gradstein SR (2005) Phytogeography of the bryophyte floras of oak forests and páramo of the Cordillera de Talamanca, Costa Rica. J Biogeogr (in press)

Holz I, Gradstein SR, Heinrichs J, Kappelle M (2002) Bryophyte diversity, microhabitat differentiation and distribution of life forms in Costa Rican upper montane *Quercus* forest. The Bryologist 105:334–348

Janzen DH (ed) (1983) Costa Rican natural history. Univ of Chicago Press, Chicago, IL

Janzen PA, Forget PM (2001) Scatter-hoarding rodents and tree regeneration. In: Bongers F, Charles-Dominique P, Forget PM, Thery M (eds) Nouragues: dynamics and plant-animal interactions in a Neotropical rainforest. Kluwer, Dordrecht, pp 275–288

Jiménez W, Chaverri A, Miranda R, Rojas I (1988) Aproximaciones silviculturales al manejo de un robledal (*Quercus* spp.) en San Gerardo de Dota, Costa Rica. Turrialba 38(3):208–214

Kappelle M (1987) A phytosociological analysis of oak forests on the western slope of the Cordillera Oriental and in the central region of the high plain of Bogotá, Colombia. University of Amsterdam, Amsterdam, Tech Rep

Kappelle M (1991) Distribución altitudinal de la vegetación del Parque Nacional Chirripó, Costa Rica. Brenesia 36:1–14

Kappelle M (1993) Recovery following clearing of an upper montane *Quercus* forest in Costa Rica. Rev Biol Trop 41(1):47–56

Kappelle M (1996) Los bosques de roble (*Quercus*) de la Cordillera de Talamanca, Costa Rica: biodiversidad, ecología, conservación y desarrollo. Instituto Nacional de Biodiversidad (INBio), Santo Domingo de Heredia

Kappelle M (2004) Tropical montane forests. In: Burley J, Evans J, Youngquist JA (eds) Encyclopedia of Forest Sciences, vol 4. Elsevier, Oxford, UK, pp 1782–1793

Kappelle M, Brown AD (eds) (2001) Bosques nublados del Neotrópico. Instituto Nacional de Biodiversidad (INBio) and World Conservation Union (IUCN), Santo Domingo de Heredia, Costa Rica

Kappelle M, Gómez LD (1992) Distribution and diversity of montane pteridophytes of the Chirripó National Park, Costa Rica. Brenesia 37:67–77

Kappelle M, Juárez ME (1995) Agroecological zonation along an altitudinal gradient in the montane belt of the Los Santos Forest Reserve in Costa Rica. Mount Res Dev 15(1):19–37

Kappelle M, Cleef AM, Chaverri A (1992) Phytogeography of Talamanca montane *Quercus* forests, Costa Rica. J Biogeogr 19(3):299–315

Kappelle M, Van Velzen HP, Wijtzes WH (1994) Plant communities of montane secondary vegetation in the Cordillera de Talamanca, Costa Rica. Phytocoenology 22(4):449–484

Kappelle M, Kennis PAF, De Vries RAJ (1995a) Changes in diversity along a successional gradient in a Costa Rican upper montane *Quercus* forest. Biodiv Conserv 4:10-34

Kappelle M, Van Uffelen JG, Cleef AM (1995b) Altitudinal zonation of montane *Quercus* forests along two transects in the Chirripó National Park, Costa Rica. Vegetatio 119:119-153

Kappelle M, Geuze T, Leal ME, Cleef AM (1996) Successional age and forest structure in a Costa Rican upper montane *Quercus* forest. J Trop Ecol 12:681–698

Kappelle M, Van Omme E, Juárez ME (2000) Lista de la flora vascular terrestre de la cuenca superior del Río Savegre, San Gerardo de Dota, Costa Rica. Acta Bot Mex 51:1–38

Keighwin LD (1982) Stable isotope stratigraphy and paleoceanography of sites 502 and 503. In: Prell WL, Gardner JV (eds) Initial Reports of the Deep Sea Drilling Project, vol LXVIII. US Government Printing Office, Washington, DC, pp 445–453

King DI, Rappole JH (2000) Winter flocking of insectivorous birds in montane pine-oak forests in Middle America. Condor 102:664–672

Kinsey AC (1936) The origin of higher categories in Cynips. Indiana University Publ, Indiana

Kroll JC (1980) Habitat requirements of the Golden-cheeked Warbler: management implications. J Range Manage 33(1):60–65

Levey DJ, Bolker BM, Tewksbury JJ, Sargent S, Haddad NM (2005) Effects of landscape corridors on seed dispersal. Science 309:146–148

López-Barrera F (2003) Edge effects in a forest mosaic: implications for the oak regeneration in the Highlands of Chiapas, Mexico. PhD Thesis, Institute of Atmospheric and Environmental Sciences, University of Edinburgh, Edinburgh

López-Barrera F, González Espinoza M (2000) Influence of litter on emergence and early growth of *Quercus rugosa*: a laboratory study. New For 21:59–70

López-Barrera F, Newton AC, Manson RH (2005) Edge effects in a tropical montane forest mosaic: experimental tests of post-dispersal acorn removal. Ecol Res 20:31–40

Monasterio M, Sarmiento G, Solbrig OT (eds) (1987) Comparative studies on tropical mountain ecosystems. Planning for research. IUBS, Paris, Biol Int Spec issue 12

Mueller GM, Bills G (2004) Introduction. In: Mueller GM, Bills G, Foster MS (eds) Biodiversity of fungi: inventory and monitoring methods. Elsevier, San Diego, CA, pp 1–4

Naranjo EJ, Vaughan C (2000) Ampliación del ámbito altitudinal del tapir centroamericano (*Tapirus bairdii*). Rev Biol Trop 48:724

Nixon KC (1993) The genus *Quercus* in Mexico. In: Ramamoorthy TP, Bye R, Lot A, Fa J (eds) Biological diversity of Mexico: origins and distribution. Oxford Univ Press, New York, pp 447–458

Oosterhoorn M, Kappelle M (2000) Vegetation structure and composition along an interior-edge-exterior gradient in a Costa Rican montane cloud forest. For Ecol Manage 126:291–307

Price MV, Jenkins SH (1986) Rodents as seed consumers and dispersers. In: Murray DR (ed) Seed dispersal. Academic Press, New York, pp 191–235

Quintana-Ascencio PF, González-Espinosa M, Ramírez-Marcial N (1992) Acorn removal, seedling survivorship, and seedling growth of *Quercus crisipipilis* in successional forests of the highlands of Chiapas, Mexico. Bull Torrey Bot Club 119:6–18

Quintana-Ascencio PF, Ramírez-Marcial N, González-Espinosa M, Martínez-Icó (2004) Sapling survival and growth of coniferous and broad-leaved trees in successional highland habitats in Mexico. Appl Veg Sci 7:81–88

Ramírez-Marcial N, González-Espinosa M, Williams-Linera G (2001) Anthropogenic disturbance and tree diversity in montane rain forests in Chiapas, Mexico. For Ecol Manage 154:311–326

Ramírez-Marcial N, Camacho-Cruz A, González-Espinosa M (2005) Potencial florístico para la restauración de bosques en Los Altos y Montañas del Norte de Chiapas. In: González-Espinosa M, Ramírez-Marcial N, Ruiz-Montoya L (eds) Diversidad biológica de Chiapas. Plaza y Valdés, México City

Rappole JH, King DI, Leimgruber P (2000) Winter habitat and distribution of the endangered Golden-cheeked Warbler (*Dendroica chrysoparia*). Animal Conserv 2:45–59

Rzedowski J (1981) Vegetación de México. Limusa, México DF, Mexico

Sáenz GP, Guariguata MR (2001) Demographic response of tree juveniles to reduced impact logging in a Costa Rican montane forest. For Ecol Manage 140:75–84

Stiles FG (1985) On the role of birds in the dynamics of Neotropical forests. In: Diamond AW, Lovejoy TE (eds) Conservation of tropical forest birds. International Council for Bird Preservation (ICPB), Cambridge, UK, Tech Publ 4:49–59

Sugden AM, Tanner EVJ, Kapos V (1985) Regeneration following clearing in a Jamaican montane rain forest: results of a ten-year study. J Trop Ecol 1:329–351

Tobler MW (2002) Habitat use and diet of Baird's tapirs (*Tapirus bairdii*) in a montane cloud forest of the Cordillera de Talamanca, Costa Rica. Biotropica 34:468–474

Valencia-A S (2004) Diversidad del género *Quercus* (Fagaceae) en México. Bol Soc Bot Méx 75:33–53

Valencia-A S, Nixon KC (2004) Encinos. In: García-Mendoza AJ, Ordóñez MJ, Briones-Salas M (eds) Biodiversidad de Oaxaca. Instituto de Biología, Universidad Nacional Autónoma de México (UNAM), Fondo Oaxaqueño para la Conservación de la Naturaleza and WWF, México DF, Mexico, pp 219–225

Van den Bergh MB, Kappelle M (1998) Diversity and distribution of small terrestrial rodents along a disturbance gradient in montane Costa Rica. Rev Biol Trop 46(2):331–338

Van der Hammen T, Gonzalez E (1963) Historia del clima y vegetación del Pleistoceno Superior y del Holoceno de la Sabana de Bogotá. Bol Geol 40(1/3):189–266

Van Omme E, Kappelle M, Juárez ME (1997) Land cover/use changes and deforestation trends over 55 years (1941-1996) in a Costa Rican montane cloud forest watershed area. In: Abstr Vol Conf Geo-Information for Sustainable Land Management. ITC, Enschede, The Netherlands, p 5.14

Van't Veer R, Hooghiemstra H (2000) Montane forest evolution during the last 650,000 yr in Colombia: a multivariate approach based on pollen record Funza-1. J Quat Sci 15:329–346

Van Velzen HP, Wijtzes WH, Kappelle M (1993) Lista de especies de la vegetación secundaria del piso montano pacífico, Cordillera de Talamanca, Costa Rica. Brenesia 39/40:147–161

Walker LR, Brokaw NVL, Lodge DJ, Waide RB (1991) Ecosystem, plant and animal responses to hurricanes in the Caribbean. Biotropica 23(4a):311–521

Wesselingh RA, Witteveldt M, Morissette J, Den Nijs JCM (1999) Reproductive ecology of understory species in a tropical montane forest in Costa Rica. Biotropica 31(4):637–645

Wesselingh RA, Burgers HCM, Den Nijs JCM (2000) Bumblebee pollination of understorey shrub species in a tropical montane forest in Costa Rica. J Trop Ecol 16(5):657–672

Wheelwright NT (1983) Fruits and the ecology of the resplendent quetzals. The Auk 100:286–301

Widmer Y (1993) Bamboo and gaps in the oak forests of the Cordillera de Talamanca, Costa Rica. Freising, Verh Gesell Ökol 22:329–332

Widmer Y (1998) Pattern and performance of understory bamboos (*Chusquea* spp.) under different canopy closures in old-growth oak forests in Costa Rica. Biotropica 30:400–415

Wolf JHD (2005) The response of epiphytes to anthropogenic disturbance of pine-oak forests in the highlands of Chiapas, Mexico. For Ecol Manage 212:376–393

Wolf JHD, Flamenco-S A (2003) Patterns in species richness and distribution of vascular epiphytes in Chiapas, Mexico. J Biogeogr 30:1689–1707

Wolf JHD, Konings CJF (2001) Toward the sustainable harvesting of epiphytic bromeliads: a pilot study from the highlands of Chiapas, Mexico. Biol Conserv 101:23–31

Taxonomic Index

For additional lichen taxa see Appendix 6.1, pages 74–80

Subject Index

Ecological Studies
Volumes published since 2001

Volume 147
Ecosystem Approaches to Landscape Management in Central Europe (2001)
J.D. Tenhunen, R. Lenz, and R. Hantschel (Eds.)

Volume 148
A Systems Analysis of the Baltic Sea (2001)
F.V. Wulff, L.A. Rahm, and P. Larsson (Eds.)

Volume 149
Banded Vegetation Patterning in Arid and Semiarid Environments (2001)
D. Tongway and J. Seghieri (Eds.)

Volume 150
Biological Soil Crusts: Structure, Function, and Management (2001, 2003)
J. Belnap and O.L. Lange (Eds.)

Volume 151
Ecological Comparisons of Sedimentary Shores (2001)
K. Reise (Ed.)

Volume 152
Global Biodiversity in a Changing Environment: Scenarios for the 21st Century (2001)
F.S. Chapin, O. Sala, and E. Huber-Sannwald (Eds.)

Volume 153
UV Radiation and Arctic Ecosystems (2002)
D.O. Hessen (Ed.)

Volume 154
Geoecology of Antarctic Ice-Free Coastal Landscapes (2002)
L. Beyer and M. Bölter (Eds.)

Volume 155
Conserving Biological Diversity in East African Forests: A Study of the Eastern Arc Mountains (2002)
W.D. Newmark

Volume 156
Urban Air Pollution and Forests: Resources at Risk in the Mexico City Air Basin (2002)
M.E. Fenn, L. I. de Bauer, and T. Hernández-Tejeda (Eds.)

Volume 157
Mycorrhizal Ecology (2002, 2003)
M.G.A. van der Heijden and I.R. Sanders (Eds.)

Volume 158
Diversity and Interaction in a Temperate Forest Community: Ogawa Forest Reserve of Japan (2002)
T. Nakashizuka and Y. Matsumoto (Eds.)

Volume 159
Big-Leaf Mahogany: Genetic Resources, Ecology and Management (2003)
A. E. Lugo, J. C. Figueroa Colón, and M. Alayón (Eds.)

Volume 160
Fire and Climatic Change in Temperate Ecosystems of the Western Americas (2003)
T. T. Veblen et al. (Eds.)

Volume 161
Competition and Coexistence (2002)
U. Sommer and B. Worm (Eds.)

Volume 162
How Landscapes Change: Human Disturbance and Ecosystem Fragmentation in the Americas (2003)
G.A. Bradshaw and P.A. Marquet (Eds.)

Volume 163
Fluxes of Carbon, Water and Energy of European Forests (2003)
R. Valentini (Ed.)

Volume 164
Herbivory of Leaf-Cutting Ants: A Case Study on *Atta colombica* in the Tropical Rainforest of Panama (2003)
R. Wirth, H. Herz, R.J. Ryel, W. Beyschlag, B. Hölldobler

Volume 165
Population Viability in Plants: Conservation, Management, and Modeling of Rare Plants (2003)
C.A Brigham, M.W. Schwartz (Eds.)

Volume 166
North American Temperate Deciduous Forest Responses to Changing Precipitation Regimes (2003)
P. Hanson and S.D. Wullschleger (Eds.)

Volume 167
Alpine Biodiversity in Europe (2003)
L. Nagy, G. Grabherr, Ch. Körner, D. Thompson (Eds.)

Volume 168
Root Ecology (2003)
H. de Kroon and E.J.W. Visser (Eds.)

Volume 169
Fire in Tropical Savannas: The Kapalga Experiment (2003)
A.N. Andersen, G.D. Cook, and R.J. Williams (Eds.)

Volume 170
Molecular Ecotoxicology of Plants (2004)
H. Sandermann (Ed.)

Volume 171
Coastal Dunes: Ecology and Conservation (2004)
M.L. Martínez and N. Psuty (Eds.)

Volume 172
Biogeochemistry of Forested Catchments in a Changing Environment: A German Case Study (2004)
E. Matzner (Ed.)

Volume 173
Insects and Ecosystem Function (2004)
W. W. Weisser and E. Siemann (Eds.)

Volume 174
Pollination Ecology and the Rain Forest: Sarawak Studies (2005)
D. Roubik, S. Sakai, and A.A. Hamid (Eds.)

Volume 175
Antarctic Ecosystems: Environmental Contamination, Climate Change, and Human Impact (2005)
R. Bargagli

Volume 176
Forest Diversity and Function: Temperate and Boreal Systems (2005)
M. Scherer-Lorenzen, Ch. Körner, and E.-D. Schulze (Eds.)

Volume 177
A History of Atmospheric CO_2 and its Effects on Plants, Animals, and Ecosystems (2005)
J.R. Ehleringer, T.E. Cerling, and M.D. Dearing (Eds.)

Volume 178
Photosynthetic Adaptation: Chloroplast to Landscape (2005)
W.K. Smith, T.C. Vogelmann, and C. Chritchley (Eds.)

Volume 179
Lamto: Structure, Functioning, and Dynamics of a Savanna Ecosystem (2005)
L. Abbadie et al. (Eds.)

Volume 180
Plant Ecology, Herbivory, and Human Impact in Nordic Mountain Birch Forests (2005)
F.E. Wielgolaski (Ed.) and P.S. Karlsson, S. Neuvonen, D. Thannheiser (Ed. Board)

Volume 181
Nutrient Acquisition by Plants: An Ecological Perspective (2005)
H. BassiriRad (Ed.)

Volume 182
Human Ecology: Biocultural Adaptations in Human Cummunities (2006)
H. Schutkowski

Volume 183
Growth Dynamics of Conifer Tree Rings: Images of Past and Future Environments (2006)
E.A. Vaganov, M.K. Hughes, and A.V. Shashkin

Volume 184
Reindeer Management in Northernmost Europe: Linking Practical and Scientific Knowledge in Social-Ecological Systems (2006)
B.C. Forbes, M. Bölter, L. Müller-Wille, J. Hukkinen, F. Müller, N. Gunslay, Y. Konstantinov (Eds.)

Volume 185
Ecology and Conservation of Neotropical Montane Oak Forests (2006)
M. Kappelle (Ed.)

Springer
the language of science

springeronline.com

New Textbook in Plant Ecology

Plant Ecology

E.-D. Schulze, Max-Planck-Institute for Biogeochemistry, Jena, Germany; **E. Beck**, University of Bayreuth, Germany; **K. Müller-Hohenstein**, University of Bayreuth, Germany

This textbook covers Plant Ecology from the molecular to the global level. It presents the following areas in unprecedented breadth and depth:
► Molecular ecophysiology ► Autecology ► Ecosystem ecology ► Synecology ► Global aspects of plant ecology
The book is carefully structured and well written: complex issues are elegantly presented and easily understandable. It contains more than 500 photographs and drawings, mostly in colour, illustrating the fascinating subject.

From the reviews of the German edition ► *"A book with scientific depth - a future standard work. Its up-to-date and global approach alone make it of interest to a large target group. ... Numerous images, tables, flow charts, graphics, and memory boxes facilitate comprehension. The comprehensive manner in which individual topics are presented make the book both a sound textbook and reference work."* ► **Forstzeitung**

TEXT BOOK

2005. IX, 702 p. 506 illus., most of them in colour. Hardcover
ISBN 3-540-20833-X ► **€ 74,95 | £ 57,50**

Easy Ways to Order ► Springer Distribution Center • Haberstr. 7 • 69126 Heidelberg, Germany • Tel.: +49 (0) 6221 - 345 - 4303 Fax: +49 (0) 6221 - 345 - 4229 • e-mail: SDC-bookorder@springer-sbm.com • or through your bookseller. All Euro and GBP prices are net-prices subject to local VAT, e.g. in Germany 7%. All prices exclusive of carriage charges. Prices and other details are subject to change without notice. d&p · BA_011675x

MIX
Papier aus verantwortungsvollen Quellen
Paper from responsible sources
FSC® C105338

FSC
www.fsc.org

If you have any concerns about our products,
you can contact us on
ProductSafety@springernature.com

In case Publisher is established outside the EU,
the EU authorized representative is:
Springer Nature Customer Service Center GmbH
Europaplatz 3, 69115 Heidelberg, Germany

Printed by Libri Plureos GmbH
in Hamburg, Germany